Mechanics

Mechanics

Tarvinder Kaur Chadha

PUBLISHING FOR ONE WORLD

NEW AGE INTERNATIONAL (P) LIMITED, PUBLISHERS

(formerly Wiley Eastern Limited)

New Delhi • Bangalore • Chennai • Guwahati • Hyderabad

Jalandhar • Kolkata • Lucknow • Mumbai • Ranchi

Visit us at **www.newagepublishers.com**

Published by New Age International (P) Ltd., Publishers
First Edition : 2004
Reprint : 2005

ISBN : 81-224-1553-9

Rs. 150.00

2 3 4 5 6 7 8 9 10

C-05-09-385

Printed in India at Pack Printer, New Delhi.
Typesetter : Spectrum Media, Delhi.

PUBLISHING FOR ONE WORLD

NEW AGE INTERNATIONAL (P) LIMITED, PUBLISHERS
(formerly Wiley Eastern Limited)
4835/24, Ansari Road, Daryaganj, New Delhi - 110002
Visit us at **www.newagepublishers.com**

PREFACE

After about twenty eight years of experience as a teacher/examiner in Mechanics to the Post graduate students in the university, knowing fully the psychology and the need of the students, the author decided to bring a book in Mechanics for M.A./M.Sc. students. While framing the syllabus of a particular subject, subject matter is selected from various books. Hence a student has to refer many textbooks, which he finds difficult and time consuming. No doubt, for thorough knowledge of the subject, a student must refer to some other books, but due to varied topics and shortage of time, it is beyond the capacity of an average student. Hence, this book is an attempt to provide students with a text book covering the entire syllabus. In this book, the author does not claim to express anything original, but has the satisfaction that after going through all necessary books, on the subject, efforts are made to make the book all-round complete. The book has been written with a view to cover the complete syllabus of Guru Nanak Dev University, Amritsar and Punjab University, Patiala but there is not even a single book available which covers the whole syllabii of these Universities. It covers all common topics of various Post Graduate courses in Mechanics. To make this book useful for other Universities and for candidates appearing for competitive examinations, some additional topics have been added.

Besides this, Mechanics is an indispensable part of Physics, as it may be regarded as a fundamental tool for various branches of Modern Physics. An engineer has also to depend upon theory of Mechanics. Consequently, I can claim that this book will be quite useful for students of Physics and Engineering.

Clarity no less than utility has been author's goal in writing this book. My aim is to present the subject matter in a straight forward and systematic way as possible to make it intelligible to an average University student. Stress has been given to basic principles and an approach has been adopted that is helpful in understanding the subject. For applications of the theory, a large number of typical examples have been solved and at the end of every chapter a large number of unsolved examples are added. The reader must attempt to solve them in order to get a better insight and understanding of the subject. Many of these problems have been selected from various University examination papers, so as to make the reader familiar with the type of questions, usually set there in.

Some quantities used in Mechanics are scalar while others are vectors. Many equations in Mechanics assume a compact form while written in vector notations. The relationship between various quantities involved in equation is revealed immediately when they are written in vector form. Moreover, these equations are independent of particular coordinate system. Once they are written in vector form, they can be used for any coordinate system depending on the nature of the problem. This approach is more convenient and precise. Most of the existing books follow the analytical approach, but in this book, effort has been made to make use of vectors as far as possible.

Due to change in style of question papers in the university examinations, a student has to give short answer of many questions in a specific time. These questions cover the whole syllabus and are compulsory. Either due to lack of practice or having no knowledge of the type of the questions to be asked, students find it very difficult to answer such questions in limited time. Hence, to meet such difficulties of the students, a !ist of such questions is given at the end of the book.

There are a number of existing books on Mechanics, but most of them were written long back, number of recent books is very small. The present book is an attempt to fulfill the requirements of the students in modern trend of life. I hope that this book will prove more suitable to those for whom it is intended.

The book covers the necessary portions of Statics, Dynamics of Particle, Dynamics of rigid body and Generalized coordinates. To make the book self contained, some results of elementary Mechanics have been recapitulated in 'Preliminary Concepts'.

I wish to dedicate this book to those I treasure above else on this Earth, and who have given meaning to my life—my husband and my children. I take the opportunity of thanking authorities of Guru Nanak Dev University, who very kindly sanctioned me sabbatical leave for writing this book. Finally, I acknowledge my thanks to the publishers for being kind to publish the book in the present form.

T.K. CHADHA

CONTENTS

PRELIMINARY CONCEPTS

1.1 INTRODUCTION

Mechanics or the science of motion of material bodies has been a matter of great interest for ages. Mechanics may be classified into two classes :

I. Dynamics: Further Dynamics can be classified into two subsections :

(a) **Kinematics:** It deals with all possible motions of material systems without reference to the agency, which causes motion. We describe motion of the bodies in terms of quantities such as displacement, velocity etc. It is a geometrical description of the motion.

(b) **Kinetics:** It deals with the motion of the body along with forces associated with it and which cause the motion. As such this subsection introduces the concept of force and explains the possibility of most favourable types of motion which takes place under the action of given forces. Hence, we investigate the properties of moving system by studying the equations of motion containing forces.

II. Statics: It deals with the systems of forces which actually give no motion to the system. For equilibrium of the system the net effect of forces is zero.

1.2 FRAME OF REFERENCE

In order to describe the motion of a body or of a paricle we need two things, (1) *a frame of reference* (2) *a time keeper.* It is not possible to describe absolute motion, but only motion relative to surrounding objects; and a suitable frame of reference which depends on the kind of motion that is desired to be described. For example, in the case of body from the surface of the Earth, a set of axes with the origin at the point of projection is suitable for the description of the motion relative to the Earth. But, for the description of the motion of planets, it would be more convenient to take a frame of axes with origin at the Sun's centre.

The motion of the objects on the surface of the Earth is described relative to some point on it. Thus earth itself is a frame of reference. There are generally two types of reference systems:

1. The systems relative to which the body, not acted upon by external force is unaccelerated. This also includes the state of rest.

2. The systems relative to which the body, not acted upon by external force is accelerated.

The frame relative to which the body, not acted upon by external force, is un-accelerated *i.e.*, at rest or moving with constant linear velocity is called *inertial frame.* We can say inertial frame is one in which law of inertia or Newton's first law is valid. The frames relative to which the body, not acted upon by external force is accelerated, are non-inertial frames. Any frame of reference moving with constant velocity relative to an inertial frame is also inertial.

A co-ordinate frame fixed in earth is not an inertial frame, since earth rotates about its axis and about the Sun. In-fact any thing which is capable of turning is not an inertial frame. To describe the motion of a body, we should always choose inerial frames, because if we choose non-inertial frames, these involve different forces and it is very complicated to formulate the laws common to all observers.

1.3 DEFINITIONS

In order to study mechanics we must be familiar with some basic concepts.

Space and time : We assume that time and space are continuous and there are universal standards of length and time.

Matter : Matter is one of the primary conceptions of the mind, it cannot be defined satisfactorily, but it is clear any definition of matter would have to embrace all things that can be perceived by sense of touch and some things which are perceived by other senses.

Mass : Mass of the particle is a measure of the quantity of the matter. It is measured variously in ib., kg, tons, etc.

Inertia : The mode by which the presence of matter is most easily perceived is through the effort to produce in it a sudden change of motion. This property of matter is called *inertia.*

Particle : It is a body so small that, for the purposes of our investigations the distance between its points may be neglected or we can say it is a quantity of having negligible dimensions.

Motion : It is change of position or displacement.

Velocity : It is the rate of change of displacement.

Acceleration : Rate of change of velocity is called acceleration.

Hence, if $\vec{x}$ denotes displacement $\frac{d\vec{x}}{dt}$ is velocity and $\frac{d^2\vec{x}}{dt^2}$ is acceleration.

Displacement, velocity and acceleration are vector quantities.

Linear Momentum : It is the product of mass and velocity. Thus if m is the mass of the particle and $\vec{v}$ is the velocity of the particle at a particular instant, then the vector $\vec{p}$.

$$\vec{p} = m\vec{v}$$

is called the *linear momentum* of the particle at that instant. It is a vector quantity.

Angular momentum : It is the moment of linear momentum about some point or some axis.

Work : A force is said to do work, when its point of application undergoes a displacement in the line of action of the force. Suppose a force $\vec{F}$ acts on a particle, which is constrained to move along a curve C joining points A and B. Let P , P′ be neighbouring points on C, such that

$$\vec{OP} = \vec{r}\ ,\ \vec{OP'} = \vec{r} + \delta\vec{r}\ ,$$

$$\vec{PP'} = \delta\vec{r} \text{ or } \delta\vec{s}$$

we choose the work in terms of arc length of C, then the work done by in moving its point of application P to P′ is

$$\vec{F}\cdot d\vec{r} = \vec{F}\cdot d\vec{s}$$

Thus the total work done in moving the particle from A to B is

$$W = \int_A^B \vec{F}\cdot d\vec{r} = \int_A^B \vec{F}\cdot d\vec{s} = \int_A^B F\cos\theta\ ds.$$

θ is the angle $\vec{F}$ makes with $\delta\vec{r}$ at P. These expressions are termed *line integrals* of $\vec{F}$. They are evaluated along the curve C and may depend on C.

Kinetic Energy : It is energy due to motion. If the particle P has constant mass *m* and velocity $\vec{v}$ at P, then

$$\vec{F}\cdot d\vec{r} = m\left(\frac{d\vec{v}}{dt}\right)\cdot d\vec{r} = m\vec{v}\cdot\frac{d\vec{v}}{dt}dt = md\left(\frac{1}{2}v^2\right)dt\ .$$

Hence, the total work done on the particle is

$$\int_A^B \vec{F}\cdot d\vec{r} = \left[\frac{1}{2}mv^2\right]_A^B = \frac{1}{2}mv_B^2 - \frac{1}{2}mv_A^2\ .$$

Thus the quantity $T = \frac{1}{2}mv^2$ has the same dimensions as work and is called *Kinetc energy* of the particle. Thus the total work done by the external force $\vec{F}$ in carrying the paticle from A to B along C is equal to the change in Kinetic energy. *Energy is defined as capacity for doing work.* Thus, when a body is in motion, work can be done in overcoming resistance before the body is brought to rest, and the amount of work is called Kinetic energy.

Potential Energy : A body may possess energy in virtue of its position, represented by the work that would be done by the forces acting on the body if it moved from its stated position to some standard position. This is called *Potential Energy.*

Power : It is the rate of doing work. The practical units of power is Horse Power and represents the doing of work of 33000 foot pounds per minute.

Since, $\vec{F} \cdot d\vec{r}$ is the work done by the force in moving the particle from P to P′, if the motion takes place in time δt the rate of working of the force at P is

$$\lim_{\delta t \to 0}\left(\vec{F} \cdot \frac{\delta \vec{r}}{\delta t}\right) = \vec{F} \cdot \frac{d\vec{r}}{dt} = \vec{F} \cdot \vec{v} .$$

Power is called *activity* of the force. Denoting it by A, we see that the work done is

$$W = \int_{t_A}^{t_B} A dt$$

Further, we note that

$$\frac{dT}{dt} = \frac{d}{dt}\left(\frac{m}{2}v^2\right) = m\vec{v} \cdot \frac{d\vec{v}}{dt} = \vec{F} \cdot \vec{v} = \text{A (scalar)}$$

Conservative forces : When the work done in moving particle from A to B does not depend on the shape of the curve along which the particle moves but only on the initial and final points, then the forces are said to be *conservative.*

Law of conservation of energy : Hence, in conservative field, energy *i.e.,* K.E. + P.E. = constant. Clearly for a system of particles in conservative field of force, the principle of conservation of energy still holds, since it holds for individual particle. In this case there exists a *potential function V* such that

$$\vec{F} = -\, grad\, V$$

$$\therefore \quad W = \int_A^B \vec{F} \cdot d\vec{r} = -\int_A^B grad\, V \cdot d\vec{r} = V_A - V_B$$

But we know that

$$\int_A^B \vec{F} \cdot d\vec{r} = \frac{1}{2} m\,(v_B^2 - v_A^2)$$

$$\therefore \quad V_A + \frac{1}{2} m v_A^2 = V_B + \frac{1}{2} m v_B^2 .$$

Impulse : The impulse of a constant force during a given interval of time is the product of force and time during which it acts *i.e., Ft, F* is the constant force and *t* the interval.

In every case, we can prove that

Impulse of a force = the change of momentum produced by it.

(i) Thus using the formula for constant force and acceleration,

Impulse of a force = $Ft = m(v - u) = mv - mu$ = change of momentum.

(ii) For a variable force.

$$X = m\ddot{x},$$

$$\text{the impulse of the force} = \int_{t_1}^{t_2} X\,dt = \int_{t_1}^{t_2} m\ddot{x}\,dt$$

$= m(\dot{x}_2 - \dot{x}_1)$, where $\dot{x}_1, \dot{x}_2$ are the speeds at the begining and end of the interval

$= m\dot{x}_2 - m\dot{x}_1$

= change of momentum produced.

Work Done by a couple : It is the product of moment of the couple and the angle through which its arm is turned. Since, all couples of the same moment in the same plane are equivalent. It is immaterial about what point the arm turns.

Efficiency : In most machines some part of the work done is expended in overcoming friction, or resistance of that nature; this portion of the work is regarded as *'wasted'* and the remaining portion is described as *'useful work'*. The ratio of the useful work to the whole work done is called the *efficiency* of the machine.

Effectiveness of the Brakes : The effect of applying brakes to one pair of a four wheeled car in motion is to alter the division of the weight between the wheels.

Note : Problems on running trains with constant propulsive force, road resistance, brake resistance may conveniently be solved by equating momentum to the impulse of the forces and equating kinetic energy to the work done, without introducing the acceleration.

Thus, if m be the mass, F the propulsive force, and R, the road resistance, the equation for getting up a speed v in a distance x and time t from rest are

Impulse $(F - R)t = mv$, the momentum generated.

work done $(F - R)x = mv^2/2$, the kinetic energy created.

Similarly, after steam is shut off and brakes are applied, the total resistance R_1, R_2 and if this brings the train to rest in a distance x and time t, then since,

momentum destroyed = impulse of the retarding force,

$$\therefore \quad mv = (R_1 + R_2)t;$$

since,

K. E. destroyed = work done by the retarding force,

$$\therefore \quad mv^2/2 = (R_1 + R_2)x.$$

1

FORCES IN THREE DIMENSIONS

1.1 PARALLELOPIPED OF FORCES

If three forces acting at a point be represented in magnitude and direction by three sides of a parallelopiped passing through a point then their resultant is given in magnitude and direction by the diagonal of the parallelopiped passing through that point.

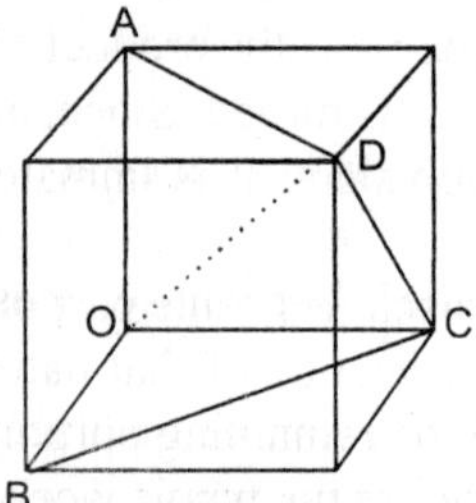

Fig. 1.1

Thus if OA, OB, OC be three co-terminous edges of parallelopiped then the resultant of forces represented by OA, OB, OC shall be represented by the diagonal OD of the parallelopiped both in magnitude and direction. The converse is also true.

If X, Y, Z be three forces mutually perpendicular at a point O then their resultant is given by

$$R^2 = X^2 + Y^2 + Z^2,$$

and if (f, g, h) be any point on the line of action of R, its equation is given by

$$\frac{x-f}{X} = \frac{y-g}{Y} = \frac{z-h}{Z}.$$

1.2 THEOREMS RELATING TO COUPLES

Couples: A system of forces and not lying in the same straight line but such that $\vec{F}_1 + \vec{F}_2 = \vec{0}$, is called a *couple.* The two forces are parallel and equal in magnitude but oppositely directed. The perpendicular distance between two parallel lines of the forces is called the *arm of the couple.* A couple cannot be reduced to a single force as the other systems.

A couple is represented by a line perpendicular to the plane of the couple and whose length is proportional to the magnitude of the moment. Thus a couple being a directed quantity, the parallelogram and parallelopiped laws will be valid for couples also.

Theorem I : The moment of a couple with respect to any axis perpendicular to its plane is independent of the position of the axis.

Theorem II : The moment of a couple is the product of the common magnitude of the forces and the perpendicular distance between them.

Theorem III : Two couples which act in the same plane and which have the same moment are equivalent.

Theorem IV : Two couples which act in parallel planes and which have the same moment are equivalent.

Theorem V : Couples are vectors.

Theorem VI : If a rigid body, acted upon by a couple, is given a small rotation $d\theta$ about any axis which is parallel to the axis of the couple, the work done by the couple is the porduct of the angular displacement and the moment of the couple.

Theorem VII : Any system of forces acting upon a rigid body, which is not equivalent to a single couple, can be replaced by two forces which are perpendicular to each other, but whose lines of action, in general, do not intersect.

Theorem VIII : A couple can be displaced and rotated in its own plane. The position of the couple has no importance as it can be transferred from one plane to any parallel plane.

1.3 AXIS OF A COUPLE

To find the axis of the couple, one of whose force acts through the origin and the other along the line

$$\frac{x-x_1}{l}=\frac{y-y_1}{m}=\frac{z-z_1}{n},$$

passing through the point $P(x_1, y_1, z_1)$.

Let F be the magnitude of the force of the couple then components of the force F at P are

$$X = F_l,\ Y = Fm, Z = Fn.$$

Similarly, the components of force $-F$ at O are $-X, -Y, -Z$.

Draw perpendicular PA from P on the plane xy and perpendicular AB on x-axis. Draw SBS' parallel to OZ. Introduce at B two equal forces Z along BS and BS', these forces being in equilibrium among themselves do not alter the equilibrium of the system. Now Z at A and $-Z$ at B form a couple of moment $y_1 Z$ about x-axis; and Z at B and $-Z$ at O form a couple of moment $-x_1 Z$ about y-axis.

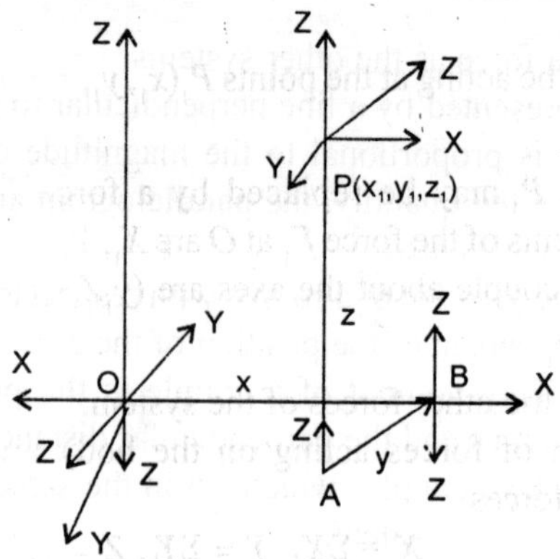

Fig. 1.2

Hence, Z at P and $-Z$ at O give two couples of moment $y_1 Z$ about x-axis and $-x_1 Z$ about y-axis.

Similarly, the couple X at P and $-X$ at O is equivalent to two couples of moment $z_1 X$ about y-axis $-y_1 X$ about z-axis and the couple Y at P and $-Y$ at O give two couples of moments $x_1 Y$ about z-axis and $-z_1 Y$ about x-axis.

Thus the couple F at P and $-F$ at O are equivalent to three couples

$$(y_1 Z - z_1 Y) = F(y_1 n - z_1 m) \text{ about } x\text{-axis},$$

$$(z_1 X - x_1 Z) = F(z_1 l - x_1 n) \text{ about } y\text{-axis},$$

$$(x_1 Y - y_1 X) = F(x_1 m - y_1 l) \text{ about } z\text{-axis},$$

Since, the plane of the couple passes through the origin, its axis is therefore

$$\frac{x}{y_1 n - z_1 m} = \frac{y}{z_1 l - x_1 n} = \frac{z}{x_1 m - y_1 l}$$

1.4 TO FIND THE RESULTANT OF A GIVEN SYSTEM OF FORCES ACTING AT GIVEN POINTS OF THE BODY

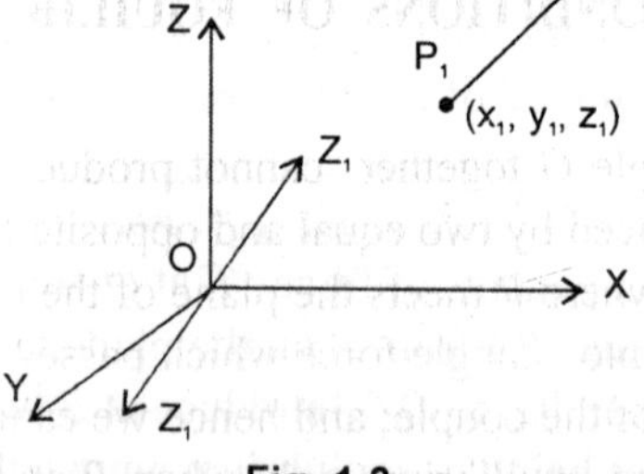

Fig. 1.3

Let us choose an arbitrary point of the body as origin and three mutually perpendicular axes OX, OY, OZ as axes of reference. Let a system of forces

$F_1, F_2, F_3, \ldots\ldots\ldots F_n$ be acting at the points $P_1(x_1, y_1, z_1), P_2(x_2, y_2, z_2), \ldots\ldots\ldots P_n$ (x_n, y_n, z_n).

The force F_1 at P_1 may be replaced by a force F_1 at O and a couple $(F_1, -F_1)$. Components of the force F_1 at O are X_1, Y_1, Z_1 along the axes and the components of the couple about the axes are $(y_1Z_1 - z_1Y_1)$, $(z_1X_1 - x_1Z_1)$ and $(x_1Y_1 - y_1X_1)$.

Similarly, for all the other forces of the system.

Thus the system of forces acting on the body are equivalent to three components of the forces

$$X = \Sigma X_1,\ Y = \Sigma Y_1,\ Z = \Sigma Z_1,$$

and three component couples

$$L = \Sigma(y_1Z_1 - z_1Y_1),\ M = \Sigma(z_1X_1 - x_1Z_1),$$

$$N = \Sigma(x_1Y_1 - y_1X_1).$$

the summation being from 1 to n.

The component forces can be compounded into a single froce R such that

$$R^2 = X^2 + Y^2 + Z^2,$$

along the line

$$\frac{x}{X} = \frac{y}{Y} = \frac{z}{Z},$$

and three component couples by a single couple G, where

$$G^2 = L^2 + M^2 + N^2,$$

and axis is

$$\frac{x}{L} = \frac{y}{M} = \frac{z}{N}.$$

Thus the system of forces reduces to a single force R and a single couple G at an arbitrary point O. Such a combination is called a *Dyname* and $X, Y, Z; L, M, N$ are the six components of the dyname (R, G).

1.5 GENERAL CONDITIONS OF EQUILIBRIUM OF A RIGID BODY

A force R and a couple G together cannot produce equilibrium. Since, the couple G can be replaced by two equal and opposite forces one of which acts through the point O where R meets the plane of the couple. This force and R can be compounded into a single force which passes through O and does not meet the other force of the couple; and hence we cannot have equilibrium.

Hence, there can be equilibrium only when R and G are separately zero. But we know that

$$R^2 = X^2 + Y^2 + Z^2,\ G^2 = L^2 + M^2 + N^2$$

Hence, for equilibrium we must have :

$$X = 0, Y = 0, Z = 0;$$
$$L = 0, M = 0, N = 0;$$

i.e. the sum of the resolved parts of the system of forces parallel to any three axes of coordinates must separately be zero and also the sums of their moments about three axes must separately be zero.

1.6 MOMENT OF A FORCE ABOUT A LINE

Let the given line be

$$\frac{x - x'}{l'} = \frac{y - y'}{m'} = \frac{z - z'}{m'}$$

and a force F acts along the line

$$\frac{x - f}{l} = \frac{y - g}{m} = \frac{z - h}{n}.$$

Now the components of F along the axis are Fl, Fm, Fn, and components of the moments about the axes are

$$F(gn - hm), \ F(hl - fn), \ F(fm - gl).$$

Now the shortest distance between two lines is given by

$$\begin{vmatrix} f - x' & g - y' & h - z' \\ l & m & n \\ l' & m' & n' \end{vmatrix} \div \sin \theta$$

where θ is the angle between two lines.

The shortest distance between F and z-axis is

$$r = \begin{vmatrix} f & g & h \\ l & m & n \\ 0 & 0 & 1 \end{vmatrix} \div \sin \theta.$$

$$\therefore \qquad r \sin \theta = fm - gl.$$

Now the moment of F about z-axis is

$$fY - gX = F(fm - gl) = Fr \sin \theta.$$

We can generalize this result:

The moment of a force about a line is equal to the product of the force, shortest distance and the sine of the included angle.

Alternatively

Resolve the force F into two components, Q parallel to the line and S perpendicular to the line; the product of S and the shortest distance between line of action of S and the given line is the moment of the force F about the required line.

If the lines are parallel then $\theta = 0$, therefore moment is also zero.

If AB, CD are portions of the line of action of F and the given line and V be the volume of the tetrahedron whose opposite edges are AB, CD.

$$V = \frac{1}{6} AB \cdot CD \cdot r \sin\theta .$$

$$\therefore \quad \text{moment} = \frac{6V}{AB \cdot CD} \cdot F .$$

1.7 CONDITIONS OF EQUILIBRIUM OF A RIGID BODY WITH ONE POINT FIXED

Let us take the fixed point O as the origin and three perpendicular lines through it as axes of coordinates. Let the external forces acting on the body reduce to component forces X, Y, Z and components couples L, M, N. Let the force of constraint at O have component X', Y', Z', parallel to the axes. Applying results of the article 1.5, we have

$$X + X' = 0, Y + Y' = 0, Z + Z' = 0 \quad \text{...(1)}$$

$$L = 0, M = 0, N = 0. \quad \text{...(2)}$$

The equations (1) give only the component reactions at O in terms of the external forces.

The equation (2) give the conditions of equilibrium, which states that the sum of moments of the external forces about any three perpendicular lines passing through the fixed point O must be separately zero.

If the external forces all act in one plane passing through O, these conditions reduce to the simpler condition that the sum of the moments of the external forces about O must vanish.

1.8 CONDITIONS OF EQUILIBRIUM OF A RIGID BODY WITH TWO POINTS FIXED

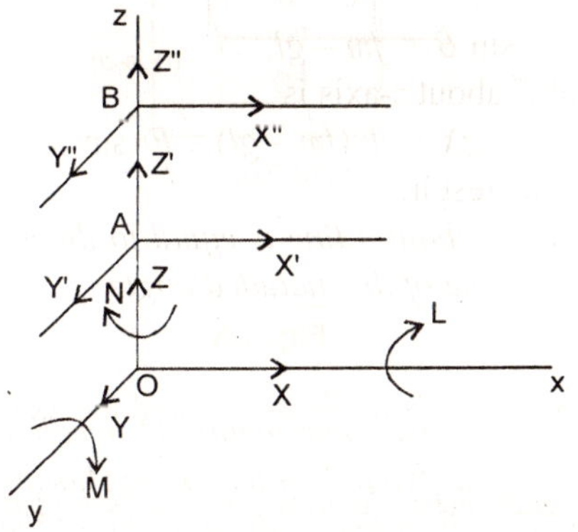

Fig. 1.4

Let two points A and B be the fixed points of the body so that the body can turn about the fixed axis AB. Let us take the fixed line AB as z-axis and any point O on it as origin.

Let $OA = z'$, $OB = z''$, X', Y', Z', X'', Y'', Z'' be components of the forces of constraints at A and B respectively.

Let the external forces acting on the body reduce to the dyname (X, Y, Z, L, M, N). The conditions of equilibrium of the forces are:

$$X + X' + X'' = 0 \quad (1)$$
$$Y + Y' + Y'' = 0 \quad (2)$$
$$Z + Z' + Z'' = 0 \quad (3)$$
$$L - Y' z' - Y'' z'' = 0 \quad (4)$$
$$M + X' z' + X'' z'' = 0 \quad (5)$$
$$N = 0 \quad (6)$$

(1) and (5) give X', X'', (2) and (4) give Y', Y'' but there is only one relation between Z' and Z'' given by (3) hence their values remain indeterminate.

Finally (6) is the only relation between external forces, so that condition of equilibrium is that the sum of the moments of the external forces about the fixed axis AB must be zero.

Examples

1. Forces Acting on a Cube

Case (i) Two equal forces R act on a cube, whose centre is fixed and whose edge is $2a$, along the diagonals of adjacent faces which do not meet; show that the moment of the couple which will keep the cube at rest is either $Ra\sqrt{3}$ or Ra according to the direction of the forces.

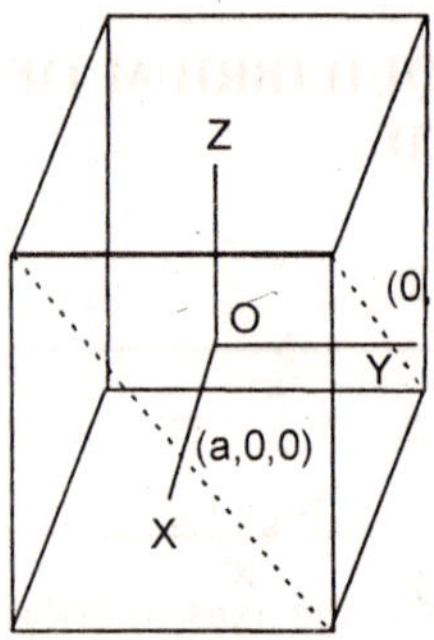

Fig. 1.5

Let the centre of the cube be the origin and the axes parallel to the edges of the cube.

Let two forces each equal to R act on the faces parallel to the yz and zx planes through the points $(a, 0, 0)$ and $(0, a, 0)$ respectively.

The resolved parts of these forces parallel to the axes are:

$$\left(0, \pm\frac{R}{\sqrt{2}}, \mp\frac{R}{\sqrt{2}}\right), \quad \left(\pm\frac{R}{\sqrt{2}}, 0 \pm\frac{R}{\sqrt{2}}\right).$$

Therefore the moments of the forces about the axes are:

$$L = \sum (y_1 Z_1 - z_1 Y_1) = \pm \frac{aR}{\sqrt{2}},$$

$$M = \sum (z_1 X_1 - x_1 Z_1) = \pm \frac{aR}{\sqrt{2}},$$

$$N = \sum (x_1 Y_1 - y_1 X_1) = \pm \frac{aR}{\sqrt{2}} \pm \frac{aR}{\sqrt{2}} = \pm \sqrt{2}\, aR \text{ or } 0,$$

according to the direction of the forces.
Therefore, either

$$G = aR\left(\frac{1}{2} + \frac{1}{2} + 2\right)^{1/2} = \sqrt{3}\, aR,$$

or
$$G = aR\left(\frac{1}{2} + \frac{1}{2}\right)^{1/2} = aR.$$

Case (ii) Six forces each equal to P, act along the edges of the cube, taken in order which do not meet a given diagonal. Show that their resultant is a couple of moment

$$2\sqrt{3}\, Pa,$$

where a is the edge of the cube.

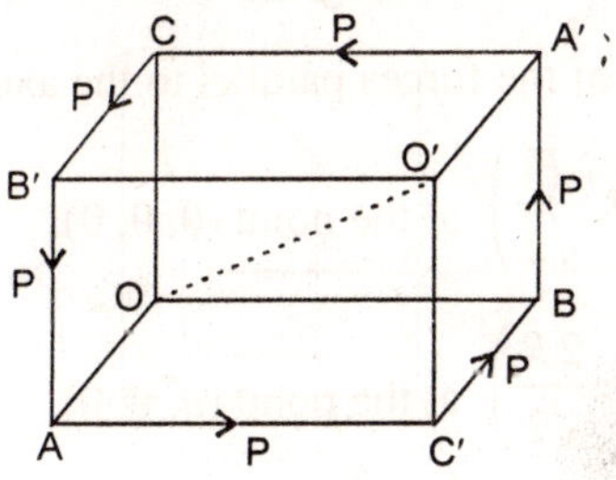

Fig. 1.6

Let OA, OB, OC, be taken as the axes of coordinates and let the six forces which do not meet the diagonal OO' be taken as shown in the figure. Then the resolved parts of the forces along the axes are:

$$X = -P + P = 0,$$
$$Y = P - P = 0,$$
$$Z = P - P = 0.$$

The moments of the couples about the axes are:

$$L = -Pa - Pa = -2Pa,$$
$$M = -Pa - Pa = -2Pa,$$

$$N = -Pa - Pa = -2Pa.$$

$$\therefore \quad G = 2\sqrt{3}\,Pa.$$

Therefore, the resultant of the forces is a couple of momemt $2\sqrt{3}\,Pa$.

Case (iii) OA, OB, OC are the edges of a cube of side a and OO', AA' BB', CC' are its diagonals; along OB', $O'A$, BC, $C'A'$ act forces equal to P, $2P$, $3P$, $4P$; show that they are equivalent to force $\sqrt{35}\,P$ at O along a line whose direction cosines are proportional to –3, –5, 6 together with a couple $\frac{Pa}{2}\sqrt{114}$ about a line whose direction cosines are proportional to 7, –2, 2.

The forces are shown in the figure.

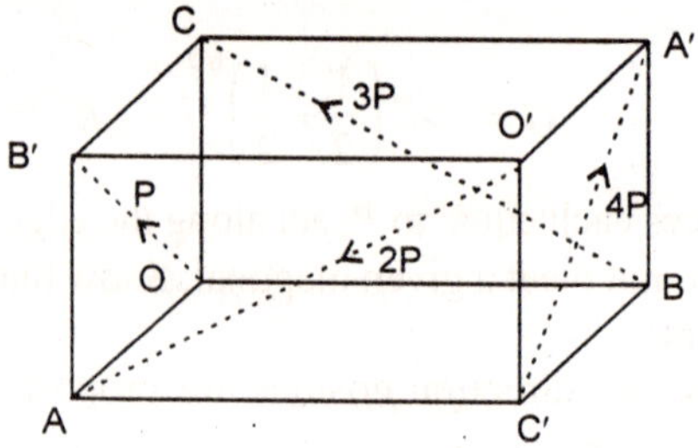

Fig. 1.7

The resolved parts of the forces parallel to the axes are:

$$\left(\frac{P}{\sqrt{2}}, 0, \frac{P}{\sqrt{2}}\right) \text{ at the point } (0, 0, 0),$$

$$\left(0, -\frac{2P}{\sqrt{2}}, -, \frac{2P}{\sqrt{2}}\right) \text{ at the ponit } (a, 0, 0),$$

$$\left(0, -\frac{3P}{\sqrt{2}}, \frac{3P}{\sqrt{2}}\right) \text{ at the point } (0, a, 0),$$

$$\left(-\frac{4P}{\sqrt{2}}, 0, \frac{4P}{\sqrt{2}}\right) \text{ at the point } (a, a, 0).$$

$$\therefore \quad X = \frac{P}{\sqrt{2}} - \frac{4P}{\sqrt{2}} = -\frac{3P}{\sqrt{2}},$$

$$Y = -\frac{2P}{\sqrt{2}} - \frac{3P}{\sqrt{2}} = -\frac{5P}{\sqrt{2}},$$

$$Z = \frac{P}{\sqrt{2}} - \frac{2P}{\sqrt{2}} + \frac{3P}{\sqrt{2}} + \frac{4P}{\sqrt{2}} = \frac{6P}{\sqrt{2}}.$$

Therefore the resultant force is

$$R = \sqrt{X^2 + Y^2 + Z^2} = \sqrt{35}P,$$

and the direction of the line of action of the resultant are proportional to X, Y, Z, *i.e.*, –3, –5, 6.

Components of couple of moments are:

$$L = \frac{3Pa}{\sqrt{2}} + \frac{4Pa}{\sqrt{2}} = \frac{7Pa}{\sqrt{2}},$$

$$M = \frac{2Pa}{\sqrt{2}} - \frac{4Pa}{\sqrt{2}} = -\frac{2Pa}{\sqrt{2}},$$

$$N = -\frac{2Pa}{\sqrt{2}} + \frac{4Pa}{\sqrt{2}} = \frac{2Pa}{\sqrt{2}}.$$

Therefore the resultant couple is

$$G = (L^2 + M^2 + N^2)^{1/2} = \frac{Pa\sqrt{114}}{2},$$

about a line whose direction cosines are proportional to L, M, N, *i.e.*, 7, –2, 2.

2. Four forces act along generators of the same system of a hyperboloid. Their magnitudes are such that if they were transferred parallel to themselves to act at one point they would be in equilibrium; show that they are in equilibrium when acting along generators.

Let P_r $(r = 1, 2, 3, 4)$ be the forces acting along four generators of the same system of the hyperboloid $\frac{x^2}{a^2} + \frac{y^2}{b^2} - \frac{z^2}{c^2} = 1$,

whose equations are:

$$\frac{x - a\cos\theta_r}{a\sin\theta_r} = \frac{y - b\sin\theta_r}{-b\cos\theta_r} = \frac{z}{c},$$

The direction cosines of these lines are:

$\frac{a\sin\theta_r}{k_r}, \frac{-b\cos\theta_r}{k_r}, \frac{c}{k_r}$, where $k_r^2 = a^2\sin^2\theta_r + b^2\cos^2\theta_r + c^2$.

Since, we are given that when the forces are transferred parallel to themselves to act at one point they are in equilibrium, we have

$$X = \sum \frac{P_r a\sin\theta_r}{k_r} = 0,$$

$$Y = \sum -\frac{P_r b\cos\theta_r}{k_r} = 0,$$

and

$$Z = \sum \frac{P_r c}{k_r} = 0.$$

Now

$$L = \sum (y_1 Z_1 - z_1 Y_1) = \sum b\sin\theta_r \cdot \frac{P_r c}{k_r} = \frac{bc}{a} \cdot X = 0,$$

$$M = \sum (z_1 X_1 - x_1 Z_1) = \sum (-a\cos\theta_r) \cdot \frac{P_r c}{k_r} = \frac{ca}{b} \cdot Y = 0,$$

$$N = \sum (x_1 Y_1 - y_1 X_1) = \sum a\cos\theta_r \left(\frac{-P_r b\cos\theta_r}{k_r}\right) - b\sin\theta_r \left(\frac{P_r a\sin\theta_r}{k_r}\right)$$

$$= \sum -\frac{abP_r}{k_r} = -\frac{ab}{c} Z = 0.$$

Hence, the forces are in equilibrium.

3. A square table stands on four legs placed respectively at the middle points of its sides; find the greatest weight that can be put at one of its corner without upsetting the table.

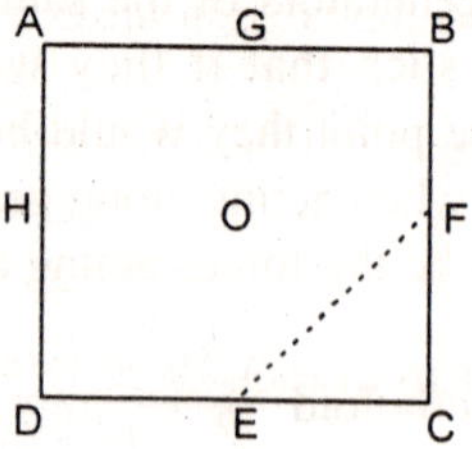

Fig. 1.8

Let $ABCD$ be the top of the table and E, F, G, H, the middle points of its sides.

Suppose the weight is placed at the corner C. If the table overturns about the line joining the feet of the legs at E and F, and the legs at G and H will loose the contact with the ground.

Let W be the weight of the table acting at its C. G, O and W' be the weight placed at C. The distance of C from EF = distance of O from EF.

The table will be just on point of turning if W and W' have equal moments about EF.

Hence, Moments of W and W' are equal if $W = W'$

Therefore the greatest weight that can be put at one corner without upsetting the table is equal to the weight of the table.

4. A round table stands upon three equidistant weightless legs at its edge, and a man sits upon its edge opposite a leg. It just upsets and falls upon its edge and two legs. He then sits upon its highest point and just tips it again. Show that the radius of the table is, $\sqrt{2}$ times the length of the leg.

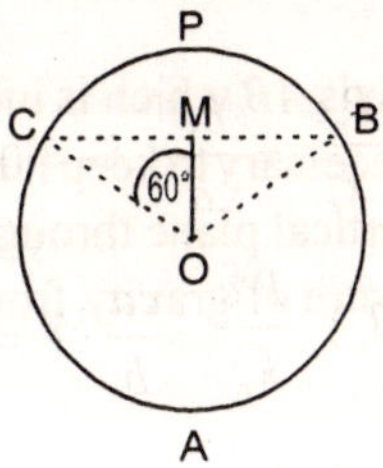

Fig. 1.9 *(i)*

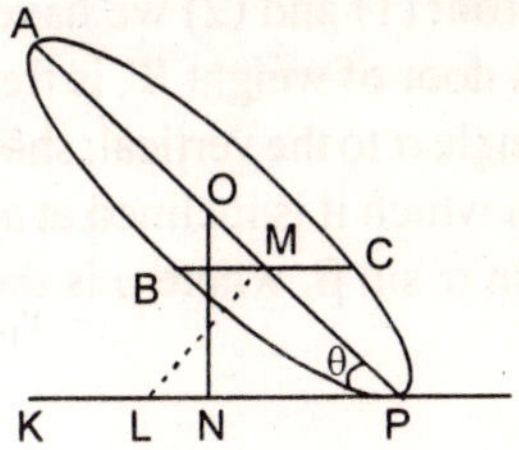

Fig. 1.9 *(ii)*

Let ABC be the top of the table, with legs at A, B, C and the man sitting at P. The length of the perpendiculars from the centre upon BC is $a \cos 60°$, *i.e.*, $\frac{1}{2}a$, where a is the redius of the circle. Hence, the lenghts of the perpendiculars both from the centre O and P on BC are equal. Considering, moments about BC, we see that the weight at P should be equal to the weight of the table, when it just upsets.

The second figure shows the table in upset position, with point P on the ground.

Let l be the length of the leg of the table and $\angle APK = \theta$, we have

$$\tan \theta = \frac{ML}{PM} = \frac{l}{a/2} = \frac{2l}{a}, \tag{1}$$

since, the perpendicular from P on BC is $a/2$.

Now $$LP = PM \sec \theta = \frac{1}{2} a \sec \theta$$

$$KP = AP \cos \theta = 2a \cos \theta.$$

$\therefore$ $$KL = 2a \cos \theta - \frac{a}{2} \sec \theta$$

$$NL = \frac{1}{2} a \sec \theta - a \cos 0.$$

The table will return to its original position, with the man going over to A, if the moment of his weight at A about L is just greater than moment of the weight of the table L.

Hence, we have

$$w.NL = w.KL$$

$$\therefore \quad \frac{1}{2} a \sec\theta - a\cos\theta = 2a\cos\theta - \frac{1}{2} a \sec\theta,$$

$$\therefore \quad \sec\theta = 3\cos\theta,$$

$$\therefore \quad \cos\theta = \frac{1}{\sqrt{3}}, \tan\theta = \sqrt{2}. \qquad (2)$$

From (1) and (2) we have $a = l\sqrt{2}$.

5. A door of weight W, is free to turn about an axis AB which is inclined at an angle α to the vertical; show that the couple necessary to keep it in a position in which it is inclined at an angle β to the vertical plane through AB is $Wa \sin\alpha \sin\beta$, where a is the distance of its centre of gravity from AB.

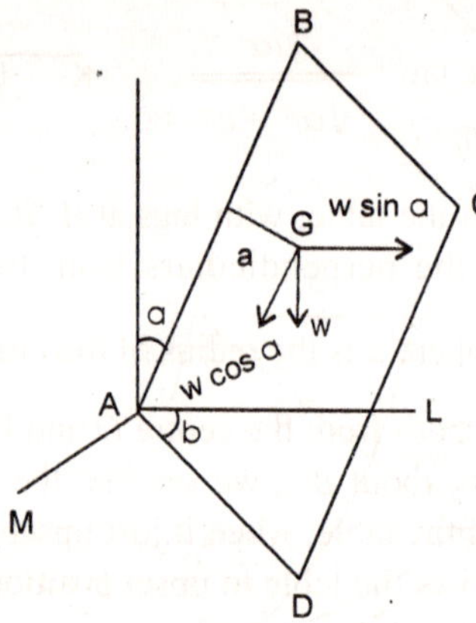

Fig. 1.10

AB is the axis of the door inclined at an angle α to the vertical and LAB is the vertical plane through AB. When the door is held in the position $ABCD$, let AD makes an angle β with AL.

The weight W of the door acting through its centre of gravity G has components $W\cos\alpha$ parallel to AB and $W\sin\alpha$ parallel to AL.

Let AM be the perpendicular to the plane $ABCD$. Let AD, AM, AB, be the axes of coordinates. Hence components of W are

$$X = W\sin\alpha\cos\beta, \ WY = W\sin\alpha\sin\beta, \ Z = W\cos\alpha$$

The moment of W about AB is $Y \cdot a$. *i.e.*, $Wa \sin\alpha \sin\beta$, which is also the moment of the couple necessary to keep the door in the position $ABCD$.

6. A rectangular gate is hung is the ordinary way on two hinges so that the line joining the hinges makes an angle α with the vertical. Show that the work which must be done to move it through an angle θ from its position of equilibrium is $Wa \sin\alpha(1 - \cos\theta)$, where W is the weight and $2a$ is the breadth of the gate.

From previous example, we see that work done in moving the gate through an angle $\delta\beta$ from the position $ABCD$ is

$$Wa \sin \alpha \sin \beta \cdot \delta\beta.$$

Therefore the work done in moving the gate through an angle θ from its position of equilibrium (in the plane BAL)

$$= \int_0^\theta Wa \sin\alpha \sin\beta \cdot \delta\beta, = Wa \sin \alpha (1 - \cos \theta)$$

7. A uniform straight rod, of length $2c$, is placed in a horizontal position as high as possible within a hollow sphere, of radius a. Show that the line joining the middle point of the rod to the centre of the sphere makes with the vertical an angle $\tan^{-1} \dfrac{\mu\alpha}{\sqrt{\alpha^2 - c^2}}$.

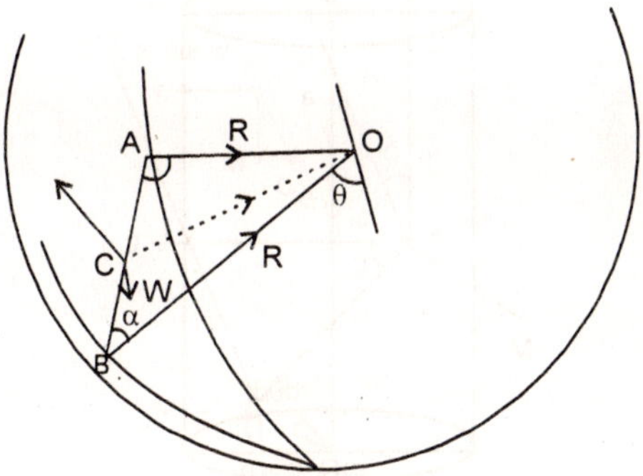

Fig. 1.11

Let AB be the rod resting horizontally inside a rough sphere such that the line joining its middle point C to the centre O of the sphere makes the greatest angle θ with the vertical.

Since, the rod tends to slip down always remaining horizontal, the point C will describe a circle of radius OC about O in the vertical plane.

The reactions at A and B are each equal to R acting along the radii AO and BO, making the same angle α with AB. Hence, their resultant is $2R \sin \alpha$ acting along CO, where $\cos \alpha = \dfrac{c}{a}$.

The frictional forces at A and B are each μR at right angles to OC in a vertical plane. Hence, their resultant is $2\mu R$ acting at C at right angles to CO, making an angle θ with the horizontal.

For equilibrium of the rod, we have

$$2R \sin \alpha \cos \theta + 2\mu R \sin \theta = W,$$

$$2R \sin \alpha \sin \theta = 2 \mu R \cos \theta.$$

The second equation gives

$$\tan\theta = \frac{\mu}{\sin\alpha} = \frac{\mu a}{\sqrt{a^2 - c^2}}.$$

8. A heavy circular cylinder rests with plane base upon a rough horizontal table; if its weight be W and the normal pressure be supposed to be uniformly distributed over the base, show that the moment of the couple about its axis which would just twist it is $\frac{2}{3}\mu Wa$, where μ is the coefficient of friction and a is the radius of the base.

It is evident that total normal reaction on the base of the cylinder is equal to the weight of the cylinder. Since this reaction is uniformly distributed over the whole base, the reactions per unit area of the base is $\frac{W}{\pi a^2}$.

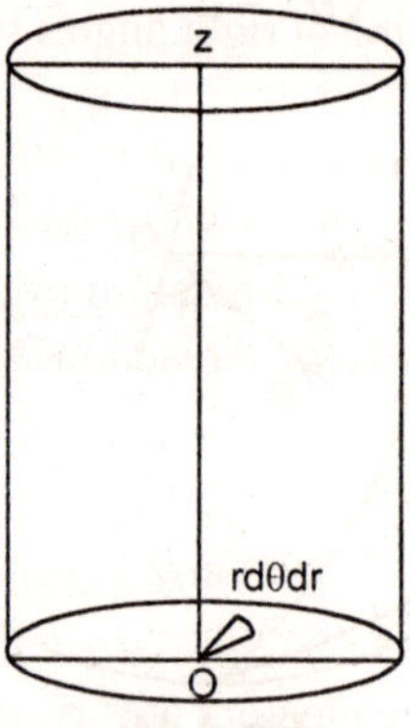

Fig. 1.12

Let us consider an elementary area $rd\theta dr$ of the base. The reaction on this area is $\left(\frac{W}{\pi a^2}\right) rd\theta dr$. The moment of the friction about the axis of the cylinder is $\left(\frac{\mu W}{\pi a^2}\right) rd\theta dr.r$.

Therefore the frictional moment on the entire base, which is equal to the couple required to trun the cylinder

$$\frac{\mu W}{\pi a^2}\int_{r=0}^{r=a}\int_{\theta=0}^{\theta=2\pi} r^2 d\theta dr = \frac{\mu W}{\pi a^2}\frac{1}{3}a^3.2\pi = \frac{2}{3}\mu Wa.$$

9. A right circular cone, of weight W and vertical angle 2α, is placed with its vertex downwards and supported by a circular hole cut in a horizontal table. If μ be the coefficient of friction and b the radius of the hole, show that the moment of the least couple that will move the cone is μWb cosec α.

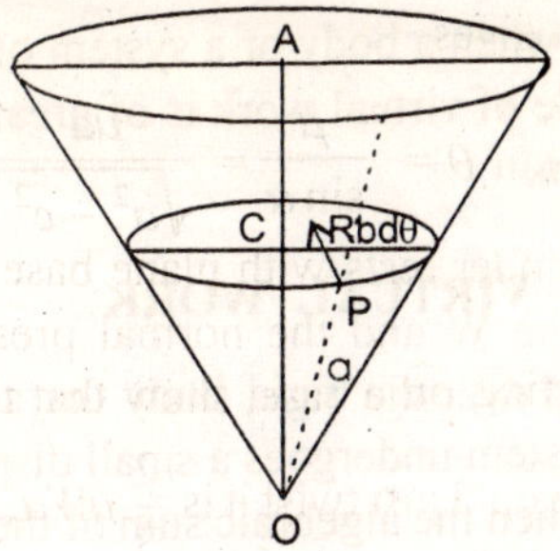

Fig. 1.13

Let us consider an element $bd\theta$ situated at the point P of the circular hole of the radius b into which the cone rests.

If R be the reaction per unit length of the arc of the hole, the reaction of this element is $Rbd\theta$, acting at right angles with generator OP of the cone through P.

The components of this reaction are $Rbd\theta \cos\alpha$ in the plane of the hole passing through its centre C and $Rbd\theta \sin \alpha$ vertical.

It is evident that the horizontal reaction of the different parts of the hole balance each other, while vertical reactions support the weight of the cone. Therefore,

$$W = \int_{\theta=0}^{\theta=2\pi} Rbd\theta.\sin\alpha = 2\pi Rb \sin \alpha.$$

The frictional force of the element at P is $\mu Rbd\theta$ along the tangent to the arc, and its moment about the axis OA of the cone is $\mu Rbd\theta\, b$. Therefore total moment.

$$= \int_0^{2\pi} \mu Rb^2 d\theta = 2\pi\mu Rb^2 = \mu bW \text{ cosec } \alpha.$$

1.9 VIRTUAL WORK

Suppose that a body is in equilibrium under the action of any system of forces. Let the body be imagined to undergo a slight displacement consistent with the geometrical conditions, such that the point A of the body is displaced to the point A′, then AA′ is called *virtual displacement* and the work done during this displacement is called *virtual work*. The word virtual implies that this displacement is not actual, it is only hypothetical or imaginary. There is nothing essentially different from real work.

If AA' makes an angle θ with the line action of the force $\vec{F}$ at A then work done is $\vec{F}.\overrightarrow{AA}$. Like actual work virtual work is also a vector quantity. Although no displacements occur when a body or a system of bodies is in equilibrium, nevertheless the principle of virtual work is of great value in determining the conditions of equilibrium.

1.10 PRINCIPLE OF VIRTUAL WORK

If a system of forces acting on a rigid body (or a system of bodies) be in equilibrium and if the system undergoes a small displacement consistent with geometrical conditions, then the algebraic sum of the virtual work done by the system of forces is zero, when small quantities of second and higher orders are neglected.

Conversely, if the algebraic sum of the virtual work for every such displacement is zero, the system is in equilibrium.

Proof :

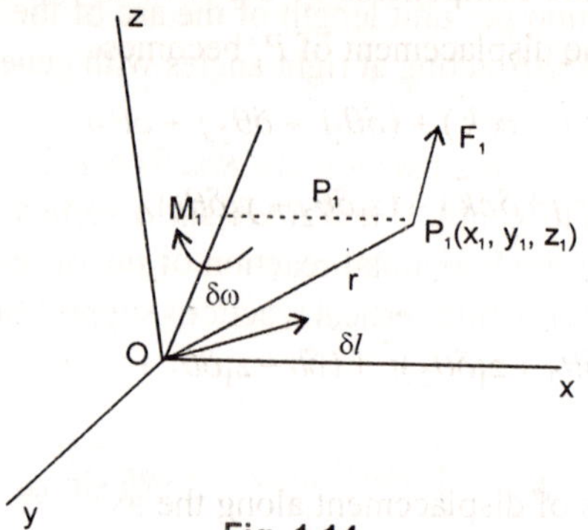

Fig. 1.14

Let $\vec{F_1}, \vec{F_2}, \ldots\ldots \vec{F_n}$ be the forces on a system at the point P_1, P_2 ......, P_n and $\delta\overrightarrow{p_1}$, $\delta\overrightarrow{p_2}$..........$\delta\overrightarrow{p_n}$. be the small virtual displacements of their points of applications along their lines of action, then virtual work done is

$$\vec{F_1}\cdot\delta\overrightarrow{p_1}+\vec{F_2}\cdot\delta\overrightarrow{p_2}+\ldots\ldots\ldots+\vec{F}\cdot\delta\overrightarrow{p_n}. \qquad (1)$$

We know that the work done by a force during any displacement is equal to the work done by its components. Let X_1, Y_1, Z_1 be the components of the force $\vec{F_1}$ parallel to the axes and δx_1, δy_1, δz_1, be the components of $\delta\overrightarrow{p_1}$ in the same directions, then

$$\therefore \quad \vec{F_1}\cdot\delta\overrightarrow{p_1} = X_1\delta x_1 + Y_1\delta y_1 + Z_1\delta z_1,$$

with similar notations for other forces.

The most general displacement of a rigid body consists of a translation and a rotation about an arbitrary point. Taking the arbitrary point O as origin, we give small translational displacement $\delta\vec{l}$, and rotational displacement $\delta\vec{\omega}$ about an axis through O.

On account of $\delta\vec{l}$, P_1 (x_1, y_1, z_1) is transferred to ($x_1 + \delta a$, $y_1 + \delta b$, $z_1 + \delta c$) where $\delta a, \delta b, \delta c$ are components of $\delta\vec{l}$. On account of rotation $\delta\vec{\omega}$ the actual displacement of P_1 in a direction perpendicular to the plane P_1OM_1 is $PM_1\ \delta\omega$ where $PM_1 = p_1$ is the perpendicular from P_1 on the axis of rotation.

Now $\qquad p_1 = r_1 \sin\theta$, where $OP_1 = r_1$ and $\angle P_1OM_1 = \theta$

Hence, displacement due to rotation becomes $r_1\,\delta\omega \sin\theta$ or in vector notation $\delta\vec{\omega}\times\vec{r}_1$.

Hence, total displacement of P_1 is

$$\delta\vec{p}_1 = \delta\vec{l} + \delta\vec{\omega}\times\vec{r}_1.$$

If $\delta\theta_1$, $\delta\theta_2$, $\delta\theta_3$ be the components of $\delta\vec{\omega}$ and $\vec{i}, \vec{j}, \vec{k}$ be the unit vectors along the axes, then the displacement of P_1 becomes

$$\begin{aligned}\delta\vec{p}_1 &= (\delta a\vec{i} + \delta b\vec{j} + \delta c\vec{k}) + (\delta\theta_1\vec{i} + \delta\theta_2\vec{j} + \delta\theta_3\vec{k})\times(x_1\vec{i} + y_1\vec{j} + z_1\vec{k})\\ &= (\delta a\vec{i} + \delta b\vec{j} + \delta c\vec{k}) + (z_1\delta\theta_2 - y_1\delta\theta_3)\vec{i} + (x_1\delta\theta_3 - z_1\delta\theta_1)\vec{j}\\ &\qquad + (y_1\delta\theta_1 - x_1\delta\theta_2)\vec{k}.\\ &= (\delta a - y_1\delta\theta_3 + z_1\delta\theta_2)\vec{i} + (\delta b - z_1\delta\theta_1 + x_1\delta\theta_3)\vec{j}\\ &\qquad + (\delta c - x_1\delta\theta_2 + y_1\delta\theta_1)\vec{k}.\end{aligned}$$

Hence, components of displacement along the axes are:

$$\begin{aligned}\delta x_1 &= \delta a - y_1\delta\theta_3 + z_1\delta\theta_2,\\ \delta y_1 &= \delta b - z_1\delta\theta_1 + x_1\delta\theta_3,\\ \delta z_1 &= \delta c - x_1\delta\theta_2 + y_1\delta\theta_1.\end{aligned}$$

$$\therefore\ \vec{F}_1\cdot\delta\vec{p}_1 = X_1(\delta a - y_1\delta\theta_3 + z_1\delta\theta_2) + Y_1(\delta b - z_1\delta\theta_1 + x_1\delta\theta_3) + Z_1(\delta c - x_1\delta\theta_2 + y_1\delta\theta_1)$$

Since, δa, δb, δc, $\delta\theta_1$, $\delta\theta_2$, $\delta\theta_3$ are the same for all the points of the body, the work done by all the forces

$$\begin{aligned}&= \delta a\sum X_1 + \delta b\sum Y_1 + \delta c\sum Z_1 + \delta\theta_1\sum(y_1Z_1 - z_1Y_1)\\ &\qquad + \delta\theta_2\sum(z_1X_1 - x_1Z_1) + \delta\theta_3\sum(x_1Y_1 - y_1X_1)\end{aligned}$$

But we know that if the system of forces is in equilibrium, then

$$\begin{aligned}\sum X_1 &= \sum Y_1 = \sum Z_1 = \sum(y_1Z_1 - z_1Y_1)\\ &= \sum(z_1X_1 - x_1Z_1) = \sum(x_1Y_1 - y_1X_1) = 0.\end{aligned}$$

Hence, the virtual work of the system is zero.

1.11 CONVERSE

If the virtual work of the system is zero, then the system is in equilibrium.

Proof. If the virtual work done by the system of forces is zero, then

$$\delta a \sum X_1 + \delta b \sum Y_1 + \delta c \sum Z_1 + \delta\theta_1 \sum (y_1 Z_1 - z_1 Y_1)$$
$$+\delta\theta_2 \sum (z_1 X_1 - x_1 Z_1) + \delta\theta_3 \sum (x_1 Y_1 - y_1 X_1) = 0.$$

for all values of δa, δb, δc, $\delta\theta_1$, $\delta\theta_2$, $\delta\theta_3$, since, these quantities are arbitrary, the necessary conditions are

$$\sum X_1 = \sum Y_1 = \sum Z_1 = \sum (y_1 Z_1 - z_1 Y_1)$$
$$= \sum (z_1 X_1 - x_1 Z_1) = \sum (x_1 Y_1 - y_1 X_1) = 0.$$

but these are the conditions of the equilibrium of forces, hence it is proved that system is in equilibrium.

1.12 WORK DONE IN SOME STANDARD CASES

While calculating the work done by a force in a certain displacement, following is the convention of sign.

The work done is positive (or negative) according as the distances of the various points from a fixed line are measured in the same (or opposite) directions in which the force is acting.

1. **Work Done by the Weight of the body**

   Let W be the weight of the body, z the height of its centre of gravity G above a fixed horizontal level XX'.

   In any small displacement, let z increases to $z + \delta z$, then the work done is $-W\delta z$, since displacement δz is in a direction opposite to that of W.

   If, however, z is the measure of depth of G below XX' in any displacement z is increased by δz, then the work done by W is $+ W\delta z$.

2. **Work done by Tension in a string**

   Let AB be the string and T, T be the tensions at its ends A and B along AB and BA.

   In any small displacement, let AB becomes $A'B'$ inclined at angle θ to AB.

   Since θ is small, therefore to first power of θ, $\cos\theta = 1$

   Drawn $A'L$ and $B'M$ perpendiculars to AB.

   Hence, work done by the tension

$$= T.AL - T.BM$$
$$= T(AL - BM)$$
$$= T(AB - A'B' \cos\theta)$$
$$= T(AB - A'B')$$
$$= -T(A'B' - AB)$$

$$= -T \times \text{Extension produced in } AB$$
$$= -T\delta l,$$

where δl the increase in the length of the string.

3. **Work done by the thrust in a rod**

If AB be a rod of length l, instead of string, the thrust T, T at its ends are along AB and BA respectively,

Then the work done by the thrust in any small displacement which is due to increas δl in l is $+ T\delta l$.

If however, the rod be in a state of tension, the work done by this tension is $-T\delta l$.

1.13 FORCES WHICH DO NOT APPEAR IN THE EQUATION OF VIRTUAL WORK

(*i*) **Tension in an inextensible string.** By (2) above the work done by the tension is $-T\delta l$.

If the string is inextensible then $\delta l = 0$, therefore the work done by T is zero.

Similarly, the thrust in a light rod does no work.

(*ii*) **Reaction of any smooth surface with which the body is in contact.** Since, the reaction at the point of contact is normal to the surface and as such is at right angle to the displacement, hence no work is done by it.

(*iii*) **Reaction of a fixed surface on which body rolls without sliding.** Since, the point of contact is instantaneously at rest, hence there is no displacement, and as such no work is done.

(*iv*) **Mutual Reaction between any two bodies which Roll upon one another.** By Newton's third law of motion, the action and reaction are equal and opposite, the virtual work done by them is zero.

(*v*) **Reaction at a fixed point or a fixed axis about which the body is constrained to rotate.** Sincc, the point of application of the force does not undergo any change and hence no work is done.

Notes :

(*a*) Sometimes, it is required to find the tension in the string in an inextensible string or the thrust in a light rod. In such cases, the string is supposed to be replaced by two tensions T, T acting inwards and the rod is supposed to be represented by two thrusts outward.

(*b*) Since, the work done by a force in direction perpendicular to its line of action is zero, hence, while writing equation of virtual work, the displacement should be so given as to be perpendicular to any particular force which is not required.

1.14 STABILITY OF EQUILIBRIUM

When a body is slightly disturbed from a position of equilibrium, there are three possibilities:

(*i*) Body may tend to return to its original position, then the equilibrium is said to be *stable.*

(*ii*) Body may recede away from the original position, then the equilibrium is called *unstable.*

(*iii*) Body may remain in equilibrium in new position as well, then the equilibrium is said to be *neutral.*

Positions of Equilbrium

If a body or a system of bodies are under the action of no other force than the weights of the bodies and mutual actions and reactions between them, the positions of equilibrium correspond to maximum and minimum velues of the height of the centre of gravity of the system above a fixed horizontal level.

Let w_1, w_2, ...be the weights of the bodies which are in equilibrium, when their C.G's are at heights z_1, z_2,...........above a fixed horizontal level, if W is the total weight and $\bar{z}$ be the height of C.G. of the system, then

$$W\bar{z} = w_1 z_1 + w_2 z_1 + \ldots\ldots$$

Since, the work done by mutual actions and reactions do not enter in the equation of virtual work, the virtual work done by the system is $Wd\bar{z}$.

If the system is in equilibrium, then the virtual work done is zero.

$$\therefore \quad Wd\bar{z} = 0$$

$$\therefore \quad d\bar{z} = 0.$$

Hence, $\bar{z}$ is maximum or minimum.

Working Rule

(*i*) Let the height $\bar{z}$ of the C.G. above a fixed horizontal level be expressed as a function of some independent variable θ (say).

(*ii*) Solve the equation $\dfrac{d\bar{z}}{d\theta} = 0$, giving $\theta = \alpha, \beta, \gamma, \ldots\ldots$

(*iii*) Examine the value of $\dfrac{d^2\bar{z}}{d\theta^2}$ for $\theta = \alpha, \beta, \gamma, \ldots\ldots\ldots\ldots$respectively.

If for $\theta = \alpha$ (say), $\dfrac{d^2\bar{z}}{d\theta^2}$ is positive then equilibrium is stable for $\theta = \alpha$. If $\dfrac{d^2\bar{z}}{d\theta^2}$ is negative for $\theta = \alpha$, then equilibrium is unstable. If $\dfrac{d^2\bar{z}}{d\theta^2}$ is zero for $\theta = \alpha$, we examine $\dfrac{d^3\bar{z}}{d\theta^3}$ for $\theta = \alpha$. If it is non-zero, the equilibrium is neutral, but if it is zero, we have to examine $\dfrac{d^4\bar{z}}{d\theta^4}$ for $\theta = \alpha$. If it is positive, the equilibrium is stable and if it is negative, equilibrium is unstable and so on.

1.15 ENERGY TEST OF STABILITY

For any conservative system, the sum of kinetic energy and potential energy is constant. In position of equilibrium, kinetic energy is zero, therefore potential energy constant or stationary, *i.e.*, potential energy is either *maximum or minimum*.

If the system be slightly displaced from the position when the potential energy is maximum, the potential energy will decrease so that kinetic energy will increase or in other words kinetic energy will be positive as in the equilibrium position it is zero. Hence, the system will move away from the position of equilibrium, therefore equilibrium will be *unstable*.

If the system is slightly displaced from the position of minimum potential energy, the potential energy will increase, so that kinetic energy will become negative, which is impossible. Hence, the system will move back to the position of minimum potential energy, so that the equilibrium is *stable*.

If potential energy is neither maximum nor minimum the equilibrium will be neutral.

Examples

1. *A* regular octahedron formed of twelve equal rods, each of weight w freely jointed together, is suspended from one corner. Show that the thrust in each rod is $\frac{3}{2}w\sqrt{2}$.

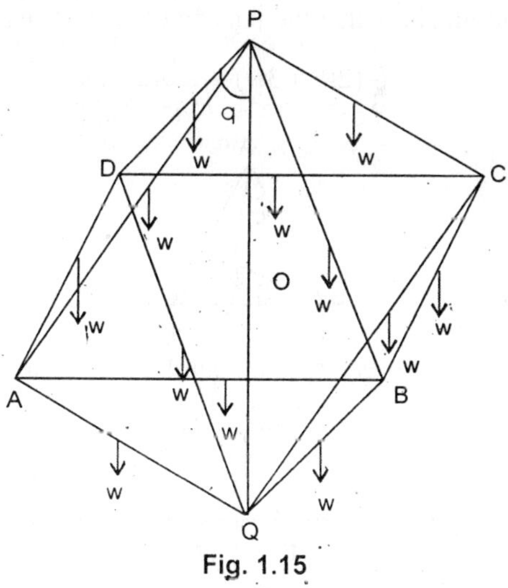

Fig. 1.15

Let the length of each rod of the octahedron (P, $ABCD$, Q) be $2a$, and let it be suspended from the vertex P.

When the angle that one of the slant rods say, AP, makes an angle θ with the vertical PQ, we have

$$OP = 2a\cos\theta,\ OA = 2a\sin\theta,$$

$$AP = \sqrt{2}\,OA = 2\sqrt{2}\ a\sin\theta.$$

Let the system be given a small displacement so that θ becomes θ + δθ, the centre of the gravity moves a little up while the figure $ABCD$ remains a square. Let the weight of each rod be w and the thrust in each horizontal rod T. The equation of the virtual work gives

$$-4w\delta(a\cos\theta) - 4w\delta(2a\cos\theta) - 4w\delta(3a\cos\theta)$$

$$-4T\delta\ (2\sqrt{2}\ a\sin\theta) = 0.$$

$$\therefore \qquad 24wa\sin\theta\delta\theta = 8\sqrt{2}\ Ta\cos\theta\delta\theta.$$

$$\therefore \qquad T = \frac{3w}{\sqrt{2}}\tan\theta\,.$$

Now, in the position of equilibrium, $AB = 2a$. Therefore $l = \sqrt{2}\sin\theta$, *i.e.*, $\theta = 45°$ Hence, we obtain

$$T = \frac{3}{2}\sqrt{2}\,w.$$

2. A tripod consists of three equal uniform bars, each of length a and weight w, which are freely jointed at one extremity, their middle points being joined by strings, each of length b. The tripod is placed with its free ends in contact with a smooth horizontal plane and a weight W is attached to the common joint; show that the tension of each string is

$$\frac{2}{3}(2W + 3w)\frac{b}{\sqrt{9a^2 - 12b^2}}.$$

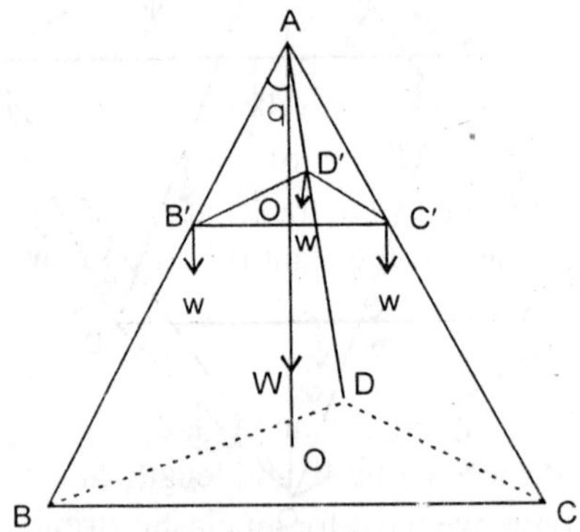

Fig. 1.16

Let the legs of tripod be AB, AC and AD each of length a and let $B'C'$, $C'D'$ and $D'B'$ be the strings each of length b. Let AO be the perpendicular from A on BCD and let $\angle OAB = \theta$, we have

$$OA = a\cos\theta,\ OB = a\sin\theta.$$

$$BC = \sqrt{3}\, a\sin\theta \text{ } (BCD \text{ is an equilateral } \Delta),$$

$$B'C' = \frac{1}{2}\sqrt{3}\, a\sin\theta.$$

Let the displacement changes θ to θ + δθ, while the lengths of rods remain unchanged.

The equation of the virtual work gives

$$-W\delta(a\cos\theta) - 3w\delta\left(\frac{1}{2}a\cos\theta\right) - 3T\delta\left(\frac{1}{2}\sqrt{3}\,a\sin\theta\right) = 0.$$

$$\therefore \quad T = \frac{2W + 3w}{3\sqrt{3}}\tan\theta.$$

Now, in the position of equilibrium $B'C' = b$.

$$\therefore \quad \frac{1}{2}\sqrt{3}a\sin\theta = b,$$

$$\therefore \quad \sin\theta = \frac{2b}{\sqrt{3}a},$$

$$\therefore \quad \tan\theta = \frac{2b}{\sqrt{3a^2 - 4b^2}}.$$

Hence, we obtain

$$\therefore \quad T = \frac{2b\,(2W + 3w)}{3\sqrt{9a^2 - 12b^2}}.$$

3. Twelve similar and uniform rods are jointed at their ends to form an octahedron and are suspended from one of the vertices, being supported by a string joining this vertex to the opposite vertex; the string is elastic and such that the total weights of the rods would stretch it to twice its length, the latter being equal to the length of the either rod. In the position of equilibrium, show that the slant rods are inclined to the vertical at an angle $\cos^{-1}\frac{3}{4}$.

See figure of Ex. 1 in which P and Q are joined by an elastic string.
Let the weight of each rod be W and length $2a$.
From Hooke's Law, we have for an elastic string,

$$\lambda = 12W.$$

When one of the slant rods make an angle θ with the vertical, the length of the string is $4a\cos\theta$ and its tension is

$$T = 12W\frac{4a\cos\theta - 2a}{2a}.$$

The equation of the virtual work is

$$-4W\delta(a\cos\theta) - 4W\delta(2a\cos\theta) - 4W\delta(3a\cos\theta) + T\delta(4a\cos\theta) = 0.$$

$$\therefore \quad 24Wa\sin\theta\delta\theta = 12W\frac{4a\cos\theta - 2a}{2a}(4a\sin\theta\delta\theta)$$

$$= 48Wa\sin\theta\,(2\cos\theta - 1)\delta\theta$$

Since, $\sin\theta \neq 0$, hence we have

$$1 = 2(2\cos\theta - 1),$$

$$\therefore \quad 4\cos\theta = 3,$$

$$\therefore \quad \theta = \cos^{-1}\frac{3}{4}.$$

4. A heavy elastic string, whose natural length is $2\pi a$, is placed round a smooth cone whose axis is vertical and whose semi-vertical angle is α. If W be the weight and λ the modulus of elasticity of the string, prove that it will be in equilibrium when in the form of a circle whose radius is $\left(1 + \frac{W}{2\pi\lambda}\cot\alpha\right)$.

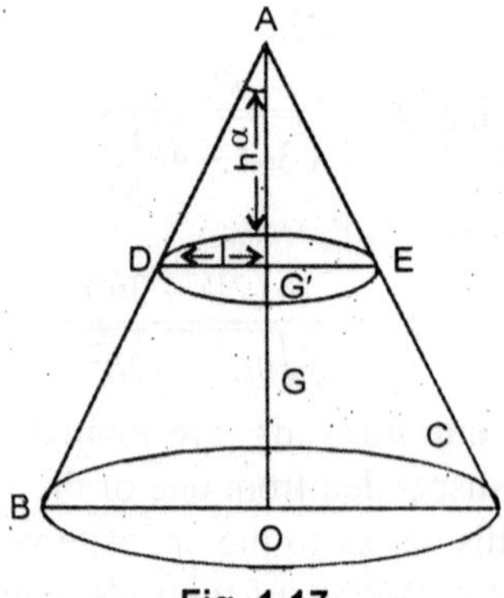

Fig. 1.17

Let ABC be the cone of semi-vertical angle α on which the elastic string of weight W rests in the position DE.

If h be the depth of the string below A, we have

$$DE = 2h\tan\alpha.$$

Therefore the extended length of the string is $2\pi h\tan\alpha$, and its tension is

$$T = \lambda\frac{2\pi h\tan\alpha - 2\pi a}{2\pi a} = \lambda\frac{h\tan\alpha - a}{a}.$$

The equation of the virtual work gives

$$W\delta h - T\delta(2\pi h\tan\alpha) = 0,$$

$$\therefore \quad W = T.\, 2\pi \tan\alpha,$$

$$\therefore \quad T = \frac{W}{2\pi}\cot\alpha = \lambda\frac{h\tan\alpha - a}{a},$$

$$\therefore \quad h\tan\alpha = a\left(1 + \frac{W}{2\pi\lambda}\cot\alpha\right).$$

This is the radius of the circle in which the string rests in equilibrium.

5. A smooth paraboloid of revolution is fixed with its axis vertical and vertex upwards; on it is placed a heavy elastic string of unstretched length $2\pi c$; when the string is in equilibrium show that it rests in the from of a circle of radius $\dfrac{4\pi ac\lambda}{4\pi a\lambda - cW}$, where W is the weight of the string, λ its modulus of elasticity, and $4a$ the latus rectum of the generating parabola.

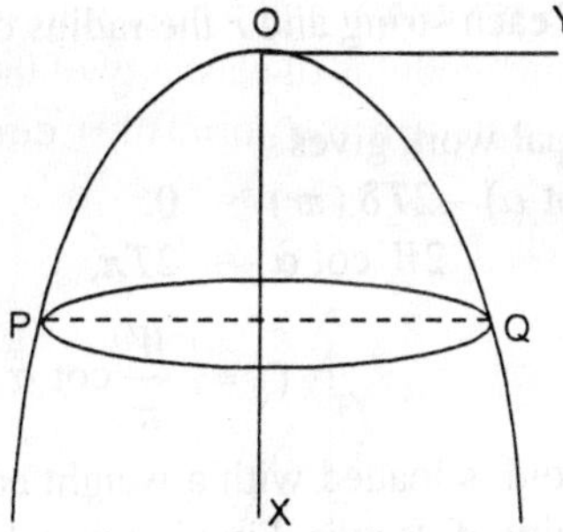

Fig. 1.18

Let POQ be the part of the paraboloid of revolution, whose generating parabola is $y^2 = 4ax$ and PQ be the elastic string resting on it as a circle of radius k, at a depth h below O.

The equation of virtual work gives

$$W\delta h - T\delta(2\pi k) = 0. \tag{1}$$

But

$$k^2 = 4ah,$$

$$\therefore \quad k\delta k = 2a\delta h. \tag{2}$$

From (1) and (2) we get

$$T = \frac{Wk}{4\pi a}.$$

From Hooke's Law

$$T = \lambda\frac{2\pi k - 2\pi c}{2\pi c} = \lambda\frac{k-c}{c}.$$

Therefore

$$\lambda\frac{k-c}{c} = \frac{Wk}{4\pi a}$$

$$\therefore \quad 4\pi a\lambda\,(k - c) = Wkc.$$

$$\therefore \quad k(4\pi a\lambda - cW) = 4\pi ac\lambda,$$

$$\therefore \quad k = \frac{4\pi ac\lambda}{4\pi a\lambda - cW}.$$

6. Two equal particles are connected by two given weightless strings, which are placed like a necklace on a smooth cone whose axis is vertical and whose vertex is uppermost; show that the tension of each string is $\frac{W}{\pi}\cot\alpha$, where W is the weight of each particle and 2α the vertical angle of the cone.

   See Figure of Ex. 4 above.

   Let the two particles D and E, each of weight W, tied together by two strings, be placed over the cone ABC of vertical angle 2α.

   If T is the tension of each string and r the radius of the circle in which it rests on the cone.

   The equation of virtual work gives

$$2W\delta(r\cot\alpha) - 2T\delta(\pi r) = 0.$$

$$\therefore \quad 2W\cot\alpha = 2T\pi,$$

$$\therefore \quad T = \frac{W}{\pi}\cot\alpha.$$

7. A solid oblate spheroid is loaded with a weight equal to n times its own weight at one extremity of its axis. Find in what different positions it can be in equilibrium resting on a smooth horizontal plane, and in which of these the equilibrium is stable.

   Show that if $e^2 < \frac{n}{2n+1}$, there are only two positions of equilibrium.

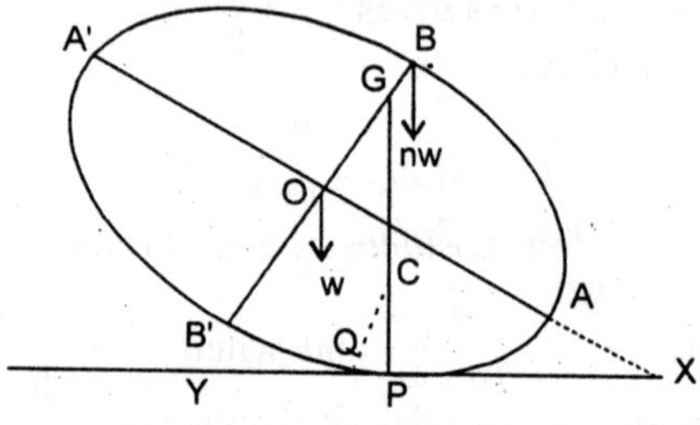

Fig. 1.19

Let $ABA'B'PA$ be the generating ellipse of the oblate spheroid, whose equation referred to the axes of x and y through the centre O is

$$\frac{x^2}{a^2} + \frac{y^2}{c^2} = 1.$$

Let w be the weight of spheroid, and a weight nw be placed at B. The C.G. of the system is G, where $nBG = OG$, so that

$$\frac{OG}{n} = \frac{BG}{1} = \frac{OG+BG}{n+1} = \frac{c}{n+1}.$$

If P be the point of contact of the spheroid and the horizontal plane, PG must be vertical for equilibrium. Therefore PG is also the normal to the ellipse at P. It follows that the positions of equilibrium can be obtained by drawing normals to the ellipse from the point G.

Since, four normals in general can be drawn to an ellipse from a given point, there are four positions of equilibrium.

If P be a point ($a \cos \phi$, $c \sin \phi$), the equation of PG is

$$ax \sec \phi - cy \operatorname{cosec} \phi = a^2 - c^2,$$

$$\therefore ax \sin \phi - cy \cos \phi = a^2 e^2 \sin \phi \cos \phi.$$

This passes through $G\left(0, \frac{-cn}{n+1}\right)$ if

$$\frac{c^2 n}{n+1} \cos \phi = a^2 e^2 \sin \phi \cos \phi,$$

so that either $\cos \phi = 0$,

$$\therefore \quad \phi = \frac{\pi}{2}, \frac{3\pi}{2},$$

or

$$\sin \phi = \frac{c^2 n}{(n+1) a^2 e^2} = \frac{(1-e^2) n}{(n+1) e^2},$$

which gives real values of ϕ when

$$(1 - e^2) n < (n + 1)e^2$$

i.e., when

$$e^2 > \frac{n}{2n+1}.$$

Hence, there are four positions of equilibrium (two symmetrical and two obliqe) when $e^2 > \frac{n}{2n+1}$, and when $e^2 \leq \frac{n}{2n+1}$ there are only two positions (both symmetrical) of equilibrium.

If C be the centre of curvature at P and Q, a neighbouring point of P, then

$$CQ = CP, GQ < GC + CQ = GC + CP = GP.$$

Therefore GP is maximum and the equilibrium at P is unstable.

It follows that when there are four positions of equilibrium, the oblique positions are unstable and the symmetrical positions are stable. When there are two positions of equilibrium (at B and B'), the one at B is clearly stable and that at B' unstable.

8. Four equal spheres rests in contact at the bottom of a spherical bowl, their centres being at the corners of a hòrizontal square. A fifth sphere of the same radius and density is placed upon them. It is required under what conditions equilibrium exists, all contacts being regarded as smooth.

Let the radius of the bowl be a and radius of the spheres be b. If equilibrium exists, each sphere is in equilibrium under the action of its weight and the reactions of the bowl and the other spheres. Since, the contacts are smooth, all the forces which act upon a given sphere pass through the centres of the other sphere. In particular, the top ball is in equilibrium under the action of its own weight and the reactions of the four lower spheres. It is assumed from symmetry that these four reactions are numerically equal. If R_1 is the magnitude of each of these reactions and W is the weight of each sphere, then

$$4R_1 \cos 45° = W;$$

$$\therefore \qquad R_1 = \frac{W}{2\sqrt{2}}.$$

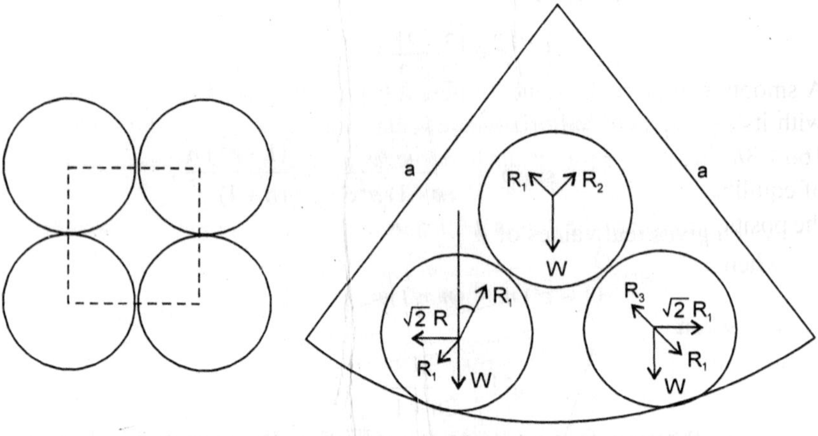

Fig. 1.20 *(i)* Fig. 1.20 *(ii)*

Now, consider the actions of the lower spheres on one another. Again, from consideration of symmetry, these actions are all equal and will be denoted by R_2. Since they pass through the centres of the spheres, the two actions on each sphere can be combined into a single horizontal force the magnitude of which is $\sqrt{2}\ R_2$, and the direction is perpendicular to the axis of the bowl. Let the magnitude of the action of the bowl on each sphere be denoted by R_3 since, they are all of the same magnitude. If the angle which this action makes with the vertical is α_3, than

$$\sin \alpha_3 = \frac{\sqrt{2}\, b}{a-b}.$$

Thus the four forces which act upon each sphere can be regarded as acting at the centre of the sphere; they lie in the same plane; their directions are known; and the magnitude of two of the forces R_1 and W are known. On resolving these forces horizontally and vertically, we obtain the following equations

$$\sqrt{2}\, R_2 = R_3 \sin \alpha_3 - R_1 \sin 45°,$$
$$R_3 \cos \alpha_3 = R_1 \cos 45° + W,$$

Hence,

$$R_3 = \frac{5}{4} W \sec \alpha_3,$$

$$R_2 = \frac{W}{4\sqrt{2}} (5 \tan \alpha_3 - 1).$$

The pressures between the four lower spheres vanish if $R_2 = 0$; *i.e.*, if $\tan \alpha_3 = 1/5$. Equilibrium can exist only if $\tan \alpha_3 \geq 1/5$, in order that this may be true, it is necessary that

$$a \leq \left(2\sqrt{13} + 1\right) b.$$

9. A smooth solid circular cone, of height h and vertical angle 2α, is at rest with its axis vertical in horizontal circular hole of radius a. Show that if $16a > 3h \sin 2\alpha$ the equilibrium is stable, and there are two other positions of equilibrium and that if $16 < 3h \sin 2\alpha$ the equilibrium is unstable, and the position in which the axis is vertical is the only position of equilibrium.

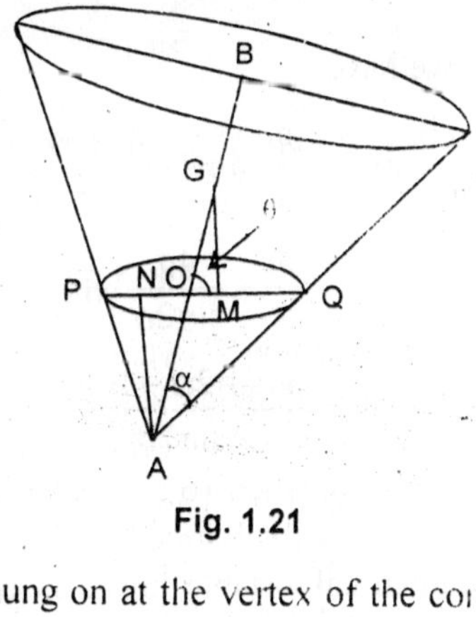

Fig. 1.21

If a weight w be hung on at the vertex of the cone, whose weight is W, prove that the corresponding condition is $16a\,(w + W) > 3Wh \sin 2\alpha$.

Considering the inclined position of the cone, let its section by the plane of the hole be an ellipse at PQ, whose axis is $PQ = 2a$. If the axis of the cone meets the horizontal plane of the hole in O, the inclination of the axis to the horizontal is $\angle BOQ = \theta$.

The height of the centre of gravity G is

$$z = GM = AG \sin\theta - AN,$$

$$= \frac{3}{4} h \sin\theta - AQ \sin(\theta - \alpha)$$

$$= \frac{3}{4} h \sin\theta - \frac{PQ \sin(\theta + \alpha)}{\sin 2\alpha} \sin(\theta - \alpha),$$

from the triangle APQ, in whicn $\angle AQP = \theta - \alpha$ and $\angle APQ = \pi - (\theta + \alpha)$.

$$\therefore \quad z = \frac{3}{4} h \sin\theta - \frac{a(\cos 2\alpha - \cos 2\theta)}{\sin 2\alpha}.$$

For equilibrium, we have

$$\frac{dz}{d\theta} = \frac{3}{4} h \cos\theta - \frac{2a \sin 2\theta}{\sin 2\alpha} = 0.$$

Therefore, either $\cos\theta = 0$, giving $\theta = \frac{\pi}{2}$,

$$\text{or} \quad \frac{3}{4} h - \frac{4a \sin\theta}{\sin 2\alpha} = 0, \therefore \sin\theta = \frac{3h}{16a} \sin 2\alpha.$$

Hence, there are three positions of equilibrium (one symmetrical in which the axis is vertical and two oblique in which the axis is inclined) provided.

$$16a > 3h \sin 2\alpha.$$

If $16a \le 3h \sin 2\alpha$, there is only one position of equilibrium which is symmetrical position.

To examine stability, we have

$$\frac{d^2 z}{d\theta^2} = -\frac{3h}{4} \sin\theta - \frac{4a \cos 2\theta}{\sin 2\alpha}$$

$$= -\frac{3h}{4} + \frac{4a}{\sin 2\alpha} \text{ when } \theta = \frac{\pi}{2}$$

$$= -\frac{3h \sin 2\alpha - 16a}{4 \sin 2\alpha}$$

$$> 0, \text{ when } 16a > 3h \sin 2\alpha.$$

Therefore z is minimum in this case. It follows that there are three positions of equilibrium, the symmetrical is stable, hence the other two are unstable.

But when $16a < 3h \sin 2\alpha$, $\frac{d^2 z}{d\theta^2} < 0$, so that z is minimum it follows that there is only one position of equilibrium, which is unstable.

When a weight w is hung from the vertex A of the cone, whose own weight is W, the centre of gravity is G' lying on AB, such that

$$wAG' = W.GG'$$

$$\therefore \qquad \frac{AG'}{W} = \frac{GG'}{w} = \frac{AG' + GG'}{w+W} = \frac{\frac{3}{4}h}{w+W},$$

$$\therefore \qquad AG' = \frac{3h}{4} \cdot \frac{W}{w+W}.$$

Hence, replacing h by $hW/(w + W)$ in the above solution, we see that the condition for three positions of equilibrium in which the symmetrical is stable, is

$$16a > \frac{3hW}{w+W} \sin 2\alpha,$$

$$\therefore \qquad 16a\,(w + W) > 3hW \sin 2\alpha.$$

10. A uniform right circular cone of height h and vertical angle 2α rests with its vertex downward and its axis vertical, between two smooth parallel rails at a distance d apart in a horizontal plane. Prove that the equilibrium is stable for angular displacements in which axis remains in a vertical plane parallel to the rails, if $h < \frac{4}{3} d \operatorname{cosec} 2\alpha$.

See the Figure of Example (9) above.

Let us consider the cone in the inclined position with its axis AB making an angle θ with the horizontal. The section of the cone by the plane of the rails is an ellipse whose major axis is PQ and the minor axis is the distance d between the rails.

Let the axis AB meet the horizontal plane of the rails in O. Then the height of the centre of gravity G of the cone above the rails is

$$z = OG \sin\theta = (AG - AO) \sin\theta.$$

We know from the property of the conics that

$AP.AQ \sin^2\alpha$ = the square of the semi-minor axis of the section $= \frac{1}{4} d^2$

Also from the triangles AOP and AOQ, we know that

$$\frac{AP}{\sin\theta} = \frac{AO}{\sin(\theta+\alpha)}$$

and

$$\frac{AQ}{\sin\theta} = \frac{AO}{\sin(\theta-\alpha)}.$$

$$\therefore \qquad AO^2 = \frac{AP.AQ}{\sin^2\theta}.\sin(\theta+\alpha)\sin(\theta-\alpha)$$

$$= \frac{d^2}{4\sin^2\alpha\sin^2\theta}\cdot\frac{1}{2}(\cos 2\alpha - \cos 2\theta)$$

$$\therefore \qquad z = \frac{3h}{4}\sin\theta - \frac{d\sqrt{(\cos 2\alpha - \cos 2\theta)}}{2\sqrt{2}\sin\alpha}$$

For equilibrium, we have

$$\frac{dz}{d\theta} = \frac{3h}{4}\cos\theta - \frac{d}{2\sqrt{2}\sin\alpha}\cdot\frac{\sin 2\theta}{\sqrt{(\cos 2\alpha - \cos 2\theta)}} = 0,$$

which gives,

$$\cos\theta = 0, \therefore \theta = \frac{\pi}{2}$$

For stability, we consider

$$\frac{d^2z}{d\theta^2} = -\frac{3h}{4}\sin\theta - \frac{d}{2\sqrt{2}\sin\alpha}$$

$$\times\left\{\frac{2\cos 2\theta}{(\cos 2\alpha - \cos 2\theta)^{1/2}} - \frac{\sin^2 2\theta}{(\cos 2\alpha - \cos 2\theta)^{3/2}}\right\}.$$

When $\theta = \frac{\pi}{2}$

$$\frac{d^2z}{d\theta^2} = \frac{-3h}{4} + \frac{d}{\sqrt{2}\sin\alpha}\cdot\frac{1}{(1+\cos 2\alpha)^{1/2}}$$

$$= -\frac{3h}{4} + \frac{d}{\sqrt{2}\sin\alpha}\cdot\frac{1}{\sqrt{2}\cos\alpha} = -\frac{3h}{4} + \frac{d}{\sin 2\alpha}.$$

Therefore, the equilibrium is stable for displacement in the vertical plane parallel to the rails if

$$\frac{3}{4}h < d\,\text{cosec}\,2\alpha.$$

$$\therefore \qquad h < \frac{4d}{3}\,\text{cosec}\,2\alpha.$$

11. A right circular cone, whose vertical angle is a right angle, is placed vertex downwards through a square hole in a horizontal plane so as to touch each side. Show that, if the height h of the cone exceeds twice the side of

the square, a position of equilibrium is possible with axis inclined to the horizon at an angle $\sin^{-1}\sqrt{\dfrac{2a}{3h-4a}}$. Prove that this position is stable.

(See the figure of Example 9)

In the inclined position the cone rests in the square hole with the major and minor axes of the section lying along the diagonals of the square. Let the axis AB of the cone make an angle θ with the horizontal and meet the plane of the hole in the point O. Let PQ be the major axis of the section.

If α, β be the semi-axes of the elliptic section, we have

$\alpha^2 + \beta^2$ = square of the radius of the director circle of the ellipse.

Now, the sides of the square are the tangents to the ellipse which are mutually perpendicular, so that the vertices of the square lie on the director circle. Therefore, the radius of the director circle is $a/\sqrt{2}$, where a is the side of the square. Thus,

$$\alpha^2 + \beta^2 = \frac{1}{2}a^2.$$

Since, the vertical angle of the cone ($\angle PAQ$) is a right angle, we have

$$AP^2 + AQ^2 = PQ^2 = 4\alpha^2$$

and

$$AP \cdot AQ \sin^2 45° = \beta^2$$

$\therefore$

$$2AP \cdot AQ = 4\beta^2.$$

$\therefore$

$$AP + AQ = 2\sqrt{\alpha^2+\beta^2} = \sqrt{2}a,$$

But

$$AP = \frac{AN}{\sin(\theta+45°)},\quad AQ = \frac{AN}{\sin(\theta-45°)}$$

$$\therefore \quad AN\left(\frac{\sqrt{2}}{\sin\theta+\cos\theta}+\frac{\sqrt{2}}{\sin\theta-\cos\theta}\right) = \sqrt{2}a,$$

$$\therefore \quad AN = \frac{a(\sin^2\theta-\cos^2\theta)}{2\sin\theta} = \frac{a(2\sin^2\theta-1)}{2\sin\theta} = a\left(\sin\theta-\frac{1}{2}\operatorname{cosec}\theta\right).$$

Hence, the height of the centre of gravity G of the cone above the hole is

$$Z = AG\sin\theta - AN$$

$$= \frac{3}{4}h\sin\theta - AN$$

$$= \frac{3}{4}h\sin\theta - a\sin\theta + \frac{1}{2}a\operatorname{cosec}\theta.$$

For equilibrium we have

$$\frac{dz}{d\theta} = \frac{3}{4}h\cos\theta - a\cos\theta - \frac{1}{2}a\cos\theta\,\mathrm{cosec}^2\theta = 0,$$

which gives either

$$\cos\theta = 0,\ \therefore\ \theta = \frac{\pi}{2}$$

$$\text{or } \frac{3}{4}h - a = \frac{1}{2}a\,\mathrm{cosec}^2\theta,\ i.e.,\ \sin^2\theta = \frac{2a}{3h-4a} \quad (1)$$

θ will be real only if

$$2a < 3h - 4a,\ i.e.,\ h > 2a.$$

If follows that the inclined position of equilibrium, given by (1) is possible if the height of the cone exceeds twice the side of the square.

For stability, we consider

$$\frac{d^2z}{d\theta^2} = -\frac{3h}{4}\sin\theta + a\sin\theta + \frac{1}{2}a(\mathrm{cosec}\,\theta + 2\cos^2\theta\,\mathrm{cosec}^3\theta)$$

When $\theta = \frac{1}{2}\pi$, $\frac{d^2z}{d\theta^2} = -\frac{3}{4}h + a + \frac{1}{2}a < 0$

Therefore, in the case when the oblique position of equilbrium exists the symmetric position is unstable, hence it follows that the oblique position is stable.

1.16 POINSOT'S CENTRAL AXIS WRENCH

Prove that a given system of forces on a rigid body reduces to a force together with a couple whose axis coincides with the direction of the force.

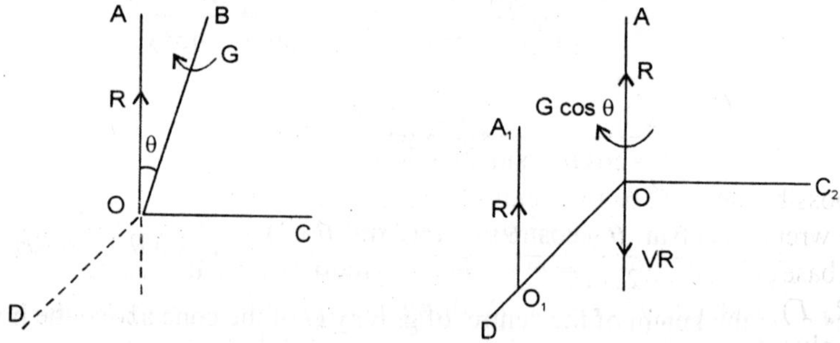

Fig. 1.22 *(i)* Fig. 1.22 *(ii)*

Let the system reduce to the dyname (R, G) at O, the axis OB of G making an angle θ with the direction OA of R. Draw OC in the plane AOB perpendicular to OA and OD perpendicular to the plane AOB. Now G can be resolved into $G\cos\theta$ about OA and $G\sin\theta$ about OC, the plane of $G\cos\theta$ is the plane *COD* and that of $G \sin \theta$ is the plane *AOD*. Again $G \sin \theta$ can be replaced by two equal and unlike parallel forces in the plane *AOD*, having the same moment. Let one of the forces be *R* acting along *AO* produced and the other be at the distance d from *O* so that

$$Rd = G \sin \theta,$$

$$\therefore \qquad d = OO_1 \frac{G\sin\theta}{R}.$$

Now, *R* and $-R$ at *O* balance, hence the system reduces to *R* along O_1A_1 and the couple $G \cos\theta$ in the plane *COD*, its axis can be transferred to O_1A_1 instead of *OA*. Thus the system reduces to a forces *R* along O_1A_1 and a couple $G \cos \theta$ about O_1A_1. Such a system is called a *wrench* and the axis O_1A_1 is the *Poinsot's central axis.* The force *R* is called *intensity* of the wrench and $G \cos\theta$ is called *Poinsot's moment* and is usually denoted by Γ.

The ratio $\frac{\Gamma}{R}$ is called the *pitch* of wrench and is denoted by *p*.

Any straight line considered with a pitch is called a *screw.* Thus screw is a definite line with a given pitch. If a force *R* acts on a screw of pitch *p*, it means a wrench of which intensity is *R* and a couple *pR*.

To prove that for a given system, the wrench is unique, *i.e.*, there is one and only one central axis.

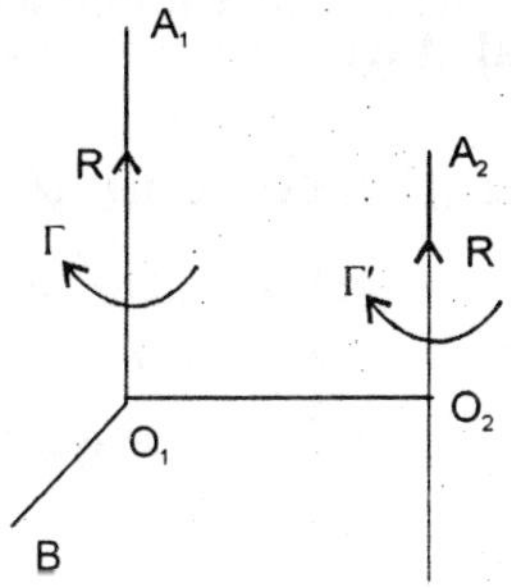

Fig. 1.23

If possible let there be two central axes O_1A_1 and O_2A_2. Since, the intensity of the wrench is always the same both in direction and magnitude, whatever be the base point, the two lines O_1A_1 and O_2A_2 must be parallel. Let the wrenches be (R, Γ) and (R, Γ'). Now *R* along O_2A_2 is equivalent to *R* along O_1A_1 plus a couple *Rl* about a line O_1B perpendicular to O_1A_1, where $O_1O_2 = l$. Thus the given system (R, Γ') is equivalent to a force *R* and two couples Γ about O_1A_1 and *Rl* about O_1B. Hence, the resultant couple about O_1A_1 cannot be Γ. Thus the two systems are not equivalent, hence, our assumption is not correct.

Invariants

First, the intensity $R = \sqrt{X^2 + Y^2 + Z^2}$ is the same whatever be the base point and the axis.

Secondly, the Poinsot's moment $G \cos\theta$ or Γ is the moment about a fixed line *i.e.*, the central axis and is therefore the same, though G and θ will vary from point to point. Thus, two invariants are R and $\Gamma = G \cos \theta$.

Now, d .c.'s. of R are $\frac{X}{R}, \frac{Y}{R}, \frac{Z}{R}$ that of G are $\frac{L}{G}, \frac{M}{G}, \frac{N}{G}$.

$$\cos\theta = \frac{LX + MY + NZ}{RG},$$

$$\therefore \quad LX + MY + NZ = RG \cos\theta = R\Gamma,$$

and therefore is invariant.

Thus for a given system

$$R^2 = X^2 + Y^2 + Z^2, \Gamma = \frac{LX + MY + NZ}{R},$$

$$\bar{P} = \frac{\Gamma}{R} = \frac{LX + MT + NZ}{X^2 + Y^2 + Z^2} = \frac{II}{I}$$

where I and II are first and second invariants.

Equation of the Central Axis

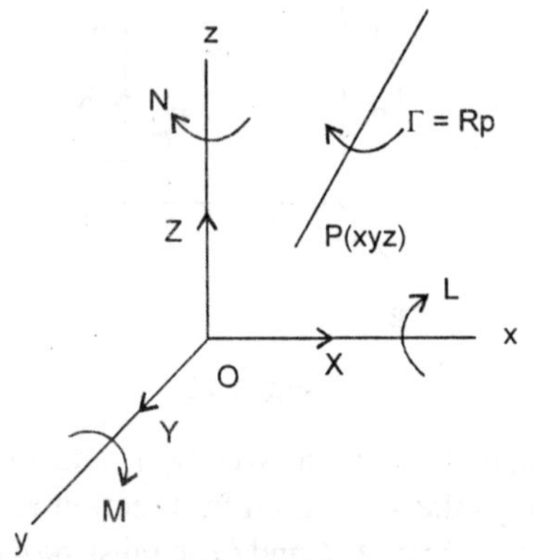

Fig. 1.24

Let the system reduce to dyname $(X, Y, Z; L, M, N)$ at O and let $P(x, y, z)$ be any point on the central axis. The system is equivalent to (R, pR) along the axis,

Now, L = Moment of the system about x-axis

= moment of R about x-axis + component of Rp along x-axis

$$= yZ - zY + Rp.\frac{X}{R}$$

because d.c'.s. of central axis are $\frac{X}{R}, \frac{Y}{R}, \frac{Z}{R}$. At P $\begin{Bmatrix} x, y, z \\ X, Y, Z \end{Bmatrix}$

$$\therefore \qquad L = yZ - zY + pX$$

$$\therefore \qquad \frac{L - yZ + zY}{X} = P.$$

Similarly,

$$\frac{M - zX + xZ}{Y} = p, \quad \frac{N - xY + yY}{Z} = p.$$

Thus equation of the central axis is given by

$$\frac{L - yZ + zY}{X} = \frac{M - zX + xZ}{Y} = \frac{N - xY + yX}{Z} = p = \frac{\Gamma}{R}.$$

1.17 TO FIND THE RESULTANT WRENCH

(a) A System of Forces

If $X, Y, Z; L, M, N$ be the components of the dyname of the system at the origin, the resultant is a wrench (R, pR), so that

$$\frac{L - yZ + zY}{X} = \frac{M - zX + xZ}{Y} = \frac{N - xY + yX}{Z} = p = \frac{\Gamma}{R}$$

$$R^2 = X^2 + Y^2 + Z^2; \; R^2p = \Gamma R = LX + MY + NZ,$$

and the equation of central axis is

$$\frac{L - yZ + zY}{X} = \frac{M - zX + xZ}{Y} = \frac{N - xY + yX}{Z} = p.$$

(b) Two Wrenches

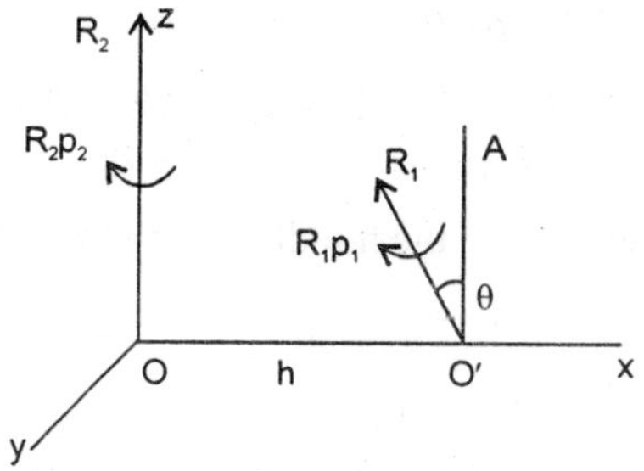

Fig. 1.25

Let the wrenches be $(R_1\, p_1\, R_1)$, $(R_2\, p_2\, R_2)$, θ the angle and h the shortest distance between the axes of the two wrenches. Let us take the axis of the wrench (R_2, $p_2 R_2$) as z-axis. Let X, Y, Z, L, M, N be the components of dyname at O

$$\begin{pmatrix} 0,0,0 \\ 0,0,R_2 \end{pmatrix}, \quad \begin{pmatrix} h,0,0 \\ 0, R_1 \sin\theta, R_1\cos\theta \end{pmatrix}$$

$$X = 0, Y = R_1 \sin\theta, \ Z = R_1 \cos\theta + R_2,$$

$$L = O, M = -hR_1 \cos\theta + R_1p_1 \sin\theta, N = hR_1 \sin\theta + p_1R_1 \cos\theta + p_2R_2.$$

The resultant is a wrench (R, pR), where

$$R^2 = X^2 + Y^2 + Z^2 = R_1^2 \sin^2\theta + (R_1 \cos\theta + R_2)^2$$

$$= R_1^2 + R_2^2 + 2R_1R_2 \cos\theta.$$

$$R^2p = LX + MY + NZ$$

$$= R_1^2 p_1 + R_2^2 p_2 + R_1R_2 \{(p_1 + p_2) \cos\theta + h \sin\theta.\}$$

Central axis is

$$\frac{L - yZ + zY}{X} = \frac{M - zX + xZ}{Y} = \frac{N - xY + yX}{Z} = p.$$

Now,

$$\frac{L - yZ + zY}{X} = \frac{0 - y(R_2 + R_1\cos\theta) + zR_1\sin\theta}{0},$$

$$\frac{M - zX + xZ}{Y} = \frac{-hR_1\cos\theta + R_1p_1\sin\theta - z.0 + x(R_2 + R_1\cos\theta)}{R_1\sin\theta},$$

$$\frac{N - xY + yX}{Z} = \frac{hR_1\sin\theta + R_1p_1\cos\theta + R_2p_2 - xR_1\sin\theta - y.0}{R_2 + R_1\cos\theta}$$

$$\therefore \quad y(R_2 + R_1 \cos\theta) = zR_1 \sin\theta. \tag{1}$$

$$(-hR_1 \cos\theta + R_1p_1 \sin\theta + xR_2 + xR_1 \cos\theta)(R_2 + R_1 \cos\theta)$$

$$= R_1 \sin\theta(hR_1 \sin\theta + R_1p_1 \cos\theta + R_2p_2 - xR_1 \sin\theta),$$

$$\therefore x(R_1^2 + R_2^2 + 2R_1R_2 \cos\theta)$$

$$= hR_1^2 + R_1R_2 \{h \cos\theta - (p_1 - p_2) \sin\theta\}$$

$$\therefore xR^2 = hR_1^2 + R_1R_2 \{h \cos\theta - (p_1 - p_2) \sin\theta\} \tag{2}$$

If c be the distance from the middle point of the shortest distance, then

$$c = x - \frac{h}{2}.$$

and (2) becomes

$$\left(c + \frac{h}{2}\right) R^2 = hR_1^2 + R_1R_2 [h \cos\theta - (p_1 - p_2) \sin\theta,]$$

$$\therefore cR^2 = \frac{h}{2}(2R_1^2 - R^2 + 2R_1R_2 \cos\theta) - R_1R_2(p_1 - p_2) \sin\theta,$$

$$= \frac{h}{2}(R_1^2 - R_2^2) - R_1 R_2 (p_1 - p_2) \sin \theta. \quad (3)$$

(c) Two Forces

This can be deduced from (b) by putting

$$p_1 = p_2 = 0.$$

$$\therefore R^2 = R_1^2 + R_2^2 + 2R_1R_2 \cos \theta,$$

$$R^2p = R_1R_2 h \cos \theta$$

and the central axis is

$$y(R_2 + R_1 \cos \theta) = zR_1 \sin \theta,$$

and

$$R^2x = hR_1 (R_1 + R_2 \cos \theta),$$

$$cR^2 = \frac{h}{2}(R_1^2 - R_2^2).$$

(d) Any number of Wrenches

R^2 will be a homogeneous function of R_1, R_2, R_3 ....

$\therefore R^2 = a_{11}R_1^2 + a_{22}R_2^2 + ...+ 2a_{12}R_1R_2 + 2a_{13}R_1R_3 + ...$

To find constans, put R_3, R_4.......... = 0, then

$$R^2 = a_{11}R_1^2 + a_{22}R_2^2 + 2a_{12}R_1R_2.$$

But in this case,

$$R^2 = R_1^2 + R_2^2 + 2R_1R_2 \cos \theta_{12}.$$

$$\therefore \quad a_{11} = 1 = a_{22}, a_{12} = \cos \theta_{12}$$

where θ_{12} is the angle between R_1 and R_2.

Hence, in general

$$a_{11} = a_{22} = a_{33} = ...= 1$$

and

$$a_{12} = \cos\theta_{12}, a_{13} = \cos\theta_{13}, \text{ etc.}$$

Thus

$$R^2 = R_1^2 + R_2^2 + R_3^2 + ... + 2R_1R_2 \cos\theta_{12} + 2R_1R_3 \cos\theta_{13} + ...$$

Similarly,

$$R^2p = b_{11}R_1^2 + b_{22}R_2^2 + 2b_{12}R_1R_2$$

$$= R_1^2p_1 + R_2^2p_2 + R_1R_2 \{(p_1 + p_2) \cos\theta_{12} + h_{12} \sin \theta_{12}\}$$

Hence,

$$b_{11} = p_1, b_{22} = p_2, 2b_{12} = (p_1 + p_2) \cos \theta_{12} + h_{12} \sin \theta_{12},$$

where h_{12} is the shortest distance between R_1 and R_2.

Hence, in general

$$R^2p = R_1^2p_1 + R_2^2p_2 + ... + R_1R_2\{(p_1+p_2)\cos\theta_{12} + h_{12}\sin\theta_{12}\}$$
$$+ R_1R_3\{(p_1+p_3)\cos\theta_{13} + h_{13}\sin\theta_{13}\} + ...$$

(e) Any number of Forces

This is obtained $R^2p = R^2{}_1p_2 + ...+R_1R_2\{(p_1+p_2)\cos\theta_{12} + h_{12}\sin\theta_{12}\} +$ from (d) by putting

$$p_1 = p_2 = ... = 0.$$

$$\therefore \quad R^2 = R_1^2 + R_2^2 + ... + 2R_1R_2\cos\theta_{12} + ...$$

$$R^2p = R_1R_2h_{12}\sin\theta_{12} + R_1R_3h_{13}\sin\theta_{13} + ...$$

1.18. GIVEN A WRENCH TO REPLACE IT BY TWO FORCES

Show that a given system of forces can be replaced by two forces, equivalent to the given system in an infinite number of ways and that the tetrahedron formed by the two forces is of constant volume.

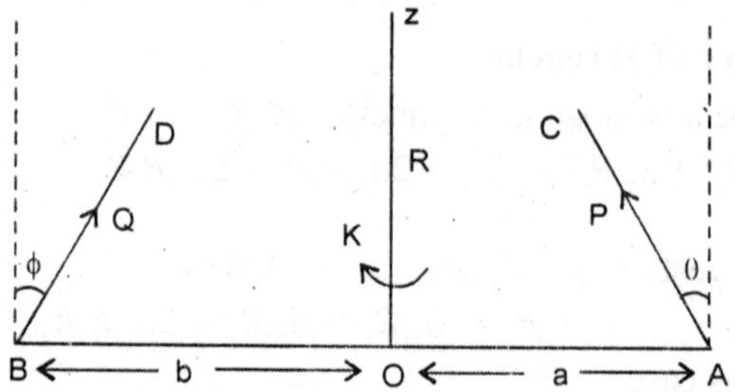

Fig. 1.26

Let the given system have Oz as its central axis and let R, K be the resultant force and the resultant couple about Oz.

On a line through O perpendicular to Oz take any two points A and B on opposite sides of O, such that $OA = a$ and $OB = b$. Take this line as x-axis.

Let us assume that system is equivalent to two forces P, Q at A and B making angles θ, φ (opposite sense) with z-axis. Let us take Oy, perpendicular to x-axis and zO as z-axis.

Let $X, Y, Z; L, M, N$ be the components of dyname at O.

$$\begin{pmatrix} a, 0, 0 \\ 0, P\sin\theta, P\cos\theta \end{pmatrix} \quad \begin{pmatrix} -b, 0, 0 \\ 0, -Q\sin\phi, Q\cos\phi \end{pmatrix}$$

$$X = 0, Y = P\sin\theta - Q\sin\phi, Z = P\cos\theta + Q\cos\phi.$$

$$L = 0, M = -aP\cos\theta + bQ\cos\phi, N = aP\sin\theta + bQ\sin\phi.$$

Since, z-axis is central axis, hence, its d.c' s are 0, 0, 1 and also proportional to X, Y, Z and L, M, N, we have

$$X = 0 = Y, L = 0 = M, Z = R, N = K.$$

$$R = P\cos\theta + Q\cos\phi \tag{1}$$

$$P \sin \theta - Q \sin \phi = 0 \tag{2}$$

$$aP \cos \theta - bQ \cos \phi = 0 \tag{3}$$

$$K = aP \sin \theta + bQ \sin \phi \tag{4}$$

(1) and (3) give

$$\frac{P \cos \theta}{b} = \frac{Q \cos\phi}{a} = \frac{R}{a+b} \tag{5}$$

(2) and (4) give

$$P \sin \theta = Q \sin \phi = \frac{K}{a+b}. \tag{6}$$

Hence,

$$P^2 = \frac{K^2 + R^2 b^2}{(a+b)^2}, Q^2 = \frac{K^2 + R^2 a^2}{(a+b)^2}, \tan \theta = \frac{K}{Rb}, \tan \phi = \frac{K}{Ra}. \tag{7}$$

Whatever be the values of a and b, we obtain real values of P, Q, θ and ϕ, so that our assumption is correct.

Also,

$$R^2 = P^2 + Q^2 + 2PQ \cos (\theta + \phi). \tag{8}$$

and

$$KR = PQ (a + b) \sin (\theta + \phi). \tag{9}$$

Let AC, BD represent P, Q in magnitude.

Now volume of the tetrahedron determined by P, Q is given by

$$\frac{1}{6} PQ (a + b) \sin (\theta + \phi) = \frac{1}{6} KR \text{ by (9)}$$
$$= \text{constant.}$$

1.19 FIND THE RESULTANT WRENCH ON SCREWS INTERSECTING AT RIGHT ANGLE CYLINDROID

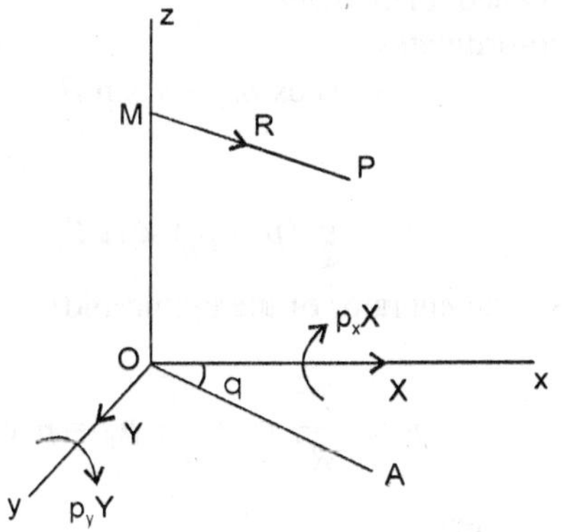

Fig. 1.27

Let screws be taken as axes of x and y and let their pitches be p_x and p_y.

Let P and Q be the intensities of the wrenches along these screws so that their corresponding couples are Pp_x, Qp_y

Hence, the components of dyname at O are

$$X = P, Y = Q, Z = 0, L = Pp_x, M = Qp_y, N = 0.$$

Therefore, resultant is a wrench (R, pR), such that

$$R^2 = P^2 + Q^2, R^2 p = LX + MY + NZ = P^2 p_x + Q^2 p_y,$$

and the central axis is

$$\frac{L - yZ + zY}{X} = \frac{M - zX + Xz}{Y} = \frac{N - xY + yX}{Z} = p.$$

$$\therefore \quad \frac{Pp_x + zQ}{P} = \frac{Qp_y - zP}{Q} = \frac{-xQ + Py}{0} = p,$$

which gives

$$xQ = Py, \therefore \frac{P}{Q} = \frac{x}{y},$$

$$p_x + \frac{zQ}{P} = p, p_y - \frac{zP}{Q} = p,$$

$$\therefore \quad p_y - p_x = z\left(\frac{P}{Q} + \frac{Q}{P}\right),$$

$$\therefore \quad p_y - p_x = z\left(\frac{x}{y} + \frac{y}{x}\right)$$

$$\therefore \quad z(x^2 + y^2) = (p_y - p_x)\,xy.$$

This surface is called *cylindroid.*

In cylindrical co-ordinates

$$x = r\cos\theta, y = r\sin\theta,$$

So that

$$z = \frac{1}{2}(p_y - p_x)\sin 2\theta.$$

Thus, central axis is a generator of the cylindroid.

Also,

$$p = \frac{P^2}{R^2}p_x + \frac{Q^2}{R^2}p_y = p_x\cos^2\theta + p_y\sin^2\theta.$$

If we take $p = \dfrac{K}{r^2}$, then

$$K = p_x x^2 + p_y y^2.$$

Thus, if we draw conic in xy plane and take the radius vector OA parallel to the resultant screw, the pitch of resultant screw is inversely proportional to the square of this radius vector. This is called *pitch conic.*

Conversely, if a wrench (R, pR) be given on a cylindroid, then

$$P = R\cos\theta, Q = R\sin\theta, L = Rp_x\cos\theta, M = Rp_y\sin\theta$$

and

$$p = p_x\cos^2\theta + p_y\sin^2\theta.$$

1.20 THE RESULTANT OF A SYSTEM OF WRENCHES ON A CYLINDROID

Let the wrenches be $(R_1, p_1 R_1)$ $(R_2, p_2 R_2)$ $(R_3, p_3 R_3)$ .... making angles $\theta_1, \theta_2, \theta_3$ .... with x-axis.

Each wrench can be resolved along the x and y axes. Now the wrench $(R_1, p_1 R_1)$ is equivalent to

A wrench $(R_1\cos\theta_1, R_1\cos\theta_1 p_x)$ along x-axis and a wrench $(R_1\sin\theta_1, R_1\sin\theta_1 p_y)$ along y-axis.

Similarly, for other wrenches. Hence, the resultant wrench along x-axis is

$$X = \sum R_1\cos\theta_1, L = \sum R_1\cos\theta_1 p_x = p_x\sum R_1\cos\theta_1 = p_x X.$$

Similarly, along y-axis

$$Y = \sum R_1\sin\theta_1, M = \sum R_1\sin\theta_1 p_y = p_y\sum R_1\sin\theta_1 = p_y Y.$$

Thus, we have two wrenches $(X, p_x X)$ and $(Y, p_y Y)$ along x-axis and y-axix respectively, therefore there is a wrench (R, pR) on the cylindroid, such that

$$R^2 = X^2 + Y^2 = \left(\sum R_1\cos\theta_1\right)^2 + \left(\sum R_1\sin\theta_1\right)^2$$

$$z = \frac{1}{2}(p_y - p_x)\sin 2\phi,$$

where

$$\tan 2\phi = \frac{Y}{X} = \frac{\sum R_1\sin\theta_1}{\sum R_1\cos\theta_1},$$

and

$$p = p_x\cos^2\phi + p_y\sin^2\phi.$$

Therefore the conditions of equilibrium of wrenches on a *cylindroid* are

$$\sum R_1\cos\theta_1 = 0, \sum R_1\sin\theta_1 = 0.$$

Thus, if by reducing wrenches at the origin, their intensities are in equilibrium, then wrenches themselves are also in equilibrium.

Since,

$$L = p_x \sum R_1 \cos\theta_1 = 0, M = p_y \sum R_1 \sin\theta_1 = 0.$$

1.21 PROVE THAT ONE AND ONLY ONE CYLINDROID CAN BE DRAWN THROUGH TWO GIVEN SCREWS

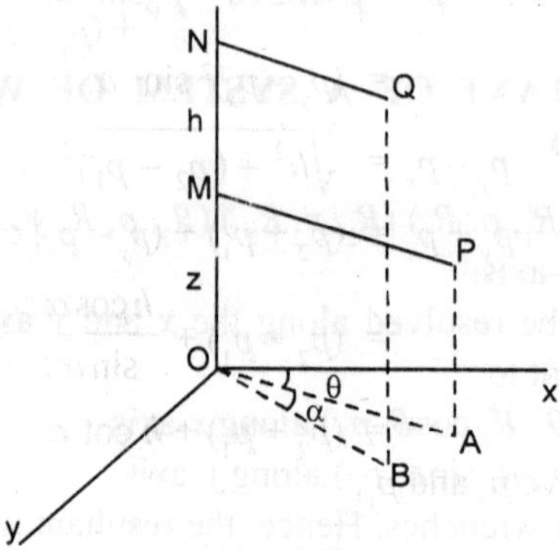

Fig. 1.28

Let us take the shortest distance (=h) between the two given screws of pitches p_1 and p_2 as z-axis, let the screws make angle θ with x-axis, and α the angle between the screws.

Then

$$p_1 = p_x \cos^2\theta + p_y \sin^2\theta,$$
$$p_2 = p_x \cos^2(\theta+\alpha) + p_y \sin^2(\theta+\alpha),$$

$$z = \frac{1}{2}(p_y - p_x)\sin 2\theta,$$

$$z + h = \frac{1}{2}(p_y - p_x)\sin 2(\theta+\alpha).$$

Here, p_1, p_2 z, $z + h$ are known, these four relations are just sufficient to determine p_x, p_y θ, z.

$$p_1 + p_2 = p_x[\cos^2\theta + \cos^2(\theta+\alpha)] + p_y[\sin^2\theta + \sin^2(\theta+\alpha)]$$

$$= \frac{p_x}{2}[1+\cos 2\theta + 1 + \cos 2(\theta+\alpha)] + \frac{p_y}{2}[1 - \cos 2\theta + 1 - \cos 2(\theta+\alpha)]$$

$$= (p_x + p_y) + \frac{1}{2}(p_x - p_y)[\cos 2\theta + \cos 2(\theta+\alpha)]$$

$$= (p_x + p_y) + (p_x - p_y)\cos\alpha\cos(2\theta+\alpha).$$

Similarly,

$$p_2 - p_1 = (p_y - p_x)\sin\alpha\sin(2\theta+\alpha).$$

$$h = (p_y - p_x) \sin \alpha \cos (2\theta + \alpha)$$

$$\therefore \quad \tan (2\theta + \alpha) = \frac{p_2 - p_1}{h},$$

this gives θ.

Now,

$$h^2 + (p_2 - p_1)^2 = (p_y - p_x)^2 \sin^2 \alpha \cos^2 (2\theta + \alpha) + (p_y - p_x)^2 \sin^2 \alpha \sin^2 (2\theta + \alpha)$$

$$= (p_y - p_x)^2 \sin^2 \alpha.$$

$$\therefore \quad p_y - p_x = \sqrt{h^2 + (p_2 - p_1)^2} \ \operatorname{cosec} \alpha,$$

$$p_y + p_x = (p_2 + p_1) + (p_y - p_x) \cos \alpha \cos (2\theta + \alpha)$$

$$= (p_2 + p_1) + \frac{h \cos \alpha \cos(2\theta + \alpha)}{\sin \alpha \cos(2\theta + \alpha)}$$

$$= (p_2 + p_1) + h \cot \alpha.$$

These relations give p_x and p_y.

1.22 SCREW DISPLACEMENT OF A RIGID BODY

Every displacement of a rigid body can be produced by a rotation about a certain axis and a translation along the same axis. The ratio of the translation to rotation is called *pitch* and such a displacement is called a *twist about a screw.*

Find the work done by a wrench (R, pR) due to a twist about a screw of pitch p_1.

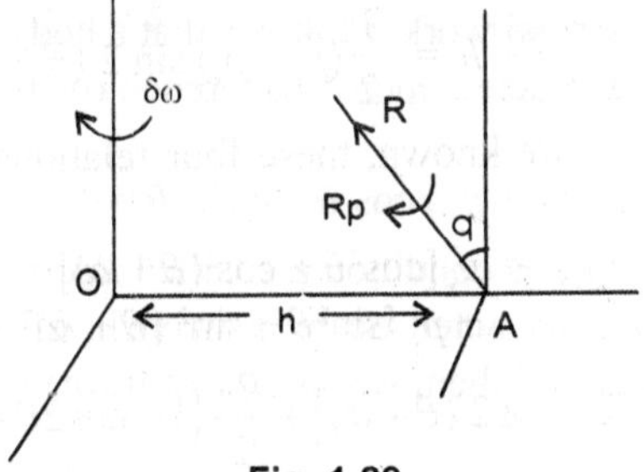

Fig. 1.29

Let Ox be the axis of pitch p_1 and MP the axis of the wrench, OM the shortest distance h and θ the angle between them. Draw MK parallel to Ox and ML perpendicular to MK and OM.

Let $\delta\omega$ be the small twist about the screw of pitch p_1. Due to this rotation translation is $p_1 \ \delta\omega$.

The force R is equivalent to $R\cos\theta$ along MK and $R\sin\theta$ along ML.

$$\begin{pmatrix} 0, 0, h \\ R\cos\theta, 0, 0 \end{pmatrix}, \begin{pmatrix} 0, 0, h \\ 0, R\sin\theta, 0 \end{pmatrix}$$

The component $R\cos\theta$ is equivalent to a force $R\cos\theta$ along Ox and a couple $R\cos\theta \,.\, h$ about Oy.

The component $R\sin\theta$ is equivalent to a force $R\sin\theta$ along Oy and a couple $-R\sin\theta \,.h$ about Ox.

The couple pR about MP is equivalent to couples $pR\cos\theta$ about MK and $pR\sin\theta$ about ML and the axes of these may be removed to Ox and Oy.

The given wrench is thus equivalent to

a force $R\cos\theta$ along Ox,
a force $R\sin\theta$ along Oy.
a couple $R(p\cos\theta - h\sin\theta)$ about Ox,
a couple $R(p\sin\theta + h\cos\theta)$ about Oy

Due to angular displacement work done by couple $R(p\cos\theta - h\sin\theta)$ is $R(p\cos\theta - h\sin\theta)\,\delta\omega$ and that by force is zero.

Due to linear displacement $p_1\delta\omega$ the work done by couple is zero and that by force $R\cos\theta p_1\delta\omega$.

Hence, the total work done due to small displacement is

$$= R\delta\omega\,[(p+p_1)\cos\theta - h\sin\theta]$$

This work is not altered if p and p_1, are interchanged so that work done is the same as it would be if we had a wrench $(R; p_1 R)$ about a screw Ox, and the body were displaced through the same angle $\delta\omega$ about a screw of the pitch p on MP as axis.

1.23 RECIPROCAL SCREWS

From the principle of virtual work it follows that a body free to move only on a screw of x-axis Ox, and acted upon by a wrench on the screw MP, it will be in equilibrium if

$$(p+p_1)\cos\theta - h\sin\theta = 0.$$

Screws which satisfy this condition are called *Reciprocal screws*, they are therefore such that a body acted on by a wrench, of any intensity and the proper pitch, about either of them will be at rest if free to move only about the other.

If the axes of the screw and wrench intersect then $h = 0$, hence from the above condition we get

$$(p+p_1)\cos\theta = 0.$$

From this either $(p+p_1) = 0$. $\therefore$ $p = -p_1$ *i.e.*, if their pitches are equal and opposite, or $\cos\theta = 0$, $\therefore\ \theta = \dfrac{\pi}{2}$ *i.e.*, if the axes of the wrench and screw are at right angle to each other.

Prove that a screw γ is reciprocal to each of two screws α and β, if it is reciprocal to any screw δ on the cylindroid determined by α and β.

Now α, β and δ are three screws on the same cylindroid, a wrench about δ is equivalent to two wrenches about α and β, if the intensity of wrench δ is equivalent to two component intensities of the wrenches α and β. Replace the wrench by these two component wrenches about α and β.

Since, the screws γ and α are reciprocal, the virtual work of the wrench about α for a displacement about γ is zero. Similarly, for the component about β.

Hence, the total work of the wrench δ for a displacement about γ is zero. Hence, δ and γ are reciprocals.

Examples

1. Two equal forces act one each along each of the straight lines

$$\frac{x \mp a\cos\theta}{a\sin\theta} = \frac{y - b\sin\theta}{\mp b\cos\theta} = \frac{z}{c};$$

show that their central axis must, for all values of θ, lie on the surface

$$y\left(\frac{x}{z} + \frac{z}{x}\right) = b\left(\frac{a}{c} + \frac{c}{a}\right).$$

Let P be the force, then at the point $(a\cos\theta, b\sin\theta, 0)$ we have a force whose components are proportional to $a\sin\theta \,.\, P$, $-b\cos\theta P$, cP, and at the $(-a\cos\theta, b\sin\theta, 0)$ a force whose components are proportional to $a\sin\theta.P$, $b\cos\theta.P$, $c.P$.

Hence,

$$X = \sum X_1 \propto 2a\sin\theta.P,$$

$$Y = 0,\ Z \propto 2cP,$$

$$L = \sum(y_1 Z_1 - z_1 Y_1) \propto 2bc\sin\theta.P$$

$$M = 0,\ N \propto -2ab.P.$$

The equation of central axis is

$$\frac{L - yZ + zY}{X} = \frac{M - zX + xZ}{Y} = \frac{N - xY + yX}{Z}.$$

Substituting the values of X, Y, Z, L, M, N we get

$$\frac{bc\sin\theta - yc}{a\sin\theta} = \frac{-ab + ya\sin\theta}{c},\ za\sin\theta = x.c.$$

Eliminating θ from these relations, we get

$$\left(bc.\frac{xc}{za}-yc\right)\cdot\frac{z}{xc} = \left(-ab+y\cdot\frac{xc}{z}\right)\cdot\frac{1}{c},$$

$$\therefore \qquad \frac{bc}{a}-\frac{yz}{x} = \frac{-ab}{c}+\frac{yx}{z},$$

$$\therefore \qquad y\left(\frac{x}{z}+\frac{z}{x}\right) = b\left(\frac{a}{c}+\frac{c}{a}\right).$$

This is the locus of central axis.

2. Two forces P and Q act along the straight lines whose equations are

$$y = x\tan\alpha,\ z = c,\ y = -x\tan\alpha,\ z = -c$$

respectively. Show that their central axis lies on the straight line

$$y = x\cdot\frac{P-Q}{P+Q}\tan\alpha,\quad \frac{z}{c} = \frac{P^2-Q^2}{P^2+2PQ\cos 2\alpha+Q^2}.$$

For all values of P and Q, prove that this line is a generator of the surface

$$(x^2+y^3)\,z.\sin 2\alpha = 2cxy.$$

Equations of the lines can be written as

$$\frac{x}{\cos\alpha} = \frac{y}{\sin\alpha} = \frac{z-c}{0};\ \frac{x}{\cos\alpha} = \frac{y}{-\sin\alpha} = \frac{z+c}{0}$$

$\therefore X = (P+Q)\cos\alpha,\ Y = (P-Q)\sin\alpha,\ Z = 0,\ L = -c(P+Q)\sin\alpha,$
$M = c(P-Q)\cos\alpha,\ N = 0.$

Equations of the central axis are given by

$$\frac{L-yZ+zY}{X} = \frac{M-zX+xZ}{Y} = \frac{N-xY+yX}{Z}.$$

Substituting the values of X, Y, Z, L, M, N we get

$$\frac{-c(P+Q)\sin\alpha+z(P-Q)\sin\alpha}{(P+Q)\cos\alpha} = \frac{c(P-Q)\cos\alpha-z(P+Q)\cos\alpha}{(P-Q)\sin\alpha}$$

$$= \frac{-x(P-Q)\sin\alpha+y(P+Q)\cos\alpha}{0} \qquad (1)$$

The last member gives

$$\frac{y}{x} = \frac{(P-Q)\tan\alpha}{(P+Q)} \qquad (2)$$

The first equality gives

$$\frac{z}{c} = \frac{(P^2+Q^2)}{(P^2+Q^2+2PQ\cos 2\alpha)}.$$

Relation (1) can be written as

$$-c\tan\alpha + \frac{z(P-Q)\tan\alpha}{(P+Q)} = c\cot\alpha - \frac{z(P+Q)\cot\alpha}{(P-Q)}.$$

Using (2), we get

$$z\left(\frac{y}{x}+\frac{x}{y}\right) = c\,(\tan\alpha + \cot\alpha),$$

$$\therefore \qquad z\,(x + y^2)\sin 2\alpha = 2cxy.$$

3. Three forces act along the lines

$$x = 0, y - z = a; y = 0, z - x = a; z = 0, x - y = a.$$

Show that they cannot reduce to a couple.

Prove also that if the system reduces to a single force its line of action must lie on the surface

$$x^2 + y^2 + z^2 - 2yz - 2zx - 2xy = a^2.$$

The equations of the given lines can be written as

$$\frac{x}{0} = \frac{y-a}{1/\sqrt{2}} = \frac{z}{1/\sqrt{2}},$$

$$\frac{x}{1/\sqrt{2}} = \frac{y}{0} = \frac{z-a}{1/\sqrt{2}},$$

$$\frac{x-a}{1/\sqrt{2}} = \frac{y}{1/\sqrt{2}} = \frac{z}{0}.$$

Let P, Q, S be the forces acting along these lines. Then the components of the forces are

$$0, \frac{P}{\sqrt{2}}, \frac{P}{\sqrt{2}} \text{ at } (0, a, 0),$$

$$\frac{Q}{\sqrt{2}}, 0, \frac{Q}{\sqrt{2}} \text{ at } (0, 0, a),$$

$$\frac{S}{\sqrt{2}}, \frac{S}{\sqrt{2}}, 0 \text{ at } (a, 0, 0).$$

$$\therefore \qquad X = \frac{1}{\sqrt{2}}(Q+S),\; Y = \frac{1}{\sqrt{2}}(S+P),\; Z - \frac{1}{\sqrt{2}}(P+Q)$$

It is obvious that R will be equal to zero, if $X = Y = Z = 0$, but it is not possible in this case unless $P = Q = S = 0$. Hence, the system cannot reduce to a single couple.

Now,

$$L = \frac{aP}{\sqrt{2}}, M = \frac{aQ}{\sqrt{2}}, N = \frac{aS}{\sqrt{2}}.$$

If the system reduces to a single force, $K = 0$.
Therefore, equations of the central axis are

$$\frac{aP - y(P+Q) + z(S+P)}{Q+S} = \frac{aQ - z(Q+S) + x(P+Q)}{S+P}$$

$$= \frac{aS - x(S+P) + y(Q+S)}{P+Q} = \frac{K}{R} = 0,$$

From these relations we get

$$(a - y + z)P - yQ + zS = 0,$$
$$xP + (a - z + x)Q - zS = 0,$$
$$-xP + yQ + (a - x + y)S = 0.$$

Eliminating P, Q, S from these relations, we get

$$\begin{vmatrix} a-y+z & -y & z \\ x & a-z+x & -z \\ -x & y & a-x+y \end{vmatrix} = 0$$

Multiplying elements of first row by x, second by y and third by z and adding to first row, we get

$$\frac{a}{x}\begin{vmatrix} x & y & z \\ x & a-z+x & -z \\ -x & y & a-x+y \end{vmatrix} = 0.$$

The determinant now can be easily reduced to a second order one, which on expansion gives

$$(a + x - y - z)(a - x + y + z) + 4yz = 0,$$
$$\therefore \quad a^2 - (x - y - z)^2 + 4yz = 0,$$
$$\therefore \quad x^2 + y^2 + z^2 - 2yz - 2zx - 2xy = a^2.$$

4. A single force is equivalent to component forces X, Y, Z along the axes of coordinates and to couples L, M, N about these axes; show that the magnitude of the single force is $\sqrt{X^2 + Y^2 + Z^2}$ and that the equation to its line of action is

$$\frac{yZ - zY}{L} = \frac{zX - xZ}{M} = \frac{xY - yX}{N} = 1.$$

The components of the force parallel to the axes are X, Y, Z. Hence,

$$R = \sqrt{X^2 + Y^2 + Z^2}$$

Also, since the resultant is single force, $K = 0$.
The equations of the central axis are

$$\frac{L - yZ + zY}{X} = \frac{M - zX + xZ}{Y} = \frac{N - xY + yX}{Z} = 0.$$

which give

$$\frac{yZ - zY}{L} = \frac{zX - xZ}{M} = \frac{xY - yX}{N} = 1.$$

5. A force P acts along the axis of x and another force $n\,P$ along a generator of the cylinder $x^2 + y^2 = a^2$; show that the central axis lies on the cylinder

$$n^2 (nx - z)^2 + (1 + n^2)^2 y^2 = n^4 a^2.$$

Let generator make an angle α with x-axis
The compents of the forces are

$$P, 0, 0 \text{ at } (0, 0, 0)$$
$$0, 0, nP \text{ at } (a \cos \alpha, a \sin \alpha, 0).$$

Therefore,

$$X = P, Y = 0, Z = nP, L = nP\,a \sin \alpha, \ M = -\,nP\,a \cos \alpha, \ N = 0.$$

The equations to the central axes are

$$\frac{nPa \sin \alpha - nyP}{P} = \frac{-nPa \cos \alpha - Pz + xnP}{0} = \frac{yP}{nP}.$$

Hence, we get

$$nx - z - na \cos \alpha = 0, \qquad (1)$$
$$n^2 (a \sin \alpha - y) = y. \qquad (2)$$

From (1) and (2) we get

$$\cos \alpha = \frac{nx - z}{na}, \ \sin \alpha = \frac{1 + n^2}{an^2} \cdot y.$$

Squaring and adding, we get

$$(1 + n^2)^2 y^2 + n^2 (nx - z)^2 = n^4 a^2.$$

This is the required locus of the central axis.

6. Two forces act, one along the line $y = 0$, $z = 0$, and other along the line $x = 0$, $z = c$. As the forces vary, show that surface generated by the axis of their equivalent wrench is $(x^2 + y^2)\,z = cy^2$.

Let the forces be P and Q of which the first acts along the x-axis ($y = 0 = z$) and the second along the line $x = 0$, $z = c$.
Therefore,

$$X = P, Y = Q, Z = 0, L = -\,cQ, M = 0, N = 0.$$

The equations of the central axes are

$$\frac{-cQ+zQ}{P} = \frac{-zP}{Q} = \frac{-xQ+yP}{0},$$

giving

$$xQ = yP \text{ or } \frac{P}{Q} = \frac{x}{y},$$

$$(z-c)\frac{Q}{P} = -z\frac{P}{Q}.$$

Eliminating P, Q the locus of the axis of the wrench is

$$(z-c)\frac{y}{x} = -z\frac{x}{y}$$

$$\therefore \quad cy^2 = z(x^2+y^2).$$

7. A force parallel to the axis of z acts at the point $(a, 0, 0)$ and an equal force perpendicular to the axis of z acts at the point $(-a, 0, 0)$. Show that the central axis of the system lies on the surface

$$z^2(x^2+y^2) = (x^2+y^2-ax)^2.$$

Let each force be equal to P, the components of the forces are

$0, 0, P$ at the point $(a, 0, 0)$.

$P\cos\alpha, P\sin\alpha, 0$ at the point $(-a, 0, 0)$.

Therefore

$$X = P\cos\alpha, Y = P\sin\alpha, Z = P, L = 0, M = -aP, N = -aP\sin\alpha.$$

The equations of the central axes are

$$\frac{-yP+zP\sin\alpha}{P\cos\alpha} = \frac{-aP-zP\cos\alpha+xP}{P\sin\alpha}.$$

$$= \frac{-aP\sin\alpha - xP\sin\alpha + yP\cos\alpha}{P}$$

$$\therefore \quad \frac{z\sin\alpha - y}{\cos\alpha} = \frac{x-z\cos\alpha-a}{\sin\alpha} = \frac{y\cos\alpha - x\sin\alpha - a\sin\alpha}{1}$$

From the equality of the first and third expressions, we get

$$z\sin\alpha - y = y\cos^2\alpha - (x+a)\sin\alpha\cos\alpha,$$

$$\therefore \quad z\sin\alpha = y(1+\cos^2\alpha) - (x+\alpha)\sin\alpha\cos\alpha. \quad (1)$$

Similarly, from first and second relations, we get

$$z\sin^2\alpha - y\sin\alpha = x\cos\alpha - z\cos^2\alpha - a\cos\alpha.$$

$$\therefore z = x\cos\alpha + y\sin\alpha - a\cos\alpha. \quad (2)$$

Multiplying (2) by $\sin\alpha$ and subtracting from (1), we get

$$0 = -2x\sin\alpha\cos\alpha + y(1+\cos^2\alpha) - y\sin^2\alpha.$$

$$\therefore \quad 2y\cos^2\alpha = 2x\sin\alpha\cos\alpha,$$

$$\therefore \quad \tan\alpha = \frac{y}{x} \quad (3)$$

Eliminating α from (2) and (3), we get

$$z = (x-a)\cdot\frac{x}{\sqrt{x^2+y^2}} + y\cdot\frac{y}{\sqrt{x^2+y^2}},$$

$$\therefore (x^2+y^2)z^2 = (x^2+y^2-ax)^2,$$

which is the locus of the central axis.

8. Along the normal at every point of an octant of an ellipsoid cut off by the principal planes acts a force proportional to the element of the surface at P. Show that these forces are equivalent to a single force acting along the line

$$a\left(x-\frac{4a}{3\pi}\right) = b\left(y-\frac{4b}{3\pi}\right) = c\left(z-\frac{4c}{3\pi}\right),$$

where $2a$, $2b$, $2c$ are the axes of the ellipsoid.

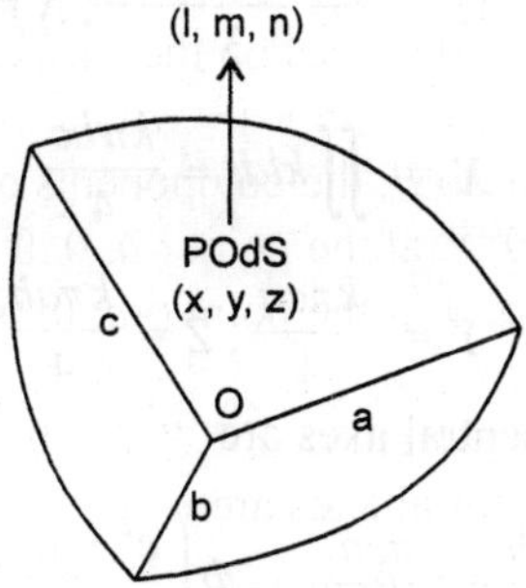

Fig. 1.30

Let l, m, n be the direction cosines of the normal at the point P (x, y, z) of the ellipsoid and let ds be the area of an element of the surface at P. Then

$$X = \iint klds,\ Y = \iint kmds,\ Z = \iint knds,$$

where kds is the force along the normal at P.

Now $\iint lds = \iint dydz$, *i.e.*, the projection of the area on the yz-plane, and

in the case of the octant it is $\dfrac{\pi bc}{4}$.

Therefore

$$L = \iint kynds - \iint kzmds$$

$$= k\left[\iint ydxdy - \iint zdxdz\right].$$

But we know that

$$\bar{x} = \frac{\iint x dx dy}{\iint dx dy} = \frac{4a}{3\pi},$$

$$\therefore \qquad \iint x dx dy = \frac{4a}{3\pi}.\frac{\pi ab}{4} = \frac{a^2 b}{3}.$$

Therefore

$$L = k\left[\frac{b^2 a}{3} - \frac{c^2 a}{3}\right] = \frac{ka(b^2 - c^2)}{3}.$$

Similarly,

$$M = \frac{kb(c^2 - a^2)}{3}, N = \frac{kc(a^2 - b^2)}{3}$$

$$X = \iint klds = \frac{k\pi bc}{4}$$

$$Y = \frac{k\pi ca}{4}, Z = \frac{k\pi ab}{4}.$$

The equations of the central axes are

$$\frac{a\left(\frac{b^2 - c^2}{3}\right) - y\frac{\pi ab}{4} + z\frac{\pi ca}{4}}{\frac{\pi bc}{4}} = \frac{b\left(\frac{c^2 - a^2}{3}\right) - z\frac{\pi cb}{4} + x\frac{\pi ab}{4}}{\frac{\pi ca}{4}}$$

$$= \frac{c\left(\frac{a^2 - b^2}{3}\right) - x\cdot\frac{\pi ab}{4} + y\frac{\pi bc}{4}}{\frac{\pi ab}{4}} = \frac{K}{R} = 0\cdot$$

since, the forces reduce to a single force.
From the first member, we get

$$a\left(\frac{b^2 - c^2}{3}\right) - y.\frac{\pi ab}{4} + z\frac{\pi ca}{4} = 0,$$

$$\therefore \qquad 4b^2 - 3\pi by = 4c^2 - 3\pi cz,$$

$$b\left(y - \frac{4b}{3\pi}\right) = c\left(z - \frac{4c}{3\pi}\right).$$

Similarly, from the three members, we get

$$a\left(x-\frac{4a}{3\pi}\right) = b\left(y-\frac{4b}{3\pi}\right) = c\left(z-\frac{4c}{3\pi}\right).$$

9. Any number of wrenches of the same pitch p act along the generators of the same system of hyperboloid $\frac{x^2}{a^2}+\frac{y^2}{b^2}-\frac{z^2}{c^2}=1$. Show that they will reduce to a single force provided their central axis is parallel to a generator of the cone.

$$\left(p+\frac{bc}{a}\right)x^2+\left(p+\frac{ca}{b}\right)y^2+\left(p-\frac{ab}{c}\right)z^2 = 0.$$

The equations of a generator of the hyperboloid $\frac{x^2}{a^2}+\frac{y^2}{b^2}-\frac{z^2}{c^2}=1$ are

$$\frac{x-a\cos\alpha}{a\sin\alpha} = \frac{y-b\sin\alpha}{-b\cos\alpha} = \frac{z}{c}.$$

Let the force P_r act at the point ($a\cos\alpha_r$, $b\sin\alpha_r$, 0). Since, pitches of all the wrenches are equal, the couple at the rth point is

$$k_r = pP_r$$

$$r = 1, 2, 3, \ldots\ldots\ldots\ldots n.$$

Thus the components of the forces are

$$\frac{P_r a\sin\alpha_r}{D_r}, \frac{-P_r b\cos\alpha_r}{D_r}, \frac{P_r c}{D_r} \text{ at } (a\cos\alpha_r, b\sin\alpha_r, 0)$$

and couples of the moments pP_r, where

$$D_r^2 = a^2\sin^2\alpha_r + b^2\cos^2\alpha_r + c^2.$$

Therefore, we have

$$X = a\sum\frac{P_r\sin\alpha_r}{D_r}, Y = -b\sum\frac{P_r\cos\alpha_r}{D_r}, Z = c\sum\frac{P_r}{D_r}$$

and

$$L = \sum\frac{P_r bc\sin\alpha_r}{D_r}+a\sum\frac{pP_r\sin\alpha_r}{D_r}$$

$$= (bc+ap)\sum\frac{P_r\sin\alpha_r}{D_r} = \left(\frac{bc}{a}+p\right)X.$$

Similarly

$$M = \left(\frac{ca}{b}+p\right)Y,$$

$$N = -ab\sum\frac{P_r\cos^2\alpha_r}{D_r}-ab\sum\frac{P_r\sin^2\alpha_r}{D_r}+c\sum\frac{pP_r}{D_r}$$

$$= -ab\sum \frac{P_r}{D_r} + c\sum \frac{pP_r}{D_r} = \left(p - \frac{ab}{c}\right)Z.$$

If the system reduce to a single force

$$LX + MY + NZ = 0,$$

$$\therefore \left(p + \frac{bc}{a}\right)X^2 + \left(p + \frac{ca}{b}\right)Y^2 + \left(p - \frac{ab}{c}\right)Z^2 = 0. \qquad (1)$$

Since, the direction cosines of the central axis are proportional to X, Y, Z, the equations to a line through the origin and parallel to the generators are

$$\frac{x}{X} = \frac{y}{Y} = \frac{z}{Z}.$$

Hence, from (1) the central axis is parallel to a generator of the cone

$$\left(p + \frac{bc}{a}\right)x^2 + \left(p + \frac{ca}{b}\right)y^2 + \left(p - \frac{ab}{c}\right)z^2 = 0.$$

10. If P and Q be two non-intersecting forces whose directions are perpendicular, show that the distances of the central axis from their lines of action are as Q^2 to P^2.

Let AB be the shortest distance between the given forces P and Q and let OC be their central axis. Suppose P makes an angle θ with the direction of OC. Then Q makes the angle $90^\circ - \theta$ with the same direction.

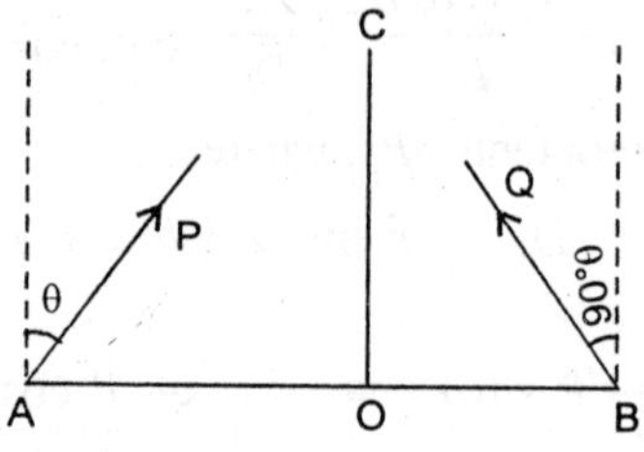

Fig. 1.31

Since, the resultant force of P and Q acts along OC, resolving along perpendicular to OC, we get

$$P \sin \theta - Q \sin (90^\circ - \theta) = 0. \qquad (1)$$

Now the resultant couple acts about OC, taking moments about a line perpendicular to OC, we get

$$P \cos \theta . AO - Q \cos (90^\circ - \theta). BO = 0. \qquad (2)$$

From (1), we have

$$\frac{P}{Q} = \frac{\cos\theta}{\sin\theta},$$

and from (2), we have

$$\frac{AO}{BO} = \frac{Q\sin\theta}{P\cos\theta} = \frac{Q^2}{P^2}.$$

11. Show that the minimum distance between two forces, which are equivalent to a given system (R, K) and which are inclined at a given angle 2α, is $\frac{2K}{R}\cot\alpha$ and two forces are then equal to $\frac{1}{2}R\sec\alpha$.

Let the system (R, K) be equivalent to forces P and Q acting at distances a, b from R and inclined to angles θ, φ angles, in opposite directions. In the notations of ∮1.8 the we, have $\theta + \phi = 2\alpha$ using the results of ∮1.18 we have

$$KR = PQ(a+b)\sin 2\alpha.$$

Now KR is constant, hence $a + b$ will be minimum when PQ is maximum. We know that

$$R^2 = P^2 + Q^2 + 2PQ\cos 2\alpha$$
$$= (P-Q)^2 + 4PQ\cos^2\alpha,$$

so that PQ is maximum when $P = Q$.
In that case

$$R = 2P\cos\alpha, \quad \therefore P = \frac{1}{2}R\sec\alpha.$$

$$(a+b) = \frac{KR}{P^2\sin 2\alpha} = \frac{K}{R}\cdot\frac{4\cos^2\alpha}{\sin 2\alpha} = \frac{K}{R}2\cot\alpha.$$

12. Wrenches of the same pitch p act along the edges of a regular tetrahedron $ABCD$ of side a. If the intensities of the wrenches along AB, DC are the same, and also along BC, DA and DB, CA, show that the pitch of the equivalent wrench is

$$p + \frac{a}{2\sqrt{2}}.$$

We know that the resultant of two equal wrenches (R_1, pR_1) acting at A and B along the lines AC and BD is the wrench (R, K), where

$$R^2 = 2R_1^2(1+\cos\alpha)$$

$$KR = 2pR_1^2(1+\cos\alpha) + R_1^2 c\sin\alpha,$$

acting at the middle point of the shortest distance between AC and BD (equal to c) bisecting the angle between them.

When AC and BD are the opposite edges of a regular tetrahedron, $\alpha = 90°$, so that the resultant wrench is (R, K) where $R = \sqrt{2}R_1$ and $KR = (2p + c)\,R_1^2$.

Let the wrenches (R_1, pR_1), (R_2, pR_2), (R_3, pR_3), act on the pairs of opposite edges AB and DC, BC and DA, CA and DB of the regular tetrahedron $ABCD$ of side a. Then since the shortest distance between the pairs of opposite edges of a regular tetrahedron meet in a common point O and bisect each other at right angles, it will be seen that the resultant wrench of the three pairs of equal wrenches act at O along three mutually perpendicular lines.

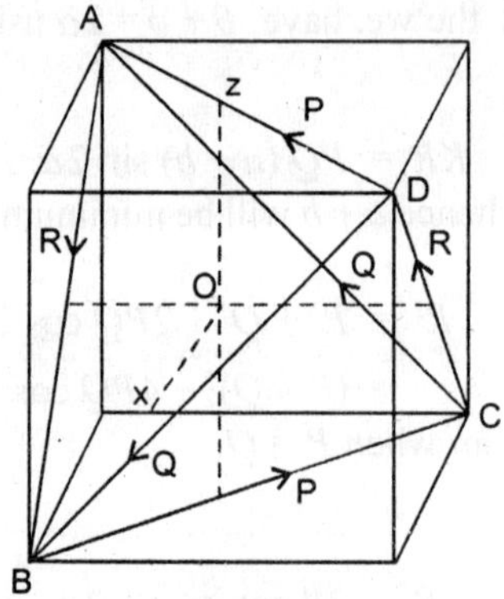

Fig. 1.32

Therefore the pitch of the final resultant wrench

$$= \frac{\sum KR}{\sum R^2} = \frac{(2p+c)\sum R_1^2}{2\sum R_1^2} = p + \frac{c}{2}.$$

Now a is the length of the faces of tetrahedron and c is the length of diagonals

$$\therefore \qquad c = \frac{a}{\sqrt{2}}$$

Substituting the value of c we get,

$$\text{Pitch of the final wrench} = p + \frac{a}{2\sqrt{2}}.$$

13. Two wrenches of pitches p, p' have axes at a distance $2a$ from one another. If the resultant wrench is of pitch ω and its axis is equidistant from the axes of the component wrenches, show that the angle between them is

$$\tan^{-1}\frac{a(2\omega - p - p')}{a^2 - (\omega - p)(\omega - p')}.$$

Let the axis of one wrench (P, pP) be taken as the axis of x and the shortest distance $OB = 2a$ between it and the axis of the other wrench $(P', p'P')$ be taken as the axis of z.

Let the axis of the resultant wrench $(R, R\omega)$ passing through the middle point A of OB make an angle θ with Ox. Then if the angle between the two wrenches α, is we have

$$R\cos\theta = P + P'\cos\alpha$$
$$R\sin\theta = P'\sin\alpha$$
$$R\,\omega\cos\theta = Pp + Pp'\cos\alpha - aP'\sin\alpha,$$
$$R\,\omega\sin\theta = P'p'\sin\alpha + aP'\cos\alpha - aP.$$

Eliminating θ between these equations, we get

$$\omega(P + P'\cos\alpha) = Pp + P'p'\cos\alpha - aP'\sin\alpha$$
$$\therefore \quad (\omega - p)P = (p'\cos\alpha - a\sin\alpha - \omega\cos\alpha)P'$$
$$\omega P'\sin\alpha = P'p'\sin\alpha + aP'\cos\alpha - aP$$
$$\therefore \quad aP = (p'\sin\theta + a\cos\alpha - \omega\sin\alpha)P'.$$

Therefore eliminating P and P', we get

$$\frac{\omega - p}{a} = \frac{(p' - \omega)\cos\alpha - a\sin\alpha}{(p' - \omega)\sin\alpha + a\cos\alpha}$$

$$\therefore [a^2 - (\omega - p)(\omega - p')]\sin\alpha = a(p + p' - 2\omega)\cos\alpha,$$

$$\therefore \tan\alpha = \frac{a(p + p' - 2\omega)}{a^2 - (\omega - p)(\omega - p')} = \frac{-a(2\omega - p - p')}{a^2 - (\omega - p)(\omega - p')}$$

$$\therefore \alpha = -\tan^{-1}\frac{a(2\omega - p - p')}{a^2 - (\omega - p)(\omega - p')}.$$

The negative sign shows that the angle α is obtuse in figure.

14. Show that a wrench, of which the pitch is ωc may be replaced by two forces inclined at angle 2θ to each other, the shortest distance between them being $2c$, and their magnitudes are

$$\frac{R}{2}\left[\sqrt{1 + \omega\tan\theta} \pm \sqrt{1 - \omega\cot\theta}\right].$$

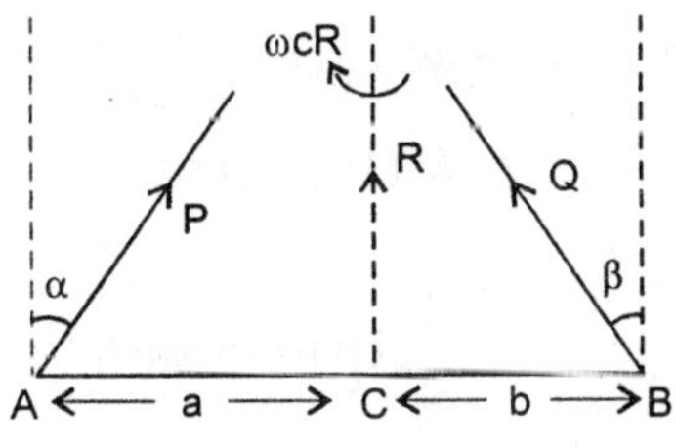

Fig. 1.33

Let the wrench $(R, \omega cR)$ acting at C be equivalent to two forces P and Q acting at A and B, inclined at angles α and β in opposite directions to R. Let AB be the shortest distance between P and Q and let $AC = a$, $BC = b$. Then we have $\alpha + \beta = 2\theta$, $a + b = 2c$.

Resolving the forces along and perpendicular to R, and also taking moments about these lines, we get

$$R = P\cos\alpha + Q\cos\beta, \qquad (1)$$

$$0 = P\sin\alpha - Q\sin\beta, \qquad (2)$$

$$\omega c R = P\sin\alpha . a + Q\sin\beta . b, \qquad (3)$$

$$0 = P\cos\alpha . a - Q\cos\beta . b. \qquad (4)$$

From (1) and (4), we get

$$\frac{P\cos\alpha}{b} = \frac{Q\cos\beta}{a} = \frac{P\cos\alpha + Q\cos\beta}{a+b} = \frac{R}{a+b},$$

From (2) and (3), we have

$$P\sin\alpha = Q\sin\beta = \frac{P\sin\alpha . a + Q\sin\beta . b}{a+b} = \frac{\omega c R}{a+b}.$$

$$\omega c R . aR = (a+b)^2 PQ \sin\alpha\cos\beta,$$

$$\omega c R . bR = (a+b)^2 PQ \cos\alpha\sin\beta.$$

Adding these we get

$$\omega c R^2 = (a+b)\sin(\alpha+\beta) = 2cPQ\sin 2\theta$$

$$\therefore \qquad 2PQ = \frac{\omega R^2}{\sin 2\theta}. \qquad (5)$$

Again from (1) and (2), by squaring and adding,

$$R^2 = P^2 + Q^2 + 2PQ\cos(\alpha+\beta)$$

$$= (P+Q)^2 - 2PQ(1 - \cos 2\theta),$$

so that by (5)

$$(P+Q)^2 = R^2 + \omega R^2 \cdot \frac{1-\cos 2\theta}{\sin 2\theta}$$

$$= R^2(1 + \omega\tan\theta).$$

Similarly,

$$(P-Q)^2 = R^2 - \omega R^2 \cdot \frac{1+\cos 2\theta}{\sin 2\theta}$$

$$= R^2(1 - \omega\cot\theta).$$

Hence, we get

$$P = \frac{1}{2}R\left[(1+\omega\tan\theta)^{1/2} + (1-\omega\cot\theta)^{1/2}\right]$$

$$Q = \frac{1}{2}R\left[(1+\omega\tan\theta)^{1/2} - (1-\omega\cot\theta)^{1/2}\right]$$

15. A system of forces given by $(X, Y, Z; L, M, N)$ is replaced by two forces, one acting along the axis of x, and another force. Show that the magnitudes of the forces are

$$\frac{LX + MY + NZ}{L}, \frac{\left[(MY + NZ)^2 + L^2(Y^2 + Z^2)\right]^{1/2}}{L}$$

Let the given system $(X, Y, Z; L; M, N)$ be equivalent to the forces P along x-axis and Q along the line

$$\frac{x - \alpha}{l} = \frac{y - \beta}{m} = \frac{z - \gamma}{n}.$$

Then we have

$$X = P + Ql, Y = Qm, Z = Qn, L = Q(n\beta - m\gamma),$$
$$M = Q(l\gamma - n\alpha), N = Q(m\alpha - l\beta).$$

$$\therefore \quad LX + MY + NZ = LP + Q^2 \sum l(n\beta - m\gamma) = LP.$$

$$P = \frac{LX + MY + NZ}{L}.$$

Again from the first three equations, we have

$$Q^2 = (X - P)^2 + Y^2 + Z^2$$
$$= \frac{(MY + NZ)^2}{L^2} + Y^2 + Z^2,$$

$$\therefore \quad Q = \frac{\left[(MY + NZ)^2 + L^2(Y^2 + Z^2)\right]^{\frac{1}{2}}}{L}.$$

16. Show that the wrench $(X, Y, Z; L, M, N)$ is equivalent to two forces, one along the line $x = y = z$, and the other along the line given by

$$Lx + My + Nz = 0,$$
$$x(Y - Z) + y(Z - X) + z(X - Y) = L + M + N,$$

and find the magnitude of the two forces.

Let the given wrench be equivalent to the forces P along the line $x = y = z$, and Q along the line

$$\frac{x - \alpha}{l} = \frac{y - \beta}{m} = \frac{z - \gamma}{n}.$$

Then we have

$$X = \frac{P}{\sqrt{3}} + Ql, \; Y = \frac{P}{\sqrt{3}} + Qm, \; Z = \frac{P}{\sqrt{3}} + Qn,$$

$$L = Q(n\beta - m\gamma), M = Q\,(l\gamma - n\alpha), N = Q\,(m\alpha - l\beta).$$

$$\therefore\ LX + MY + NZ = \frac{P(L + M + N)}{\sqrt{3}} + Q^2 . \sum l(n\beta - m\gamma)$$

$$= \frac{P(L + M + N)}{\sqrt{3}}.$$

$$\therefore \qquad P = \sqrt{3}.\frac{LX + MY + NZ}{L + M + N}.$$

Also, we have

$$Ql = X - \frac{P}{\sqrt{3}} = X - \frac{LX + MY + NZ}{L + M + N}$$

$$= \frac{M(X - Y) + N(X - Z)}{L + M + N}$$

Similarly

$$Qm = \frac{N(Y - Z) + L\,(Y - X)}{L + M + N},\ Qn = \frac{L(Z - X) + M(Z - Y)}{L + M + N}$$

$$\therefore \qquad Q = \frac{\sqrt{\sum [M(X - Y) + N(X - Z)]^2}}{L + M + N}$$

Again from the above equations, we have

$$L\alpha + M\beta + N\gamma = Q\,.\sum \alpha\,(n\beta - m\gamma) = 0,$$

and

$$(Y - Z)\,\alpha + (Z - X)\,\beta + (X - Y)\gamma = Q\,\{\alpha\,(m - n) + \beta(n - l) + \gamma(l - m)\}$$
$$= L + M + N.$$

Therefore the locus of (α, β, γ), *i.e.*, the line of action of Q is given by the equations

$$Lx + My + Nz = 0,$$
$$(Y - Z)x + (Z - X)\,y + (X - Y)\,z = L + M + N.$$

EXERCISE

1. Four forces acting through the angular points of a tetrahedron perpendicular to the opposite faces and proportional to them are in equilibrium. Show that they are in equilibrium if they act either all inwards or all outwards.
2. A circular uniform table, of 120 kg weight rests on four equal legs placed symmetrically round its edge; find the least weight which hung upon the edge of the table will just overturn it.
3. A rectangular table is supported in a horizontal position by four legs at its four angles, and a given weight is placed at a given point of it; show that

the thrust on each leg is indeterminate, and find the greatest and least value of it which it can have for a given position of the weight.

4. Two smooth planes, each inclined at angle α to the vertical, intersect in a horizontal line. A uniform rod, of weight W and length $2a$, is placed between them in a horizontal position making an angle θ with their line of intersection. Show that the horizontal couple required to maintain the equilibrium is $Wa \cos\theta \cot\alpha$.

5. A uniform rod, of weight W, can turn freely about a hinge at the end, and rests with the other end against a rough vertical wall making an angle α with the wall. Show that this end may rest anywhere on an are of a circle of angle 2 $\tan^{-1}$ ($\mu \tan\alpha$), and that in either of the extreme position the normal reaction of the wall is $\frac{1}{2}W\left[\cot^2\alpha+\mu^2\right]^{-\frac{1}{2}}$ wher μ is the co-efficient of friction.

6. A hemisphere, whose surface is rough and whose centre is O, is fixed with its base on a horizontal plane. One end of a straight uniform rod is freely jointed to a fixed point A in the plane and the other rests on the surface of the hemisphere at P so that the rod is just on the point of slipping. Show that the plane through O and the rod makes with the vertical plane through OA an angle $\tan^{-1}$ ($\mu \cos\alpha$ cosec β), where μ is the co-efficient of friction, α is the angle OAP, and β is the angle OPA.

7. A uniform triangular table ABC has three equal legs at A, B, and C rests on a rough horizontal plane. Find the least couple that will cause the table to move.

8. A parallelopiped is formed of twelve weightless rods which are jointed at their ends, and is in equilibrium under the action of four stretched elastic strings which connect the four pairs of opposite vertices. Show that the tensions of the strings and the actions in the rods are proportional to their lengths.

9. A heavy elastic string of natural length a, modulus λ and weight W is placed upon a smooth vertical cone, the angle of the vertex of which is 2α. What is the distance of its plane of equilibrium from the vertex?

10. A square board of side a and weight W is supported by four light rods of length l which are vertical and which are freely hinged at points in the ceiling and at the corners of the board. The board is rotated through an angle θ and a string is tied around the four mid points of the rods (which also form a square). What is the tension in the string, if the system is in equilibrium?

11. The legs of a tripod are of length l and make an angle θ with the vertical. However much the legs of the tripod are spread, they always form an equilateral triangle. The tripod rests on a smooth plane and its legs are

kept from spreading by an endless string around the feet. If the total weight of the tripod and the load which it supports is w, show that the tension in the string is

$$T = \frac{w}{3\sqrt{3}} \tan\theta.$$

12. The axis of z being vertically upwards, and the origin being A, a uniform square board $ABCD$, of weight W and side $2a$, is mounted so that it can turn freely about AB which is fixed in the direction whose direction cosines are ($\sin\theta$, 0, $\cos\theta$). A weightless string, fastened to the board at C, passes through a smooth fixed ring at (0, $2a$, 0), and carries hanging weight w at its other end. Prove that there is equilibrium when the angle ϕ, which the board makes with the plane of xz satisfies the equation

$$\frac{\cot\phi}{\sqrt{(3-2\sin\phi)}} = \frac{W}{2w}\sin\theta.$$

Investigate the stability of the equilibrium.

13. A parabolic curve is placed in vertical plane with its axis vertical and vertex downward. Its latus rectum is $4l$. A smooth heavy beam of length b rests against the concave arc and a peg at the focus. Find the inclination of the beam to the vertical.

14. A rectangular beam, the sides of which are a and b, is placed horizontally with one edge on a rough floor and a second edge against a smooth wall. Prove that equilibrium is not possible unless the height of the highest edge of the beam exceeds

$$\frac{a^2 + 2\mu ab + b^2}{\sqrt{a^2 + (2\mu a + b)^2}}.$$

15. Two rough cylinders lie in contact with each other on a rough horizontal plane. A third similar cylinder is placed on top of them with its axis parallel to the other two axes. If equilibrium exists, show that the co-efficient of friction exceeds $2 - \sqrt{3}$.

16. Three equal spheres are held together on a horizontal plane by a rubber band around them. If a cube of weight W is placed upon them with one diagonal vertical and the three lower faces of the cube in contact with the spheres, and if all of the contacts are smooth, show that the tension of the string of the band is increased by 1/3.

17. Three forces each equal to P, act on a body; one at the point (a, 0, 0) parallel to Oy, the second at the point (0, b, 0) parallel to Oz, and the third at the point (0, 0, c) parallel to Ox; the axes being rectangular, find the resultant wrench in magnitude and direction.

18. Equal forces act along two perpendicular diagonals of opposite faces of a cube of side a; show that they are equivalent to a single force R acting along a line through the centre of the cube, and a couple $\frac{1}{2}aR$ with the same line for axis.

19. $OBDC$ is a rectangle such that $OB = b$ and $OC = c$; also OA is a perpendicular to its plane; along OA, CD and BD act forces X, Y and Z respectively. Show that the component force R and a couple K of the resultant wrench are $\sqrt{X^2+Y^2+Z^2}$ and $\frac{X(Zb-Yc)}{\sqrt{X^2+Y^2+Z^2}}$. Show also that with OA, OB and OC as axes of x, y, and z the equation to the central axis is

$$\frac{x}{X} = \frac{y}{Y} - \frac{ZK}{XYR} = \frac{z}{Z} + \frac{KY}{XZR}.$$

20. Forces P, Q, R act along three non-intersecting edges of a cube; find the central axis.

21. Forces X, Y, Z act along the three straight lines

$$y = b,\ Z = -c\ ;\ z = c,\ x = -a;\ x = a,\ y = -b,$$

respectively; show that they will have a single resultant if

$$\frac{a}{X} + \frac{b}{Y} + \frac{c}{Z} = 0;$$

and that the equation of its line of action are any two of the three

$$\frac{y}{Y} - \frac{z}{Z} - \frac{a}{X} = 0,\ \frac{z}{Z} - \frac{x}{X} - \frac{b}{Y} = 0,\ \frac{x}{X} - \frac{y}{Y} - \frac{c}{Z} = 0.$$

22. If a force (X, Y, Z) acts along a generator of the hyperbolic paraboloid $\frac{x^2}{a^2} - \frac{y^2}{b^2} = 2\frac{z}{c}$, and be equivalent to an equal force (X, Y, Z) at the origin together with a couple (L, M, N), show that

$$aL \pm bM = 0,\ bX \pm aY = 0,\ cN \pm abZ = 0.$$

23. A force F acts along the axis of x, and a force mF along a straight line, intersecting the axis of x at a distance c from the origin and parallel to the plane of yz. Show that as this straight line turns round the axis of x, the central axis of the forces generates the surface

$$\{m^2 z^2 + (m^2 - 1)\, y^2\}\ \{c - x\}^2 = x^2 z^2.$$

24. Two forces P and Q are such that their central axis is given in position and the line of action of P is given. Show that the locus of the line of action of Q is a conicoid.

25. Forces act along the edges of a regular tetrahedron, viz. P along BC and DA, Q along CA and DB, and R along AB and DC. Show that pitch of the equivalent wrench is $\frac{1}{2\sqrt{2}}$ of the edge of the tetrahedron.
26. At all points on a given straight line are drawn the axes of principal moment corresponding to any given system of forces; show that these axes lie on a hyperbolic paraboloid and that their ends lie on another line.
27. On three given screws, whose axes are mutually perpendicular and concurrent, their act wrenches of p_1, p_2 and p_3 whose resultant is on a given screw of pitch p. Show that the locus of this latter screw is on the hyperboloid

$$(p - p_1)\, x^2 + (p - p_2)\, y^2 + (p - p_3)\, z^2 + (p - p_1)(p - p_2)(p - p_3) = 0,$$

the axes of co-ordinates being the axes of the given screws.

2

KINEMATICS OF PARTICLES AND RIGID BODIES

In this chapter we will confine our consideration to the motion of particle and rigid bodies moving in a curve (other than straight line) apart from the forces that cause the motion.

2.1 TANGENTIAL AND NORMAL COMPONENTS OF VELOCITY AND ACCELERATION

The investigation in this case will be applicable to the case of motion in space.

When the axes of reference are taken in the body, which is in motion, then axes will be moving when body moves. Suppose that moving axes at any time are parallel to the unit tangent, unit principal normal and unit bi-normal $\hat{t}, \hat{n}, \hat{b}$.

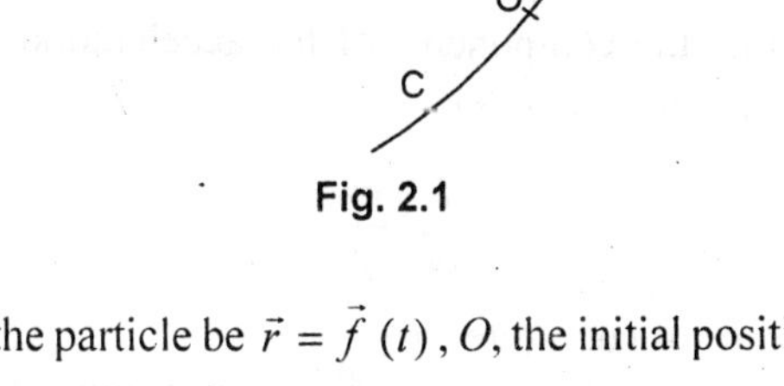

Fig. 2.1

Let the path of the particle be $\vec{r} = \vec{f}(t)$, O, the initial position of the particle of P at time t, let arc $OP = s$, $\overrightarrow{AP} = \vec{r}$.

If $\vec{v}$ is the velocity of P at time t, then

$$\vec{v} = \frac{d\vec{r}}{dt} = \frac{d\vec{r}}{ds} \cdot \frac{ds}{dt} = \frac{ds}{dt}\hat{t} \quad ...(1)$$

This shows that the speed v is equal to the absolute value of the scalar $\frac{ds}{dt}$ and the direction of the velocity of *P* is the same as that of the tangent to the path and the sense is the same or opposite to that of the unit tangent vector $\hat{t}$, according as $\frac{ds}{dt}$ is positive or negative, *i.e.*, according as, *s* increases or decreases with time *t*.

If $\vec{f}$ is the acceleration of *P* at time *t*, then

$$\vec{f} = \frac{d\vec{v}}{dt} = \frac{d}{dt}\left(\frac{ds}{dt}\hat{t}\right)$$

$$= \frac{d^2s}{dt^2}\hat{t} + \frac{ds}{dt}\cdot\frac{d\hat{t}}{dt}$$

$$= \frac{d^2s}{dt^2}\hat{t} + \frac{ds}{dt}\cdot\frac{d\hat{t}}{ds}\cdot\frac{ds}{dt}$$

$$= \frac{d^2s}{dt^2}\hat{t} + \left(\frac{ds}{dt}\right)^2 \kappa\hat{n} + 0.\hat{b}$$

where κ is the curvature of the path.

Thus $\frac{d^2s}{dt^2}, \left(\frac{ds}{dt}\right)^2 \kappa, 0$ are the components of acceleration along the vectors $\hat{t}, \hat{n}, \hat{b}$ respectively.

We can state that the components of the acceleration along the tangent, principal normal and bi-normal are

$$\frac{dv}{dt}, \frac{v^2}{\rho}, 0$$

where ρ is the radius of curvature.

Since, component along bi-normal is zero, hence components of acceleration along the tangent and normal are

$$\frac{d^2s}{dt^2}, \frac{v^2}{\rho}.$$

Hence, *the acceleration vector at any point of the path lies in the osculating plane at that point.*

Since,

$$\frac{d^2s}{dt^2} = \frac{d}{dt}\left(\frac{ds}{dt}\right) = \frac{dv}{dt} = \frac{dv}{ds}\cdot\frac{ds}{dt} = v\frac{dv}{ds}\cdot$$

Hence, tangential component of acceleration can be written as $v\dfrac{dv}{ds}$ also.

Note: *Mechanical significance of tangential and normal components of acceleration.*

Whenever a particle moves in a plane along a curve the *resulant tangential component* of acceleration changes the *magnitude of velocity*, while *normal component* of acceleration changes the direction of velocity. Hence, if the tangential component is zero, the particle will move with a constant speed. If normal component of acceleration is zero, the motion will be along a straight line, and the particle cannot move in a curve.

Example

A particle moves along a helix so that its co-odinates at time t are ($a\cos\theta$, $a\sin\theta$, $a\,\theta\tan\alpha$), where a, α are constants and $0<\alpha<\dfrac{\pi}{2}\cdot$ The speed increases linearly with time from 0 at t = 0 to V at time $t = T$. Find the acceleration at any time $t < T$. The motion taking place in the sense of θ increasing, and starting from the point (a, 0, 0).

At time t ($0 < t < T$), the position vector of the particle at the point P ($a\cos\theta$, $a\sin\theta$, $a\theta\tan\alpha$) is

$$\vec{r} = a(\cos\theta\,\hat{i} + \sin\theta\,\hat{j} + \theta\tan\alpha\,\hat{k})$$

$$\therefore \frac{d\vec{r}}{dt} = \frac{d\vec{r}}{d\theta}\frac{d\theta}{dt} = a\theta(-\sin\theta\,\hat{i} + \cos\theta\,\hat{j} + \tan\alpha\,\hat{k})$$

The magnitude of $-\sin\theta\,\hat{i} + \cos\theta\,\hat{j} + \tan\alpha\,\hat{k}$ is

$$(\sin^2\theta + \cos^2\theta + \tan^2\alpha)^{1/2} = \sec\alpha$$

$$\therefore \frac{d\vec{r}}{dt} = a\dot{\theta}\sec\alpha\,(-\sin\theta\cos\alpha\,\hat{i} + \cos\theta\cos\alpha\,\hat{j} + \sin\alpha\,\hat{k})$$

$$\therefore \left(\frac{Vt}{T}\right)\hat{t} = a\dot{\theta}\sec\alpha\,(-\sin\theta\cos\alpha\,\hat{i} + \cos\theta\cos\alpha\,\hat{j} + \sin\alpha\,\hat{k})$$

Hence,

$$\frac{Vt}{T} = a\dot{\theta}\sec\alpha \qquad ...(1)$$

$$\hat{t} = -\sin\theta\cos\alpha\,\hat{i} + \cos\theta\cos\alpha\,\hat{j} + \sin\alpha\,\hat{k} \qquad ...(2)$$

From (1), we have

$$\dot{\theta} = \frac{Vt}{aT} \cos \alpha.$$

Integrating, we get

$$\theta = \frac{Vt^2}{2aT} \cos \alpha + A,$$

where A is a constant of integeration.

The point ($a \cos \theta$, $a \sin \theta$, $a\,\theta \tan \alpha$) becomes starting point (a, 0, 0) when $\theta = 0$. Since, initial value of θ is zero. So initial condition is

$t = 0, \theta = 0.$ $\therefore$ $A = 0$

$$\therefore \quad \theta = \frac{Vt^2}{2aT} \cos \alpha$$

From (2), we get

$$\frac{d\hat{t}}{ds} = \frac{d\hat{t}}{d\theta} \cdot \frac{d\theta}{ds} = (-\cos\theta \cos\alpha\, \hat{i} - \sin\theta \cos\alpha \hat{j}) \frac{d\theta}{ds} \quad ...(3)$$

Now

$$\frac{d\theta}{ds} = \frac{d\theta}{dt} / \frac{ds}{dt} = \frac{Vt \cos \alpha}{aT} / \frac{Vt}{T} = \frac{\cos \alpha}{a}$$

Substituting the value of $\frac{d\theta}{ds}$ in (3), we get

$$\frac{d\hat{t}}{ds} = -(\cos\theta \cos\alpha \hat{i} + \sin\theta \cos\alpha \hat{j}) \frac{\cos\alpha}{a}$$

$$= -(\cos\theta \hat{i} + \sin\theta \hat{j}) \frac{\cos^2 \alpha}{a}$$

$$= \kappa \hat{n}$$

where

$$\kappa = \text{curvature} = \frac{\cos^2 \alpha}{a}$$

$$\hat{n} = \text{normal unit vector} = -(\cos\theta \hat{i} + \sin\theta \hat{j})$$

Thus at any time t ($0 < t < T$) the particle moves with acceleration having components V / T along the tangent and $\kappa v^2 = \frac{V^2 t^2 \cos^2 \alpha}{aT^2}$ along the normal.

2.2 RADIAL AND TRANSVERSE COMPONENTS OF VELOCITY AND ACCELERATION, WHEN THE PARTICLE IS MOVING IN A PLANE OR VELOCITY AND ACCELERATION IN POLAR CO-ORDINATES

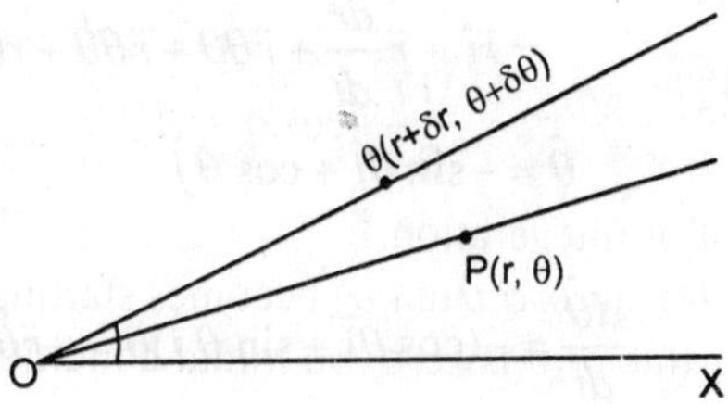

Fig.2.2

Let at time t the partricle be at the point $P\ (r,\ \theta)$ and at time $t + \delta t$ be at the point $Q\ (r + \delta r,\ \theta + \delta\theta)$. Let

$$\overrightarrow{OP} = r\hat{r}$$

where

$$\hat{r} = \cos\theta\hat{i} + \sin\theta\hat{j}$$

is a unit vector in the radial direction.

$$\vec{v} = \frac{d\vec{r}}{dt} = \frac{d}{dt}(r\hat{r}) = \dot{r}\hat{r} + r\frac{d\hat{r}}{dt}$$

$$\frac{d\hat{r}}{dt} = \frac{d\hat{r}}{d\theta}\frac{d\theta}{dt} = \frac{d}{d\theta}(\cos\theta\hat{i} + \sin\theta\hat{j})\,\dot{\theta}$$

$$= (-\sin\theta\ddot{i} + \cos\theta\hat{j})\,\dot{\theta}$$

$$= \left[\cos(90+\theta)\hat{i} + \sin(90+\theta)\hat{j}\right]\dot{\theta} = \dot{\theta}\hat{\theta},$$

where

$$= \hat{\theta} = \cos(90+\theta)\hat{i} + \sin(90+\theta)\hat{j}$$

is a unit vector in the transverse direction.

Substituting the value of $\frac{d\hat{r}}{dt}$ in (1), we get

$$v = \dot{r}\hat{r} + r\dot{\theta}\hat{\theta}$$

Thus radial and transverse components of velocity are $\dot{r}, \dot{r}\dot{\theta}$ in the sense of r and θ increasing.

Acceleration

$$\vec f = \frac{d\vec v}{dt} = \frac{d}{dt}(\ddot r\hat r + \ddot r\theta\hat\theta)$$

$$= \ddot r\hat r + \ddot r\frac{d\hat r}{dt} + \ddot r\dot\theta\hat\theta + \ddot r\theta\hat\theta + r\dot\theta\frac{d\hat\theta}{dt} \qquad (3)$$

$$\hat\theta = -\sin\theta\hat i + \cos\theta\hat j$$

$$\frac{d\hat\theta}{dt} = -(\cos\theta\hat i + \sin\theta\hat j)\dot\theta = -\ddot r\hat\theta$$

Substituting the values of $\frac{d\hat r}{dt}$ and $\frac{d\hat\theta}{dt}$ we get

$$\vec f = \ddot r\,\hat r + 2\ddot r\,\theta\hat\theta + \ddot r\,\theta\,\hat\theta - \ddot r\,\theta^2\hat r$$

$$= (\ddot r - \ddot r\,\theta^2)\,\hat r + (\ddot r\,\theta + 2\ddot r\,\theta)\hat\theta\,.$$

Hence, components of acceleration along radial and transverse directions are $\ddot r - r\,\dot\theta^2$ and $r\ddot\theta + 2\dot r\,\dot\theta$ in the sense of r and θ increasing.

2.3 RELATIVE VELOCITY

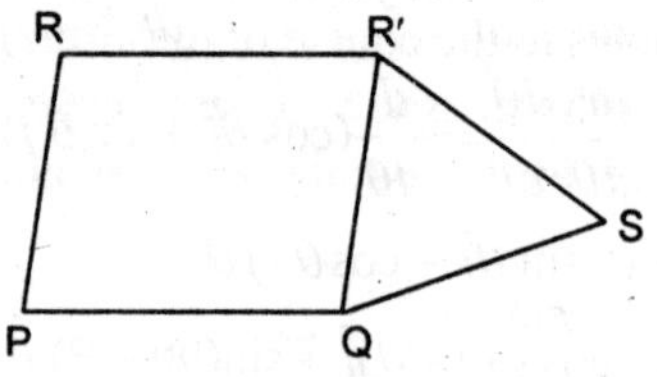

Fig. 2.3

Let two points P, Q move in the same plane with velocities represented by $\overrightarrow{PR}, \overrightarrow{QS}$. Complete the parallelogram $PQR'R$. Then $\overrightarrow{QR'}$ is a vector representing a velocity equal in magnitude and direction to that of P. But $\overrightarrow{QS} = \overrightarrow{QR'} + \overrightarrow{R'S}$, therefore $\overrightarrow{R'S}$, represents a velocity which when compounded with velocity of P gives the velocity of Q. Hence, $\overrightarrow{R'S}$, is the velocity of Q relative to P.

Let (x_p, y_p, z_p), (x_q, y_q, z_q) be the co-odinates of P and Q at time t referred to fixed axes through origin O.

Let
$$\vec{r}_p = x_p\vec{i} + y_p\vec{j} + z_p\vec{k}$$
$$\vec{r}_q = x_q\vec{i} + y_q\vec{j} + z_q\vec{k}$$

Since $\dfrac{d\hat{i}}{dt} = \vec{0}$, etc.

$$\therefore \quad \dot{\vec{r}}_p = \dot{x}_p\hat{i} + \dot{y}_p\hat{j} + \dot{z}_p\hat{k}$$
$$\ddot{\vec{r}}_p = \ddot{x}_p\hat{i} + \ddot{y}_p\hat{j} + \ddot{z}_p\hat{k}$$

Similar expressions hold for $\dot{\vec{r}}_q, \ddot{\vec{r}}_q$

Now the position of P relative to Q is $\vec{r}_p - \vec{r}_q$ so the velocity of P relative to Q is
$$\dot{\vec{r}}_p - \dot{\vec{r}}_q = (\dot{x}_p - \dot{x}_q)\hat{i} + (\dot{y}_p - \dot{y}_q)\hat{j} + (\dot{z}_p - \dot{z}_q)\hat{k}$$
and acceleration of P relative to Q is
$$\ddot{\vec{r}}_p - \ddot{\vec{r}}_q = (\ddot{x}_p - \ddot{x}_q)\hat{i} + (\ddot{y}_p - \ddot{y}_q)\hat{j} + (\ddot{z}_p - \ddot{z}_q)\hat{k}.$$

Example

An aircraft persues a straight course with consant velocity V and is being chased by a guided missile moving with constant speed $2V$ and fitted with a homing device to ensure that its motion is always directed at the target. Initially missile is at right angles to the course of the aircraft and distant R from it. Find the polar equation of the missile's persuit curve relative to the target, taking the course of the target as the initial line $\theta = 0$, and find the time taken by it to strike the target.

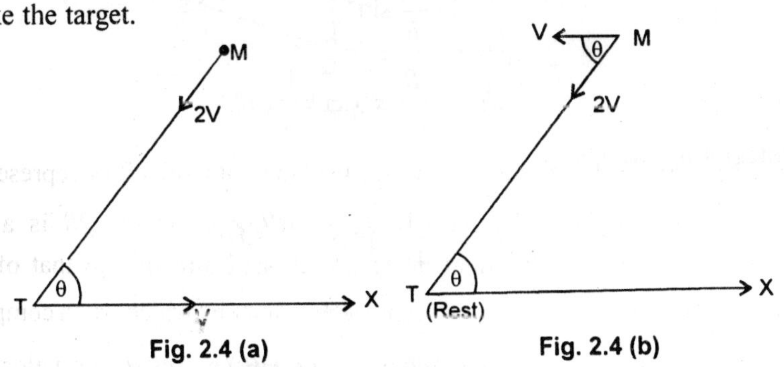

Fig 2.4 *(a)* shows the situation at time t.
Fig. 2.4 *(b)* shows the motion of missile relative to the target T.
Let us take
$$TM = R, \quad \angle MTX = \theta$$

In Fig. 2.4(a) resolving velocities along radial and transverse directions, we get

$$\dot{r} = -V(2+\cos\theta)$$

$$r\dot{\theta} = V\sin\theta$$

$$\therefore \quad \frac{dr}{r} = -\left(\frac{2}{\sin\theta}+\cot\theta\right)d\theta\cdot$$

Integrating, we get

$$\frac{A}{r} = \tan^2\frac{\theta}{2}\sin\theta\cdot$$

Initially $\theta = \frac{\pi}{2}, r = R$

$$\therefore \quad \frac{A}{R} = \tan^2\frac{\pi}{4}\sin\frac{\pi}{2} = 1\cdot$$

$$\therefore \quad A = R$$

Hence,

$$\frac{R}{r} = \sin\theta\tan^2\frac{\theta}{2} = 2\sin^3\frac{\theta}{2}/\cos\frac{\theta}{2}\cdot$$

Thus

$$r = 0, \text{ when } \theta = \pi$$

$$\therefore \quad \dot{\theta} = \frac{V\sin\theta}{r} = \frac{V}{R}\sin^2\theta\tan^2\frac{\theta}{2}$$

$$= \frac{4V}{R}\sin^4\frac{\theta}{2}$$

$$\therefore \quad dt = \frac{R}{4V}\operatorname{cosec}^4\frac{\theta}{2}d\theta.$$

Integrating, we get

$$\int_0^t dt = \frac{R}{4V}\int_{\frac{\pi}{2}}^{\pi}\operatorname{cosec}^4\frac{\theta}{2}d\theta$$

$$t = \frac{R}{4V}\int_{\frac{\pi}{2}}^{\pi}\left(1+\cot^2\frac{\theta}{2}\right)\operatorname{cosec}^2\frac{\theta}{2}d\theta$$

$$= \frac{2R}{3V}\cdot$$

2.4 ANGULAR VELOCITY

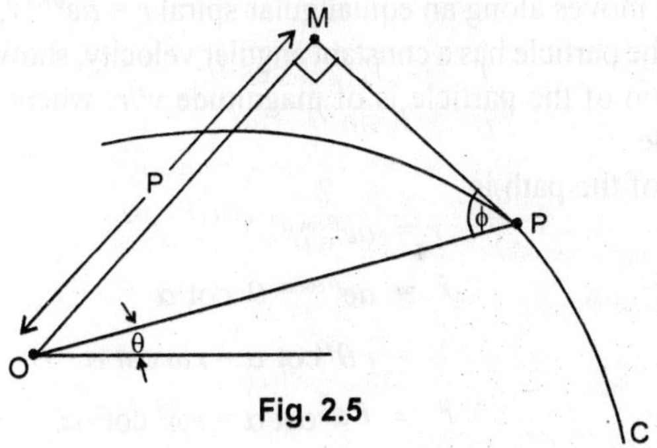

Fig. 2.5

If a point is moving in a plane, the angular velocity of P about a fixed pint O in the plane is the rate of increase of the angle that the line OP makes with a fixed direction in the plane. Thus if a point P be moving in any direction with velocity V, making an angle θ with a fixed direction Ox, the angular velocity of P about O is $\frac{d\theta}{dt}$. But if $OP = r$ and the tangent to the path makes an angle ϕ with OP, we have

$$r\frac{d\theta}{ds} = \sin\phi$$

$$\therefore \quad \frac{d\theta}{dt} = \frac{d\theta}{ds}\frac{ds}{dt} = \frac{V\sin\phi}{r}$$

Hence,

Angular velocity = (Component of the velocity at right angles to OP)/OP.

We can write

$$r\dot{\theta} = V\sin\phi$$

$$\therefore r^2\dot{\theta} = Vr\sin\phi = Vp$$

where p is the length of perpendicular drawn frum origin O on the tangent, *i.e.*,

$$r^2\theta = \text{moment of the velocity about the origin.}$$

Again a sectorial element of area of a plane curve is $\frac{1}{2}r^2\delta\theta$, hence, $\frac{1}{2}r^2\dot{\theta}$ is the rate of increase of a sectorial area as P moves along the curve, and therefore $r^2\dot{\theta}$ or Vp is twiçe the rate at which the radius vector OP sweeps out area. Generally $r^2\dot{\theta}$ is denoted by h, *i.e.*, $h = r^2\dot{\theta}$.

Examples

1. A particle moves along an equiangular spiral $r = ae^{\theta \cot \alpha}$, and if the radius vector to the particle has a constant angular velocity, show that the resultant acceleration of the particle is of magnitude v^2/r, where v is the speed of the particle.

   Equation of the path is

$$r = ae^{\theta \cot \alpha}$$

$$\therefore \quad \dot{r} = ae^{\theta \cot \alpha}\,\dot{\theta} \cot \alpha$$

$$= r\dot{\theta} \cot \alpha = r\omega \cot \alpha$$

$$\therefore \quad \ddot{r} = \dot{r}\omega \cot \alpha = r\omega^2 \cot^2 \alpha.$$

   Let f_r, f_θ be the radial and transverse components of acceleration, then

$$f_r = \ddot{r} - r\dot{\theta}^2 = r\omega^2 \cot^2 \alpha - r\omega^2 = r\omega^2 (\cot^2 \alpha - 1)$$

$$f_\theta = r\ddot{\theta} + 2\dot{r}\dot{\theta} = 0 + 2\dot{r}\omega = 2\dot{r}\omega = 2r\omega^2 \cot \alpha.$$

   Hence, resultant acceleration

$$= \sqrt{f_r^2 + f_\theta^2}$$

$$= \sqrt{r^2\omega^4 (\cot^2 \alpha - 1)^2 + 4r^2\omega^4 \cot^2 \alpha}$$

$$= r\omega^2 (\cot^2 \alpha + 1) = r\omega^2 \operatorname{cosec}^2 \alpha$$

   Now,

$$v^2 = \dot{r}^2 + r^2\dot{\theta}^2 = r^2\omega^2 \cot^2 \alpha + r^2\omega^2$$

$$= r^2\omega^2 \operatorname{cosec}^2 \alpha$$

   Hence, resultant acceleration $= \dfrac{v^2}{r}$.

   Let Ψ be the angle which the resulant acceleration makes with the radius vector, then

$$\tan \Psi = \frac{f_\theta}{f_r} = \frac{2r\omega^2 \cot \alpha}{r\omega^2 (\cot^2 \alpha - 1)} = \frac{2\cot \alpha}{(\cot^2 \alpha - 1)}$$

$$= \frac{2\cos \alpha \sin \alpha}{\cos^2 \alpha - \sin^2 \alpha} = \frac{\sin 2\alpha}{\cos 2\alpha} = \tan 2\alpha$$

$$\therefore \quad \Psi = 2\alpha$$

2. A point is moving in a circle. Show that its angular velocity about a point on the circumference of the circle is half its angular velocity about the centre.

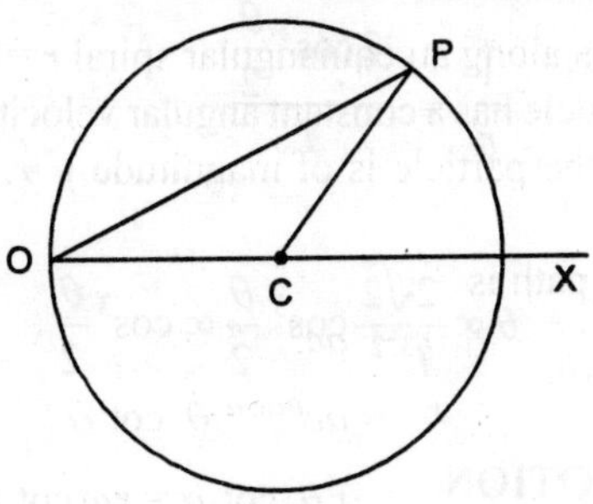

Fig. 2.6

Let C be the centre of the circle and O any point on the circumference of the circle, then

$$\angle POC = \frac{1}{2} \angle PCX$$

$$\therefore \quad \frac{d}{dt}(\angle POC) = \frac{1}{2}\frac{d}{dt}(\angle PCX)$$

i.e., angular velocity about a point on the circumference $= \frac{1}{2}$ angular velocity about the centre.

3. A particle moves in a parabola with constant speed. Show that the angular velocity about the focus varies as $\cos^3 \frac{\theta}{2}$, where θ is the angular distance from the focus.

Equation of the parabola is

$$p^2 = ar, \text{ when focus is origin}$$

$$\therefore \quad p = a^{1/2} r^{1/2}.$$

We know that

$$r^2\dot{\theta} = Vp$$

$$\dot{\theta} = \frac{Vp}{r^2} = \frac{Va^{1/2}r^{1/2}}{r^2} = \frac{Va^{1/2}}{r^{3/2}}$$

$$\therefore \quad \dot{\theta} \propto \frac{1}{r^{3/2}}.$$

Also we know that for parabola

$$\frac{l}{r} = 1 + \cos\theta = 2\cos^2\frac{\theta}{2}$$

$$\therefore \qquad \frac{1}{r} = \frac{2\cos^2\dfrac{\theta}{2}}{l}.$$

Hence,

$$\dot{\theta} \propto \frac{2\sqrt{2}}{l^{3/2}} \cos^3\frac{\theta}{2} \propto \cos^3\frac{\theta}{2}.$$

2.5 CIRCULAR MOTION

At time t, let v be the velocity of a point P moving in a cicle of radius r, centre O. Let θ be the angle that OP makes with a fixed direction Ox.

Let s be the arc AP measured from a point A on Ox. Then the velocity

$$v = \dot{s} = r\dot{\theta}.$$

Also the acceleration components are $\ddot{s}$ or $r\ddot{\theta}$ along the tangent and $\dfrac{v^2}{r}$ or $r\dot{\theta}^2$ towards the centre.

If the motion be uniform *i.e.*, v is constant, then there is no acceleration along the tangent and there is acceleration v^2/r or $r\dot{\theta}^2$ towards the centre.

2.6 RELATIVE ANGULAR VELOCITY

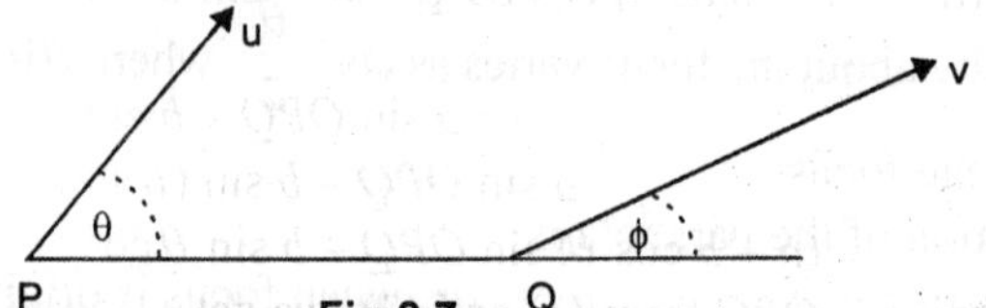

Fig.2.7

It two points P, Q are moving in a plane with velocities u, v, then either point in general posses an angular velocity relative to the other, which may be represented thus:

Let the directions of velocities make angles θ, ϕ with the line PQ.

The component velocity of Q relative to P at right angles to PQ is $v \sin \phi - u \sin \theta$, therefore the angular velocity of Q relative to P is

$$\frac{v\sin\phi - u\sin\theta}{\text{PQ}}$$

When $v \sin \phi = u \sin \theta$, then PQ is moving parallel to itself.

Example

Two points describe two concentric circles with velocities varying inversely as the square roots of the radius of the circles, find the position in which the relative angular velocity vanishes.

Let O be the centre of the circles, a, b the radii u, v the velocities. If P, Q are the positions of the points when the angular velocity vanishes, then velocities resolved at right angle to PQ must be equal.

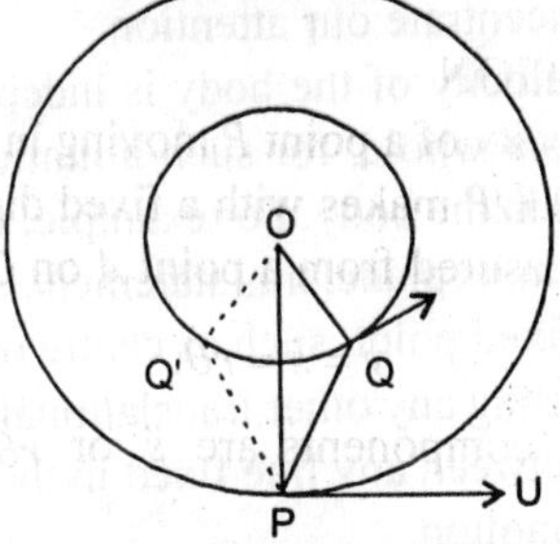

Fig.2.8

Let

$$\angle POQ = \theta.$$

Therefore

$$u \cos OPQ + v \cos OQP = 0$$

$$\therefore \quad a^{-1/2} \cos OPQ - b^{-1/2} \cos (\theta + OPQ) = 0$$

$$\therefore \quad (a^{-1/2} - b^{-1/2} \cos \theta) \cos OPQ + b^{-1/2} \sin \theta \sin OPQ = 0 \qquad ...(1)$$

But

$$a \sin OPQ - b \sin OQP = 0$$

$$\therefore \quad a \sin OPQ - b \sin (\theta + OPQ) = 0$$

$$\therefore \quad (a - b \cos \theta) \sin OPQ - b \sin \theta \cos OPQ = 0. \qquad ...(2)$$

Eliminating $\angle OPQ$ from (1) and (2), we get

$$\frac{b \sin \theta}{a - b\cos \theta} = \frac{-(a^{-1/2} - b^{-1/2} \cos \theta)}{b^{-1/2} \sin \theta}$$

$$\therefore \quad (a^{-1/2} - b^{-1/2} \cos \theta)(a - b \cos \theta) + b^{1/2} \sin^2 \theta = 0$$

$$\therefore \quad a^{1/2} + b^{1/2} \cos^2 \theta - ab^{-1/2} \cos \theta - a^{-1/2} b \cos \theta + b^{1/2} \sin^2 \theta = 0$$

$$\therefore \quad \cos \theta = \frac{a^{1/2} b^{1/2}}{(a - a^{1/2} b^{1/2} + b)}.$$

2.7 ANGULAR VELOCITY OF A BODY

When a plane body is moving in its plane, we may speak of 'angular velocity of a rigid body' without specifying any particular line or point of reference, meaning there by the rate of increase of the angle between any line AB fixed in the body and any line Ox in the plane of motion. *i.e.*, if at time t, the angle between AB and Ox be θ, then $\dot{\theta}$ is the angular velocity of the body. Also if any other line CD fixed in the body makes an angle $\theta + \alpha$ with Ox at the same instant, then α is a constant angle between AB and CD and remains unchanged by the motion, so that $\dot{\theta}$ is also the rate of increase of the angle betwen CD and Ox. Hence, in measuring angular velocity, it is immaterial upon what line fixed in the body, we concentrate our attention.

Further, the angular velocity of the body is independent of any motion of translation of the body as a whole, for such a motion would not change the direction of any line fixed in the body. For example, when we say that a wheel has an angular velocity ω in its plane, this statement is independent of whether wheel is turning about a fixed point, such as centre or rolling along in contact with a fixed line or possessing any other translational motion, ω being the rate of increase of the angle between any line fixed in the wheel in its plane and a line fixed in the plane of motion.

Examples

1. A circle rolls along in contact with a straight line, find the velocity and acceleration of any point on the circumference.

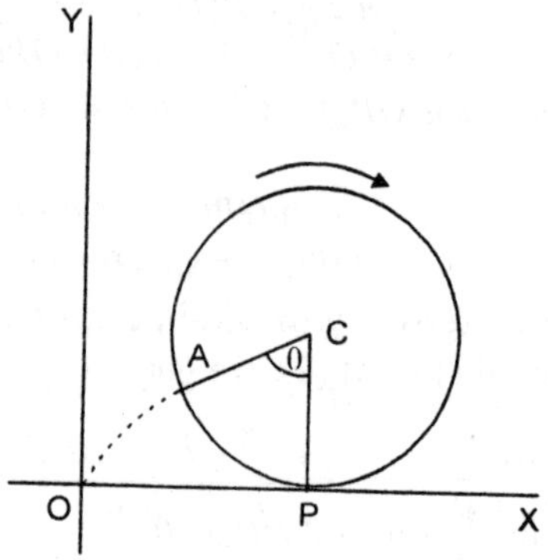

Fig. 2.9

Let us take the given line as x-axis. Let C be the centre of the circle, a the radius, P the point of contact with Ox at time t. Let A be the point on the cicumference of the circle whose velocity and acceleration are required.

Let $\angle ACP = \theta$, then $\dot{\theta}$ is the angular valocity of the circle. For convenience, we take O as origin so that $OP = \text{arc } AP$ *i.e.*, O is the position of the point

A in the rolling motion. Taking an axis Oy at right angle to Ox, then the co-ordinates of A are (x, y) which are given by

$$x = OP - AC \sin\theta = a\theta - a\sin\theta = a(\theta - \sin\theta)$$
$$y = CP - AC\cos\theta = a(1-\cos\theta)$$

Hence, velocity components are given by

$$\dot{x} = a(1-\cos\theta)\dot{\theta}$$

$$\dot{y} = a\dot{\theta}\sin\theta.$$

$$\therefore \quad v = \sqrt{\dot{x}^2+\dot{y}^2}$$

$$= a\dot{\theta}\sqrt{(1-\cos\theta)^2+\sin^2\theta}$$

$$= \dot{\theta}\, 2a\sin\frac{\theta}{2} = \dot{\theta}\, AP$$

If Ψ is the angle which the resultant velocity makes with Ox, then

$$\tan\Psi = \frac{a\dot{\theta}\sin\theta}{a\dot{\theta}(1-\cos\theta)} = \cot\frac{\theta}{2} = \tan\left(90^\circ+\frac{\theta}{2}\right)$$

$$\therefore \quad \Psi = 90^\circ + \theta.$$

i.e., in the direction of perpendicular PI in the sense of θ increasing.

Acceleration: Acceleration components are

$$\ddot{x} = a\ddot{\theta}(1-\cos\theta) + a\dot{\theta}^2\sin\theta$$

$$\ddot{y} = a\ddot{\theta}\sin\theta + a\dot{\theta}^2\cos\theta$$

$$f = \sqrt{[a\ddot{\theta}(1-\cos\theta)+a\dot{\theta}^2\sin\theta]^2+[a\ddot{\theta}\sin\theta+a\dot{\theta}^2\cos\theta]^2}$$

$$= \sqrt{\begin{matrix} a^2\ddot{\theta}^2(1-\cos\theta)^2 + a^2\dot{\theta}^4\sin^2\theta + 2a^2\dot{\theta}^2\ddot{\theta}\sin\theta(1-\cos\theta) \\ +a^2\ddot{\theta}^2\sin^2\theta + a^2\dot{\theta}^4\cos^2\theta + 2a^2\dot{\theta}^2\ddot{\theta}\sin\theta\cos\theta \end{matrix}}$$

$$= a\sqrt{4\ddot{\theta}^2\sin^2\frac{\theta}{2}+\dot{\theta}^4+2\dot{\theta}^2\ddot{\theta}.\sin\theta}.$$

If ϕ is the angle which the direction of acceleration makes with x-axis, then

$$\tan\phi = \frac{\ddot{y}}{\ddot{x}} = \frac{\ddot{\theta}\sin\theta+\dot{\theta}^2\cos\theta}{\ddot{\theta}(1-\cos\theta)+\dot{\theta}^2\sin\theta}.$$

Special cases: As a special case, we may put $\theta = 0$, so that A is the point of contact of the circle and the line, then we get

$$\dot{x} = 0 = \dot{y}.$$

Showing that the point of contact of the rolling has no velocity *i.e.*, it is the instantaneous centre of rotation.
Acceleration components are

$$\ddot{x} = 0, \ \ddot{y} = a\dot{\theta}^2$$

showing that the point of contact has an acceleration $a\dot{\theta}^2$ towards the centre.

2. A circle A of radius a turns round its axis with uniform angular velocity ω. A circle B of radius b rolls on the circle A and its uniform angular velocity ω'. Find the time taken.
   (*i*) for the point of contact to make a complete circuit of A,
   (*ii*) for the centre of B to return to former position.
   Determine the acceleration of the common point of two circles and the greatest acceleration on the circle B.

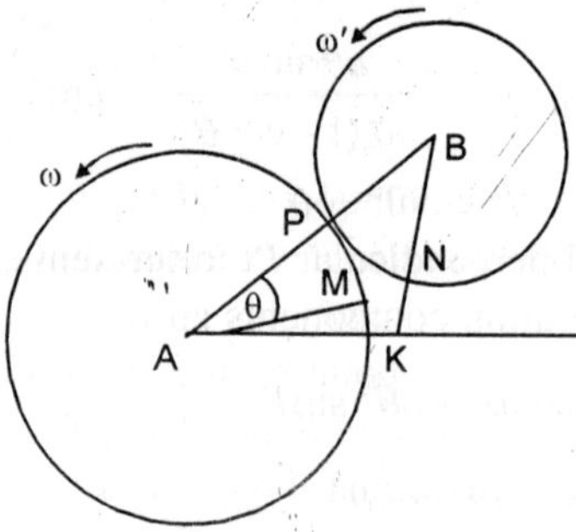

Fig. 2.10

Suppose that at time $t = 0$, the points M, N are in contact and on the fixed line Ax through the centre A of the circle A. Let P be the point of contact at time t. Since, AM and BM are lines fixed in relation to the circles, in time t they turn through angles ωt, $\omega' t$. Therefore if BN meets Ax in K, the angle

$$BKx = \omega' t \quad MAx = \omega t.$$

Again arc PM = arc PN, so that if $\angle PAM = \theta$ then $\angle PBN = \dfrac{a\theta}{b}$.

Hence,

$$\omega' t - \theta = \omega t + \frac{a\theta}{b}$$

$$\therefore \qquad \theta = \frac{b(\omega' - \omega)t}{a+b}, \ \dot{\theta} = \frac{b(\omega' - \omega)}{a+b}.$$

The time taken for P to make a circuit of the circle A is

$$\frac{2\pi}{\dot{\theta}} = \frac{2\pi\,(a+b)}{b\,(\omega'-\omega)}$$

Again the time taken for the point to describe a cicle A depends on the angular velocity of the line AB, which is $\dot{\theta}+\omega$.

Hence, the time required $= 2\pi/(\dot{\theta}+\omega)$.

For the acceleration we have the acceleration of the point P on the circle A

$$a\left(\frac{d}{dt}PAx\right)^2 \text{ along } PA$$

$$= a\,(\dot{\theta}+\omega)^2$$

$$= a\left[\frac{b\,(\omega'-\omega)}{a+b}+\omega\right]^2 = \frac{a\,(a\omega+b\omega')^2}{(a+b)^2}$$

and the acceleration of the point on the circle B

= acceleration relative to centre B + acceleration of B.

$$- b\omega'^2 + (a\omega+b\omega')^2/(a+b)$$

Also the acceleration of any point on the circle B is compounded of its acceleration relative to the centre B, *i.e.*, $b\omega'^2$ along the radius towards B, and the acceleration of B, so that the greatest acceleration will be when these components coincide in direction, *i.e.*,

$$b\omega'^2 + (a+b\omega')^2/(a+b).$$

2.8 INSTANTANEOUS ANGULAR VELOCITY AND INSTANTANEOUS AXIS OF ROTATION

Suppose the rotation of the body about an axis OC is $\delta\theta$ in time, δt, then

$$\lim_{\delta t\to 0}\frac{\delta\theta}{\delta t} = \frac{d\theta}{dt} = \omega \text{ (say)}$$

is called the *instantaneous angular velocity at time t*.

Any rotation of a rigid body about a fixed point O is equivalent to a rotation about a definite axis through O. The axis about which the body instantaneously rotates is called *instantaneous axis of rotation.*

2.9 VECTOR ANGULAR VELOCITY

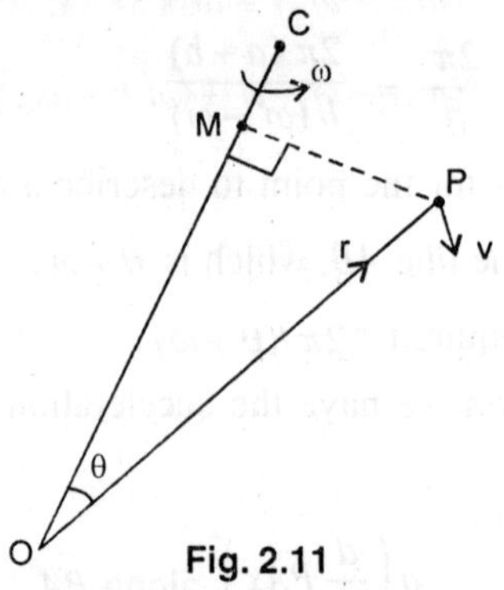

Fig. 2.11

Let a rigid body rotate about a fixed point O and let ω be its angular velocity about the instantaneous axis OC at time t. If O be a fixed point of the body such that $\overrightarrow{OP} = \vec{r}$. Draw PM perpendicular from P on OC. Now $PM = r \sin \theta$, hence P describes a circle of radius PM round M, therefore its speed is = radius $\times \dot{\theta}$ = $r\omega \sin \theta$ and direction is $\hat{a} \times \vec{r}$, where $\hat{a}$ is a unit vector along OC, so chosen that a right handed screw advancing along it turns in the sense of ω.

$$\vec{v} = (\omega r \sin\theta)\hat{t}$$

where $\hat{t}$ is the unit vector in the sense $\hat{a} \times \vec{r}$

$$\therefore \quad \vec{v} = \omega\hat{a} \times \vec{r} = \vec{\omega} \times \vec{r}$$

where

$$\vec{\omega} = \omega\hat{a}$$

$\vec{\omega}$ is called the *instantaneous vector angular velocity* of the body about O. Its magnitude ω is the physical angular velocity about the instantaneous axis OC and $\hat{a}$ specifies the direction of this instantaneous axis.

Cor. *Prove that for a rigid body*

$$\vec{\omega} = \frac{1}{2} curl\,\vec{v}.$$

We know that

$$\vec{v} = \vec{\omega} \times \vec{r}.$$

Let

$$\vec{\omega} = \omega_1\hat{i} + \omega_2\hat{j} + \omega_3\hat{k}$$

$$\vec{r} = r\hat{i} + y\hat{j} + z\hat{k}.$$

$$\therefore \quad \vec{v} = (\omega_1\hat{i} + \omega_2\hat{j} + \omega_3\hat{k}) \times (x\hat{i} + y\hat{j} + z\hat{k})$$

$$= (\omega_2 z - \omega_3 y)\,\hat{i} + (\omega_3 x - \omega_1 z)\,\hat{j} + (\omega_1 y - \omega_2 x)\hat{k}.$$

If

$$\vec{v} = v_1\hat{i} + v_2\hat{j} + v_3\hat{k}.$$

then

$$v_1 = \omega_2 z - \omega_3 y, \quad v_2 = \omega_3 x - \omega_1 z, \quad v_3 = \omega_1 y - \omega_2 x$$

$$curl\,\vec{v} = \left(\frac{\partial v_3}{\partial y} - \frac{\partial v_2}{\partial z}\right)\hat{i} + \left(\frac{\partial v_1}{\partial z} - \frac{\partial v_3}{\partial x}\right)\hat{j} + \left(\frac{\partial v_2}{\partial x} - \frac{\partial v_1}{\partial y}\right)\hat{k}$$

$$= \left[\frac{\partial}{\partial y}(\omega_1 y - \omega_2 x) - \frac{\partial}{\partial z}(\omega_3 x - \omega_1 z)\right]\hat{j}$$

$$+ \left[\frac{\partial}{\partial x}(\omega_3 x - \omega_1 z) - \frac{\partial}{\partial y}(\omega_2 z - \omega_3 y)\right]\hat{k}$$

$$= 2\omega_1\hat{i} + 2\omega_2\hat{j} + 2\omega_3\hat{k}$$

$$= 2(\omega_1\hat{i} + \omega_2\hat{j} + \omega_3\hat{k}) = 2\vec{\omega}$$

$$\therefore \quad \vec{\omega} = \frac{1}{2}\,curl\,\vec{v}.$$

2.10 GENERAL MOTION OF A RIGID BODY

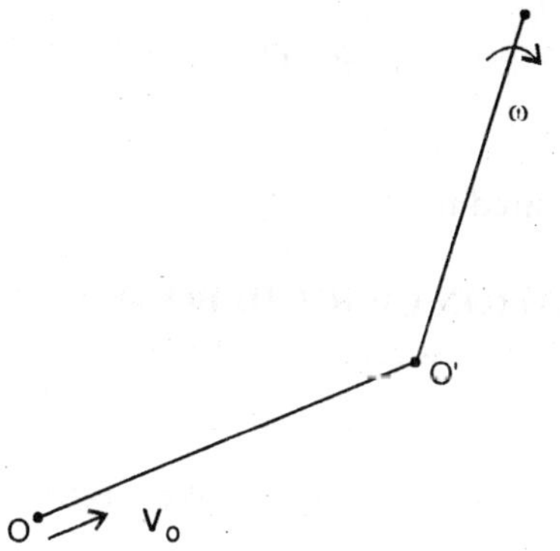

Fig. 2.12

In general case when no point of the rigid body is fixed, then body's motion can be regarded as the displacement of one point of it to a new point O', followed by an appropriate turning about a certain axis through O' until the body takes up required orientation. Thus the general body motion may be described as a translation and a rotation.

When we take the limit as $O' \to O$, the motion is specified by vector $\vec{v}_0$ giving the instantaneous velocity of O and an angular velocity $\vec{\omega}$, being the instantaneous angular velocity about O.

Now suppose P is any other point of the body and let its velocity be $\vec{v}_p$. If $\overrightarrow{OP} = \vec{r}$, then the velocity of P relative to O is $\vec{\omega} \times \vec{r}$.

$$\therefore \quad \vec{v}_p - \vec{v}_0 = \vec{\omega} \times \vec{r}$$

$$\therefore \quad \vec{v}_p = \vec{v}_0 + \vec{\omega} \times \vec{r}.$$

Cor. Prove that $\vec{\omega}$ is independent of the position of O.

Let $\overrightarrow{OP} = \vec{r}, \overrightarrow{OO'} = \vec{s}, \overrightarrow{O'P} = \vec{r}'$.

Then

$$\overrightarrow{OP} = \overrightarrow{OO'} + \overrightarrow{O'P}$$

$$\therefore \quad \vec{r} = \vec{s} + \vec{r}'.$$

Also

$$\vec{v}_p = \vec{v}_0 + \vec{\omega} \times \vec{r}$$

$$\vec{v}_{o'} = \vec{v}_0 + \vec{\omega} \times \vec{s}$$

$$\therefore \quad \vec{v}_p - \vec{v}_0{}' = \vec{\omega} \times (\vec{r} - \vec{s})$$

$$= \vec{\omega} \times \vec{r}'$$

$$\therefore \quad \vec{v}_p = \vec{v}_0{}' + \vec{\omega} \times \vec{r}'.$$

This proves the required result

2.11 GENERAL MOTION OF RIGID BODY AS A SCREW MOTION

With notations of the previous article, we have

$$\vec{v}' = \vec{v} + \vec{\omega} \times \vec{s}. \tag{1}$$

If point O' is such that $\vec{v}'$ and $\vec{\omega}$ are parallel vectors, than

$$\vec{\omega} \times \vec{v}' = \vec{0}.$$

Taking vector product of (1) with $\vec{\omega}$, we get

$$\vec{\omega} \times \vec{v}' = \vec{\omega} \times [\vec{v} + (\vec{\omega} \times \vec{s})] = \vec{\omega} \times \vec{v} + \vec{\omega} \times (\vec{\omega} \times \vec{s})$$

$$\therefore \quad \vec{0} = \vec{\omega} \times \vec{v} + (\vec{\omega} . \vec{s})\, \vec{\omega} - (\vec{\omega} . \vec{\omega})\, \vec{s} = \vec{\omega} \times \vec{v} + (\vec{\omega} . \vec{s})\, \vec{\omega} - \omega^2 \vec{s} \cdot$$

If $\vec{\omega} \neq 0$, then

$$\vec{s} = \frac{\vec{\omega} \times \vec{v}}{\omega^2} + \frac{(\vec{\omega} . \vec{s})\, \vec{\omega}}{\omega^2} = \frac{\vec{\omega} \times \vec{v}}{\omega^2} + \lambda \vec{\omega} \cdot \quad (2)$$

where

$$\lambda = \frac{(\vec{\omega} . \vec{s})}{\omega^2} \cdot$$

Now as λ varies it is clear from (2) that the point O' having position vector $\vec{s}$ traces out a straight line through the point represented by $\vec{s} = \dfrac{\vec{\omega} \times \vec{v}}{\omega^2}$ and having the same direction as that of $\vec{\omega}$.

At the particular instant considered, every pont on the line (2) has a velocity parallel to $\vec{\omega}$. Hence, this line is called *axis of screw* or *central axis*. Every point on this axis moves along it and the body turns about this axis.

Invariants

The velocity $\vec{V}$ at any point P, where $\overrightarrow{OP} = \vec{r}$, is

$$\vec{V} = \vec{v} + \vec{\omega} \times \vec{r} \, .$$

Taking dot product with $\vec{\omega}$, we get

$$\begin{aligned} \vec{\omega} . \vec{V} &= \vec{\omega} \cdot \vec{v} + \vec{\omega} \cdot (\vec{\omega} \times \vec{r}) \\ &= \vec{\omega} \cdot \vec{v} + (\vec{\omega} \times \vec{\omega}) \cdot \vec{r} \\ &= \vec{\omega} \cdot \vec{v} \, . \end{aligned}$$

Thus scalar quantity $\Gamma = \vec{\omega} . \vec{v}$ is invariant for the system at any instant.

Let $p = \dfrac{\Gamma}{\omega^2}$, then p is also an invariant. At any point on the axis of the screw with position vector $\vec{v}'$, since, $\vec{v}'$ and $\vec{\omega}$ are parallel, we see that $p = \dfrac{\vec{v}' . \vec{\omega}}{\omega^2} = \dfrac{v'}{\omega}$. This latter ratio is the distance, which the body advances along the axis of the screw per unit angle of the turn. Hence, p is called the *pitch* of the screw.

2.12 COMPOSITION OF ANGULAR VELOCITIES.

Let O and P be two fixed points in a body $\overrightarrow{OP} = \vec{r}$, and the body turns through a small angle $\delta\theta_1$ in the positive sense about an axis through O, specified by unit vector $\hat{a}_1$.

The movement of P is $(\hat{a}_1 \delta\theta_1) \times \vec{r}$; hence, the new position of P relative to O is

$$\vec{r}_1 = \vec{r} + (\hat{a}_1 \delta\theta_1) \times \vec{r}.$$

Now suppose the body rotates through another small angle $\delta\theta_2$ in the positive sense about an axis through O specified by unit vector $\hat{a}_2$. Since $(\hat{a}_2 \delta\theta_2) \times \vec{r}_1$ is the movement of P, the new position of P relative to O is now

$$\vec{r}_{12} = \vec{r}_1 + (\hat{a}_2 \delta\theta_2) \times \vec{r}_1$$

Substituting the value of $\vec{r}_1$, we get

$$\begin{aligned} \vec{r}_{12} &= \vec{r} + (\hat{a}_1 \delta\theta_1) \times \vec{r} + (\hat{a}_2 \delta\theta_2) \times (\vec{r} + \delta\theta_1 \hat{a}_1 \times \vec{r}) \\ &= \vec{r} + (\delta\theta_1 \hat{a}_1 + \delta\theta_2 \hat{a}_2) \times \vec{r}, \text{ to first order.} \end{aligned}$$

If the operations were performed in the reverse order, we will get for the displacement of P relative to O,

$$\vec{r}_{21} = \vec{r} + (\delta\theta_2 \hat{a}_2 + \delta\theta_1 \hat{a}_1) \times \vec{r}, \text{ to first order.}$$

Thus to first order $\vec{r}_{12} = \vec{r}_{21}$, but this result will not hold for finite rotations.

If δt be the time in which these operations take place, the velocity of P relative to O at time t is

$$\begin{aligned} \lim_{\delta t \to 0} \frac{\vec{r}_{12} - \vec{r}}{\delta t} &= (\dot{\theta}_1 \hat{a}_1 + \dot{\theta}_2 \hat{a}_2) \times \vec{r} \\ &= (\vec{\omega}_1 + \vec{\omega}_2) \times \vec{r} \end{aligned}$$

where $\vec{\omega}_1 = \dot{\theta}_1 \hat{a}_1$ = vector angular velocity about $\hat{a}_1$, $\vec{\omega}_2 = \dot{\theta}\, \hat{a}_2$ = vector angular velocity about $\hat{a}_2$.

Thus the vector angular velocities about a fixed point of the rigid body can be added in the same way as the other vector quantities, provided they are infinitesimal. Hence, it is permissible to resolve such quantities into components.

Because of essentially vectorial character of both linear and angular velocities, we can generalise the result for infinitesimal angular velocities.

If a rigid body moves in such a way that a velocity $\vec{v}_1$ of a point O fixed in it together with spin $\vec{\omega}_1$ is to be combind with a velocity $\vec{v}_2$ of O and spin $\vec{\omega}_2$ of the body, then the net motion is a velocity $\vec{v}_1 + \vec{v}_2$ and a spin $\vec{\omega}_1 + \vec{\omega}_2$ of the body.

Examples

1. The points $(a, 2a, -a)$, $(-a, -a, a)$, (a, a, a) of a rigid body have instantaneous velocities $\left[\sqrt{3}v/2, 0, \sqrt{3}v/2\right], \left[-v/\sqrt{3}, 0, -v/\sqrt{3}\right]$, $\left[0, -v/\sqrt{3}, v/\sqrt{3}\right]$. Show that the body has the line through the origin having direction cosines $\left[1/\sqrt{3}, -1/\sqrt{3}, -1/\sqrt{3}\right]$ as the instantaneous axis of the rotation and that the magnitude of the angular velocity is $v/2a$.

Let A, B, C be the point $(a, 2a, -a)$, $(-a, -a, a)$, (a, a, a) whose velocities are $\vec{v}_1, \vec{v}_2, \vec{v}_3$ and $\vec{\omega} = \omega_1\hat{i} + \omega_2\hat{j} + \omega_3\hat{k}$ be the angular velocity of the body, then

$$\vec{r}_1 = a\hat{i} + 2a\hat{j} - a\hat{k}, \; \vec{r}_2 = -a\hat{i} - a\hat{j} + a\hat{k}, \; \vec{r}_3 = a\hat{i} + a\hat{j} + a\hat{k},$$

and

$$\vec{v}_1 = \vec{\omega} \times \vec{r}_1, \vec{v}_2 = \vec{\omega} \times \vec{r}_2, \vec{v}_3 = \vec{\omega} \times \vec{r}_3$$

From these relations, we get

$$-\omega_2 a - 2a\omega_3 = \frac{\sqrt{3}v}{2} \quad (1)$$

$$\omega_3 a + \omega_1 a = 0 \quad (2)$$

$$\omega_1 2a + \omega_2 a = \frac{\sqrt{3}v}{2} \quad (3)$$

$$\omega_2 a + \omega_3 a = -\frac{v}{\sqrt{3}} \quad (4)$$

$$\omega_2 a - \omega_1 a = -\frac{v}{\sqrt{3}} \quad (5)$$

$$\omega_2 a - \omega_3 a = 0. \quad (6)$$

Solving these, we get

$$\omega_1 = \frac{v}{2a\sqrt{3}}, \; \omega_2 = \frac{-v}{2a\sqrt{3}}, \; \omega_3 = \frac{-v}{2a\sqrt{3}}$$

$$\therefore \quad \vec{\omega} = \frac{v}{2a}\left(\frac{1}{\sqrt{3}}\hat{i} - \frac{1}{\sqrt{3}}\hat{j} - \frac{1}{\sqrt{3}}\hat{k}\right).$$

Hence, d.c.'s of the instantaneous axis of rotation are $\frac{1}{\sqrt{3}}, \frac{-1}{\sqrt{3}}, \frac{-1}{\sqrt{3}}$ and magnitude is $v/2a$.

2. Show that two equal and opposite rotations of a rigid body about distinct parallel axes are equivalent to a translation of a rigid body.

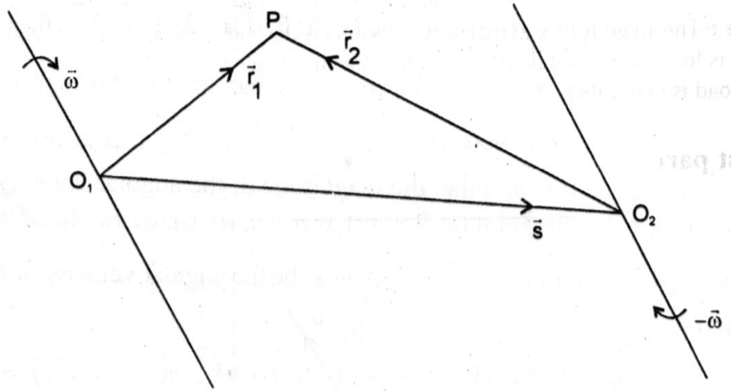

Fig. 2.13

Let O_1, O_2 be any two points chosen on the axes of rotation about which the angular velocities are ω, $-\omega$ respectively. Let

$$\overrightarrow{O_1O_2} = \vec{s}, \overrightarrow{O_1P} = \vec{r}_1, \overrightarrow{O_2P} = \vec{r}_2,$$

where P is any point of the rigid body. Let $\vec{v}_p$ be the velocity of P.

Now, we have

$$\begin{aligned}\vec{v}_p &= (\vec{\omega} \times \vec{r}_1) + (-\vec{\omega} \times \vec{r}_2) \\ &= \vec{\omega} \times (\vec{r}_1 - \vec{r}_2) = \vec{\omega} \times \vec{s}\,.\end{aligned}$$

This is independent of the position of P and is perpendicular to the plane of $\vec{\omega}$, $-\vec{\omega}$, $\vec{s}$. Thus every point of the rigid body moves instantaneously with this velocity, so the total velocity is one of translation only.

3. A car of width b is moving with constant velocity V close to the edge of a straight road. If a pedestrian steps on the road at a point distant d in front of a car, what is the least uniform velocity at which he must be able to walk in order to cross the road in safety?

If the car is at rest, but moves off with constant acceleration f at the same instant that the pedestrian starts to cross the road from a point on the edge of the road distant d in front of car, show that the uniform velocity at which he must be able to walk is

$$f\left[\sqrt{d^2+b^2}-d\right]^{1/2}$$

and that if he walks with this velocity, his direction must be inclined to the edge of the road at an angle.

$$\cot^{-1} \frac{\sqrt{b^2 + d^2} - d}{b}.$$

Note : Throughout this question it is assumed that the pedestrian's position on the road is located by a single point and that the distance between the car and the edge of the road is negligible.

First part

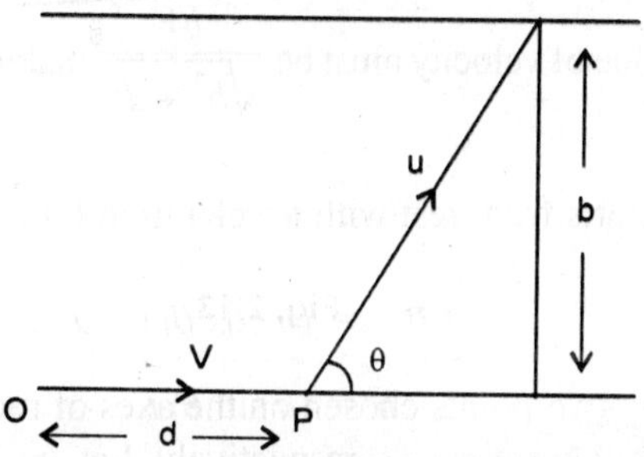

Fig. 2.14

Let the car be at the point O and man at the point P, when man steps on the road, u be the velocity of man making an angle θ with the road, we have

$$Vt = (u \cos \theta)\, t + d \qquad (1)$$

$$b = (u \sin \theta)\, t \qquad (2)$$

$$\therefore \quad t^2 u^2 = (Vt - d)^2 + b^2$$

$$\therefore \quad u = \frac{1}{t} \sqrt{(Vt - d)^2 + b^2}. \qquad (3)$$

In order that the man be able to cross the road in safety, his velocity must be greater than least value of u. Hence, we will find the least value of u.

$$\frac{du}{dt} = \frac{Vt\,(Vt - d)\,(Vt - d)^2 - b^2}{t^2 \sqrt{(Vt - d)^2 + b^2}}$$

For minimum value of u

$$\frac{du}{dt} = 0$$

$$\therefore \quad Vt\,(Vt - d) - (Vt - d)^2 - b^2 = 0$$

$$\therefore \quad t = \frac{b^2 + d^2}{dV}.$$

Putting this value of t in (3), we get

$$u = \frac{bV}{\sqrt{b^2 + d^2}}.$$

Now

$$\tan\theta = \frac{b}{Vt - d} = \frac{d}{b}.$$

Hence, least value of velocity must be $\dfrac{bV}{\sqrt{b^2 + d^2}}$ making an angle $\tan^{-1}\dfrac{d}{b}$.

Second part

When the car starts from rest with acceleration f, then we have

$$\frac{1}{2}ft^2 = (u\cos\theta)\,t + d \quad (4)$$

$$b = (u\sin\theta)\,t$$

$$\therefore \quad u^2 t^2 = \left(\frac{1}{2}ft^2 - d\right)^2 + b^2$$

$$\therefore \quad u = \frac{1}{t}\sqrt{\left(\frac{1}{2}ft^2 - d\right)^2 + b^2}$$

$$\therefore \quad \frac{du}{dt} = \frac{\left(\frac{1}{2}ft^2 - d\right)^2 + b^2 - ft^2\left(\frac{1}{2}ft^2 - d\right)}{t^2\left(\sqrt{\left(\frac{1}{2}ft^2 - d\right)^2 + b^2}\right)}$$

$$\frac{du}{dt} = 0, \text{ gives}$$

$$\left(\frac{1}{2}ft^2 - d\right)^2 + b^2 = ft^2\left(\frac{1}{2}ft^2 - d\right),$$

$$\therefore \quad t^2 = \frac{2}{f}\sqrt{b^2 + d^2}$$

$$\therefore \quad t = \left[\frac{2}{f}\sqrt{b^2 + d^2}\right]^{1/2}$$

Putting this value of t in (5), we get

$$u = \left[f\sqrt{b^2+d^2} - d\right]^{1/2}$$

$$\therefore \quad \tan\theta = \frac{b}{\frac{1}{2}ft^2 - d} = \frac{b}{\sqrt{b^2+d^2} - d}$$

$$\therefore \quad \cot\theta = \frac{\sqrt{b^2+d^2} - d}{b},$$

$$\therefore \quad \theta = \cot^{-1}\frac{\sqrt{b^2+d^2} - d}{b}$$

4. A point P moves so that it always has the velocity component V_1 parallel to the fixed line and V_2 in the transverse direction perpendicular to OP, O is a fixed point. Prove that its orbit is a conic having O as focus and discuss the cases $V_1 > V_2$, $V_1 = V_2$, $V_1 < V_2$.

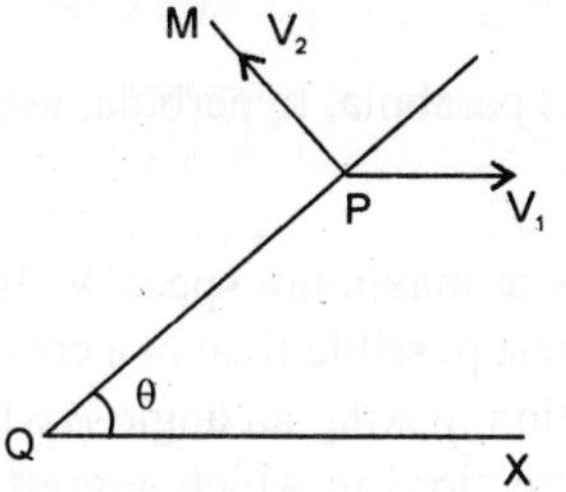

Fig. 2.15

Let O be fixed point $OP = r$, $\angle POX = \theta$, PM is perpendicular to OP. Resolving velocities V_1 and V_2 along radial and transverse directions, we get

$$\frac{dr}{dt} = V_1\cos\theta \quad (1)$$

$$r\frac{d\theta}{dt} = V_2 - V_1\sin\theta \quad (2)$$

Dividing (1) by (2), we get

$$\frac{dr}{r} = \frac{V_1\cos\theta}{V_2 - V_1\sin\theta}$$

Integration gives

$$\log \frac{r}{A} = -\log (V_2 - V_1 \sin \theta),$$

where A is a constant of integration.

$$\therefore \qquad \frac{A}{r} = V_2 - V_1 \sin \theta_1,$$

$$\therefore \qquad \frac{A}{V_2 r} = 1 - \frac{V_1}{V_2} \sin \theta$$

Let

$$B = \frac{A}{V_2}$$

Hence,

$$\frac{B}{r} = 1 - \frac{V_1}{V_2} \sin \theta$$

This is equation of conic referred to O as focus, latus rectum B, eccentricity $\frac{V_1}{V_2}$. Conic is ellipse, parabola, hyperbola, as $\frac{V_1}{V_2} < i, \frac{V_1}{V_2} = 1, \frac{V_1}{V_2} > 1$ *i.e.*, $V_1 < V_2, V_1 = V_2, V_1 > V_2$.

5. An aeroplane travels at maximum speed V. It has to travel from A to B and back in the shortest possible time in a cross wind blowing at constant velocity v in a direction making an angle ϕ with BA. Find the time taken and determine the directions in which aircraft must steer on the outward and return journey.

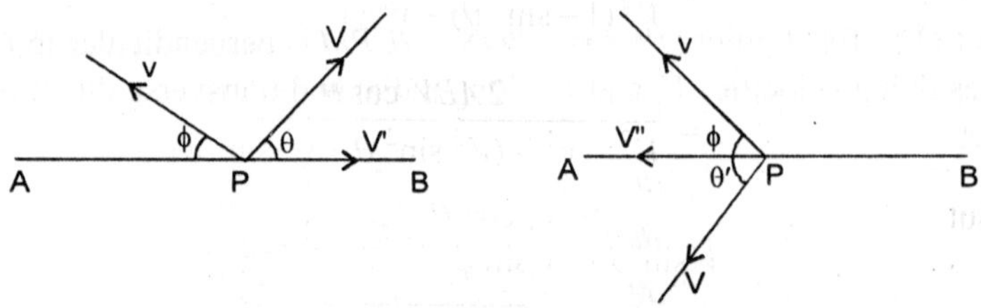

Fig. 2.16 *(a)* Fig. 2.16 *(b)*

Time will be shortest when distance travelled is shortest. Hence, the aeroplane must travel along straight line AB and back along BA.
Let the direction of the velocity of aeroplane make an angle θ with AB, while on outward journey and θ with BA while back from B to A. Let V'

and V'' be the resultant velocities along AB and BA. Let t_1 and t_2 be the times taken by aeroplane travelling from A to B and B to A respectively. Since, resultant velocities are along AB and BA respectively, there is no resultant velocity perpendicular to AB. Hence, we have

$$V \sin \theta - v \sin \phi = 0, \tag{1}$$

$$V' = V \cos \theta - v \cos \phi \tag{2}$$

$$V \sin \theta' - v \sin \phi = 0 \tag{3}$$

$$V'' = V \cos \theta' + v \cos \phi \tag{4}$$

$$\therefore \quad \sin \theta = \frac{v}{V} \sin\phi = \sin \theta'$$

$$\therefore \quad \theta = \theta'.$$

Hence, directions of velocities of aeroplane make same angle with AB or BA.

Now

$$t_1 = \frac{AB}{V'} = \frac{AB}{V \cos\theta - v \cos\phi}$$

$$t_2 = \frac{AB}{V''} = \frac{AB}{V \cos\theta + v \cos\phi}.$$

$$\text{Total time} = t_1 + t_2 = AB\left[\frac{1}{V\cos\theta - v\cos\phi} + \frac{1}{V\cos\theta + v\cos\phi}\right]$$

$$= \frac{2ABV\cos\theta}{V^2\cos^2\theta - v^2\cos^2\phi}$$

$$= \frac{2ABV\cos\theta}{V^2(1-\sin^2\theta) - v^2(1-\sin^2\phi)}$$

$$= \frac{2ABV\cos\theta}{V^2 - v^2 - (V^2\sin^2\theta - v^2\sin^2\phi)}.$$

But

$$V \sin \theta = v \sin \phi$$

$$\therefore \quad V\cos\theta = \sqrt{V^2 - v^2\sin^2\phi}.$$

$$\therefore \quad t_1 + t_2 = \frac{2AB\,(V^2 - v^2\sin\phi)^{1/2}}{-v^2 + V^2}.$$

2.13 HODOGRAPH

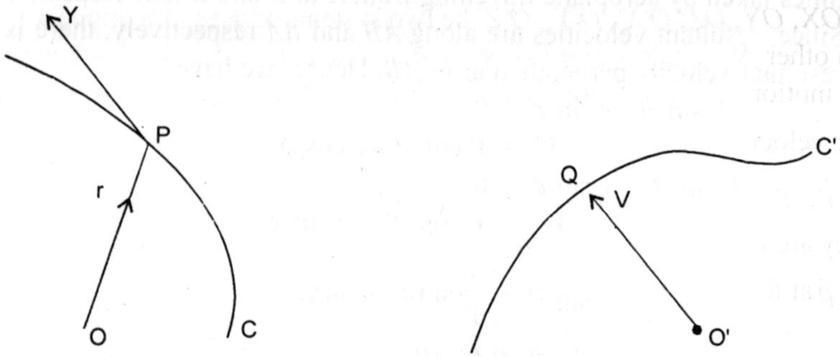

Fig. 2.17 *(a)* Fig. 2.17 *(b)*

Suppose $\overrightarrow{OP} = \vec{r}(t)$ at some time *t*, *O* being fixed. Then *P* moves along a curve *C* Fig. 2.17 (a).

Let $\vec{V} = \dfrac{d\vec{r}}{dt}$ be the velocity of *P*. In Fig. 2.17 (*b*) through a fixed point *O′* we draw $\overrightarrow{O'Q} = \vec{V}$. Then the locus *C′* of Q is called *hodograph* of the velocity of *P*.

Example

Prove that the hodograph of ellipse is similar and similarly situated to the original ellipse.

Let the point *P* describes the ellipse

$$x = a\cos nt, y = b\sin nt$$

where *t* is the time.

$$\therefore \qquad \dot{x} = -an\sin nt, \ \dot{y} = bn\cos nt,$$

so that the locus of $(\dot{x}, \dot{y})$ in the hodograph plane is

$$\left(\frac{\dot{x}}{an}\right)^2 + \left(\frac{\dot{y}}{bn}\right)^2 = 1,$$

which is an ellipse with semi-major and semi-minor axes *na*, *nb* and with centre of the original ellipse.

2.14 MOVING AXES

It is often convenient in the solution of dynamical problems to refer the motion of a body to a frame of reference which is iself moving in a known manner.

The motion of the body referred to a fixed frame is then its motion referred to moving frame compounded with the motion of that frame.

Let OX, OY, OZ; $O'X'$, $O'Y'$, $O'Z'$ be two systems of axes, moving relative to each other. We shell regard first system as fixed and second as moving.

The motion of the system $OX'Y'Z'$ is given by the velocity of O' and the angular velocity $\vec{\omega}$ about an axis through O'.

Let $\hat{i}'$, $\hat{j}'$, $\hat{k}'$ denote unit vectors along $O'X'$, $O'Y'$, $O'Z'$ respectively, so that they are functions of time. Then if p', q', r' be the components of variable vector $\vec{p}$ at time t relative to the position of moving axes at time t, we have

$$\vec{p} = p'\hat{i}' + q'\hat{j}' + r'\hat{k}'.$$

Hence, the rate of change of $\vec{p}$ relative to the fixed axes, is given by

$$\frac{d\vec{p}}{dt} = \sum \frac{dp'}{dt}\hat{i}' + \sum p'\frac{d\hat{i}'}{dt}.$$

But $\hat{i}'$ is the position vector of same point A' on OX' at unit distance from O', $\frac{d\hat{i}'}{dt}$ is the velocity of A' relative to O', hence,

$$\frac{d\hat{i}'}{dt} = \vec{\omega} \times \hat{i}'.$$

Similarly

$$\frac{d\hat{j}'}{dt} = \vec{\omega} \times \hat{j}', \quad \frac{d\hat{k}'}{dt} = \vec{\omega} \times \hat{k}'$$

$$\therefore \quad \frac{d\vec{p}}{dt} = \sum \frac{dp'}{dt}\hat{i}' + \sum p'\,\vec{\omega} \times \hat{i}'$$

$$= \sum \frac{dp'}{dt}\hat{i}' + \vec{\omega} \times \sum p'\hat{i}'$$

$$= \sum \frac{dp'}{dt}\hat{i}' + \vec{\omega} \times \vec{p}.$$

Here, $\sum \frac{dp'}{dt}\hat{i}'$ is the rate of change of $\vec{p}$ relative to the moving axes.

Denoting this by $\frac{\partial \vec{p}}{\partial t}$, we get

$$\frac{d\vec{p}}{dt} = \frac{\delta \vec{p}}{\delta t} + \vec{\omega} \times \vec{p}.$$

Thus we have the following result:

If $\frac{d}{dt}$ denotes the rate of change relative to a fixed frame and $\frac{\partial}{\partial t}$ relative to frame moving with angular velocity $\vec{\omega}$, then for any vector $\vec{p}$,

$$\frac{d\vec{p}}{dt} = \frac{\delta \vec{p}}{\delta t} + \vec{\omega} \times \vec{p}.$$

Cor. Prove that

$$\frac{d\vec{\omega}}{dt} = \frac{\partial \vec{\omega}}{\partial t}.$$

Replacing $\vec{p}$ by $\vec{\omega}$ in the above result, we get

$$\frac{d\vec{\omega}}{dt} = \frac{\partial \vec{\omega}}{\partial t} + \vec{\omega} \times \vec{\omega}$$
$$= \frac{\partial \vec{\omega}}{\partial t}.$$

2.15 VELOCITY AND ACCELERATION IN MOVING AXES

Replacing $\vec{p}$ by $\vec{r}$ in relation

$$\frac{dp}{dt} = \frac{\partial p}{\partial t} + \vec{\omega} \times \vec{p}$$

we get velocity

$$\vec{v} = \frac{d\vec{r}}{dt} = \frac{\partial \vec{r}}{\partial t} + \vec{\omega} \times \vec{r}$$

and replacing $\vec{p}$ by $\vec{v}$, we get acceleration

$$\vec{f} = \frac{d\vec{f}}{dt} = \frac{\partial \vec{v}}{\partial t} + \vec{\omega} \times \vec{v}$$

I. Two Dimensional Motion

(*a*) Cartesian Co-ordinates

Velocity: Let $P(x, y)$ be any point of the body, whose position vector is given by

$$\vec{r} = x\hat{i} + y\hat{j},$$

and $\vec{\omega} = \omega\hat{k}$ be the angular velocity about an axis passing through O and normal to xy plane, and velocity of P be

$$\vec{v} = u\hat{i} + v\hat{j},$$

Now we have

$$\vec{v} = \frac{\partial \vec{r}}{\partial t} + \vec{\omega} \times \vec{r}$$

Sub-stituting the values of $\vec{r}, \vec{\omega}, \vec{v}$ and equating the co-efficients of $\hat{i}, \hat{j}$ we get

$$u = \dot{x} - \omega y, \; v = \dot{y} + \omega x.$$

Hence,

$$\vec{v} = (\dot{x} - \omega y)\,\hat{i} + (\dot{y} + \omega x)\,\hat{j}.$$

Acceleration: Acceleration $\vec{f} = f_1\hat{i} + f_2\hat{j}$ is given by

$$\vec{f} = \frac{d\vec{v}}{dt} = \frac{\partial \vec{v}}{\partial t} + \vec{\omega} \times \vec{v}.$$

Substituting the value of $\vec{v}, \vec{f}, \vec{\omega}$ and equating the coeficients of $\hat{i}, \hat{j}$, we get

$$f_1 = \ddot{x} - 2\omega\dot{y} - \omega^2 x, \; f_2 = \ddot{y} + 2\omega\dot{x} - \omega^2 y.$$

(*b*) Polar Co-ordinates

Put $x = r, y = 0$, in results of previous case we get

$$u = \dot{r}, \; v = r\omega$$

$$f_1 = \ddot{r} - \omega^2 r, \; f_2 = 2\omega\dot{r}.$$

II. Three Dimensional Motion

(*a*) Cartesian Co-ordinates

Let $\vec{r}$ be the position vector of a point P (x, y, z) and $\vec{\omega} = \omega_1\hat{i} + \omega_2\hat{j} + \omega_3\hat{k}$ be the angular velocity of the body. We have

$$\vec{r} = x\hat{i} + y\hat{j} + z\hat{k}.$$

Velocity of P is given by

$$\vec{v} = \frac{\partial \vec{r}}{\partial t} + \vec{\omega} \times \vec{r}.$$

$$= \frac{\partial}{\partial t}(x\hat{i} + y\hat{j} + z\hat{k}) + (\omega_1\hat{i} + \omega_2\hat{j} + \omega_3\hat{k}) \times (x\hat{i} + y\hat{j} + z\hat{k})$$

$$= \dot{x}\hat{i} + \dot{y}\hat{j} + \dot{z}\hat{k} + (\omega_2 z - \omega_3 y)\hat{i} + (\omega_3 x - \omega_1 z)\hat{j} + (\omega_1 y - \omega_2 x)\hat{k}.$$

If $\vec{v} = u\hat{i} + v\hat{j} + w\hat{k}$, then

$$u = \dot{x} + \omega_2 z - \omega_3 y,$$

$$v = \dot{y} + \omega_3 x - \omega_1 z,$$
$$w = \dot{z} + \omega_1 y - \omega_2 x.$$

Acceleration: Acceleration is given by

$$\vec{f} = \frac{\partial \vec{v}}{\partial t} + \vec{\omega} \times \vec{v},$$

$$= \frac{\partial}{\partial t}(u\hat{i} + v\hat{j} + z\hat{k}) + (\omega_1\hat{i} + \omega_2\hat{j} + \omega_3\hat{k}) \times (u\hat{i} + v\hat{j} + w\hat{k}),$$

$$= \dot{u}\hat{i} + \dot{v}\hat{j} + \dot{w}\hat{k} + (\omega_2 w - \omega_3 v)\hat{i} + (\omega_3 u - \omega_1 w)\hat{j} + (\omega_1 v - \omega_2 u)\hat{k}.$$

If $\vec{f} = f_1\hat{i} + f_2\hat{j} + f_3\hat{k}$, then

$$f_1 = \dot{u} + \omega_2 w - \omega_3 v,$$
$$f_2 = \dot{v} + \omega_3 u - \omega_1 w,$$
$$f_3 = \dot{w} + \omega_1 v - \omega_2 u.$$

Further putting the value of u, v, w we can get these components in terms of x, y, z and their derivatives.

(*b*) Cylindrical Co-ordinates

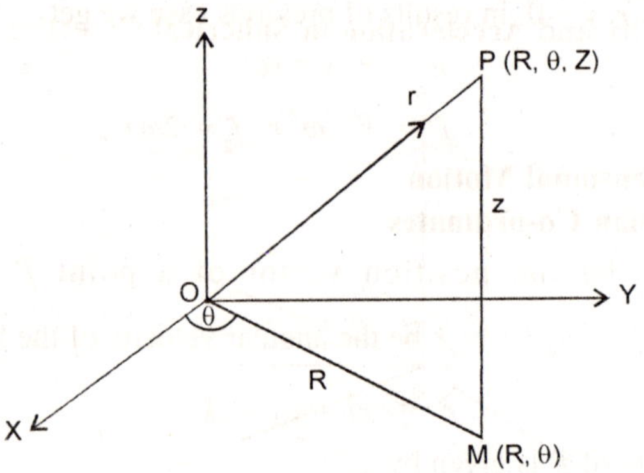

Fig. 2.18

Let *OX*, *OY*, *OZ* be fixed mutually perpendicular axes and cylindrical co-ordinates of *P* ,e (R, θ, z) and *P* vary with time. Let $\overrightarrow{OP} = \vec{r}$; $\hat{R}, \hat{\theta}, \hat{z}$ be the unit vectors in the directions of R, θ, z increasing and unit vectors constitute a right handed frame such that $\hat{R} \times \hat{\theta} = \hat{z}$, etc.

Now

$$\vec{r} = R\hat{R} + 0.\hat{\theta} + z\hat{z} = R\hat{R} + z\hat{z}.$$

Then since, P is in the frame specified by these unit vectors and moving with angular velocity $\vec{\omega} = \theta \hat{z}$, the velocity of P is given by

$$\vec{v} = \frac{\partial \vec{r}}{\partial t} + \vec{\omega} \times \vec{r}$$

$$= \frac{\partial}{\partial t}(R\hat{R} + z\hat{z}) + \vec{\omega} \times (R\hat{R} + z\hat{z})$$

$$= \dot{R}\hat{R} + \dot{z}\hat{z} + \dot{\theta}\hat{z}x\,(R\hat{R} + z\hat{z})$$

$$= \dot{R}\hat{R} + \dot{z}\hat{z} + R\dot{\theta}\hat{\theta}.$$

The acceleration of P is

$$\vec{f} = \frac{\partial \vec{v}}{\partial t} + \vec{\omega} \times \vec{v}$$

$$= \frac{\partial}{\partial t}(\dot{R}\hat{R} + \dot{z}\hat{z} + R\dot{\theta}\hat{\theta}) + \dot{\theta}\hat{z} \times (\dot{R}\hat{R} + \dot{z}\hat{z} + R\dot{\theta}\hat{\theta})$$

$$= (\ddot{R} - R\dot{\theta}^2)\,\hat{R} + (R\ddot{\theta} + 2\dot{R}\dot{\theta})\,\hat{\theta} + \ddot{z}\hat{z}.$$

N.B. If z = constant, then we get the radial and transverse components of plane motion.

(c) Velocity and Acceleration in Spherical Co-ordinates

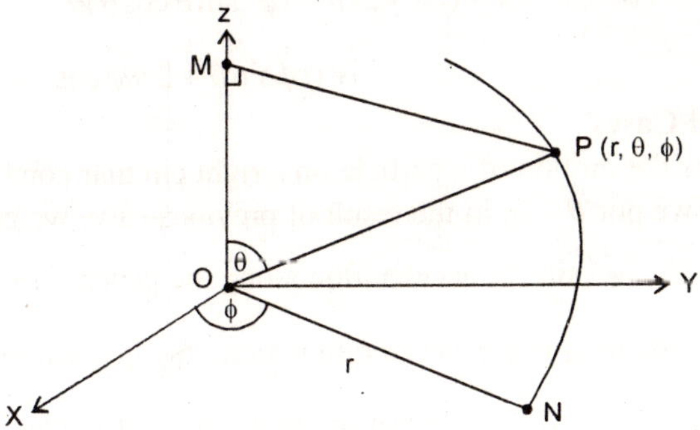

Fig. 2.19

Let OX, OY, OZ be fixed mutually perpendicular axes. Point P is varying with time, whose spherical co-ordinates are r, θ, ϕ and $\hat{r}$, $\hat{\theta}$, $\hat{\phi}$ are unit vectors in the directions of r, θ, ϕ, increasing respectively, and these constitute a right handed fame for which $\hat{r} \times \hat{\theta} = \hat{\phi}$, etc. The angular velocity of the frame is

$$\vec{\omega} = \dot{\theta}\hat{\phi} + \dot{\phi}\hat{k} = \dot{\theta}\hat{\phi} + \dot{\phi}(\cos\theta\,\hat{r} - \sin\theta\,\hat{\theta})$$

$$= \dot{\phi}\cos\theta\,\hat{r} - \dot{\phi}\sin\theta\,\hat{\theta} + \dot{\theta}\hat{\phi}.$$

Hence, velocity of P is

$$\vec{v} = \frac{\partial\vec{r}}{\partial t} + \vec{\omega}\times\vec{r}$$

$$= \frac{\partial}{\partial t}(r\hat{r}) + (\dot{\phi}\cos\theta\,\hat{r} - \dot{\phi}\sin\theta\,\hat{\theta} + \dot{\theta}\hat{\phi})\times r\hat{r}$$

$$= \dot{r}\hat{r} + r\dot{\theta}\hat{\theta} + r\dot{\phi}\sin\theta\,\hat{\phi}.$$

Acceleration is

$$\vec{f} = \frac{\partial\vec{v}}{\partial t} + \vec{\omega}\times\vec{v}$$

$$\frac{\partial}{\partial t}[\dot{r}\hat{r} + r\dot{\theta}\hat{\theta} + r\dot{\phi}\sin\theta\,\hat{\phi}] + (\dot{\phi}\cos\theta\,\hat{r} - \dot{\phi}\sin\theta\hat{\theta} + \dot{\theta}\hat{\phi})$$

$$\times(\dot{r}\hat{r} + r\dot{\theta}\hat{\theta} + r\dot{\phi}\sin\theta\,\hat{\phi})$$

$$= \ddot{r}\hat{r} + \dot{r}\dot{\theta}\hat{\theta} + r\ddot{\theta}\hat{\theta} + \dot{r}\dot{\phi}\sin\theta\,\hat{\phi} + r\ddot{\phi}\sin\theta\,\hat{\phi} + r\dot{\phi}\dot{\theta}\cos\theta\,\hat{\phi}$$

$$-(r\dot{\phi}^2\sin^2\theta + r\dot{\theta}^2)\,\hat{r} + (\dot{r}\dot{\theta} - r\dot{\phi}^2\sin\theta\cos\theta)\hat{\theta} + (r\dot{\theta}\dot{\phi}\cos\theta + \dot{r}\dot{\phi}\sin\theta)\hat{\phi}$$

$$= (\ddot{r} - r\dot{\theta}^2 - r\dot{\phi}^2\sin^2\theta)\hat{r} + (r\ddot{\theta} + 2\dot{r}\dot{\theta} - r\dot{\phi}^2\sin\theta\cos\theta)\hat{\theta}$$

$$+ (r\ddot{\phi}\sin\theta + 2r\dot{\theta}\dot{\phi}\cos\theta + 2\dot{r}\dot{\phi}\sin\theta)\hat{\phi}.$$

(*d*) Special Cases

(*i*) For the motion of a particle on a right circular cone.

If we put $\theta = \alpha$, in the result of previous case, we get

$\ddot{r} - r\dot{\phi}^2\sin^2\alpha$, acceleration along the generator OP,

$r\dot{\phi}^2\sin\alpha\cos\alpha$, acceleration along the inward normal,

$r\ddot{\phi}\sin\alpha + 2\dot{r}\dot{\phi}\sin\alpha$, acceleration along at right angles to the plane OPG.

(*ii*) For the motion of a sphere.

If we put $r = a$ in the results of spherical co-ordinates, we get

$a(\dot{\theta}^2 + \dot{\phi}^2\sin^2\theta)$ for the acceleration towards the centre,

$a(\ddot{\theta} - \dot{\phi}^2\sin\theta\cos\theta)$, along the meridian curve CP

$a(\ddot{\phi}\sin\theta + 2\dot{\phi}\dot{\theta}\cos\theta)$, perpendicular to the plane COP.

Examples

1. A rigid plane lamina is moving in its plane and one point A in it has velocity $\overline{u}$ relative to a fixed point O. If $\vec{\omega}$ is the angular velocity of the lamina, show that point P in it has the velocity $\vec{u}+\vec{\omega}\times\vec{r}$, relative to O, where $\vec{r}=\overrightarrow{AP}$. Hence or otherwise prove that the position vector $\vec{r}\,'$ of the instantaneous centre of the lamina, relative to A is given by $\vec{r}\,'=(\vec{\omega}\,.\,\vec{u})/\omega^2$.

   Show also that the acceleration of the particle in the lamina which instantaneously coincides with I has component $\dfrac{du}{dt}-\dfrac{u}{\omega}\dfrac{d\omega}{dt}$ parallel to $\vec{u}$.

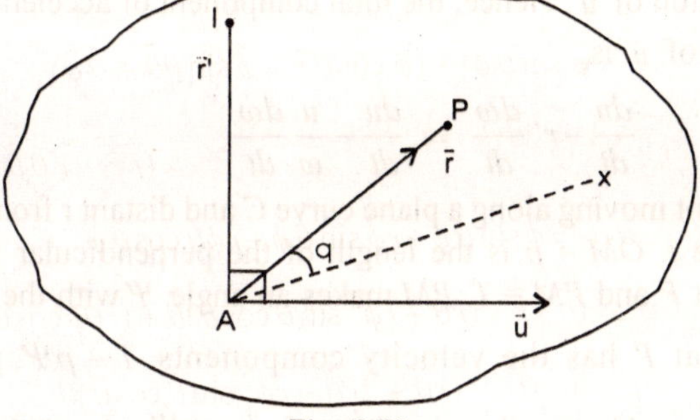

Fig. 2.20

Velocity of P relative to A is $=\vec{\omega}\times\overrightarrow{AP}$,

Hence, velocity of P relative to O is $=\vec{u}+\vec{\omega}\times\vec{r}$.

The velocity of I is

$$\vec{u}+\vec{\omega}\times\vec{r}\,'.$$

I is instantaneously at rest, hence

$$\vec{u}+\vec{\omega}\times\vec{r}\,' = 0.$$

Taking cross product with $\vec{\omega}$, we get

$$\vec{\omega}\times\vec{u}+\vec{\omega}\times(\vec{\omega}\times\vec{r}\,') = \vec{0}.$$

$$\therefore \quad \vec{\omega}\times\vec{u}+(\vec{\omega}\cdot\vec{r}\,')\,\vec{\omega}-\omega^2\vec{r}\,' = \vec{0}.$$

Since $\vec{\omega},\vec{r}\,'$ are at right angles, hence

$$\vec{\omega}\,.\,\vec{r}\,' = 0,$$

$$\therefore \quad \vec{r}\,' = (\vec{\omega}\times\vec{u})/\omega^2.$$

Relative to A, the acceleration of P has radial and transverse components $\ddot{r}-r\dot{\theta}^2$ and $r\ddot{\theta}+2\dot{r}\dot{\theta}$ where θ is the angle that AP makes with a direction

AX, chosen to be fixed in the space. Putting $\dot{\theta} = \omega$ these components become $\ddot{r} - r\omega^2, r\dot{\omega} + 2\dot{r}\omega$. When $P \equiv I$, $\vec{r} = \vec{r}'$ and so the component of acceleration of I relative to A is the transverse component $r'\dot{\omega}$ in the direction of $-\vec{u}$, since $\vec{r}' = 0$, the velocity of I relative to A being perpendicular to AI.

Now $$u^2 = \vec{u}^2 = \vec{u} \cdot \vec{u}$$

$$\therefore \quad 2u\frac{du}{dt} = 2\vec{u}\cdot\frac{d\vec{u}}{dt}$$

$$\therefore \quad \hat{u}\cdot\frac{d\vec{u}}{dt} = \frac{du}{dt}.$$

But $\frac{d\vec{u}}{dt}$ is the acceleration of A in space and $\hat{u}\cdot\frac{d\vec{u}}{dt}$ is the component in the direction of $\hat{u}$. Hence, the total component of acceleration of I in the direction of $\vec{u}$ is

$$\frac{du}{dt} - r'\frac{d\omega}{dt} = \frac{du}{dt} - \frac{u}{\omega}\frac{d\omega}{dt}.$$

2. P is a point moving along a plane curve C and distant r from a fixed origin O at time t. $OM = p$ is the length of the perpendicular from O on the tangent at P and $PM = T$. PM makes an angle Ψ with the fixed line OX. Show that P has the velocity components $\dot{T} - p\dot{\Psi}, \dot{p} + T\dot{\Psi}$ along $\overrightarrow{MP}, \overrightarrow{OM}$ and hence, that $\dot{p} + T\dot{\Psi} = 0$, $\dot{T} - p\dot{\Psi} = \dot{s}$, s being the length of arc along C from some fixed point on it to the position P. Show further that $r\dot{r} = p\dot{p} + T\dot{T}$, and hence derive the formula $\rho = \left| r\frac{dr}{dp} \right|$ for the radius of curvature $\rho = \left| \frac{ds}{d\psi} \right|$ of the curve C at P.

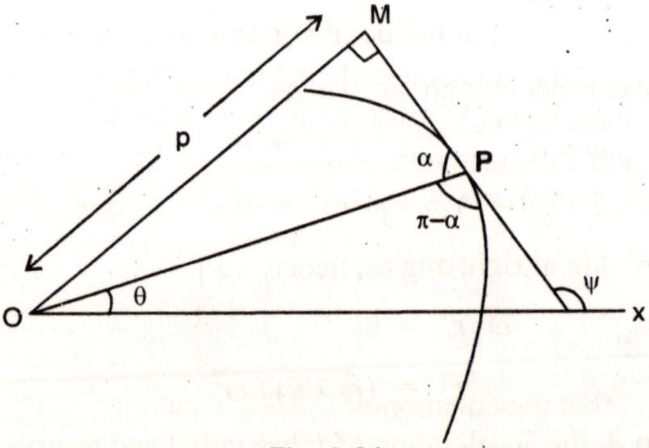

Fig. 2.21

Let OP make an angle θ with the fixed line OX and $\angle OPM = \alpha$.

Now $$OM = p,\ OP = r,\ PM = T,\ \angle\Psi = \pi - \alpha + \theta.$$

$$\therefore \quad \dot{\psi} = \dot{\theta}$$

We have

$$p = r\sin\alpha,\ \mathrm{T} = r\cos\alpha.$$

Components of velcity of P are $\dot{r}$ along OP and $r\dot{\theta}$ or $r\dot{\psi}$ along perpendicular to OP. Hence

Velocity along $MP = \dot{r}\cos\alpha - r\dot{\psi}\sin\alpha = \dot{T} - p\dot{\psi}$.

Velocity along $OM = \dot{r}\sin\alpha + r\dot{\theta}\cos\alpha = \dot{p} + T\dot{\theta} = \dot{p} + T\dot{\psi}$.

But velocity of P along the tangent MP is $\dot{s}$, hence its component along $OM = 0$.

$$\therefore \quad \dot{T} - p\dot{\psi} = \dot{s},\ \dot{p} + T\dot{\psi} = 0.$$

From ΔOPM

$$r^2 = p^2 + T^2$$

$$\therefore \quad r\dot{r} = p\dot{p} + T\dot{T}.$$

Radius of curvature of the curve C at P is

$$\rho = \left|\frac{ds}{d\psi}\right| = \left|\frac{ds}{dt}\frac{dt}{d\psi}\right| = \left|\frac{\dot{s}}{\dot{\psi}}\right| = \left|\frac{\dot{T} - p\dot{\psi}}{\dot{\psi}}\right|$$

$$= \left|-p + \frac{\dot{T}}{\dot{\psi}}\right| = \left|-p - \frac{T\dot{T}}{\dot{p}}\right| = \left|p + \frac{T\dot{T}}{\dot{p}}\right|$$

$$= \left|\frac{p\dot{p} + T\dot{T}}{\dot{p}}\right| = \left|\frac{r\dot{r}}{\dot{p}}\right| = \left|r\frac{dr}{dt}\frac{dt}{dp}\right| = \left|r\frac{dr}{dp}\right|$$

3. A cyclist rides along a straight road with uniform speed V at a given instant, the cyclist is at the point O on the road and a dog is at A, a point distant a from the road and such that OA is perpendicular to the road. The dog runs with speed V towards the cyclist. Taking OA and the road as axes of x and y respectively, and the co-ordinates of the dog as x, y, show that

$$x\frac{d^2y}{dx^2} = \left[1 + \left(\frac{dy}{dx}\right)^2\right]^{1/2},$$

and hence, that the equation of the dog's path is

$$y = \frac{1}{2}\left[\frac{x^2}{2a} - a\log\frac{x}{a} - \frac{a}{2}\right].$$

Show further that the distance between the cyclist and dog tends to $a/2$.

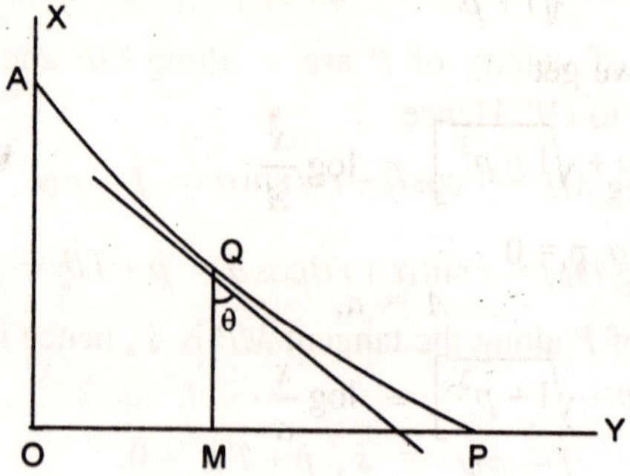

Fig. 2.22

Let the path of the dog be along the curve AP and at time t, the dog is at the point $Q(x, y)$. Draw QM perpendicular to Oy and let the tangent at Q make an angle θ with the negative x axis,

$$\therefore \qquad \tan\theta = \frac{\dot{y}}{\dot{x}} = -\frac{y - Vt}{x} \qquad (1)$$

Initially, $x = a$, $y = 0$, $t = 0$

$$\therefore \qquad \theta = 0$$

Equation (1) can be written as

$$\frac{dy}{dx} = \frac{y - Vt}{x}$$

$$\therefore \qquad x\frac{dy}{dx} = y - Vt.$$

$$\therefore \qquad x\frac{d^2y}{dx^2} + \frac{dy}{dx} = \frac{dy}{dx} - V\frac{dt}{dx}.$$

$$\therefore \qquad x\frac{d^2y}{dx^2} = -V\frac{dt}{dx} = -\frac{ds}{dt}\frac{dt}{dx} = \frac{ds}{dx} = \sqrt{1 + \left(\frac{dy}{dx}\right)^2}.$$

Let

$$\frac{dy}{dx} = p,$$

$$\therefore \qquad x\frac{dp}{dx} = \sqrt{1+p^2}\,.$$

$$\therefore \qquad \frac{dp}{\sqrt{1+p^2}} = \frac{dx}{x}\,.$$

Integrating, we get

$$\log\left[p+\sqrt{1+p^2}\right] = \log\frac{x}{A}\cdot$$

Initially, $x = a$, p = 0

$$\therefore \qquad A = a,$$

$$\therefore \qquad \log\left[p+\sqrt{1+p^2}\right] = \log\frac{x}{a}\cdot$$

$$\therefore \qquad \frac{x}{a} = p+\sqrt{1+p^2}\,.$$

$$\therefore \qquad \left(\frac{x}{a}-p\right)^2 = 1+p^2,$$

$$\therefore \qquad \frac{x^2}{a^2}-2p\frac{x}{a}-1 = 0$$

$$\therefore \qquad p = \frac{x}{2a}-\frac{a}{2x},$$

$$\therefore \qquad \frac{dy}{dx} = \frac{x}{2a}-\frac{a}{2x}$$

Integrating, we get

$$y = \frac{x^2}{4a}-\frac{a}{2}\log x + B\,.$$

But $x = a$, when $y = 0$.

$$\therefore \qquad B = -\frac{a}{4}+\frac{a}{2}\log a\,.$$

$$\therefore \qquad y = \frac{x^2}{4a}-\frac{a}{2}\log x-\frac{a}{4}+\frac{a}{2}\log a\,.$$

$$= \frac{x^2}{4a}-\frac{a}{2}\log\frac{x}{a}-\frac{a}{4}$$

$$= \frac{1}{2}\left(\frac{x^2}{2a} - a\log\frac{x}{a} - \frac{a}{2}\right).$$

Let s be the distance between man and the dog, then

$$s^2 = x^2 + (y - Vt)^2$$

$$= x^2 + \left(x\frac{dy}{dx}\right)^2 = x^2\left[1 + \left(\frac{dy}{dx}\right)^2\right]$$

$$= x^2\left[1 + \left(\frac{x}{2a} - \frac{a}{2x}\right)^2\right] = x^2\left[1 + \frac{x^2}{4a^2} + \frac{a^2}{4x^2} - \frac{1}{2}\right]$$

$$= \frac{a^2}{4} + \frac{x^4}{4a^2} + \frac{x^2}{2}.$$

When $x \to 0$, $s \to \frac{a^2}{4}$, $\therefore s \to \frac{a}{2}$.

EXERCISE

1. A wheel is rolling without slipping along a plane with angular velocity ω. A and B are taken at different distances from the centre on two different spokes. Show that at any time the velocity of A relative to B is $\vec{\omega} \times \overrightarrow{AB}$ in a direction perpendicular to AB.
2. The end P of a straight rod PQ with angular velocity ω describes a circle whose centre is at O, while the other end Q moves on a fixed line through O in the plane of the circle. The end Q' of an equal straight rod PQ' moves on the same fixed line through O. Prove that the velocities of Q and Q' are in the ratio $QO : OQ'$.
3. The velocity of a stream between parallel banks at a distance $2a$ apart is zero at the edges and increases uniformly to the middle, where it is u. A boat is rowed with constant velocity v ($> u$) relative,to the water, and goes in a line straight across. How are the bows pointed at any point of the path and how long will it take to get it across?
4. The line joining two points A, B is of constant length a and the velocities of A and B are in directions making angles α and β respectively with AB. Prove that the angular velocity of AB is $\frac{u\sin(\alpha - \beta)}{\alpha\cos\beta}$, where u is the velocity of A.

5. The instantaneous velocitie, of particles at points $(a, 0, 0)$, $\left(0, \frac{a}{\sqrt{3}}, 0\right)$, $(0, 0, 2a)$ of a rigid body are $[u, 0, 0]$, $[u, 0, v]$ $[u+v, -v\sqrt{3}, v/2]$ respectively referred to a rectangular cartesian frame. Find the magnitude and direction of spin of the body and the point at which the central axis cuts the xz-plane.
6. A bar AB moves in the plane of the rectangular axes Ox, Oy so that it always touches the parabola $y^2 = 4ax$ and the end of A of the bar moves along the x-axis. Find the co-ordinates of instantaneous centre in terms of θ, the angle between AB and Oy, and prove that the equation of the space centroid is $ay^2 + 4x(a-x)^2 = 0$. Prove also that, referred to A as pole, and AB as the initial line, the equation of the body centroid is $r = 2a \sin\theta \sec^3\theta$.
7. A point moves on the surface of the anchor ring

$$x = (c + a\sin\theta)\cos\phi,\ y = (c + a\sin\theta)\sin\phi,\ z = a\cos\theta.$$

Denoting by $\hat{a}, \hat{b}, \hat{c}$ the unit vectors in the normal, meridian and transverse directions at P, so that $\hat{a} \times \hat{b} = \hat{c}$ etc., show that the position vector $\vec{r}$ of P is

$$\vec{r} = (a + c\sin\theta)\,\hat{a} + c\cos\theta\,\hat{b},$$

and that the vector angular velocity $\vec{\omega}$ of the $(\hat{a}, \hat{b}, \hat{c})$ frame is

$$\vec{\omega} = \dot{\phi}\cos\theta\,\hat{a} - \dot{\phi}\sin\theta\,\hat{b} + \dot{\theta}\,\hat{c}.$$

Hence or otherwise deduce that the acceleration components of P are

$$[-\dot{\phi}^2 \sin\theta\,(c + a\sin\theta) - a\dot{\theta}^2,\ a\ddot{\theta} - \dot{\phi}^2\,(c + a\sin\theta)\cos\theta,$$
$$2a\dot{\theta}\dot{\phi}\cos\theta + (c + a\sin\theta)\,\ddot{\phi}]$$

8. Two points are describing concentric circles of radius a and a' with angular velocities ω and ω' respectively. Prove that angular velocity of the body joining them when its length is r is

$$[(r^2 + a^2 - a'^2)\,\omega + (r^2 - a^2 - a'^2)\,\omega']\,/\,2r^2.$$

9. An aeroplane has sspeed of v km. per *A*our and a range of action (out and home) of R km. in a calm weather. Prove that a north wind of w km. per hour has its range of action

$$R\,(v^2 - w^2)\,/\,v\,(v^2 - w^2\sin^2\phi)^{1/2},$$

in a direction whose true bearing is ϕ. If $R = 800$ km., $v = 80$ km./per hour and $w = 30$ km. per hour, find the direction in which its range is a maximum and the value of maximum range.

10. A particle moves in the curve $y = a \log \sec \frac{x}{a}$, in such a way that the tangent to the curve rotates uniformly, prove that the resultant acceleration of the particle varies as the square of radius of cur,ature.

11. A circle and a tangent to it are given. A rod moves so that it touches the circle and one end is upon the tangent. Show that the loci of the instantaneous centre in space and relative to the rod are both parabolas.

12. Assuming the formula $\vec{v}_P = \vec{v}_A + \vec{\omega} \times \overrightarrow{AP}$ relating the velocities of particles A, P of the lamina L moving in its own plane with spin $\vec{\omega}$, and that $\dot{\overrightarrow{AP}} = \vec{\omega} \times \overrightarrow{AP}$, show that the acceleration vectors $\vec{f}_A, \vec{f}_P$ are related by the equation

$$\vec{f}_P = \vec{f}_A + \dot{\vec{\omega}} \times \overrightarrow{AP} - \omega^2 \overrightarrow{AP}.$$

Prove that if J is the particle of L given by

$$\overrightarrow{AJ} = (\dot{\vec{\omega}} \times \vec{f}_A + \omega^2 \vec{f}_A) / (\omega^4 + \dot{\omega}^2),$$

the acceleration of J is instantaneously zero (instantaneous centre of acceleration).

If the spin is constant $\vec{f}_A = a(\cos pt\,\hat{i} + \sin pt\,\hat{j})$, where a, p are constants and $\hat{i}, \hat{j}$ are constant orthogonal unit vectors, show that the instantaneous centre of acceleration J moves relative to A uniformly round a circle of radius a/ω^2 with period $2\pi/p$.

3

NEWTON'S LAWS OF MOTION

3.1 NEWTON'S LAWS OF MOTION

Laws of motion enunciated by Newton in seventeenth century form the foundation of a science of dynamics, which has been developed by many other mathematicians. This dynamics, now commonly called 'Newtonian Mechanics' is the basis of all the theoretical work in applied mechanics or engineering. These laws are ingenious generalizations based on observations.

Newton's laws of motion

1. Every particle continues in its state of rest or of uniform motion in a straight line unless compelled to do otherwise by a force acting on it.
2. The rate of change of momentum is proportional to the impressed force and takes place in the direction of force.
3. To every action there is an equal and opposite reaction.

First law defines force as the cause of motion change, second gives the measure of the force as rate of change of momentum and third describes the equal and opposite force interactions of two bodies.

However, (2) needs careful interpretation. When mass varies $\frac{d\vec{p}}{dt}$ is not necessarily a force, but if m is constant, then the force

$$\vec{F} = m\frac{d\vec{v}}{dt} = m\vec{f},$$

where $\vec{f}$ is the acceleration of the particle and $\vec{p}$ the momentum.

Example

A smooth wire bent in the form of a parabola is fixed with its axis vertical and vertex downwards. A particle of mass m oscillates on the wire coming to rest at the extremities of the latus rectum. Show that the reaction of the wire on the particle when passing through the vertex is $2mg$.

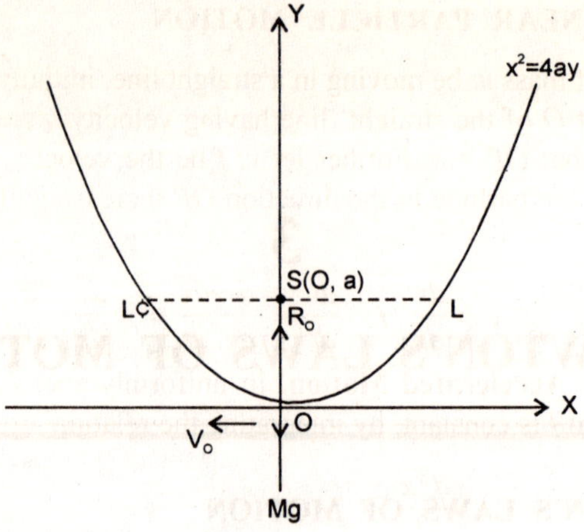

Fig. 3.1

Let the equation of the parabola be

$$x^2 = 4ay.$$

Let S (0, a) be the focus and LL' latus rectum. Suppose particle starts from L from rest and has velocity v_0 at the vertex O.

Loss in potential energy is mga and gain in K. E. is $\frac{1}{2}mv_0^2$.

We know that gain in K. E. = loss in P. E.

$$\therefore \quad \frac{1}{2}mv_0^2 = mga,$$

$$\therefore \quad v_0^2 = 2mga.$$

Let R_0 be the inward normal reation at O and ρ_0 the radius of curvature at O.

Total inward force towards O is R_0 – mg and inward acceleration is v_0^2 / ρ_0.

Equation of motion is

$$R_0 - mg = mv_0^2 / \rho_0.$$

$$\therefore \quad R_0 = mg + 2mga/\rho_0. \qquad (1)$$

$$\rho_0 = [(1 + y'^2)^{3/2}/y'']_{y=0}. \qquad (2)$$

Now

$$y = \frac{x^2}{4a}, y' = \frac{x}{2a}, y'' = \frac{1}{2a}$$

Using these values in (2) and simplifying, we get $\rho_0 = 2a$.

Using value of ρ_0 in (1) we have

$$R_0 = mg + (2m)a / 2a = 2mg.$$

3.2 RECTILINEAR PARTICLE MOTION

Let a particle of mass m be moving in a straight line, initially the particle is at the fixed point O of the straight line having velocity u and at time t at the point P, such that $OP = x$. Further let v, f be the velocity and acceleration respectively of the particle in the direction OP increasing at time t. Then we have

$$v = \frac{dx}{dt}, f = \frac{dv}{dt} = \frac{dv}{dx}\frac{dx}{dt} = v\frac{dv}{dx}.$$

(*i*) **Uniformly Accelerated Motion.** In uniformly accelerated motion the acceleration f is constant, by integrating the relation

$$\frac{d^2x}{d^2t} = f,$$

we get

$$\frac{dx}{dt} = ft + C.$$

Integrating again, we have

$$x = \frac{1}{2}ft^2 + Ct + C'$$

where C and C' are constants which can be determined by using initial conditions. Let at $t = 0$ the velocity be u (called initial velocity) and $x = 0$. Using these initial conditions we get $C = u$ and $C' = 0$

$$\therefore \qquad v = u + ft, \tag{1}$$

$$x = ut + \frac{1}{2}ft^2. \tag{2}$$

Eliminating t between (1) and (2) we get

$$v^2 = u^2 + 2fx. \tag{3}$$

This last result can be obtained by integrating the alternative expression for acceleration, $v\frac{dv}{dx} = f$, which gives

$$v^2 = 2fx + C'',$$

Using initial conditions

$$v = u, x = 0.$$

we get

$$C'' = u^2$$

$$\therefore \qquad v^2 = u^2 + 2fx.$$

(*ii*) **Resisted Motion.** Suppose a force F acts in the direction of OP and a resistance R in the direction PO. Generally R will be a function of velocity. The resultant force in the direction of OP is F-R, so the equation of motion will be

$$F - R = mf.$$

Applications. The law of acceleration in a particular problem may be given by expressing the acceleration as a function of (*i*) the time t, or (*ii*) the distance x, or (*iii*) the velocity v.

The problem of further investigating the motion can be solved as follows:

(*i*) Acceleration a given function of time, say $\phi(t)$:
We have

$$\dot{v} = \phi(t).$$

$$\therefore \quad v = \int \phi(t)\, dt + C_1$$

$$\therefore \quad \dot{x} = \Psi(t) + C_1,$$

where

$$\Psi(t) = \int \phi(t)\, dt$$

Integrating again we get

$$x = \int \Psi(t)\, dt + C_1 t + C_2$$

Constants C_1, C_2 can be determined from initial conditions.

(*ii*) Acceleration as given function of distance, say $\phi(x)$.
For this case we can proceed in two ways
(*a*) we have

$$v \frac{dv}{dx} = \phi(x),$$

Integration gives

$$v^2 = 2\int \phi(x)\, dx + C',$$

Let

$$\Psi(x) = 2\int \phi(x)\, dx,$$

We have

$$\dot{x} = \pm \sqrt{\psi(x) + C'}.$$

$$\therefore \pm \int \frac{dx}{\sqrt{\psi(x) + C''}} = t + C''.$$

(*b*) we can write

$$\ddot{x} = \phi(x).$$

Multiplying both sides by $2\dot{x}$ and integrating w.r.t.t, we get

$$\dot{x}^2 = 2\int \phi(x)\,dx + C,$$

further we can proceed as in (*a*)

3.3 RESISTANCE DEPENDING ON VELOCITY

Generally resistance is a function of velocity. It can be proportional to some power of velocity, e.g. kv^2, then we have equation of motion given by

$$m\frac{dv}{dt} = F - kv^2, \qquad (1)$$

where F is the force in the direction of the motion. When F is constant, we can write

$$\frac{mdv}{F - kv^2} = dt$$

which on integration gives

$$\frac{m}{2\sqrt{Fk}}\log\frac{\sqrt{F} + v\sqrt{k}}{\sqrt{F} - v\sqrt{k}} = t + Const.$$

Alternatively, we can write

$$mv\frac{dv}{dx} = F - kv^2,$$

$$\therefore \qquad \frac{mdv^2}{F - kv^2} = 2dx,$$

Integrating this we get

$$-\frac{m}{k}\log(F - kv^2) = 2x + Const.$$

If the force F is not costant, but working at constant rate R, we have $Fv = R$, where R is constant.

$$\therefore \qquad m\frac{dv}{dt} = \frac{R}{v} - kv^2,$$

$$\therefore \qquad mv\frac{dv}{dt} = R - kv^3.$$

This can be integrated by substitution $v = \frac{1}{z}$.

Similarly, in case of distance, we get

$$mv\frac{dv}{dx} = \frac{R}{v} - kv^2,$$

$$\therefore \qquad \frac{mdv^3}{R - kv^3} = 3dx,$$

integration gives

$$-\frac{m}{k}\log(R - kv^3) = 3x + const.$$

Examples

1. A body of mass M, travelling in a straight line, is supplied with constant power P and is subjected to a resistance Mkv^2, where v is the speed and k is a constant. Prove that the speed of the body cannot exceed a certain value and that, if it starts from rest, it acquires half the maximum speed after travelling a distance $\frac{1}{3k}\log\frac{8}{7}$. If the power is then cut off and an additional retarding force of constant value F is imposed, find the subsequent time which elapses before the body comes to rest.

   Equation of motion is

$$Mv\frac{dv}{dx} = \frac{P}{v} - Mkv^2,$$

$$\therefore \qquad v^2\frac{dv}{dx} + kv^3 = \frac{P}{M}. \qquad (1)$$

Maximum speed V is attained when $v\frac{dv}{dx} = 0$.

Hence, we get

$$V = \left(\frac{P}{Mk}\right)^{1/3}$$

$$\therefore \qquad k = \frac{P}{MV^3}.$$

Putting this value of k in (1), we get

$$v^2\frac{dv}{dx}+\frac{Pv^3}{MV^3} = \frac{P}{M},$$

$$\therefore \quad v^2\frac{dv}{dx} = \frac{P}{M}\left(1-\frac{v^3}{V^3}\right),$$

$$\therefore \quad \frac{v^2dv}{V^3-v^3} = \frac{P}{MV^3}dx\cdot$$

Integrating we get,

$$-\frac{1}{3}\log(V^3-v^3) = \frac{P}{MV^3}x+C\cdot$$

Initially $x = 0$, $v = 0$. $\therefore C = -\frac{1}{3}\log V^3\cdot$

$$\therefore \quad -\frac{1}{3}\log(V^3-v^3) = \frac{Px}{MV^3}-\frac{1}{3}\log v^3$$

$$\therefore \quad \frac{V^3-v^3}{V^3} = -e^{\frac{3Px}{MV^3}} = e^{-3kx}.$$

Hence, $v < V$.
Let $v = V/2$, when $x = x_1$

$$\therefore \quad e^{-3kx_1} = \frac{7}{8}$$

$$\therefore \quad x_1 = \frac{1}{3k}\log\frac{8}{7}.$$

Second part. When power is cut off and an additional force F of constant value is imposed, the equation of motion is

$$M\frac{dv}{dt} = -F-Mkv^2.$$

$$\therefore \quad \frac{dv}{\frac{F}{M}+kv^2} = -dt.$$

Let $t = t_1$. when $v = V/2$ and $t = t_2$ when $v = 0$.
Hence,

$$\int_{\frac{V}{2}}^{0}\frac{dv}{\frac{F}{M}+kv^2} = -\int_{t_1}^{t_2}dt = \int_{t_2}^{t_1}dt.$$

$$\therefore \quad \frac{1}{k}\int_{V/2}^{0}\frac{dv}{\frac{F}{Mk}+v^2} = \int_{t_2}^{t_1} dt.$$

$$\therefore \quad t_2 - t_1 = \sqrt{\frac{Mk}{F}}\tan^{-1}\left\{\sqrt{\frac{M}{kF}}\cdot\frac{1}{2}\left(\frac{P}{Mk}\right)^{\frac{1}{3}}\right\}.$$

2. A car of mass m starts with its engine exerting at a constant pull mP. At any subsequent time it can be switched over to work at constant rate mH. Show that if possible speed V $(>H/P)$ is to be obtained from rest against resistance mkv^2, where v is the speed, then the distance covered will be a maximum if the engine is switched over, when the speed is H/P. Find the maximum distance.

When a constant pull mP is being exerted, the equation of motion is

$$mv\frac{dv}{dx} = mP - mkv^2.$$

$$\therefore \quad \frac{vdv}{P-kv^2} = dx.$$

Integrating this and using initial conditions $v = 0$, $x = 0$, we get

$$\log\frac{P-kv^2}{P} = -2kx.$$

When engine is switched over let $v = V_1$ and $x = x_1$, Hence,

$$P - kV_1^2 = Pe^{-2kx_1}$$

$$\therefore \quad x_1 = \frac{1}{2k}\log\frac{P}{P-kV_1^2}$$

Now equation of motion is

$$v\frac{dv}{dx} = \frac{H}{v} - kv^2.$$

$$\therefore \quad \frac{v^2dv}{H-kv^3} = dx.$$

Integrating this and using the conditions $v = V_1$, $x = 0$, we get

$$\frac{H-kv^3}{H-kV_1^3} = e^{-3kx}$$

Let $x = x_2$, when $v = V$.

$$\therefore \quad \frac{H-kV^3}{H-kV_1^3} = e^{-3kx_2}$$

$$\therefore \quad x_2 = \frac{1}{3k}\log\frac{H-kV_1^3}{H-kV^3}.$$

Hence, distance travelled

$$X = x_1 + x_2$$

$$= \frac{1}{2k}\log\frac{P}{P-kV_1^2} + \frac{1}{3k}\log\frac{H-kV_1^3}{H-kV^3}$$

$$= \frac{1}{6k}\log\frac{P^3(H-kV_1^3)^2}{(P-kV_1^2)^3(H-kV^3)^2}.$$

$$\therefore \quad \frac{dX}{dV_1} = \frac{-6kV_1^2}{H-kV_1^3} + \frac{6kV_1}{P-kV_1^2} = 0,$$

gives

$$(P-kV_1^2)V_1 = H-kV_1^3.$$

$$\therefore \quad V_1 = \frac{H}{P}.$$

$$\therefore \quad X = \frac{1}{6k}\log\frac{P^3(P-kV_1^2V_1^2)}{(P-kV_1^2)^3(H-kV^3)^2}.$$

$$= \frac{1}{6k}\log\frac{P^3H^2}{(P^3-kH^2)(H-kV^3)^2}.$$

3. A particle, moving in a straight line, is subject to a retardation of amount mkv^n per unit mass, where v is the speed at time t. Show that if $n < 1$, the particle will come to rest at a distance $\dfrac{u^{2-n}}{k(2-n)}$ from the point of projection at time $t = \dfrac{u^{1-n}}{k(1-n)}$, where u is the initial speed. What happens when (*i*) $1 < n < 2$ (*ii*) $n > 2$.

Equation of motion is

$$v\frac{dv}{dx} = -kv^n$$

$$\therefore \qquad v^{1-n}\,dv = -\,kdx.$$

Integrating it and using initial condintions $v = u$, $x = 0$, we get

$$\frac{v^{2-n} - u^{2-n}}{2-n} = -\,kx.$$

The particle will come to rest, when $v = 0$.

$$\therefore \qquad x = \frac{u^{2-n}}{k(2-n)}.$$

Equation of motion can also be written as

$$\frac{dv}{dt} = -\,kv^{n};$$

$$\therefore \qquad v^{-n}\,dv = -\,kdt.$$

Integrating it and using initial conditions $t = 0$, $v = u$, we get

$$\frac{v^{1-n} - u^{1-n}}{1-n} = -\,kt.$$

The particle will come to rest when $v = 0$.

$$\therefore \qquad t = \frac{u^{1-n}}{k(1-n)}.$$

(*i*) When $1 < n < 2$, x is finite but time tends to infinity. Hence, particle will comes to rest after a distance $\dfrac{u^{2-n}}{k(2-n)}$ and after an infinite time.

(*ii*) When $n > 2$, x and t both tends to infinity. Hence, particle never come to rest after a finite time or finite distance.

4. A particle moves under gravity in a medium which offers a resistance $kv^2/(a + y)$ per unit mass, where v is the speed of the particle, y is the height above a fixed point O and a and k ($\neq -1/2$) are constatans. If particle is projected vertically upwards from O with speed u, show that it will come to rest instantaneously, when $y = h$, where

$$(a+h)^{2k+1} = a^{2k+1}\{1 + u^2(2k+1)/2ag\}.$$

Find also the speed with which the particle will return to O, and show that if u^2 and ku^2 are both small, compared with ag, the speed will be approximately $u\,(1 - ku^2/2ag)$.

Upward Motion

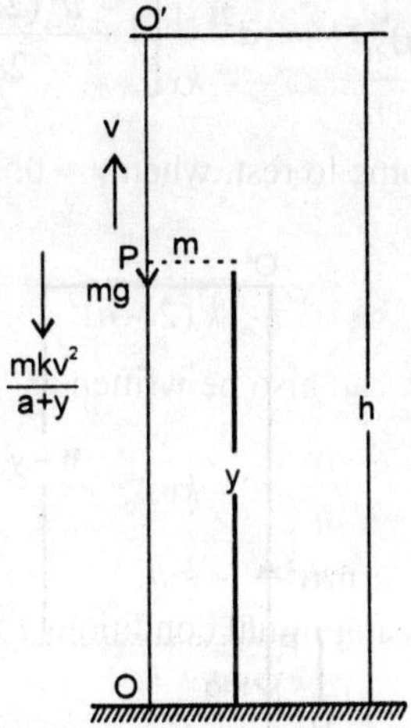

Fig.3.2 *(a)*

Equation of motion is

$$v\frac{dv}{dy} = \frac{-kv^2}{a+y} - g.$$

$$\therefore \qquad \frac{1}{2}\frac{dv^2}{dy} + \frac{kv^2}{a+y} + g = 0,$$

$$\therefore \qquad \left(D + \frac{2k}{a+y}\right)v^2 = -2g,$$

where $D = \frac{d}{dy}\cdot$ Its solution is

$$v^2(a+y)^{2k} = \frac{-2g\,(a+y)^{2k+1}}{2k+1} + C_1,\ k \neq -\frac{1}{2}\cdot$$

Using initial conditions $y = 0$, $v = u$, we get

$$C_1 = u^2 a^{2k} + \frac{2ga^{2k+1}}{2k+1},$$

$$\therefore \qquad v^2(a+y)^{2k} = \frac{-2g}{2k+1}\left[(a+y)^{2k+1} - a^{2k+1}\right] + u^2 a^{2k}.$$

Let particle comes to rest when $v = 0$, and $y = h$,

$$\therefore (a+h)^{2k+1} = a^{2k+1}\left[1+\frac{u^2(2k+1)}{2ag}\right] \quad (1)$$

Downward Motion

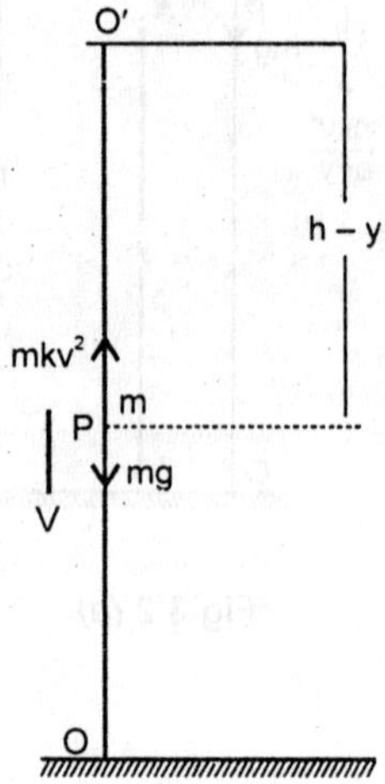

Fig. 3.2. *(b)*

Measuring distance from the lowest point O
Equation of motion is

$$\frac{d^2}{dt^2}(h-y) = \frac{-kv^2}{a+y}+g,$$

$$\therefore \quad \ddot{y} = \frac{kv^2}{a+y}-g.$$

As in the first part, its solution is

$$v^2(a+y)^{-2k} = \frac{2g}{2k-1}(a+y)^{1-2k}+C_2$$

When $y = h$, $v = 0$

$$\therefore \quad C_2 = \frac{2g}{1-2k}(a+h)^{1-2k}$$

$$\therefore \quad v^2(a+y)^{-2k} = \frac{2g}{2k-1}\left[(a+y)^{1-2k}-(a+h)^{1-2k}\right]$$

Let $v = V_1$, when, $y = 0$,

$$\therefore \quad V_1^2 a^{-2k} = \frac{2g}{2k-1}\left[a^{1-2k} - (a+h)^{1-2k}\right]$$

$$\therefore \quad V_1^2 = \frac{2ga}{2k-1}\left[1 - \left(1+\frac{h}{a}\right)^{1-2k}\right]$$

$$\therefore \quad V_1 = \left[\frac{2ga}{2k-1}\left\{1 - \left(1+\frac{h}{a}\right)^{1-2k}\right\}\right]^{1/2}.$$

Substituting value of $\left(1+\frac{h}{a}\right)$ from (1), we get

$$V_1^2 = \frac{2ga}{2k-1}\left[1 - \left\{1+\frac{u^2(2k+1)}{2ga}\right\}^{\frac{1-2k}{1+2k}}\right]$$

$$= \frac{2ga}{2k-1}\left[1 - 1 + (2k-1)\frac{u^2}{2ga} + \frac{(1-2k)u^4 k}{2g^2a^2}\right] \text{ approx.}$$

$$= \left(u^2 - \frac{ku^4}{ga}\right) = u^2\left(1 - \frac{ku^2}{ga}\right)$$

$$\therefore \quad V_1 = u\left(1 - \frac{ku^2}{ga}\right)^{1/2} = u\left(1 - \frac{ku^2}{2ga}\right) \text{ approx.}$$

3.4 SIMPLE HARMONIC MOTION

When a particle moves in a straight line so that its acceleration is always directed towards a fixed point in the line and is proportional to the distance from the point, its motion is called *Simple Harmonic Motion.*

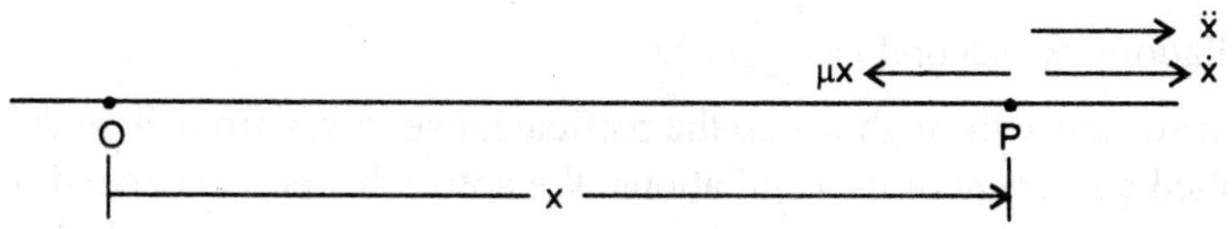

Fig. 3.3

Let O be the fixed point and P the position of the particle at the time t, such that $OP = x$, then acceleration is μx in magnitude, where μ is constant, Acceleration is towards O, therefore the acceleration in the positive direction is $-\mu x$. Hence, equation of motion is

$$\ddot{x} = -\mu x \tag{1}$$

Multiplying both sides by $2\dot{x}$ and integrating, we get

$$\ddot{x} = C - \mu x^2.$$

where C is a constant.

$$\therefore \qquad \dot{x} = \pm\sqrt{C - \mu x^2}.$$

$$\therefore \qquad \frac{dx}{\sqrt{C - \mu x^2}} = \pm\, dt,$$

where the positive or negative sign must be taken according as x is increasing or decreasing with t.

On intergration, we get

$$\sin^{-1}\sqrt{\frac{\mu}{C}}\,x = \pm\sqrt{\mu}\,t + \alpha,$$

where α is a constant,

$$\therefore \qquad x = \sqrt{\frac{C}{\mu}}\sin\left(\pm\sqrt{\mu}\,t + \alpha\right).$$

The solution contains two arbitrary constants C and α and may be written in the form

$$x = A\sin\sqrt{\mu}\,t + B\cos\sqrt{\mu}\,t \tag{2}$$

This is the general form of solution of the equation (1) and therefore contains all the special cases.

Periodicity : Solutions obtained are periodic because t is increased by $2\pi/\sqrt{\mu}$ or any multiple of $2\pi/\sqrt{\mu}$, the value of x and $\dot{x}$ remain unaltered. The interval $2\pi/\sqrt{\mu}$ is called the *periodic time* of the harmonic motion or oscillation; and the *frequency* of the oscillations means the number of oscillations per second i.e. $\sqrt{\mu}/2\pi$.

The distance through which the particle moves away from O on either side is called *amplitude* of the oscillations. We notice that period depends only the constant μ, so the period is independent of the amplitude. Further when we take

$$x = a\cos(\sqrt{\mu}\, t + \varepsilon),$$

$\sqrt{\mu}\, t + \varepsilon$ is called *phase* at that instant from which t is measured.

3.5 DAMPED SIMPLE HARMONIC MOTION

Suppose that in addition to the force varying as the distance, there is a force in the line of action of motion producing a constant acceleration f in the positive direction of x-axis, the equation of motion is given by

$$\ddot{x} = -\mu x + f,$$

$$\therefore \quad \ddot{x} = -\mu\left(x - \frac{f}{\mu}\right).$$

If we put $x = x' + \dfrac{f}{\mu}$, which is equivalent to moving the origin to the point of equilibrium $x = f/\mu$, the above equation becomes

$$\ddot{x}' = -\mu x'.$$

This shows that the effect of the constant force is merely to move the centre of the simple harmonic motion a definite distance f/μ in the direction of the constant force; *i.e.*, the centre of simple harmonic motion is the point of equilibrium, or the point where the variable force is balanced by the constant force.

3.6 FORCED OSCILLATIONS

The equation $\ddot{x} = -\mu x + f\cos pt$ (1)

represents harmonic motion disturbed by a periodic force proportional to cospt.

To find a particular integral of this equation, substituting

$$x = C\cos pt, \text{ we get}$$

$$-Cp^2\cos pt = -\mu C\cos pt + f\cos pt,$$

so that $x = C\cos pt$ is a solution provided that

$$C = \frac{f}{\mu - p^2}$$

$$\therefore \quad x = \frac{f\cos pt}{\mu - p^2}$$

is a particular integral.

Hence, complete solution of equation (1) is

$$x = A\sin\sqrt{\mu}t + B\cos\sqrt{\mu}t + \frac{f\cos pt}{\mu - p^2} \tag{2}$$

The two parts of the solution represent 'free' and 'forced' oscillations, the latter being of the same period $2\pi/p$ as the disturbing force. Free or forced oscillations can exist independently and the motion may be a combination of both, depending on the initial circumstances.

When p^2 is nearly to μ *i.e.*, when the forced and free oscillations have same period, the amplitude $f/(\mu - p^2)$ of the forced oscillations is large. When $p^2 = \mu$, the equation (1) no longer has solution of the form $x = C \cos pt$; but if we substitute $x = Ct \sin pt$ in (1), we get

$$C(-p^2 t \sin pt + 2p \cos pt) = -\mu Ct \sin pt + f \cos pt,$$

which gives $C = f/2p$, so that now forced oscillation is represented by

$x = \dfrac{f}{2p} t \sin pt$; showing that the amplitude of the forced oscillation increases continuously with t.

The phenomenon of large amplitude vibration of a body forced to vibrate with period nearly equal to that of its free vibrations is a familiar one in the theory of sound and electromagnetism and is known as *resonance*. The reason why troops are ordered to fall out of step in crossing bridge is lest by forcing upon the structure vibrations of period nearly the same as that of free vibration, displacements of large amplitude should be set up and cause the bridge to break.

3.7 DAMPED HARMONIC OSCILLATIONS

When a particle moves in a straight line under the action of force towards a fixed point in the line, varying as the distance, and the motion is resisted by a force $2k\dot{x}$ proportional to the velocity, we may write the equation of motion

$$\ddot{x} = -\mu x - 2k\dot{x},$$

$$\therefore \quad \ddot{x} + 2k\dot{x} + \mu x = 0. \tag{1}$$

It is easily seen that $x = e^{mt}$ is a solution of this equation, provided that

$$m^2 + 2km + \mu = 0.$$

So that if m_1 and m_2 are the roots of this equation, the complete solution of (1) is

$$x = Ae^{m_1 t} + Be^{m_2 t}$$

where

$$m_1, m_2 = -k \pm \sqrt{k^2 - \mu}.$$

There are three cases:

(*i*) If $\mu < k^2$ then solution is of the form

$$x = Ae^{-kt\sqrt{(k^2-\mu)t}} + Be^{-kt\sqrt{(k^2-\mu)t}}, \tag{2}$$

and the motion is non-oscillatory.

(*ii*) If $\mu = k^2$, then solution is of the form

$$x = e^{-kt}(A + Bt), \tag{3}$$

which again does not represent oscillations.

(*iii*) If $\mu < k^2$, then solution is

$$x = e^{-kt}(A\cos nt + B\sin nt), \tag{4}$$

where

$$n^2 = \mu - k^2,$$

$$\therefore \quad x = Ce^{-kt}\cos(nt + \varepsilon),$$

where C and ε are arbitrary constants. This solution represents harmonic oscillations with amplitude Ce^{-kt} decreasing steadily with time. They are called *damped oscillations* and e^{-kt} is the damping co-efficient. The oscillations die away as t increases.

3.8 DAMPED FORCED OSCILLATIONS

The equation $\ddot{x} + 2k\dot{x} + \mu x = f\cos pt$ represents damped forced oscillations. The complete solution of this equation can be obtained by adding particular integral to the solution of previous article.

Particular Integral : Substitute $x = E\cos pt + F\sin pt$, as a trial solution and we find that it satisfies the differential equation, if $E(-p^2\cos pt - 2kp\sin pt + \mu\cos pt) + F(-p^2\sin pt + 2kp\cos pt + \mu\sin pt) = f\cos pt$. Hence, we must chose E and F so as to make the last equation as identity for all values of t, i.e., equating to zero, the co-efficients of $\cos pt$ and $\sin pt$. Therefore

$$(\mu - p^2)E + 2kpF - f = 0,$$

$$-2kpE + (\mu - p^2)F = 0;$$

hence,

$$\frac{E}{\mu - p^2} = \frac{F}{2kp} = \frac{f}{(\mu - p^2)^2 + 4k^2p^2}.$$

Hence, particular integral is

$$x = f.\frac{(\mu - p^2)\cos pt + 2kp\sin pt}{(\mu - p^2)^2 + 4k^2p^2},$$

$$\therefore \quad x = f.\frac{\cos(pt + \alpha)}{\left\{(\mu - p^2)^2 + 4k^2p^2\right\}^{1/2}}$$

where

$$\tan\alpha = \frac{2kp}{p^2 - \mu}.$$

Here, x does not vanish permanently at any time, however large t is, it is called *steady state.*

If we take $R = \left[(\mu - p^2)^2 + 4k^2p^2\right]^{1/2}$, then amplitude of steady state is f/R.

Now

$$R^2 = p^4 - 2\left(\mu - \frac{k^2}{2}\right)p^2 + \mu^2 = \left[p^2 - \left(\mu - \frac{k^2}{2}\right)\right]^2 + \mu^2 - \left(\mu - \frac{k^2}{2}\right)^2$$

Thus if $\mu > \frac{k^2}{2}$, R^2 is minimum, when $p = \sqrt{\mu - \frac{k^2}{2}}$.

The amplitude of steady state is maximum for this value of p, and the corresponding value $p/2\pi$ is called the *resonance frequency* of the periodic force.

The complete solution is

$$x = Ce^{-kt}\cos(nt + \varepsilon) + \frac{f\cos(pt + \alpha)}{\left[(\mu - p^2)^2 + 4k^2p^2\right]^{1/2}},$$

where

$$n^2 = \mu - k^2.$$

The two terms represent the free and forced oscillations, and owing to the damping, the free oscillations die away and the forced oscillations alone persist.

Examples

1. A, B, C, D, E, F are the vertices of a regular hexagon having O as centre. A particle of mass m, initially in equilibrium at O, is displaced to point P. If, for any position, the particle is subjected to forces represented completely by $\mu\vec{PA}, \mu\vec{PB}, \mu\vec{PC}, \mu\vec{PD}, \mu\vec{PE}, \mu\vec{PF}$, where $\mu > 0$, show that when released P executes a S.H.M. about O of period $2\pi(m/6\mu)^{1/2}$.

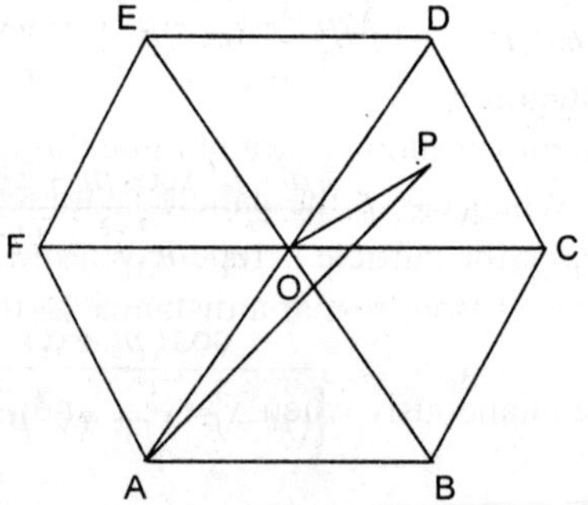

Fig. 3.4

Now

$$\mu\overrightarrow{PA} = \mu\overrightarrow{PO} + \mu\overrightarrow{OA},$$
$$\mu\overrightarrow{PB} = \mu\overrightarrow{PO} + \mu\overrightarrow{OB},$$
$$\mu\overrightarrow{PL} = \mu\overrightarrow{PO} + \mu\overrightarrow{OL},$$
$$\mu\overrightarrow{PD} = \mu\overrightarrow{PO} + \mu\overrightarrow{OD},$$
$$\mu\overrightarrow{PE} = \mu\overrightarrow{PO} + \mu\overrightarrow{OE},$$
$$\mu\overrightarrow{PF} = \mu\overrightarrow{PO} + \mu\overrightarrow{OF},$$

Adding these we get

$$\mu(\overrightarrow{PA} + \overrightarrow{PB} + \overrightarrow{PC} + \overrightarrow{PD} + \overrightarrow{PE} + \overrightarrow{PF})$$
$$= 6\mu\overrightarrow{PO} + \mu(\overrightarrow{OA} + \overrightarrow{OB} + \overrightarrow{OC} + \overrightarrow{OD} + \overrightarrow{OE} + \overrightarrow{OF})\cdot$$

When m is at O, it is in equilibrium,

$$\therefore \quad m\,(\overrightarrow{OA} + \overrightarrow{OB} + \overrightarrow{OC} + \overrightarrow{OD} + \overrightarrow{OE} + \overrightarrow{OF}) = 0\cdot$$
$$\therefore \quad \overrightarrow{OA} + \overrightarrow{OB} + \overrightarrow{OC} + \overrightarrow{OD} + \overrightarrow{OE} + \overrightarrow{OF} = 0\cdot$$
$$\therefore \quad \mu\,(\overrightarrow{PA} + \overrightarrow{PB} + \overrightarrow{PC} + \overrightarrow{PD} + \overrightarrow{PE} + \overrightarrow{PF} = 6\mu\,\overrightarrow{PO} = -6\mu x,$$

if $\overrightarrow{OP} = \vec{x}\cdot$

$$\therefore \quad \overrightarrow{PA} + \overrightarrow{PB} + \overrightarrow{PC} + \overrightarrow{PD} + \overrightarrow{PE} + \overrightarrow{PF} = -6\vec{x}\cdot$$

Equation of motion of m at P is

$$m\ddot{\vec{x}} = \mu\,(\overrightarrow{PA} + \overrightarrow{PB} + \overrightarrow{PC} + \overrightarrow{PD} + \overrightarrow{PE} + \overrightarrow{PF}) = -6\mu\vec{x}\cdot$$

Hence, P executes a S.H.M. about O, if $\mu > 0$.

$$\text{Period} = 2\pi / \sqrt{\frac{6\mu}{m}} = 2\pi\left(\frac{m}{6\mu}\right)^{1/2}.$$

2. A particle of mass m moves along a straight line through a point O and at any instant $OP = x$. When $x > a$, the paticle is attracted towards O by a force mk/x^2, when $x < a$ the particle is repelled from O by a force mak/x^3. If the particle is released from rest at a distance $2a$ from O, show that it will come to rest instantaneously when $x = \frac{a}{\sqrt{2}}$ and find the time that the particle takes to travel fro $x = a$ to $x = \frac{a}{\sqrt{2}}$.

When $x > a$, the equation of motion of P is

$$\ddot{x} = -\frac{k}{x^2},$$

$$\therefore \quad v\frac{dv}{dx} = -\frac{k}{x^2}$$

$$\therefore \quad vdv = -\frac{k}{x^2}dx.$$

Integrating, we get

$$\therefore \quad \frac{v^2}{2} = \frac{k}{x} + A.$$

Initially, when $v = 0$, $x = 2a$.

$$\therefore \quad A = -\frac{k}{2a}.$$

$$\frac{v^2}{2} = \frac{k}{x} - \frac{k}{2a}.$$

Let $v = V_1$, When $x = a$.

$$\therefore \quad \frac{V_1^2}{2} = \frac{k}{a} - \frac{k}{2a} = \frac{k}{2a}.$$

$$\therefore \quad V_1^2 = \frac{k}{a}.$$

When $x < a$, the equation of motion is

$$\ddot{x} = \frac{ak}{x^3}$$

$$\therefore \quad v\frac{dv}{dx} = \frac{ak}{x^3}$$

$$\therefore \quad vdv = \frac{ak}{x^3}dx$$

Integrating we get

$$\frac{v^2}{2} = -\frac{ak}{2x^2} + D.$$

When $x = a$, $v = V_1$

$$\therefore \qquad D = \frac{V_1^2}{2} + \frac{ak}{2a^2} = \frac{k}{2a} + \frac{k}{2a} = \frac{k}{a}.$$

$$\therefore \qquad \frac{v^2}{2} = -\frac{ak}{2x^2} + \frac{k}{a}.$$

$$\therefore \qquad x^2v^2 = \frac{k}{a}(2x^2 - a^2)\cdot$$

The particle will come to rest, when $v = 0$

$$\therefore \qquad x = \frac{a}{\sqrt{2}}.$$

Now from

$$xv = \sqrt{\frac{k}{a}(2x^2 - a^2)},$$

$$\frac{xdx}{\sqrt{2x^2 - a^2}} = \sqrt{\frac{k}{a}}\, dt.$$

Integrating, we get

$$-\int_a^{\frac{a}{\sqrt{2}}} \frac{xdx}{\sqrt{2x^2 - a^2}} = \int_{t_1}^{t_2} \sqrt{\frac{k}{a}}\, dt$$

$$\therefore \qquad \frac{a}{2} = \sqrt{\frac{k}{a}}(t_2 - t_1).$$

$$\therefore \qquad t_2 - t_1 = \frac{a^{3/2}}{2\sqrt{k}}$$

3. A particle of unit mass moves in a straight line and is attracted to an origin O in the line by a force n^2x, where x is the displacement from O. There is also a damping force $2ku$, where u is the speed. Find the condition that motion should be periodic, and show that if this condition is satisfied, the successive maximum displacements from the origin form a geometrical progression with common ratio $e^{2k\pi/\lambda}$, where $\lambda = \sqrt{n^2 - k^2}$. If the particle is also acted on by a force $F \cos pt$ and $n^2 > 2k^2$, find for what values of p the magnitude of the forced oscillation has a maximum value. Denoting these particular values of p and A by p_0 and A_0, show that

$$4\left(\frac{A_0^2}{A^2}-1\right) = \left(\frac{p^2}{p_0^2}-1\right)^2\left(\frac{\lambda}{k}-\frac{k}{\lambda}\right)^2,$$

where A is the amplitude of the forced oscillation for any value of p.
Equation of motion is

$$\ddot{x} = -n^2x - 2k\dot{x}.$$

When $k = 0$, the equation becomes

$$\ddot{x} = -n^2x,$$

which represents S.H.M.
But when $k \# 0$, then three cases arise:

(*i*) When $k^2 > n^2$ the solution is

$$x = ae^{-kt}\sin\left(t\sqrt{n^2-k^2}+\alpha\right),$$

$$= ae^{-kt}\sin(\lambda t + \alpha),$$

where $\lambda = \sqrt{n^2 - k^2}$.

Hence, motion is periodic, and period is $\frac{2\pi}{\lambda}$.

(*ii*) When $k^2 < n^2$, then solution is given by:

$$x = Ae^{-(\lambda+k)t} + Be^{-(k-\lambda)t}$$

which is a periodic or dead beat, both exponential decay with time.

(*iii*) When $k^2 = n^2$. then the solution is given by.

$$x = e^{-nt}(A + Bt).$$

Again motion is aperiodic and the system is said to be critically damped.

Second part

When $k^2 > n^2$, then

$$x = ae^{-kt}\sin(\lambda t + \alpha).$$

Periods are given by $\frac{2\pi}{\lambda}, \frac{4\pi}{\lambda}, \frac{6\pi}{\lambda}$.........

Let $x = x_1$, when $t = 0$, $x = x_2$, when $t = \frac{2\pi}{\lambda}$, $x = x_3$, when $t = \frac{4\pi}{\lambda}$.........

$\therefore$

$$x_1 = a\sin\alpha.$$

$$x_2 = e^{-k\frac{2\pi}{\lambda}}\sin\alpha$$

$$x_3 = ae^{-k\frac{4\pi}{\lambda}}\sin\alpha$$

....

....

$$\therefore \quad \frac{x_2}{x_1} = \frac{x_3}{x_2} = \ldots\ldots\ldots\ldots\ldots = e^{2k\pi\,\lambda}$$

Hence $x_1, x_2, x_3 \ldots\ldots\ldots\ldots\ldots, x_n$ form a geometrical progression with common ratio

$$e^{-2k\pi/\lambda}.$$

Third part

Equation of motion is

$$\ddot{x} = -n^2 x - 2k\dot{x} + F\cos pt.$$

Amplitude of forced vibrations is $\frac{F}{R}$, where

$$R^2 = (n^2 - p^2)^2 + 4k^2p^2.$$

$$\therefore \quad 2R\frac{dR}{dp} = -4\,p(n^2 - p^2) + 8k^2p = 0,$$

$\frac{F}{R}$ is maximum when R is minimum and for R to be minimum $\frac{dR}{dp} = 0.$

$$8k^2p - 4p\,(n^2 - p^2) = 0.$$

$$\therefore \quad p\,[2k^2 - (n^2 - p^2)] = 0.$$

$$\therefore \quad \text{either } p = 0, \text{ or } 2k^2 = n^2 - p^2.$$

But if $p = 0$, then force becomes constant, hence, $p \neq 0$.

$$\therefore \quad p_0^2 = n^2 - 2k^2 = \lambda^2 - k^2.$$

$$\therefore \quad R_0^2 = 4k^4 + 4k^2(\lambda^2 - k^2) = 4k^2\lambda^2.$$

$$\therefore \quad R_0 = 2k\lambda.$$

Now $$A = \frac{F}{R},$$

$$\therefore \quad A_0 = \frac{F}{R_0} = \frac{F}{2k\lambda}.$$

$$\frac{A_0}{A} = \frac{R}{2k\lambda}.$$

$$\therefore \quad R^2 = (n^2 - p^2)^2 + 4k^2p^2$$

$$= (p_0^2 + 2k^2 - p^2)^2 + 4k^2p^2$$

$$= (p_0^2 - p^2)^2 + 4k^4 + 4(p_0^2 - p^2)k^2 + 4k^2p^2$$

$$= (p^2 - p_0^2)^2 + 4k^2(p_0^2 + k^2)$$

$$= (p^2 - p_0^2)^2 + 4k^2\lambda^2.$$

$$\therefore \quad R^2 - 4k^2\lambda^2 = (p^2 - p_0^2)^2.$$

Now $$4\left(\frac{A_0^2}{A^2} - 1\right) = 4\left(\frac{R^2}{4k^2\lambda^2} - 1\right) = \frac{R^2 - 4k^2\lambda^2}{k^2\lambda^2} = \frac{(p^2 - p_0^2)^2}{k^2\lambda^2}$$

$$= \left(\frac{p^2}{p_0^2} - 1\right)^2 \frac{p_0^4}{k^2\lambda^2} = \left(\frac{p^2}{p_0^2} - 1\right)^2 \left(\frac{\lambda^2 - k^2}{\lambda k}\right)^2$$

$$= \left(\frac{p^2}{p_0^2} - 1\right)^2 \left(\frac{\lambda}{k} - \frac{k}{\lambda}\right)^2.$$

3.9 ELASTIC STRINGS AND SPIRAL SPRINGS

Hook's Law : *It states that tension of the string is proportional to the extension.* If l_0 is the unstretched length and l is the stretched length of the string, T the tension, then

$$T = \lambda \frac{l - l_0}{l_0},$$

λ is called the modulus of elasticity of the string.

Work done in Stretching an Elastic String : Let T be the tension, when the length is x, then

$$T = \lambda \frac{x - l_0}{l_0}.$$

The work done, when a force T moves its point of application through a distance x in its line of action, is Tdx, therefore in increasing a length from l to l' is

$$\int_l^{l'} T dx = \frac{\lambda}{l_0} \int_l^{l'} (x - l_0)\, dx$$

$$= \frac{\lambda}{2l_0} \left[(l' - l_0)^2 - (l - l_0)^2\right]$$

$$= \frac{1}{2}(l' - l)\left[\frac{\lambda(l' - l_0)}{l_0} + \frac{\lambda(l - l_0)}{l_0}\right]$$

= increase in length × mean of initial and final tension.

Example

A heavy particle is supported in equilibrium by two equal elastic strings with their other ends being attached to two points in a horizontal plane and each inclined at an angle of 60° to the vertical. The modulus of elasticity is such that when the particle is suspended from any portion of the string its extension is equal to its natural length. The particle is displaced vertically a small distance and then released. Prove that period of its small oscillations is $2\pi\sqrt{2l/5g}$, where l is the stretched length of either string in equilibrium.

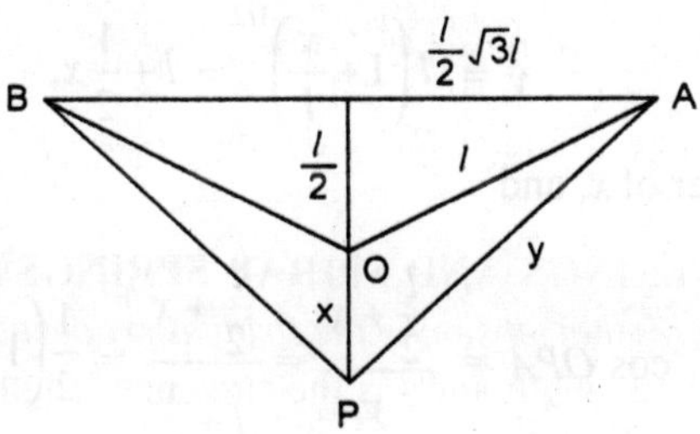

Fig. 3.5

Let m be the mass of the particle and λ the modulus of elasticity. Then by supposing the particle to be suspended from any portion of the string, since extended length is double the natural length, we find that

$$\lambda = mg.$$

If l_0 is the nature of either of the string, we have in equilibrium position

$$mg = 2\lambda\frac{l-l_0}{l_0}\cos 60° = \lambda\frac{l-l_0}{l_0};$$

but

$$\lambda = mg.$$

$$\therefore \qquad l_0 = \frac{l}{2}.$$

Let x denotes verticle dislacement and y the length of either of the string at time t. To find the period of small oscillations, we want to have an equation of the form

$$\ddot{x} = -\mu x,$$

where μ is a constant. It will therefore be sufficient for our purpose to write down the equation of motion at time t and neglect all powers of x higher then the first.

We have

$$m\ddot{x} = mg - 2\lambda\frac{y-l_0}{l_0}\cos OPA$$

where P is the particle at time $t = 0$, O is its equilibrium position and $PA = PB = y$ are the strings.

Now

$$y^2 = \left(x+\frac{l}{2}\right)^2 + \frac{3}{4}l^2 = l^2 + lx + x^2,$$

neglecting x^2.

$$y = l\left(1+\frac{x}{l}\right)^{1/2} = l+\frac{1}{2}x,$$

correct to first power of x, and

$$\cos OPA = \frac{\frac{l}{2}+x}{y} = \frac{\frac{l}{2}+x}{l+\frac{x}{2}} = \frac{1}{2}\left(1+\frac{2x}{l}\right)\left(1-\frac{x}{2l}\right)$$

$$= \frac{1}{2}\left(1+\frac{3x}{2l}\right)$$

to the first power of x.

Hence,

$$m\ddot{x} = mg - \frac{2\lambda\left(l+\frac{x}{2}-\frac{l}{2}\right)}{\frac{l}{2}}\cdot\frac{1}{2}\left(1+\frac{3x}{2l}\right).$$

$$= mg - 2\lambda\left(\frac{1}{2}+\frac{x}{2l}\right)\left(1+\frac{3x}{2l}\right)$$

$$= mg - mg\left(1+\frac{x}{l}\right)\left(1+\frac{3x}{2l}\right)$$

$$\ddot{x} = g - g\left(1+\frac{5x}{2l}\right) = \frac{5xg}{2l},$$

which represents a S.H.M. of period $2\pi\sqrt{\dfrac{2l}{5g}}$.

EXERCISE

1. A car of mass M moves on a level road, the engine working at a constant rate R. There is a constant frictional resistance and the maximum speed attainable is W. If the car starts from rest, and S, T, are respectively, the distance traversed and the time taken, before the car attains a speed V show that

$$S = (MW^3/R)\,(\log\,[W/(W - V) - (V/W) - V^2/2W^2])$$
$$T = (S/W) + (MV^2/2R).$$

2. A particle of mass m is placed on a horizontal board which is made to execute simple vertical hormonic oscillations of period T and amplitude a. If $a < (gT^2/4\pi^2)$, show that the particle does not lose contact with the board at any time.

3. A small stone of mass m is thrown vertically upward with speed V. If the air resistance at speed v is mkv^2, where k is a positive constant, show that the stone returns to its starting point with speed $V/(1+kV^2/g)^{1/2}$.
If the stone has the same mass and initial speed, but if the air rsistance is mkV, show that the stone returns to its starting point with speed U given by the equation

$$g - kU = (g + kV)\exp\{-k(U + V)/g\}$$

4. A particle of mass m oscillates in a line with natural period $2\pi/n$. If an applied periodic force $F\cos pt$ now acts in the line so that the particle is instantaneously at rest at zero time at a distance d from the centre of oscillation, prove that the displacement of the particle at subsequent time t is

$$d\cos nt + F(\cos pt - \cos nt)\,/\,(n^2 - p^2)m.$$

5. In the absence of external forces, a particle of mass m would execute a S.H.M. When a disturbing force $F\cos pt$ acts, the maximum speed attained is V. Show that the angular frequency of the free oscillations is $[(F + mp)p\,/\,mV]^{1/2}$.

6. A particle is projected vertically upwards under gravity. The resistance of the air produces an acceleration opposite to the velocity and numerically equal to kv^2, where v is the velocity and k a constant. If the initial velocity is V, and the square of kV^2/g can be neglected, show that the particle reaches its highest point in time $\dfrac{V}{g} - \dfrac{kV^4}{3g^2}$ and the greatest

altitude reached is

$$\frac{V^2}{2g} - \frac{kV^4}{4g^2}.$$

If the initial velocity, in addition to the vertical component V, has a small horizontal component U, and the resistance follows the same law, show that, when the particle returns to the original level its horizontal velocity is approximately $Ue^{kv^2/g}$.

7. A mass m hangs from a fixed point by means of a light string, which obeys Hooke's law, the mass being given a small vertical displacement. If n is the number of oscillations per second in ensuing simple harmonic motion, and If l is the length of the spring, when the system is in equilibrium, find the natural length of the spring, and show that, when the spring is extended to double its length, the tension is

$$m\,(4\pi^2 n^2 l - g)$$

8. A heavy particle of mass m is attached to one end of the elastic string of natural length a, whose other end is fixed at O. The particle is let fall from rest at O. Show that part of the motion is simple hoarmonic, and that, if the greatest depth of the particle below O is $a \cot^2\frac{\theta}{2}$, the modulus of velocity of the string is $\frac{1}{2}mg\tan^2\frac{\theta}{2}$, and that the particle attains this depth in time

$$\sqrt{\frac{2a}{g}}\,[l + (\pi - \theta)\cot\theta],$$

where θ is a positive acute angle.

9. A particle of mass is at rest on a smooth horizontal plane and is connected by three elastic strings, each of modulus λ and natural length a, to three points on the plane at the corners of equilateral triangle of side $2\sqrt{3a}$. Prove that if the particle is displaced in a direction of one of the strings and then released, it will perform oscillations of period.

$$(4\pi/3)\sqrt{\frac{am}{\lambda}}.$$

10. An elastic string of natural length a is fixed at its ends to points A, B on a smooth horizontal table distance $2a$ apart. A heavy particle is attached to the mid-point of the string, and is projected along the table with velocity u and at right angles to the string. The modulus of elasticity is equal to the weight of the table. Show that when the particle first comes to instantaneous

rest, the length of the string is $2ka$, where k is the positive root of the equation

$$k^2 - k = \frac{u^2}{4ag}.$$

Show that if $u^2/4ag$ is small, the motion is approximately simple harmonic of period

$$\pi\sqrt{\frac{2a}{g}}.$$

4

DYNAMICS OF A PARTICLE

4.1 EQUATIONS OF MOTION IN TWO AND THREE DIMENSIONS

Suppose a particle of constant mass m is situated at the point $P(x, y, z,)$ at time t and let

$$\overrightarrow{OP} = \vec{r} = x\hat{i} + y\hat{j} + z\hat{k}$$

where $\hat{i}, \hat{j}, \hat{k}$ are unit vectors along axes of co-ordinates. Let force

$$\vec{F} = X\hat{i} + Y\hat{j} + Z\hat{k},$$

By Newton's Second law of motion

$$\text{Force} = \text{mass} \times \text{Acceleration}$$

$$\therefore \quad X\hat{i} + Y\hat{j} + Z\hat{k} = m(\ddot{x}\hat{i} + \ddot{y}\hat{j} + \ddot{z}\hat{k}).$$

Hence,

$$X = m\ddot{x}, \; Y = m\ddot{y}, \; Z = m\ddot{z}.$$

These are equations of motion in three dimensions. In case of two dimensions we have only first two equations.

4.2 PROJECTILE MOTION UNDER GRAVITY

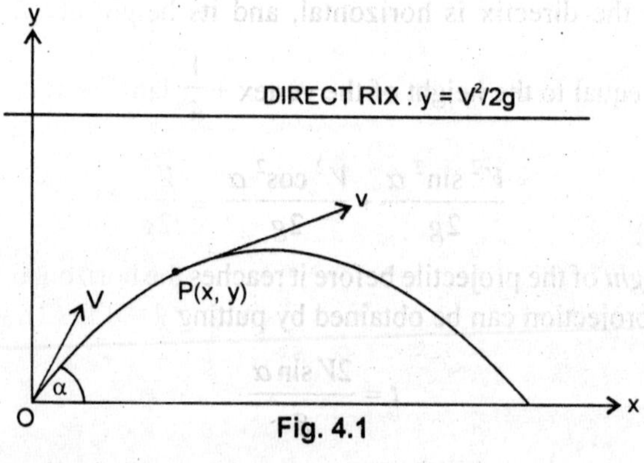

Fig. 4.1

Let a particle of mass m be projected freely under gravity in non-resisting medium. Let us take the horizontal and upward verticle lines through the point of projection O as Ox and Oy co-ordinate axes and let V be the velocity of projection in a direction making and angle α with the horizontal.

Equations of motion are

$$\therefore \quad \ddot{x} = 0, \ddot{y} = -g. \tag{1}$$

Initial conditions are

$$x = y = 0, \dot{x} = V\cos a, \ \dot{y} = V\sin\alpha, t = 0.$$

Integrating (1) and using initial conditions, we get

$$\dot{x} = V\cos\alpha, \dot{y} = V\sin\alpha - gt,$$

$$x = Vt\cos\alpha, y = Vt\sin\alpha - \frac{1}{2}gt^2. \tag{2}$$

Eliminating t, we obtain the equation of motion of the path of the particle,

$$y = x\tan\alpha - \frac{gx^2}{2V^2\cos^2\alpha}. \tag{3}$$

This equation represents a parabola and by writing this equation in the form

$$\left(x - \frac{V^2}{g}\sin\alpha\cos\alpha\right)^2 = -\frac{2V^2\cos^2\alpha}{g}\left(y - \frac{V^2\sin^2\alpha}{2g}\right),$$

we see that the latus rectum is $\frac{2V^2\cos^2\alpha}{g}$, the vertex is at the point $\left(\frac{V^2}{2g}\sin 2\alpha, \frac{V^2\sin^2\alpha}{2g}\right)$ and the axis vertically downwards.

Therefore the directix is horizontal, and its height above the point of projection is equal to the height of the vertex $+\frac{1}{4}$ latus rectum, *i.e.*

$$\frac{V^2\sin^2\alpha}{2g} + \frac{V^2\cos^2\alpha}{2g} = \frac{V^2}{2g}$$

Time of flight of the projectile before it reaches the horizontal plane through the point of projection can be obtained by putting $y = 0$ in (2), which gives

$$t = \frac{2V\sin\alpha}{g}.$$

The range of the projectile on the plane through the point of projection can be obtained by putting for t, the time of flight in the expression of x in (2), *i.e.*,

$$\frac{V^2 \sin 2\alpha}{g},$$

and the greatest range for a given V is therefore $\frac{V^2}{g}$ by taking $\alpha = \frac{\pi}{2}$.

Since, $\sin 2\left(\frac{\pi}{2} - \alpha\right) = \sin 2\alpha$, therefore in general there are two directions of projection with a given velocity which give the same horizontal range, viz., those of inclination α and $\frac{\pi}{2} - \alpha$, The two coinciding when $\alpha = \frac{\pi}{4}$.

Again we can show that the velocity at any point of the path is, in magnitude, the velocity that would be acquired in falling freely from the directrix.

If v is the velocity at any time t, then

$$\begin{aligned} v^2 &= \dot{x}^2 + \dot{y}^2 \\ &= V^2 \cos^2 \alpha + (V \sin \alpha + gt)^2 \\ &= V^2 - 2gtV \sin\alpha + g^2 t^2 \\ &= 2g\left[\frac{V^2}{2g} - \left(Vt \sin \alpha - \frac{1}{2} gt^2\right)\right] \\ &= 2g\left(\frac{V^2}{2g} - y\right) \\ &= 2g \times \text{depth below the directrix.} \end{aligned}$$

4.3 RANGE ON AN INCLINED PLANE THROUGH THE POINT OF PROJECTION.

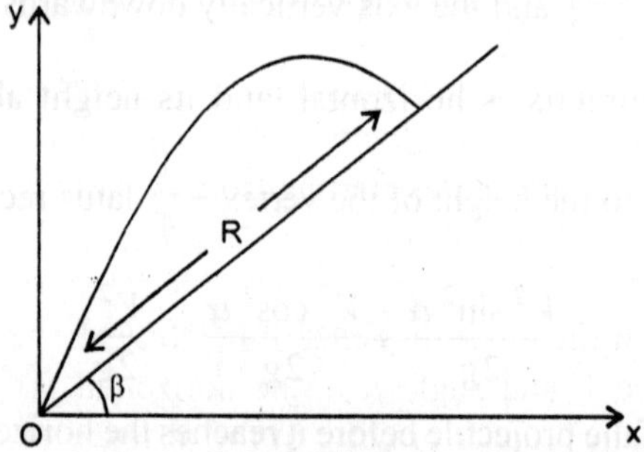

Fig. 4.2

Let the inclination of the plane to the horizontal be β, then the range on the plane can be obtained from the equation (3) of the previous article by putting $y = x \tan \beta$, hence we get

$$x \tan \beta = x \tan \alpha - \frac{gx^2 \sec^2 \alpha}{2V^2},$$

when $x \neq 0$, then

$$x = \frac{2V^2 \cos^2 \alpha}{g}(\tan \alpha - \tan \beta).$$

$$\text{The range} = x \sec \beta$$

$$= \frac{2V^2 \cos^2 \alpha}{g}(\tan \alpha - \tan \beta) \sec \beta$$

$$= \frac{2V^2 \cos \alpha}{g} \sin(\alpha - \beta) \sec^2 \beta$$

$$= \frac{V^2}{g} \sec^2 \beta [\sin(2\alpha - \beta) - \sin \beta]$$

For a given values of V and β this expression is maximum, when $2\alpha - \beta = \frac{\pi}{2}$, and this makes the maximum range

$$R = \frac{V^2}{g} /(1 + \sin \beta),$$

and we know that the polar equation of a parabola with the focus as pole is

$$\frac{l}{r} = 1 + e \cos \theta,$$

where θ is measured from the vertex and l is the semi latus rectum.

Comparing last two equations, we see that if we construct a parabola with focus at O, latus rectum $\frac{2V^2}{g}$, axis vertical and vertex A upwards, any radius OP drawn from O will give the maximum range in the direction OP for a particle projected from O with velocity V. This parabola is an outer boundary that can be reached by projectiles starting from point O with velocity V.

Examples

1. A short putter at the top A of a cliff of height h above sea level throws a short at a speed V and angle α to the horizontal. The shot describes a parabolic trajectory under gravity, in the absence of air resistance, and enters the sea at B. If the horizontal distance R of B from the foot of the cliff is to be maximum find R, α in term of V.

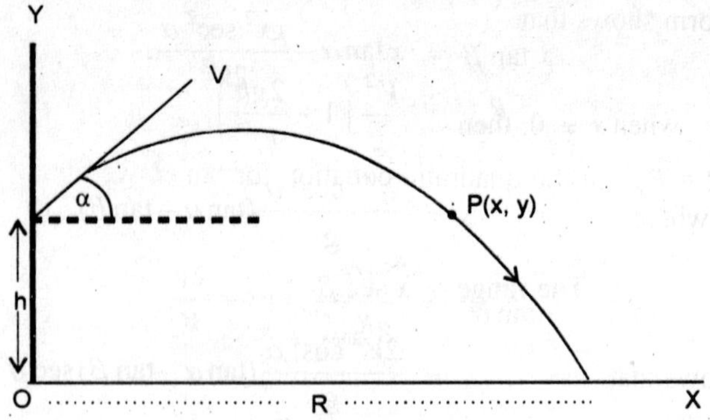

Fig. 4.3

At time t suppose the shot has reached to the point $P(x, y,)$. The equations of motion are

$$\ddot{x} = 0, \ddot{y} = -g.$$

Initial conditions are

$$t = 0, \dot{x} = V\cos\alpha, y = V\sin\alpha, x = 0, y = h$$

Integrating above differential equations twice and using the initial conditions, we get

$$\dot{x} = V\cos\alpha, x = Vt\cos\alpha,$$

$$\dot{y} = V\sin\alpha - gt, y = h + Vt\sin\alpha - \frac{1}{2}gt^2.$$

Eliminating t from the forms for x, y, we get

$$y = h + x\tan\alpha - \frac{gx^2}{2V^2}\sec^2\alpha.$$

When $x = R$, $y = 0$, gives

$$\frac{gR^2}{2V^2}\tan^2\alpha - R\tan\alpha - \left(\frac{gR^2}{2V^2} - h\right) = 0.$$

Regarding this equation as a quadratic for tan α, roots will be real if

$$R^2 \geq \frac{gR^2}{V^2}\left(\frac{gR^2}{V^2} - 2h\right),$$

$$\therefore \quad R \leq \frac{V^2}{g}\left(1 + \frac{2gh}{V^2}\right)^{1/2}.$$

The last form shows that

$$R_{max.} = \frac{V^2}{g}\left(1+\frac{2gh}{V^2}\right)^{1/2}.$$

Writing $R = R_{max.}$ in the quadratic equation for tan α, we obtain the criticlal value α^*, where

$$\tan\alpha^* = \frac{V^2}{gR_{max}} = \left(1+\frac{2gh}{V^2}\right)^{-1/2}.$$

which shows that

$$\tan\,\alpha^* < 1,\ \alpha^* < \frac{\pi}{4},\ \text{provided h} > 0.$$

The special case of $h = 0$, gives

$$R_{max} = \frac{V^2}{g},\ \alpha^* = \frac{\pi}{4}.$$

It is instructive to see how the value of R as given by quadratic equation.

$$\frac{gR^2}{2V^2}\sec^2\alpha - R\tan\alpha - h = 0$$

varies with R and α for a parabolic case. The appropriate solution is

$$R = \frac{V^2\sin 2\alpha}{2g}\left[1+\left(1+\frac{2gh}{V^2\sin^2\alpha}\right)^{1/2}\right].$$

When $2gh < V^2\sin^2\alpha$, then

$$R \cong \frac{V^2\sin 2\alpha}{2g}\left[1+1+\frac{gh}{V^2\sin^2\alpha}\right]$$

$$\cong \frac{V^2\sin 2\alpha}{g} + h\cot\alpha.$$

Clearly $\dfrac{V^2\sin 2\alpha}{g}$ is the dominant term in this expression and so R will be much more sensitive in V than in α.

2. An aeroplane is flying with uniform speed v in an arc of a vertical circle of radius a, whose centre is at a height h vertically above O of the ground. If a bomb is dropped from the aeroplane when at a height Y and strikes the ground at O, show that Y satisfies the equation

$$KY^2 + Y(a^2 - 2hk) + K(h^2 - a^2) = 0,$$

where

$$K = h + \frac{a^2 g^2}{2v^2}.$$

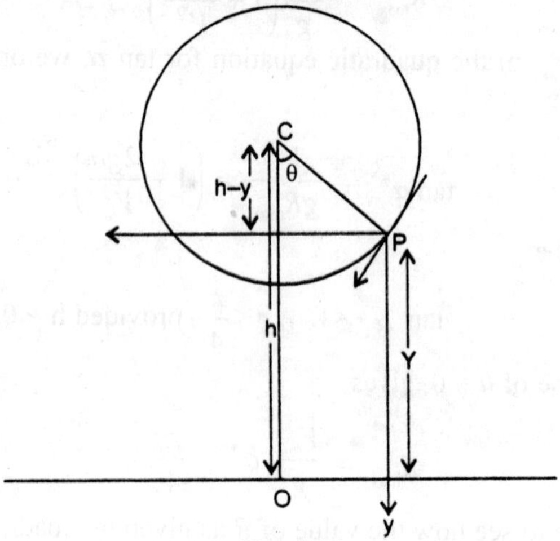

Fig. 4.4

Let C be the centre of the circle, and bomb is dropped when the aeroplane is at the point P. Let the horizontal and vertical lines (as shown in the figure) through P be taken as axes of references and CP makes an angle θ with the vertical CO.

The equation of the path of bomb is given by

$$y = x \tan\theta + \frac{gx^2}{2v^2 \cos^2\theta} \qquad \text{... (1)}$$

The co-ordinate of O are $(a \sin\theta, Y)$, if the bomb strikes the ground at O, the equation (1) will satisfy these co-ordinates. Hence, we have

$$Y = a \sin\theta \tan\theta + \frac{ga^2 \sin^2\theta}{2v^2 \cos^2\theta}$$

$$= a \sin\theta \tan\theta + \frac{ga^2}{2v^2} \tan^2\theta.$$

But $\tan\theta = \dfrac{\sqrt{a^2 - (h-Y)^2}}{h-Y}$

$$\therefore \quad Y = (h-Y)\frac{a^2 - (h-Y)^2}{(h-Y)^2} + \frac{ga^2}{2v^2} \cdot \frac{a^2 - (h-Y)^2}{(h-Y)^2}$$

$$= \frac{a^2 - (h - Y)^2}{(h - Y)^2}\left[h - Y + \frac{ga^2}{2v^2}\right] \qquad (A)$$

But

$$K = h + \frac{ga^2}{2v^2}.$$

Hence, we get

$$Y = \frac{\left[a^2 - (h - y)^2\right][K - Y]}{(h - Y)^2}$$

$$\therefore \quad (h - Y)^2 (Y + K - Y) = a^2 (K - Y)$$

$$\therefore \quad K(h - Y)^2 = a^2 (K - Y)$$

$$\therefore \quad K(h^2 + Y^2 - 2hY) = a^2 (K - Y)$$

$$KY^2 + Y(a^2 - 2hK) + K(h^2 - a^2) = 0.$$

3. A particle is projected under gravity with velocity V in a direction making an angle θ with an upward line of greatest slope of plane of inclination α, the point of projection being on the plane. The motion takes place in vertical plane through the line of greatest slope. Show that the particle strikes the plane at an angle α with the horizontal then θ is given by equation

$$(1 + 3\tan^2\alpha)\tan\theta = 2\tan\alpha.$$

In this case show that the velocity with which the particle strikes the plane is

$$V(1 + 8\sin^2\alpha)^{-1/2}.$$

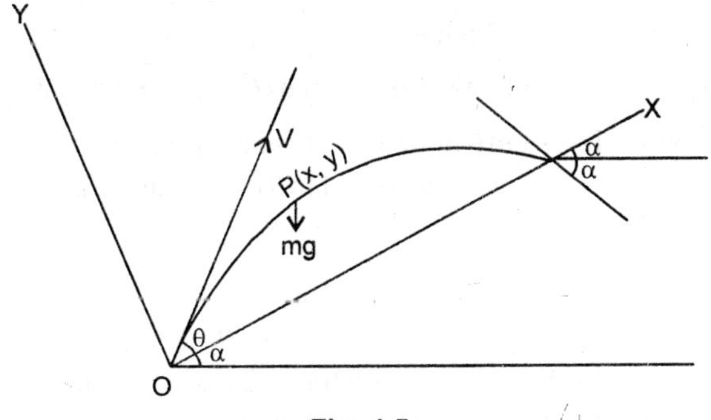

Fig. 4.5

Let us take the line of greatest slope as x-axis and a line perpendicular to it y-axis. Let at time t the particle be at the point $P(x, y)$.

The equations of motion are

$$\ddot{x} = -g\sin\alpha, \ddot{y} = -g\cos\alpha. \qquad \text{.... (1)}$$

Initial conditions are $t = 0, x = 0, y = 0, \dot{x} = V\cos\theta, \dot{y} = V\sin\theta$.

Integrating (1) and using intial conditions, we get

$$\dot{x} = -gt\sin\alpha + V\cos\theta, x = -\frac{1}{2}gt^2\sin\alpha + Vt\cos\theta.$$

$$\dot{y} = -gt\cos\alpha + V\sin\theta, \ y = -\frac{1}{2}gt^2\cos\alpha + Vt\sin\theta.$$

When particle strikes the plane $y = 0$.

Hence, we get,

$$\text{When } t \# 0, t = \frac{2V\sin\theta}{g\cos\alpha}.$$

Let

$\dot{x}_0$ = Velocity of the particle along the plane when it strikes the plane

$$= V\cos\theta - \frac{2V\sin\theta\sin\alpha}{\cos\alpha} = V(\cos\theta - 2\sin\theta\tan\alpha)$$

$\dot{y}_0$ = Velocity of the particle perpendicular to the plane when it strikes the plane

$$= V\sin\theta - 2V\sin\theta = -V\sin\theta.$$

Now resultant velocity makes an angle $\pi - 2\alpha$ with positive direction of x-axis

$$\therefore \quad \tan(\pi - 2\alpha) = \frac{\dot{y}_0}{\dot{x}_0}$$

$$\therefore \quad \tan 2\alpha = \frac{-V\sin\theta}{V(\cos\theta - 2\sin\theta\tan\alpha)}$$

$$\therefore \quad \frac{2\tan\alpha}{1-\tan^2\alpha} = \frac{\sin\theta}{(\cos\theta - 2\sin\theta\tan\alpha)}$$

$$\therefore \quad 2\tan\alpha(1 - 2\tan\theta\tan\alpha) = (1-\tan^2\alpha)\tan\theta$$

$$\therefore \quad \tan\alpha = \tan\theta(1 + 3\tan^2\alpha).$$

Let

$$V_0^2 = \dot{x}_0^2 + \dot{y}_0^2$$

$$= V^2(\cos\theta - 2\sin\theta\tan\alpha)^2 + V^2\sin^2\theta$$

$$= V^2(\cos^2\theta + 4\sin^2\theta\tan^2\alpha - 4\sin\theta\cos\theta\tan\alpha) + V^2\sin^2\theta$$

$$= V^2(1 + 4\sin^2\theta\tan^2\alpha - 4\sin\theta\cos\theta\tan\alpha).$$

Now

$$\tan\theta = \frac{2\tan\alpha}{1+3\tan^2\alpha}$$

$$\therefore \sin\theta = \frac{2\tan\alpha}{\sqrt{1+10\tan^2\alpha+9\tan^4\alpha}},\ \cos\theta = \frac{1+3\tan^2\alpha}{\sqrt{1+10\tan^2\alpha+9\tan^4\alpha}}$$

$$\therefore V_0^2 = V^2\left[1+\frac{16\tan^4\alpha}{1+10\tan^2\alpha+9\tan^4\alpha}-\frac{8\tan^2\alpha(1+3\tan^2\alpha)}{1+10\tan^2\alpha+9\tan^4\alpha}\right]$$

$$= V^2\left[\frac{1+2\tan^2\alpha+\tan^4\alpha}{1+10\tan^2\alpha+9\tan^4\alpha}\right]$$

$$= \frac{V^2\sec^4\alpha}{1+\tan^2\alpha+9\tan^2\alpha\sec^2\alpha}$$

$$= \frac{V^2\sec^4\alpha}{\sec^2\alpha(1+9\tan^2\alpha)}$$

$$= \frac{V^2}{1+8\sin^2\alpha}$$

$$\therefore V_0 = V(1+8\sin^2\alpha)^{-1/2}$$

4. A particle moves in a plane so that its velocity is the vector sum of magnitude V perpendicular to its radius vector and of a velocity of magnitude U in a fixed direction, where U and V are constants. Show that the acceleration of the particle is always directed towards a fixed point and varies as the inverse square of the distance from the point. Express the eccentricity of the orbit in terms of U, V.

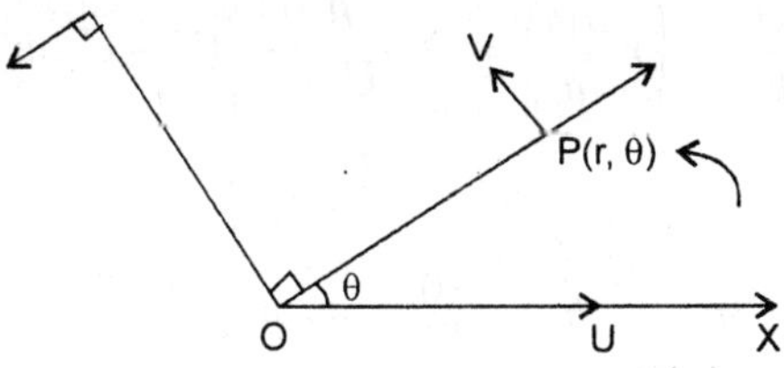

Fig. 4.6

Let us take the fixed direction as x-axis. Let at time t the particle is at the point $P(r, \theta)$, then resolving velocities along radial and transverse directions,

$$\frac{dr}{dt} = U\cos\theta$$

$$\frac{rd\theta}{dt} = V - U\sin\theta.$$

$$\therefore \quad \frac{dr}{rd\theta} = \frac{U\cos\theta}{V - U\sin\theta}$$

$$\therefore \quad \frac{dr}{r} = \frac{U\cos\theta}{V - U\sin\theta}d\theta$$

$$\therefore \quad \log\frac{r}{A} = -\log(V - U\sin\theta)$$

$$\therefore \quad r = \frac{A}{V - U\sin\theta}$$

$$\therefore \quad \frac{(A/V)}{r} = 1 - \frac{U}{V}\sin\theta.$$

This is the equation of the conic referred to focus,

with semi-latus rectunm $l = \frac{A}{V}$ and eccentricity

$$e = \frac{U}{V}.$$

Now

$$r\left(1 - \frac{U}{V}\sin\theta\right) = \frac{A}{V} = B.$$

We have

$$r^2\frac{d\theta}{dt} = r\left(r\frac{d\theta}{dt}\right) = \frac{B}{1 - \frac{U}{V}\sin\theta}(V - U\sin\theta) = BV.$$

$$\therefore \quad \frac{d}{dt}\left(r^2\frac{d\theta}{dt}\right) = 0.$$

This transverse component of acceleration is zero.

$$\text{Radial acceleration} = \frac{d^2r}{dt^2} - r\left(\frac{d\theta}{dt}\right)^2$$

$$= -\dot{\theta}U\sin\theta - \frac{(V - U\sin\theta)^2}{r}$$

$$= \frac{-U\sin\theta(V - U\sin\theta)}{r} - \frac{(V - U\sin\theta)^2}{r}$$

$$= \frac{1}{r}\left[-UV\sin\theta + U^2\sin^2\theta - V^2 + 2UV\sin\theta - U^2\sin^2\theta\right]$$

$$= \frac{1}{r}\,[UV\sin - V^2]$$

$$= \frac{-V^2}{r}\left(1 - \frac{U}{V}\sin\theta\right)$$

$$= -\frac{BV^2}{r^2} = -\frac{\mu}{r^2}$$

where

$$\mu = BV^2$$

is a constant.

Thus acceleration varies as inverse square of the distance from the fixed point and is directed towards fixed point.

4.4. MOTION IN RESISTING MEDIUM

(*i*) A particle of unit mass is projected with velocity V and inclination α in a medium whose resistance is $k \times$ velocity. Show that if k is small, the equation of the path is approximately

$$y = x\tan\alpha - \frac{gx^2}{2V^2\cos^2\alpha} - \frac{kgx^3}{3V^3\cos^3\alpha}.$$

The particle is projected in the same medium from a point on a plane inclined at angle β to the horizontal. Prove that range is approximately

$$\left[\frac{2V^2\sin(\alpha-\beta)\cos\alpha}{g\cos^2\beta}\right]\left[1 - \frac{4kV\sin(\alpha-\beta)}{3g\cos\beta}\right].$$

First Part.

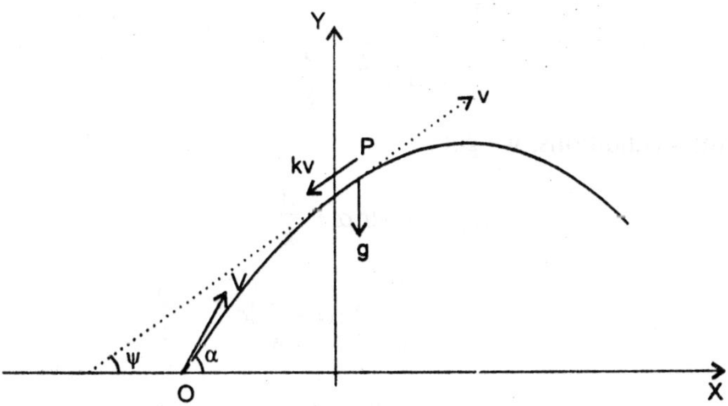

Fig. 4.7

Let the paticle be at the point $P(x, y)$ at time t. The forces acting are g downwards and kv along the tangent in the direction opposite to v. The equations of motion are

$$\ddot{x} = -kv \sin\psi = -k\dot{x} \tag{1}$$

$$\ddot{y} = -g - kv \sin \psi = -g - k\dot{y} \tag{2}$$

where ψ is the angle that tangent at P makes with the horizontal axis OX.

Initial conditions are

$$t = 0, \dot{x} = V\cos\alpha, x = 0. \dot{y} = V\sin\alpha, y = 0.$$

Equation (1) can be written as

$$(D + k)\dot{x} = 0, \text{ where } D = \frac{d}{dt}.$$

Its solution is

$$\dot{x} = A_1 e^{-kt}.$$

Using initial conditions, we get

$$A_1 = V\cos\alpha.$$

$$\therefore \quad \dot{x} = V\cos\alpha . e^{-kt}.$$

Integrating again, we get

$$x = A_2 - \frac{V\cos\alpha}{k} e^{-kt}.$$

Using initial conditions, we have

$$A_2 = \frac{V\cos\alpha}{k}$$

$$\therefore \quad x = \frac{V\cos\alpha}{k}(1 - e^{-kt}). \tag{3}$$

Equation (2) can be written as

$$(D + k)\dot{y} = -g.$$

$$\therefore \quad \dot{y} = A_3 e^{-kt} - \frac{g}{k}$$

Using initial conditions, we get.

$$A_3 = V\sin\alpha + \frac{g}{k}$$

$$\therefore \quad \dot{y} = \left(V\sin\alpha + \frac{g}{k}\right)e^{-kt} - \frac{g}{k}$$

$$\therefore \quad y = -\frac{gt}{k} - \frac{1}{k}\left(V\sin\alpha + \frac{g}{k}\right)e^{-kt} + A_4.$$

Using initial conditions, we have

$$A_4 = \frac{1}{k}\left(V\sin\alpha + \frac{g}{k}\right).$$

$$\therefore \quad y = \frac{1}{k}\left(V\sin\alpha + \frac{g}{k}\right)(1 - e^{-kt}) - \frac{gt}{k}. \tag{4}$$

From (3)

$$-t = \frac{1}{k}\log\left[1 - \frac{kx}{V\cos\alpha}\right].$$

Substituting value of t in (4), we get

$$y = x\tan\alpha + \frac{gx}{kV\cos\alpha} + \frac{g}{k^2}\log\left[1 - \frac{kx}{V\cos\alpha}\right]$$

$$= x\tan\alpha + \frac{gx}{kV\cos\alpha} - \frac{g}{k^2}\left[\frac{kx}{V\cos\alpha} + \frac{k^2x^2}{2V^2\cos^2\alpha} + \frac{k^3x^3}{3V^3\cos^3\alpha} + \ldots\ldots\right]$$

$$\cong x\tan\alpha - \frac{gx^2}{2V^2\cos^2\alpha} - \frac{kgx^3}{3V^3\cos^3\alpha}. \tag{5}$$

Second Part.

Let R be the range on the plane inclined at β to OX. Putting

$$x = R\cos\beta,\ Y = R\sin\beta$$

In (5), we get

$$R\sin\beta = R\cos\beta\tan\alpha - \frac{gR^2\cos^2\beta}{2V^2\cos^2\alpha} - \frac{kgR^3\cos^3\beta}{3V^3\cos^3\alpha}$$

$$\therefore \left[\frac{gR\cos^2\beta}{2V^2\cos^2\alpha}\right]\left[1 + \frac{2kR\cos\beta}{3V\cos\alpha}\right] = \frac{\sin(\alpha-\beta)}{\cos\alpha}$$

$$\therefore R = \left[\frac{2V^2\cos\alpha\sin(\alpha-\beta)}{g\cos^2\beta}\right]\left[1 + \frac{2kR\cos\beta}{3V\cos\alpha}\right]^{-1}$$

$$\cong \left[\frac{2V^2\cos\alpha\sin(\alpha-\beta)}{g\cos^2\beta}\right]\left[1 - \frac{2kR\cos\beta}{3V\cos\alpha}\right]$$

$$\therefore R\left[1 + \frac{4kV\sin(\alpha-\beta)}{3g\cos\beta}\right] \cong \frac{2V^2\cos\alpha\sin(\alpha-\beta)}{g\cos^2\beta}$$

$$\therefore R \cong \left[\frac{2V^2 \cos\alpha \sin(\alpha-\beta)}{g\cos^2\beta}\right]\left[1+\frac{4kV\sin(\alpha-\beta)}{3g\cos\beta}\right]^{-1}$$

$$\cong \left[\frac{2V^2 \cos\alpha \sin(\alpha-\beta)}{g\cos^2\beta}\right]\left[1-\frac{4kV\sin(\alpha-\beta)}{3g\cos\beta}\right].$$

Note: If we put $k = 0$, we get the results of trajectory in non-resisting medium.

(*ii*) Motion of a heavy particle in a medium the resistance of which varies as the square of the velocity.

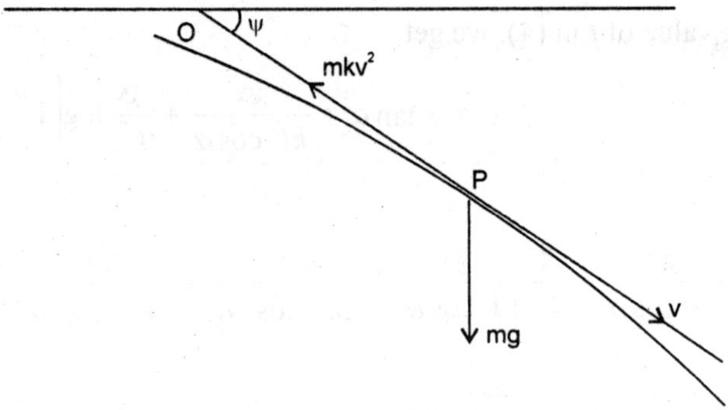

Fig. 4.8

Measuring ψ downwards from a horizontal line, the equations of motion are

$$v\frac{dv}{ds} = g\sin\psi - kv^2,$$

$$\frac{v^2}{\rho} = g\cos\psi.$$

The second equation shows that for a given velocity, the curvature of the path is independent of the resistance, a theorem which is true for motion in any resisting medium under any forces.

Eliminating v, we get

$$\cos\psi\frac{d\rho}{d\psi} - 3\rho\sin\psi + 2k\rho^2\cos\psi = 0$$

$$\therefore \quad \frac{d}{d\psi}\left(\frac{1}{\rho}\right) + \frac{3}{\rho}\tan\psi = 2k.$$

This equation has an integrating factor $\sec^3\psi$ and gives on integration

$$\frac{\sec^3 \psi}{\rho} = k\{\tan\psi \sec\psi + \log(\tan\psi + \sec\psi)\} + C.$$

This equation gives the curvature of the path in terms of its inclination to the horizontal.

If ρ_0 is the radius of curvature at the highest point, $C = 1/\rho_0$, and if ρ, ρ' are the radii of curvatures at the two points at which the tangent to the path is inclined at the same angle ψ to the horizontal, we have

$$\frac{1}{\rho} + \frac{1}{\rho'} = \frac{2\cos^3\psi}{\rho_0}.$$

4.5 PARABOLA OF SAFETY

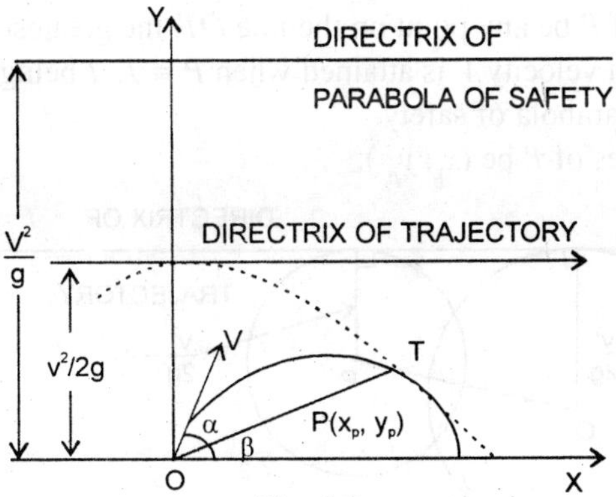

Fig. 4.9

Equation of trajectory

$$y = x\tan\alpha - \frac{gx^2}{2V^2\cos^2\alpha}$$

can be written as

$$\left(\frac{gx^2}{2V^2}\right)\tan^2\alpha - x\tan\alpha + \left[y + \frac{gx^2}{2V^2}\right] = 0 \tag{1}$$

Now suppose V be fixed; then as α varies, a family of trajectory is generated in the (x, y) plane, each passing through the point $O(0, 0)$. Now (1) is a quadratic equation in tan α, the envelope of this family is obtained by equating the discriminant of the quadratic equation to zero, giving

$$x^2 - \frac{gx^2}{V^2}\left[2y + \frac{gx^2}{V^2}\right] = 0, \tag{2}$$

when $x \neq 0$.

$$x^2 = \frac{2V^2}{g}\left[\frac{V^2}{2g} - y\right]$$

This is a parabola, and is known as *parabola of safety*.

When this parabola is rotated about y-axis to form a paraboloid of revolution, it determines the entire range of space accessible from O for a given velocity of projection V

From (2), we see that vertex of the parabola is $\left(0, \frac{V^2}{2g}\right)$, focus is (0,0) and the directrix is at the height V^2/g above the point of projection O.

Every trajectory, for given V, must touch the parabola of safety at some point T. The connection between two curves can be seen from the figure 4.9.

Cor. Prove that if P be any point on the line OT, the greatest value of OP for a given projection velocity V is attained when $P \equiv T$, T being the intersection of the line with parabola of safety.

Let co-ordinates of P be (x_p, y_p).

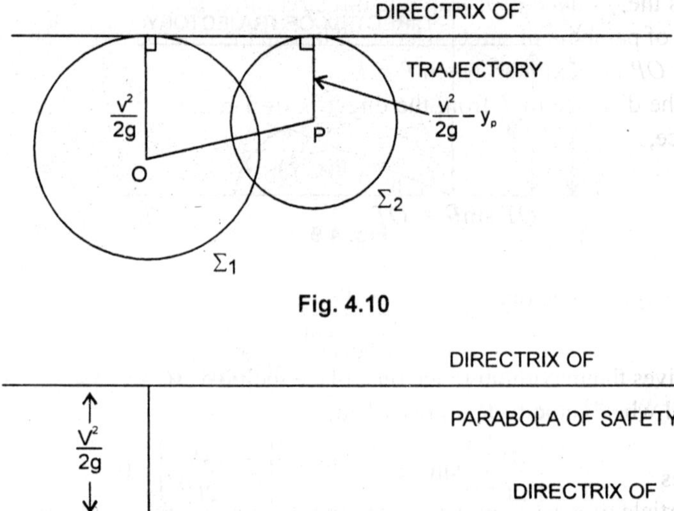

Fig. 4.10

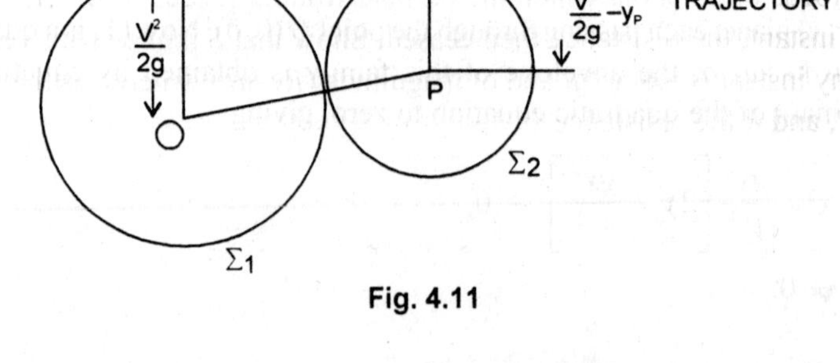

Fig. 4.11

The focus S of the trajectory lies on the circle Σ_1 having centre O and radius $\frac{V^2}{2g}$, since $\frac{V^2}{2g}$ is the distance of O from the directrix. Distance of P form the directrix is $\left(\frac{V^2}{2g} - y_p\right)$, S is on the circle Σ_2 with centre P and radius $\frac{V^2}{2g} - y_p$. Circles Σ_1, Σ_2 may intersect in a pair of distinct real points, (Fig. 4.10) a pair of imaginary points, or co-incident pair. When there is no real intersection, P cannot be reached from O with the given projection velocity V. When there are two intersections, P may be attained via two paths. When circles touch, (Fig. 4.11) the distance OP is maximum.

In Fig. (4.11)

$$OP = \left(\frac{V^2}{2g} - y_p\right) + \frac{V^2}{2g} = \frac{V^2}{g} - y_p.$$

This is the distance of P from the directrix of parabola of sefety. But O is the focus of parabola of safety. Hence, P lies on the parabola of safety, and so the range OP is maximum when $P \equiv T$,

Now the distance of T from the directrix of parabola of safety is eqaul to OT. Hence,

$$\frac{V^2}{g} - OT \sin\theta = OT$$

$$\therefore \qquad OT = \frac{V^2}{g(1 + \sin\theta)}.$$

This gives the maximum range on a plane inclined at a fixed angle θ to the horizontal when velocity V is prescribed.

Examples

1. A particle of mass m moves under gravity in a resisting medium, and S is the focus of parabola which the particle would proceed to describe if at any instant, the resistance then ceased. Show that S moves with velocity at any instant is along SP and of magnitude (Rv/mg), where v is the speed of P, and R the resistance.

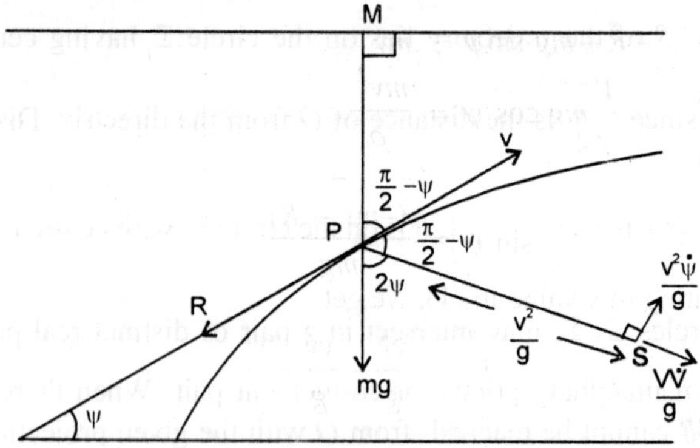

Fig. 4.12

S is the focus of the parabolic trajectory, which would be described when $R = 0$. PM is perpendicular from P on the directrix of the parabola. Then if v be the speed of the particle at P, $PS = \dfrac{v^2}{2g}$.

Let the tangent to the actual trajectory, (for which $R = 0$), make an angle ψ with horizontal at P. Since, PT instantaneously coincides with the tangent to the parabola at P, it bisects $\angle MPS$, and so

$$\angle MPS = \pi - 2\psi$$

and PS makes angle 2ψ with the downward vertical. It follows that the velocity components of S relative to P along PS and in the direction perpendicular to PS and in the sense of 2ψ increasing are respectively,

$$\frac{d}{dt}\left(\frac{v^2}{2g}\right) = \frac{v\dot{v}}{g};$$

$$\frac{v^2}{2g}\cdot 2\dot{\psi} = \frac{v^2\dot{\psi}}{g}$$

Since, I has tangential velocity v, the actual velocity components in the direction of PS and perpendicular to it in the sense of 2ψ increasing are V_1, V_2 where

$$V_1 = \frac{v\dot{v}}{g} + v\sin\psi, \tag{1}$$

$$V_2 = \frac{v^2\dot{\psi}}{g} + v\cos\psi. \tag{2}$$

For the motion of the particle in the tangential and normal directions equaltions are

$$-R - mg\sin\psi = m\dot{v} \tag{3}$$

$$mg\cos\psi = \frac{mv^2}{\rho}. \tag{4}$$

From (3),

$$\sin\psi = \frac{(-m\dot{v} - R)}{mg} = \frac{-\dot{v}}{g} - \frac{R}{mg}.$$

Substituting this value in (1), we get

$$V_1 = \frac{v\dot{v}}{g} - \frac{v\dot{v}}{g} - \frac{Rv}{mg} = -\frac{Rv}{mg} \tag{5}$$

From (4)

$$\cos\psi = \frac{v^2}{g\rho}.$$

Using this value of $\cos\psi$ in (2), we get

$$V_2 = \frac{v^2\dot{\psi}}{g} + \frac{v^3}{g\rho}. \tag{6}$$

Since, the curve is convex upwards, ψ decreases as s increases.

$$\rho = \frac{-ds}{d\psi} = \frac{-ds}{dt} \div \frac{d\psi}{dt} = \frac{-v}{\dot{\psi}}.$$

Putting value of ρ in (6), we get

$$V_2 = \frac{v^2\dot{\psi}}{g} - \frac{v^3}{g}\cdot\frac{\dot{\psi}}{v} = 0.$$

Thus the motion of S is along $\overrightarrow{SP}$ and velocity is $-Rv/mg$.

2. A particle is projected at very high velocity, whose horizontal component is u, in a medium whose resistance, per unit mass is $k \times (\text{velocity})^2$. If the angle of elevation is small, show that the approximate equation of the trajectory is

$$y = x\tan\alpha - \frac{g}{4k^2u^2}(e^{2kx} - 2kx - 1).$$

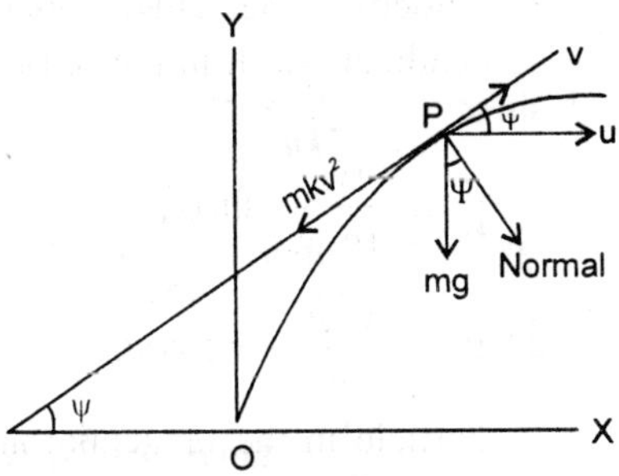

Fig. 4.13

Let v be the velocity of the particle at $P(x, y)$ at time t, and the tangent makes an angle ψ with x-axis, then equations of motion are

$$m\ddot{x} = -mkv^2 \cos\psi \tag{1}$$

$$m\ddot{y} = -mg - mkv^2 \sin\psi \tag{2}$$

First equation can be written as

$$\ddot{x} = -kv\dot{x}$$

$$\therefore \quad (\ddot{x} + k\dot{s}\dot{x}) = 0$$

$$\therefore \quad \frac{d}{dt}(\dot{x}e^{ks}) = 0$$

$$\therefore \quad \dot{x}e^{ks} = const.$$

Initially,

$$s = 0,\ \dot{x} = u = V\cos\alpha,$$

$$\therefore \quad const. = u$$

$$\therefore \quad \dot{x}e^{ks} = u = V\cos\alpha,$$

but $s \cong x$ if angle of elevation α is small.

$$\therefore \quad \dot{x}e^{kx} = u = V\cos\alpha \tag{3}$$

Similarly from (2), we get

$$\frac{d}{dt}(\dot{y}e^{ks}) = -ge^{ks}$$

$$\therefore \quad \frac{d}{dx}(\dot{y}e^{ks})\cdot\frac{dx}{dt} = -ge^{ks}$$

$$\therefore \quad \frac{d}{dx}(\dot{y}e^{ks}) = -\frac{g}{\dot{x}}e^{ks} = -\frac{g}{u}e^{2ks}$$

$$\therefore \quad \left[\dot{y}e^{kx}\right]_0^x \cong -\frac{g}{u}\int_0^x e^{2kx}\,dx \cong -\frac{g}{u}\left[\frac{e^{2kx}}{2k}\right]_0^x$$

$$\therefore \quad \dot{y}e^{kx} - V\sin\alpha = -\frac{g}{2ku}(e^{2kx} - 1)$$

$$\dot{y}e^{kx} = V\sin\alpha - \frac{g}{2ku}(e^{2kx} - 1) \tag{4}$$

Dividing (4) by (3) we get

$$\frac{\dot{y}}{\dot{x}} = \frac{dy}{dx} = \tan\alpha - \frac{g}{2ku^2}(e^{2kx} - 1)$$

$$\int_0^y dy = \int_0^x \left[\tan\alpha - \frac{g}{2ku^2}(e^{2kx} - 1)\right]dx$$

$$\therefore \quad [y]_0^y = \left[x\tan\alpha - \frac{g}{2ku^2}\left(\frac{e^{2kx}}{2k} - x\right)\right]_0^x$$

$$\therefore \quad y = x\tan\alpha - \frac{g}{2ku^2}\left[\frac{e^{2kx}-1}{2k} - x\right]$$

$$\therefore \quad y \cong x\tan\alpha - \frac{g}{4k^2u^2}\left(e^{2kx} - 1 - 2kx\right).$$

3. A particle is projected under gravity in a medium whose resistanace per unit mass is kv^n, where v is the speed. If ψ is the angle between the direction of motion of P and the horizontal, show that (*i*) the horizontal velocity u is determined by

$$u^{-n} = u_0^{-n} - \frac{nk}{g}\int_{\psi_0}^{\psi} \sec^{n+1}\psi d\psi,$$

where u_0, ψ_0 are the values of u and ψ at a point P_0 of the curve,

(ii) $$\frac{d^2y}{dx^2} = -\frac{g}{u^2},$$

where x, y are the horizontal and vertical co-ordinates of P, positive y being taken upwards. If $n = 4$. Show that

$$\frac{d^3y}{dx^3} = -2gk\left[1 + \left(\frac{dy}{dx}\right)^2\right]^{3/2}.$$

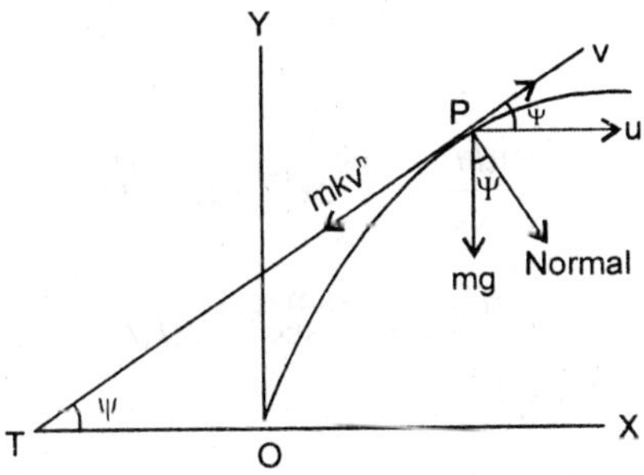

Fig. 4.14

Let the particle be projected from O, P be its position at time t, u is the horizontal component of velocity v.

$$\therefore \quad u = v \cos \psi \tag{1}$$

Equation of motion along normal and parallel to x-axis are

$$\frac{mv^2}{\rho} = mg \cos \psi$$

$$\therefore \quad \frac{v^2}{\rho} = g \cos \psi \tag{2}$$

and

$$m \frac{du}{dt} = -mkv^n \cos \psi$$

$$\therefore \quad \frac{du}{dt} = -kv^n \cos \psi. \tag{3}$$

Now ψ decreases as s increases

$$\therefore \quad \rho = -\frac{ds}{d\psi}.$$

Substituting value of ρ in (2), we get

$$v^2\left(-\frac{d\psi}{ds}\right) = g \cos \psi$$

$$\therefore \quad v\frac{d\psi}{dt} = -g \cos \psi \tag{4}$$

From (3), we get

$$\frac{du}{d\psi} \cdot \frac{d\psi}{dt} = -kv^n \cos \psi$$

Using (4), we get

$$\frac{du}{d\psi} = \frac{k}{g} v^{n+1}$$

Using (1)

$$\frac{du}{d\psi} = \frac{k}{g} u^{n+1} \sec^{n+1} \psi$$

$$\therefore \quad \frac{du}{u^{n+1}} = \frac{k}{g} \sec^{n+1} \psi d\psi.$$

Integrating, we get $\displaystyle\int_{u_0}^{u} \frac{du}{u^{n+1}} = \frac{k}{g}\int_{\psi_0}^{\psi} \sec^{n+1} \psi d\psi$

$$\therefore \qquad u^{-n} = u_0^{-n} - \frac{nk}{g}\int_{\psi_0}^{\psi} \sec^{n+1}\psi d\psi.$$

Now

$$\frac{dy}{dx} = \tan\psi$$

$$\therefore \qquad \frac{d^2y}{dx^2} = \sec^2\psi\cdot\frac{d\psi}{dx} = \sec^2\psi\cdot\frac{d\psi}{dt}\cdot\frac{dt}{dx}$$

$$= \frac{1}{u}\sec^2\psi\frac{d\psi}{dt}$$

$$= -\frac{1}{u}\sec^2\psi\frac{g}{v}\cos\psi = -\frac{g}{u}\cdot\frac{1}{v\cos\psi} = -\frac{g}{u^2}$$

When $n = u$

$$\frac{du}{d\psi} = \frac{k}{g}u^5\sec^5\psi.$$

Now

$$\frac{d^2y}{dx^2} = -\frac{g}{u^2}$$

$$\therefore \qquad \frac{d^3y}{dx^3} = \frac{2g}{u^3}\cdot\frac{du}{dx} = \frac{2g}{u^3}\cdot\frac{du}{d\psi}\cdot\frac{d\psi}{dx}$$

$$= \frac{2g}{u^3}\cdot\frac{du}{d\psi}\cdot\frac{d\psi}{dt}\cdot\frac{dt}{dx}$$

$$= -2gk\sec^3\psi$$

$$= -2gk\left[1+\left(\frac{dy}{dx}\right)^2\right]^{3/2}.$$

4.6 CONSTRAINED MOTION

Now we consider the motion of a single particle constrained to move on curves or surfaces. The constraint may be one sided, as for example when a heavy particle slides on the inside of a spherical surface and is free to break contact with the surface on the inside of the sphere but cannot get outside. Then there will be a normal pressure inwards exerted by the sphere on the particle as long as contact persists, and the pressure will vanish at the point where the particle leaves the surface. On the other hand if the constraint is two sided, for example when the particle moves on a table or a bead moves along a wire, then the normal reaction may vanish and change the sign, but particle continues to move along the prescribed path. When the particle is smooth, there is no friction between it and the constraint, there is a normal reaction which does no work during the motion of the particle. In these circumstances, if the external forces are conservative the principle of conservation of energy can be applied.

Examples

1. A horizontal smooth fixed plane, which has the shape of catenary $y = c\cosh\frac{x}{c}$ and carries a small ring which is attracted towards the directrix with a force proportional to the distance y from the directrix. Prove that the motion of the ring is S. H. M. and that the normal reaction varies inversely as y^2.

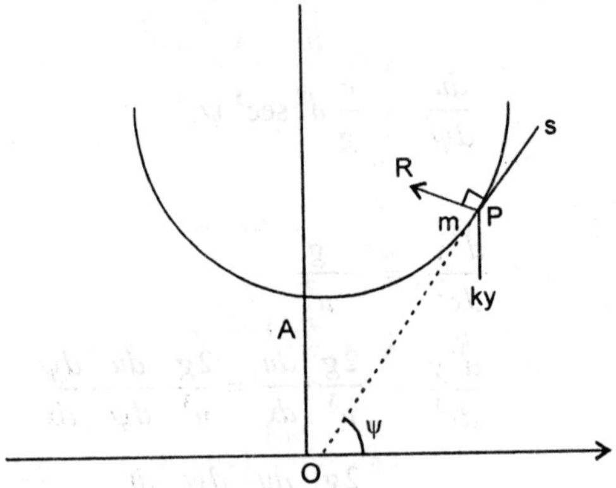

Fig. 4.15

Let A be the vertex of the catenary and the ring be at P (x, y), and ky the attraction towards the directrix, R the normal reaction of the wire on the ring at P.

Resolving tangentially

$$m\ddot{s} = -ky\sin\psi \tag{1}$$

We know for catenary

$$s = c\tan\psi, y = c\sec\psi,$$

$$\therefore \quad y\sin\psi = (c\sec\psi)\sin\psi = c\tan\psi = s$$

Putting this value in (1), we get

$$m\ddot{s} = -ks$$

$$\therefore \quad \ddot{s} = -\frac{k}{m}s \tag{2}$$

Since, $k > 0$, this represents S. H. M. of period $2\pi\sqrt{\frac{m}{k}}$.

Resolving normally, we get

$$m\frac{\dot{s}^2}{\rho} = R - ky\cos\psi \qquad (3)$$

Multiplying (2) by $2\dot{s}$ and integrating, we get

$$\dot{s}^2 = A - \frac{k}{m}s^2.$$

(3) gives

$$R = \frac{m\left[A - \frac{k}{m}s^2\right]}{\rho} + kc$$

$$= kc + \frac{mA}{\rho} - \frac{kc^2\tan^2\psi}{\rho}.$$

Now

$$\rho = \frac{ds}{d\psi} = c\sec^2\psi$$

$$\therefore \qquad R = kc + \frac{mA\cos^2\psi}{c} - kc\sin^2\psi$$

$$= \left(kc + \frac{mA}{c}\right)\cos^2\psi$$

$$= \left(kc + \frac{mA}{c}\right)\frac{c^2}{y^2}.$$

2. Discuss the motion of a heavy particle, placed on the outside of a smooth circle in a verticle plane.

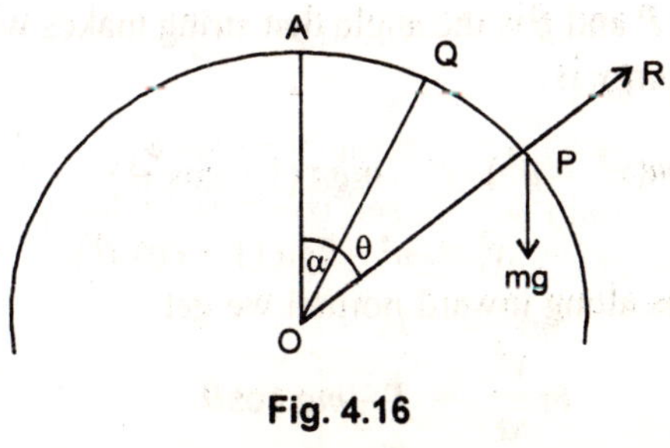

Fig. 4.16

If the particle starts from point Q at an angular distance α from the highest point A and let a be the radius of the circle and v be the velocity at P whose angular distance from A is θ. Using principle of conservation of energy, we get

$$v^2 = 2ga(\cos\alpha - \cos\theta)$$

Resolving forces along the inward normal, we have

$$\frac{mv^2}{a} = mg\cos\theta - R,$$

where R is the outward reaction of the curve.

$$\therefore \quad R = mg\cos\theta - 2mg(\cos\alpha - \cos\theta)$$

$$= mg(3\cos\theta - 2\cos\alpha).$$

$$R = 0, \text{ when } \cos\theta = \frac{2}{3}\cos\alpha$$

i.e., the particle flies off the curve when $\theta = \cos^{-1}\left(\frac{2}{3}\cos\alpha\right)$.

3. Discuss the motion in a vertical plane of a heavy particle attached by a fine string to fixed point.

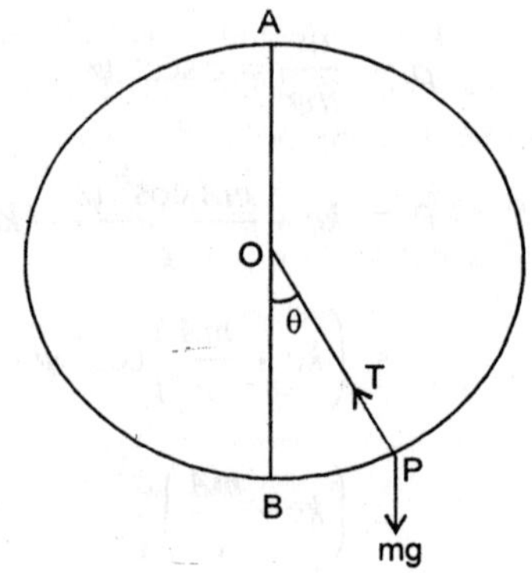

Fig. 4.17

Suppose the particle starts with velocity u from its lowest point B. If v is the velocity at P and θ is the angle that string makes with the vertical, the equation of energy is

$$\frac{1}{2}m(v^2 - u^2) = -mga(1 - \cos\theta)$$

$$\therefore \quad v^2 = u^2 - 2ga(1 - \cos\theta) \qquad (1)$$

Resolving forces along inward normal we get

$$m\frac{v^2}{a} = T - mg\cos\theta \qquad (2)$$

where T is the tension of the string.

From (1) and (2), we get

$$T = m\left(3g\cos\theta - 2g + \frac{u^2}{a}\right) \tag{3}$$

To find the height of ascent, we put $v = 0$ in (1) and get

$$2ga\cos\theta = 2ga - u^2 \tag{4}$$

and by putting $T = 0.$ in (3), we get

$$3ga\cos\theta = 2ga - u^2 \tag{5}$$

We have the following cases:

(*i*) If $u^2 < 2ga$, the string does not reach the horizontal position and the tension does not vanish.

(*ii*) If $u^2 = 2ga$, the string just reaches the horizontal position, the tension vanishes for $\theta = \frac{\pi}{2}$ and the particles swings through a quadrant on each side of the vertical.

(*iii*) If $2ga < u^2 < 5ga$, we find that there is a value of θ, an obtuse angle, given by (5), smaller than that given by (4), so that string becomes slack before the velocity vanishes and the particle will fall away from the circular path and move in a parabola till the string again becomes taut.

(*iv*) If $u^2 = 5ag$, the tension just vanishes in the highest position, but v does not vanish, so that circular motion persists.

(*v*) If $u^2 > 5ga$, neither v nor T vanishes.

This is an example of one sided constraint, if instead of the problem of a particle attached to a string, we consider that of a bead sliding on a wire, we find that for all values of u^2 greater than $4ga$, the bead will describe the complete circle.

. A fixed wire is in the shape of the cardioid $r = a(1 + \cos\theta)$, the initial line being downward vertical. A small ring of mass m can slide on the wire and is attached to the point $r = 0$ of the cycloid by an elastic string of natural length a and modulus $4mg$. If the particle is released from rest, when the string is horizontal, show that

$$a\dot{\theta}^2(1 + \cos\theta) - g\cos\theta(1 - \cos\theta) = 0.$$

Show further that $\theta = \pi/3$ is a position of stable equilibrium and that the period of small oscillations about this position is $2\pi\sqrt{\frac{2a}{g}}$.

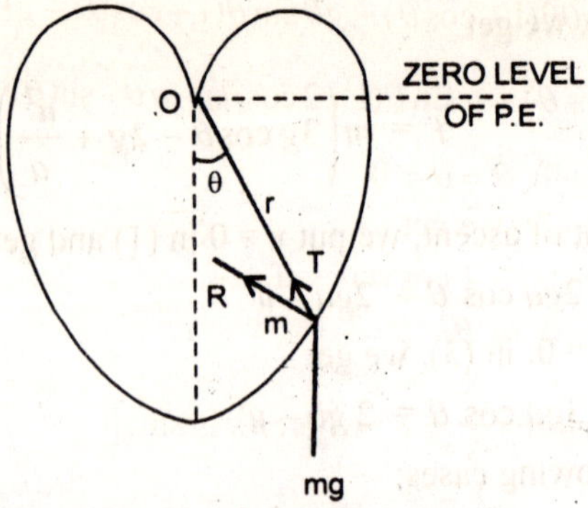

Fig. 4.18

Let the ring be at the point $P(r, \theta)$ at time t, the velocity components are $\dot{r}, r\dot{\theta}$ in the radial and transverse directions.

K. E. of the bead $= \frac{1}{2}m(\dot{r}^2 + r^2\dot{\theta}^2)$

$$= \frac{1}{2}ma^2\left[\sin^2\theta\dot{\theta}^2 + (1+\cos\theta)^2\dot{\theta}^2\right]$$

$$= ma^2\dot{\theta}^2(1+\cos\theta).$$

The gravitational P.E. of the particle (referred to the horizontal through C as zero level) is

$$= -mgr\cos\theta = -mga\cos\theta(1+\cos\theta).$$

The elastic P.E. of the string is

$$\frac{1}{2}(4mg)\frac{(r-a)^2}{a}$$

$$= 2mga\cos^2\theta.$$

Since, the forces are conservative,

$$\text{K.E.} + \text{P.E.} = \text{const.}$$

Hence,

$$-mga\cos\theta\,(1+\cos\theta) + 2mga\cos^2\theta + ma^2\,\dot{\theta}^2\,(1+\cos\theta) = \text{const.}$$

$$\therefore \quad -g\cos\theta(1-\cos\theta) + a\dot{\theta}^2(1+\cos\theta) = \text{const.}$$

Initially, $\theta = \frac{\pi}{2}, \dot{\theta} = 0,$

$\therefore$ const. $= 0$.

Hence,

$$a\dot{\theta}^2\,(1+\cos\theta) - g\cos\theta\,(1-\cos\theta) = 0. \quad (1$$

Differentiating w.r. to t, we get

$$-a\dot{\theta}^3 \sin\theta + 2a\dot{\theta}\ddot{\theta}(1+\cos\theta) + g\dot{\theta}\sin\theta(1-\cos\theta) - g\dot{\theta}\cos\theta\sin\theta = 0. \quad (2)$$

$$\therefore 2a\ddot{\theta}(1+\cos\theta) + g\sin\theta(1-2\cos\theta) - a\dot{\theta}^2\sin\theta = 0.$$

For equilibrium $\dot{\theta} = \ddot{\theta} = 0$.

Hence, from (2), we get

$$\sin\theta(1-2\cos\theta) = 0.$$

$\therefore$ either $\theta = 0$, or $\theta = \pi/3$.

In (2), we put $\theta = \frac{\pi}{3} + \varepsilon$, wherer ε is small.

$$\cos\theta = \cos\left(\frac{\pi}{3}+\varepsilon\right) = \cos\frac{\pi}{3}\cos\varepsilon - \sin\frac{\pi}{3}\sin\varepsilon \cong \frac{1}{2} - \frac{\sqrt{3}}{2}\varepsilon,$$

$$\sin\theta = \sin\left(\frac{\pi}{3}+\varepsilon\right) = \sin\frac{\pi}{3}\cos\varepsilon + \cos\frac{\pi}{3}\sin\varepsilon \cong \frac{\sqrt{3}}{2} + \frac{\varepsilon}{2}.$$

Retaining terms of first order of small quantity in $\varepsilon, \dot{\varepsilon}, \ddot{\varepsilon}$, we get from (2)

$$2a\ddot{\varepsilon}\left(\frac{3}{2} - \frac{\sqrt{3}}{2}\varepsilon\right) + g\left(\frac{\sqrt{3}}{2} + \frac{\varepsilon}{2}\right)\frac{\sqrt{3}\varepsilon}{2} = 0$$

$$\therefore \quad \ddot{\varepsilon} + \frac{g}{2a}\varepsilon = 0,$$

$$\therefore \quad \ddot{\varepsilon} = -\frac{g}{2a}\varepsilon. \quad (3)$$

This shows that for small perturbations about $\theta = \pi/3$, the motion is approximately equal to S.H.M. Thus $\theta = \pi/3$, is a stable equilibrium, the period of small oscillations being $2\pi(a/g)^{1/2}$.

4.7 ANGULAR MOMENTUM OF A PARTICLE

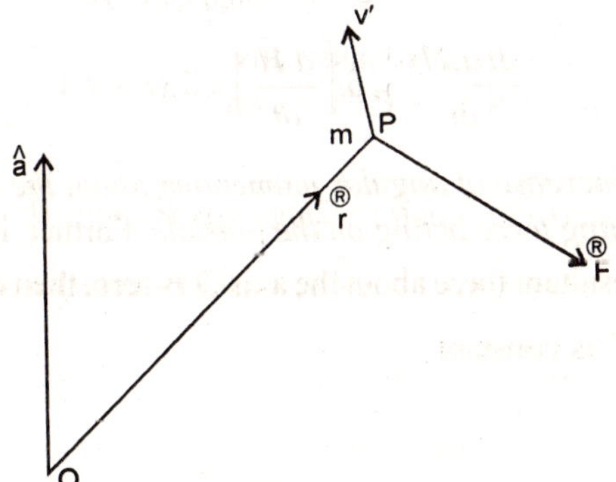

Fig. 4.19

Let a particle of mass m be moving with velocity v at time t, and it is at point P, such that for fixed origin O, $\overrightarrow{OP} = \vec{r}$, so that $\vec{v} = \dfrac{d\vec{r}}{dt}$. Then the (vector) *angular momentum or moment of momentum* of the particle is defined to be

$$\vec{H} = \vec{r} \times (m\vec{v})\cdot$$

If $\vec{F}$ is the force acting on m for m constant

$$\frac{d\vec{H}}{dt} = \vec{r} \times \left(m\frac{d\vec{v}}{dt}\right) + \vec{v} \times (m\vec{v})$$

$$= \vec{r} \times \vec{F}, \text{ since } \vec{F} = m\frac{d\vec{v}}{dt}\cdot$$

Thus the *rate of change of angular momentum of a particle is equal to the moment of the resulting force which acts on it.*

In the absence of external forces $\vec{F} = \vec{0}$

$$\therefore \qquad \frac{d\vec{H}}{dt} = \vec{0}$$

$$\therefore \qquad \vec{H} = \text{const.}$$

Thus *we see that when a particle is moving under no forces, angular momentum is constant.* This is known as *law of conservation of angular momentum.*

If $\hat{a}$ specifies unit vector in some axes drawn through O, then (physical) angular momentum about this axis is the resolute of $\vec{H}$ about it *i.e.*, $\hat{a}.\vec{H}$. Hence, the rate of increase of angular momentum about the axis $\hat{a}$ is

$$\frac{d(\hat{a}.\vec{H})}{dt} = \hat{a}.\left(\frac{d\vec{H}}{dt}\right) = \hat{a}.(\vec{r} \times \vec{F}).$$

Thus *rate of increase of angular momentum about the axis is equal to the moment of resulting force acting on the particle.* Further, it follows that if the moment of the resultant force about the axis $\hat{a}$ is zero, then the physical angular momentum $\hat{a}.\vec{H}$ is constant.

Examples

1. The interior surface of a bowl is given by $a^3z = a^2r^2 + r^4$, where a is positive constant and r denotes the distance from OZ which is fixed verically upwards. A particle of unit mass moves on this surface under gravity and surface reaction. Show that in steady motion in a horizontal circle of radius a, the moment of momentum about OZ is $(6ag)^{1/2}$. If the particle suffers a small disturbance while describing this path, determine the period of small oscillations produced.

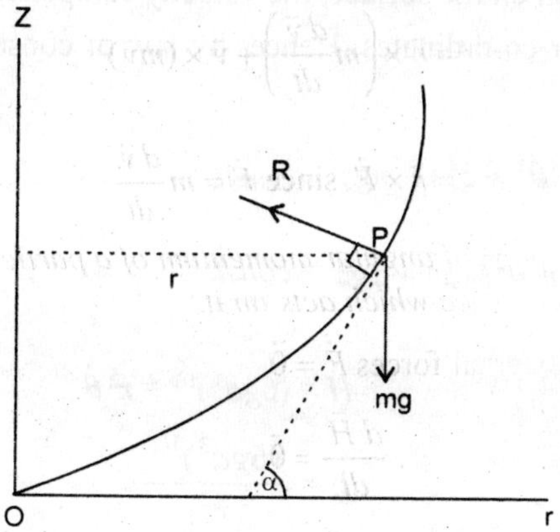

Fig. 4.20

Fig. 4.20 shows the meridian section of the surface through P.
The surface is

$$a^3 z = a^2 r^2 + r^4,$$

$$\therefore \qquad a^3 \frac{dz}{dt} = 2a^2 r \frac{dr}{dt} + 4r^3 \cdot \frac{dr}{dt}$$

$$\therefore \qquad \frac{\dot{z}}{\dot{r}} = \frac{2(a^2 r + 2r^3)}{a^3} = \tan\alpha,$$

where α is the angle which the tangent at P makes with the horizontal, when $r = a$, $z = 2a$.

$$\therefore \qquad \tan\alpha = 6. \qquad (1)$$

For this case resolving forces along radius of the circle, we get

$$R \sin\alpha = \frac{mu^2}{a}. \qquad (2)$$

Resolving forces along OZ, we have

$$R \cos\alpha = mg. \qquad (3)$$

From (2) and (3), we get

$$\tan\alpha = \frac{u^2}{ag} \tag{4}$$

From (1) and (4)

$$u^2 = 6ag.$$

Moment of momentum about OZ is

$$H = a.(6ag)^{1/2} = (6ga^3)^{1/2}. \tag{5}$$

In general motion on the surface, the velocity components are $(\dot{r}, r\dot{\theta}, \dot{z})$ in cylindrical polar co-ordinates. Hence, by law of conservation of energy, we have

$$\frac{1}{2}m(\dot{r}^2 + r^2\dot{\theta}^2 + \dot{z}^2) + mgz = const.$$

$$\therefore \quad \dot{r}^2 + r^2\dot{\theta}^2 + \dot{z}^2 + 2gz = const. \tag{6}$$

Now

$$H = (6ga^3)^{1/2} = r^2\dot{\theta}$$

$$\dot{\theta} = \frac{(6ga^3)^{1/2}}{r^2}.$$

Putting this value of $\dot{\theta}$ in (6), we get

$$\dot{r}^2 + \frac{6ga^3}{r^2} + \dot{z}^2 + 2gz = const.$$

Putting

$$z = \frac{a^2r^2 + r^4}{a^3}$$

we get

$$\dot{r}^2 + \frac{6ga^3}{r^2} + \frac{(2a^2r + 4r^3)^2}{a^6}\cdot\dot{r}^2 + \frac{2g(a^2r^2 + r^4)}{a^3} = const.$$

$$\therefore \quad (a^6 + 4a^4r^2 + 16r^6 + 16a^2r^4)\dot{r}^2/a^6 + \frac{6ga^3}{r^2} + 2g(a^2r^2 + r^4)/a^3 = const.$$

Differentiating w.r. to t, we get

$$\frac{(a^6 + 4a^4r^2 + 16r^6 + 16a^2r^4)}{a^6}\cdot\ddot{r} + \frac{(8a^4r + 96r^5 + 64a^2r^3)}{a^6}\dot{r}^2 - \frac{6ga^3}{r^3} + \frac{2g(2r^3 + a^2r)}{a^3} = 0$$

When particle suffers a small disturbance, let

$$r = a + \varepsilon, \text{ where } \varepsilon \text{ is small,}$$

$$\therefore \qquad \dot{r} = \dot{\varepsilon}, \ddot{r} = \ddot{\varepsilon}$$

Retaining only first order terms of $\varepsilon, \dot{\varepsilon}, \ddot{\varepsilon}$, from last equation, we get

$$37\ddot{\varepsilon} = \frac{-32g}{a}\varepsilon, \text{ approx.}$$

This represents S.H.M. so period of small oscillation is given by

$$T = 2\pi\sqrt{\frac{37a}{32g}}.$$

2. A particle of mass m can slide freely on a light wire bent into the shape of the curve $a^3y = x^4$, which can rotate about the y-axis, which is vertically upwards. If a steady motion takes place with the particle at $x = a$, find the angular velocity Ω of the wire. The particle is slightly disturbed from its motion at $x = a$. Prove that the period of oscillation about this position is 2π $(17a'/8g)^{1/2}$, If the wire is constrained to rotate with angular velocity Ω, find the period of oscillation if the wire is not constrained but moves freely about y-axis.

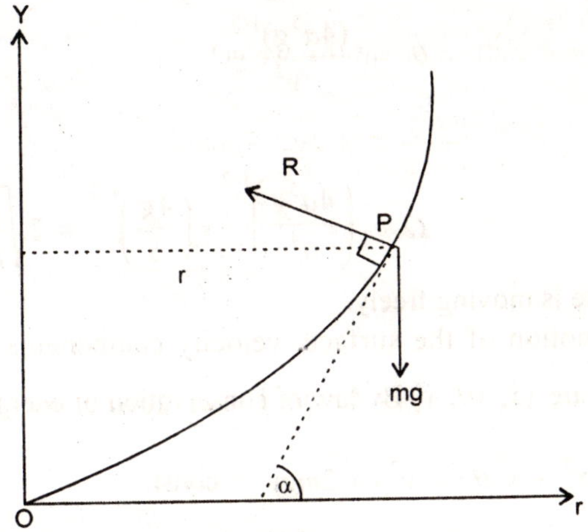

Fig. 4.21

Curve is

$$a^3 y = x^4$$

$$\therefore \qquad a^3\frac{dy}{dx} = 4x^3$$

$$\therefore \qquad \frac{dy}{dx} = \frac{4x^3}{a^3}$$

When $x = a, y = a$

$$\tan\alpha = 4 \qquad (1)$$

For this case, resolving forces along radius of the circle and y-axis, we get

$$R\cos\alpha = mg \qquad (2)$$

$$R\sin\alpha = m\frac{u^2}{a} \qquad (3)$$

$$\therefore \qquad \tan\alpha = \frac{u^2}{ag}$$

But from (1)

$$\therefore \qquad \tan\alpha = 4.$$

$$\therefore \qquad u^2 = 4ag \qquad (4)$$

Angular momentum

$$H = ma\left(\sqrt{4ag}\right) = m(4a^3g)^{1/2} \qquad (5)$$

But

$$H = mx^2\dot{\theta} = m(4a^3g)^{1/2}$$

$$\therefore \qquad \dot{\theta} = \frac{(4a^3g)^{1/2}}{x^2}. \qquad (6)$$

At $x = a$

$$\Omega = \left(\frac{4a^3g}{a^4}\right)^{1/2} = \left(\frac{4g}{a}\right)^{1/2} = 2\sqrt{\frac{g}{a}} \qquad (7)$$

When particle is moving freely.

In general motion of the surface, velocity components in cylindrical co-ordinates are $(\dot{x}, x\dot{\theta}, \dot{y})$ By law of conservation of energy

$$\frac{1}{2}m(\dot{x}^2 + x^2\dot{\theta}^2 + \dot{y}^2) + 2mgy = \text{const.}$$

$$\therefore \qquad \dot{x}^2 + x^2\dot{\theta}^2 + \dot{y}^2 + 2gy = \text{const.}$$

Substituting the value of $\dot{\theta}$ from (6), we get

$$\dot{x}^2 + \frac{4ga^3}{x^2} + \dot{y}^2 + 2gy = \text{const.}$$

$$\therefore \qquad \dot{x}^2 + \frac{4ga^3}{x^2} + \frac{16x^6}{a^6}\dot{x}^2 + \frac{2gx^4}{a^3} = \text{const.}$$

$$\therefore \quad \left(1+\frac{16x^6}{a^6}\right)\dot{x}^2+\frac{4a^3g}{x^2}+\frac{2gx^4}{a^3} = \text{const.}$$

Differentiating w.r. to t, we have

$$\left(1+\frac{16x^6}{a^6}\right)\ddot{x}+\frac{48x^5}{a^6}\dot{x}^2-\frac{4a^3g}{x^3}+\frac{4gx^3}{a^3} = 0 \qquad (8)$$

When the particle is slightly disturbed, let

$$x = a+\varepsilon, \text{ where } \varepsilon \text{ is small.}$$

$$\therefore \quad \dot{x} = \dot{\varepsilon}, \ddot{x}=\ddot{\varepsilon}$$

Retaining first order terms of ε, $\dot{\varepsilon}$, $\ddot{\varepsilon}$, (8) becomes

$$17\ddot{\varepsilon}-4g\left(1-\frac{3\varepsilon}{a}\right)+4g\left(1+\frac{3\varepsilon}{a}\right) = 0$$

$$\therefore \quad \ddot{\varepsilon} = -\frac{24g}{17a}\varepsilon$$

This represents S.H. M. of period $2\pi\sqrt{\dfrac{17a}{24g}}$.

When the particle is constrained to move on the curve with constant velocity Ω, the law of conservation of energy gives

$$\dot{x}^2-x^2\Omega^2+\dot{y}^2+2gy = \text{const.}$$

$$\therefore \quad \dot{x}^2-x^2\Omega^2+\frac{16x^6}{a^6}\dot{x}^2+\frac{2gx^4}{a^3} = \text{const.}$$

$$\therefore \quad \left(1+\frac{16x^6}{a^6}\right)\dot{x}^2-x^2\Omega^2+\frac{2gx^4}{a^3} = \text{const.}$$

Differentiating w. r. to t, we get

$$\left(1+\frac{16x^6}{a^6}\right)\ddot{x}+\frac{48x^5}{a^6}\dot{x}^2-x\Omega^2+\frac{4gx^3}{a^3} = 0 \qquad (9)$$

When the particle is disturbed, let

$$x = a+\varepsilon, \text{ where } \varepsilon \text{ is small.}$$

Retainting only first order terms of ε, $\dot{\varepsilon}$, $\ddot{\varepsilon}$, (9) reduces to

$$17\ddot{\varepsilon}-\Omega^2(a+\varepsilon)+4g\left(1+\frac{3\varepsilon}{a}\right) = 0.$$

Substituting the value of Ω from (7), we get

$$17\ddot{\varepsilon} - 4g\left(1+\frac{\varepsilon}{a}\right) + 4g\left(1+\frac{3\varepsilon}{a}\right) = 0.$$

$$\therefore \qquad \ddot{\varepsilon} = -\frac{8g}{17a}$$

This represents S.H. M. of period $2\pi\sqrt{\dfrac{17a}{8g}}$.

3. A particle of mass m moves on a smooth horizontal table and is connected by a light inextensible string passing through a smooth hole O in the table to a particle B of mass m, which moves along the vertical through O. Initially B is at rest and A is distant a from O, moving with velocity $\sqrt{\dfrac{ga}{3}}$ perpendicular to OA. Show that in subsequent motion the distance $r = OA$ lies between a and $a/2$. Show also that the tension in the string is $\dfrac{1}{6}mg\left(3+\dfrac{a^3}{r^3}\right)$.

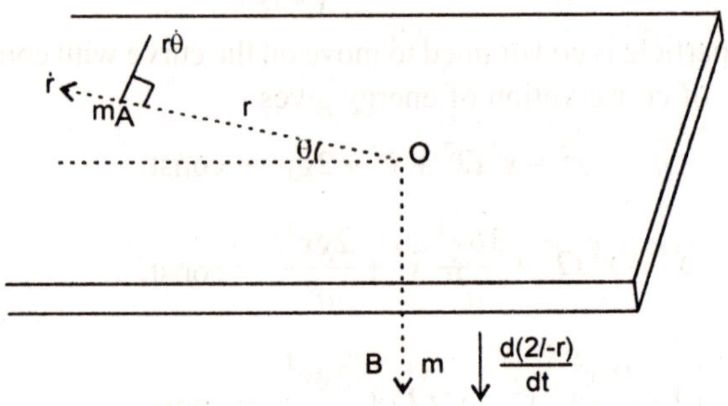

Fig. 22

Let T be the tension of the string and $OA = r$ at any instant. The table being smooth, there is no frictional force at A perpendicular to OA and tension is the only force in the direction AO.
Equations of motion for A are

$$\frac{m}{r}\frac{d}{dt}\left(r^2\frac{d\theta}{dt}\right) = 0$$

$$\therefore \qquad r^2\frac{d\theta}{dt} = \text{const.} = \text{initial value}$$

$$\therefore \qquad r^2 \frac{d\theta}{dt} = a\sqrt{\frac{ga}{3}} \qquad (1)$$

and

$$m\left[\frac{d^2r}{dt^2} - r\left(\frac{d\theta}{dt}\right)^2\right] = -T \qquad (2)$$

If l is the length of the string, equation of motion for B is

$$m\frac{d^2(l-r)}{dt^2} = mg - T$$

$$\therefore \qquad \frac{d^2r}{dt^2} = T - mg \qquad (3)$$

Subtracting (2) and (3) and using (1), we have

$$2T - mg = mr\left(\frac{d\theta}{dt}\right)^2 = mr \cdot \frac{a^3g}{3} \cdot \frac{1}{r^4} = \frac{mga^3}{3r^3}$$

$$\therefore \qquad T = \frac{1}{6}mg\left(3 + \frac{a^3}{r^3}\right).$$

Again adding (2) and (3) and using (1), we get

$$2m\frac{d^2r}{dt^2} - mr\left(\frac{d\theta}{dt}\right)^2 = -mg$$

$$2\frac{d^2r}{dt^2} - \frac{ga^3}{3r^3} = -g.$$

Multiplying this equation by $\frac{dr}{dt}$ and integrating, we get

$$\left(\frac{dr}{dt}\right)^2 + \frac{ga^3}{6r^2} = -gr + C.$$

Initially

$$r = a, \frac{dr}{dt} = 0,$$

$$\therefore \qquad C = \frac{7ga}{6}.$$

$$\therefore \quad \therefore \left(\frac{dr}{dt}\right)^2 + gr + \frac{ga^3}{6r^2} = \frac{7}{6}ag.$$

For maximum value of r, $\frac{dr}{dt} = 0$

$$\therefore \qquad \frac{ga^3}{6r^2} + gr - \frac{7}{6}ag = 0$$

$$6r^3 - 7ar^2 + a^3 = 0$$

$$\therefore \quad (r-a)(2r-a)(3r+a) = 0.$$

$$\therefore \quad r = a,\ r = a/2,\ r = -a/3.$$

But $r = -a/3$. is not possible, hence either $r = a/2$ or $r = a$. Hence, in subsequent motion r lies between a and $a/2$.

4. A particle P is moving in a plane under an attraction of magnitude $\frac{2}{r^3}$ per unit mass towards fixed point O in the plane, and at $t = 0$, $r = 2$, the radial and transverse components of velocity are $\sqrt{\frac{3}{2}}$ and l respectively. Show that $\ddot{r} = \frac{2}{r^3}$ and r as function of t.

Let

$$r = f(\theta).$$

Equations of motion along radial and transverse directions are

$$\frac{d^2r}{dt^2} - r\left(\frac{d\theta}{dt}\right)^2 = -\frac{2}{r^3} \tag{1}$$

$$\frac{1}{r}\frac{d}{dt}\left(r^2\frac{d\theta}{dt}\right) = 0. \tag{2}$$

From (2),

$$r^2\frac{d\theta}{dt} = const. = A.$$

Initially

$$t = 0,\ r = 2,\ \frac{dr}{dt} = \sqrt{\frac{3}{2}},\ r\dot{\theta} = 1$$

$$\therefore \quad A = 2,$$

$$\therefore \quad r^2\frac{d\theta}{dt} = 2,$$

$$\therefore \quad \dot{\theta} = \frac{2}{r^2}.$$

Substituting this value of $\dot{\theta}$ in (1), we get

$$\frac{d^2r}{dt^2} = \frac{2}{r^3}.$$

Multiplying by $2\frac{dr}{dt}$ and integrating, we obtain

$$\left(\frac{dr}{dt}\right)^2 = -\frac{2}{r^2} + B,$$

Initially

$$r = 2, \frac{dr}{dt} = \sqrt{\frac{3}{2}},$$

$\therefore$ $$B = 2$$

$\therefore$ $$\left(\frac{dr}{dt}\right)^2 = 2\left(1 - \frac{1}{r^2}\right),$$

$\therefore$ $$\frac{rdr}{\sqrt{r^2 - 1}} = \sqrt{2}dt$$

$\therefore$ $$\sqrt{r^2 - 1} = t\sqrt{2} + C.$$

Initially

$$t = 0, r = 2,$$

$$C = \sqrt{3}$$

$\therefore$ $$\therefore \sqrt{r^2 - 1} = t\sqrt{2} + \sqrt{3}$$

$\therefore$ $$r^2 = 1 + \left(t\sqrt{2} + \sqrt{3}\right)^2$$

$\therefore$ $$r = \sqrt{1 + (t\sqrt{2} + \sqrt{3})^2}$$

4.8 MOTION OF PARTICLE ON A ROUGH PLANE CURVE

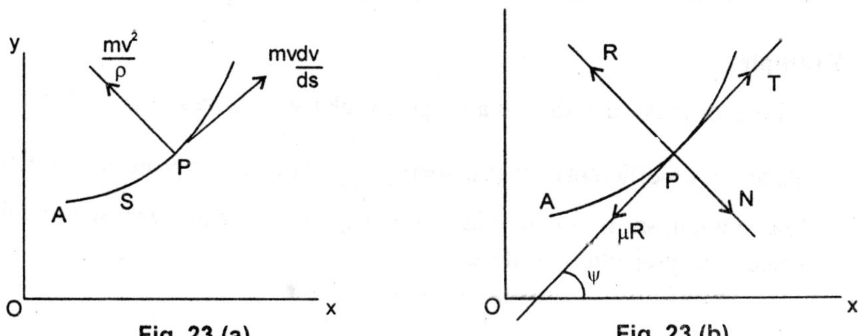

Fig. 23 (a) Fig. 23 (b)

Let s be the arc measured from a fixed point A on the curve C in the direction of motion, to the position of the particle P whose mass is m. Let R be the pressure measured inwards, μ the co-efficient of friction and T, N the components of the external force on the particle along the tangent and outward normal. The equations of motion are

$$mv\frac{dv}{ds} = T - \mu R, \tag{1}$$

$$m\frac{v^2}{\rho} = R - N. \tag{2}$$

In the case in which a heavy particle is sliding upwards on a curve in a vertical plane. Then equations reduce to

$$mv\frac{dv}{ds} = -mg\sin\psi - \mu R \tag{3}$$

$$m\frac{v^2}{\rho} = R - mg\cos\psi \tag{4}$$

where ψ is the angle which the tangent at the point P to the curve C makes with the horizontal.

Eliminating R from (3) and (4), we get

$$v\frac{dv}{ds} + \mu\frac{v^2}{\rho} = -g(\sin\psi + \mu\cos\psi),$$

$$\therefore \quad \frac{dv^2}{d\psi} + 2\mu v^2 = -2g\rho(\sin\psi + \mu\cos\psi) \tag{5}$$

From the intrinsic equation of the curve $s = f(\psi)$, we get $\rho = f'(\psi)$, and the equation (5) has an integrating factor $e^{2\mu\psi}$.

Example

A bead of mass m slides on a rough circular wire of radius a in a vertical plane, the co-efficient of friction being $\frac{1}{\sqrt{2}}$. The bead is projected from the lowest point, show that in order that it may just reach the highest point, the velocity of projection must be

$$\sqrt{\frac{1}{3}\sqrt{6}gae^{\sqrt{2}\left(\frac{\pi}{2}+\alpha\right)}},$$

where α is the acute angle $\tan^{-1}\sqrt{2}$.

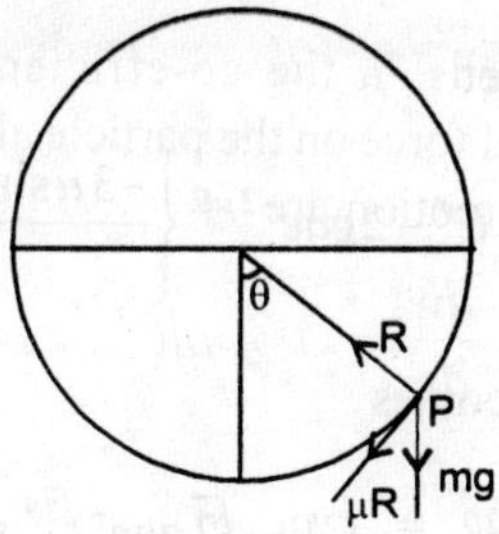

Fig. 4.24

If v is the velocity when θ is the angular distance of the bead from its lowest point, in the early stage the equations of motion are

$$mv\frac{dv}{ds} = -mg\sin\theta - \mu R \tag{1}$$

$$m\frac{v^2}{a} = R - mg\cos\theta. \tag{2}$$

Eliminating R, we get

$$\frac{dv^2}{d\theta} + 2\mu v^2 = -2ga\,(\sin\theta + \mu\cos\theta). \tag{3}$$

On integrating this equation, we get

$$v^2 e^{2\mu\theta} = C - 2gae^{2\mu\theta}\left\{\frac{3\mu\sin\theta + (2\mu^2 - 1)\cos\theta}{4\mu^2 + 1}\right\} \tag{4}$$

Putting $\mu = \frac{1}{\sqrt{2}}$, this becomes

$$v^2 e^{\sqrt{2}\theta} = v_0^2 - \sqrt{2}ga.e^{\sqrt{2}\theta}\sin\theta. \tag{5}$$

where v_0 is the value of v, when $\theta = 0$.
We see that v decreases when $\theta = \pi$, so there will be an angle θ between 0 and π for which $\cos\theta = -\frac{v^2}{ag}$, and from (2), for this value of θ pressure R vanishes and changes sign. From this point onward (3) ceases to hold, but (1) still holds good, and instead of (2), we have

$$m\frac{v^2}{a} = -R - mg\cos\theta \tag{6}$$

which with the help of (1) leads to

$$\frac{dv^2}{d\theta} - 2\mu v^2 = -2ga\,(\sin\theta + \mu\cos\theta). \tag{7}$$

On integrating this, we get

$$v^2 e^{-2\mu\theta} = C' - 2gae^{-2\mu\theta}\left\{\frac{-3\mu\sin\theta + (2\mu^2 - 1)\cos\theta}{4\mu^2 + 1}\right\} \quad (8)$$

Putting $\mu = \frac{1}{\sqrt{2}}$, this becomes

$$v^2 e^{-\sqrt{2}\theta} = C' + \sqrt{2}gae^{-\sqrt{2}\theta}\sin\theta.$$

But v vanishes, when $\theta = \pi$, $\therefore C' = 0$

and in the latter stage of motion

$$v^2 = \sqrt{2}ga\sin\theta \quad (9)$$

But the motion being continuous, the values of v^2 given by (5) and (9) are the same at the point at which R vanishes, *i.e.*, when v^2 is also $-ga\cos\theta$. Substituting this value in (9), we get

$$-\cos\theta = \sqrt{2}\sin\theta$$

$$\therefore \quad \frac{\cos\theta}{-\sqrt{2}} = \frac{\sin\theta}{1} = \frac{1}{\sqrt{3}},$$

so that $\theta = \frac{\pi}{2} + \alpha$, where α is the actute angle $\tan^{-1}\sqrt{2}$. Then from (5) and (9), we get

$$v_0^2 = \sqrt{2}gae^{\sqrt{2}\theta}\sin\theta = \frac{2}{3}\sqrt{6}gae^{\sqrt{2}\left(\frac{\pi}{2}+\alpha\right)}.$$

4.9 MOTION ON CYCLOID

The base of a rough cycloid arc is horizontal and vertex downwards. Discuss the motion of a bead down the arc.

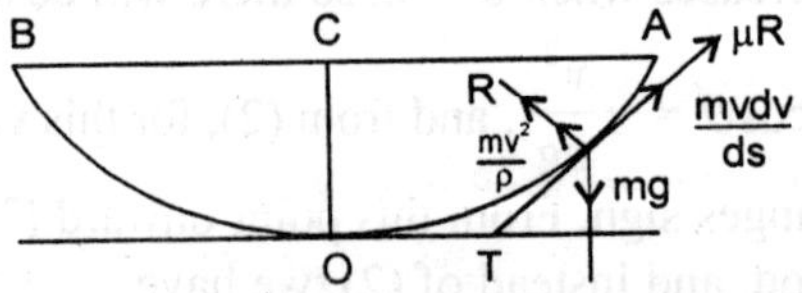

Fig. 4.25

Intrinsic equation of a cycloid is

$$s = 4a\sin\psi.$$

$$\rho = \frac{ds}{d\psi} = 4a\cos\psi.$$

Length of the arc is $8a$. For cusp $\psi = \frac{\pi}{2}$ and for vertex $\psi = 0$.

Since, the bead is sliding down the arc the frictional force acts upwards. Hence, equations of motion along tangent and inward normal are

$$mv\frac{dv}{ds} = \mu R - mg\sin\psi, \tag{1}$$

$$m\frac{v^2}{\rho} = R - mg\cos\psi. \tag{2}$$

Eliminating R between (1) and (2), we get

$$mv\frac{dv}{ds} = \mu(m\frac{v^2}{\rho} + mg\cos\psi) - mg\sin\psi$$

$$\therefore \quad \frac{1}{2}\frac{dv^2}{ds} = \mu\frac{v^2}{\rho} + g(\mu\cos\psi - \sin\psi).$$

Putting $\rho = \frac{ds}{d\psi} = 4a\cos\psi$, we get

$$\frac{dv^2}{ds}\cdot\frac{ds}{d\psi} - 2\mu v^2 = 2g4a\cos\psi(\mu\cos\psi - \sin\psi)$$

$$\therefore \quad \frac{dv^2}{d\psi} - 2\mu v^2 = 8ag\cos\psi(\mu\cos\psi - \sin\psi). \tag{3}$$

This is a linear differential equation, whose I.F. is $e^{-2\mu\psi}$. Multiplying both sides by I.F. and integrating, we get

$$v^2e^{-2\mu\psi} = 8ag\int e^{-2\mu\psi}\cos\psi(\mu\cos\psi - \sin\psi)d\psi + C. \tag{4}$$

Put

$$e^{-\mu\psi}(\mu\cos\psi - \sin\psi) = z.$$

$$\therefore \left[-\mu e^{-\mu\psi}(\mu\cos\psi - \sin\psi) + e^{-\mu\psi}(-\mu\sin\psi - \cos\psi)\right]d\psi = dz.$$

$$\therefore -e^{-\mu\psi}\left[1+\mu^2\right]\cos\psi d\psi = dz.$$

$$\therefore \quad -e^{-\mu\psi}\cos\psi d\psi = -\frac{dz}{1+\mu^2}. \tag{5}$$

Hence, from (4) and (5), we get

$$v^2 e^{-2\mu\psi} = 8ag\int e^{-\mu\psi}\cos\psi . e^{-\mu\psi}(\mu\cos\psi - \sin\psi)d\psi + C$$

$$= 8ag\int z\cdot\frac{-dz}{1+\mu^2} + C$$

$$= -\frac{8ag}{1+\mu^2}\cdot\frac{z^2}{2} + C.$$

Putting for z, we get

$$v^2 e^{-2\mu\psi} = -\frac{4ag}{1+\mu^2}.e^{-2\mu\psi}(\mu\cos\psi - \sin\psi)^2 + C. \tag{6}$$

The constant of integration C can be found from initial conditions.

Examples

1. A bead slides on a smooth cycloidal wire

$$x = a(2\psi + \sin 2\psi), \quad y = a(1 - \cos\psi),$$

which is fixed with the (x, y) plane vertical and the axes of x, y inclined at angles α and $\frac{\pi}{2} - \alpha$ with the upwards vertical. The bead moves from rest at the vertex. Prove that it does not reach the cusp, if $2\tan\alpha > \pi$, but moves to rest when

$$\sin\alpha + 2\psi\sin\alpha + \sin(2\psi - \alpha) = 0.$$

From equations

$$x = a(2\psi + \sin 2\psi),\ y = a(1 - 2\cos\psi),$$

we get

$$(ds)^2 = 4a^2\left[(1+2\cos 2\psi)^2 + \sin^2 2\psi\right]$$

$$= 16a^2\cos^2\psi$$

$$\therefore \qquad ds = 4a\cos\psi.$$

Integrating, we get

$$s = 4a\sin\psi.$$

Now the particle will move from vertex O from rest to the left of O if

$$s = -4a\sin\psi. \tag{1}$$

Equation of motion along the tangent is

$$\ddot{s} = -g\sin\left(\frac{\pi}{2} - \alpha + \psi\right) = -g\cos(\psi - \alpha).$$

$$\therefore \quad v\frac{dv}{ds} = -g\cos(\psi-\alpha),$$

$$\therefore \quad v\frac{dv}{d\psi}\cdot\frac{d\psi}{ds} = -g\cos(\psi-\alpha),$$

$$\therefore \quad v\frac{dv}{d\psi} = -g\cos(\psi-\alpha)\frac{ds}{d\psi}. \qquad (2)$$

Substituting the value of $\frac{ds}{d\psi}$ in (2), we get

$$v\frac{dv}{d\psi} = 4ga\cos\psi\ \cos(\psi-\alpha),$$

$$= 2ag[\cos(2\psi-\alpha)+\cos\alpha].$$

Integrating, we have

$$v^2 = 2ag[\sin(2\psi-\alpha)+2\psi\cos\alpha] + const.$$

Inititally

$$v = 0,\ \psi = 0,$$

$$\therefore \quad const. = 2ag\sin\alpha.$$

$$\therefore \quad v^2 = 2ag[\sin(2\psi-\alpha)+2\psi\cos\alpha+\sin\alpha] \qquad (3)$$

At cusp

$$\psi = -\frac{\pi}{2}$$

$$\therefore \quad v^2 = 2ag[\sin(-\pi-\alpha)-\pi\cos\alpha+\sin\alpha,]$$

$$= 2ag[2\sin\alpha-\pi\cos\alpha].$$

If $2\sin\alpha > \pi\cos\alpha$ *i.e.*, $2\tan\alpha > \pi$, then v changes its direction, so bead will come back before reaching cusp. Now the dead will come to rest, if $v = 0$, *i.e.*,

$$\sin(2\psi-\alpha)+2\psi\cos\alpha+\sin\alpha = 0.$$

2. A rough cycloid has its plane vertical and the line joining the cusps horizontal. A heavy particle slides down the curve from rest at a cusp and comes to rest again at the point on the other side of the vertex where the tangent is inclined at 45° to the vertical. Show that the co-efficient of friction satisfies the equation

$$3\mu\pi + 4\log(1+\mu) = 2\log 2.$$

From equation (6) of Art. 4.9, we have

$$v^2 e^{-2\mu\psi} = \frac{-4ag}{1+\mu^2} e^{-2\mu\psi} (\mu \cos\psi - \sin\psi)^2 + C.$$

We are given that $v = 0$, when $\psi = \frac{\pi}{2}$ at the cusp. Also $v = 0$, when $\psi = -45°$ as we are given that tangent is inclined at 45° to the horizontal and since it is on the other side of the vertex, ψ is to be measured in clock wise direction and hence we have taken $\psi = -45°$.

$$\therefore 0 = \frac{-4ag}{1+\mu^2} e^{-\mu\pi} (0-1)^2 + C. \qquad \text{for } \psi = \frac{\pi}{2}.$$

$$0 = \frac{-4ag}{1+\mu^2} e^{\mu\pi/2} (\mu \frac{1}{\sqrt{2}} + \frac{1}{\sqrt{2}})^2 + C. \qquad \text{for} = -\frac{\pi}{4}$$

Subtracting these relations, we get

$$e^{-\mu\pi}.1 = e^{\mu\pi/2} \frac{(\mu+1)^2}{2},$$

$$\therefore \qquad 2 = e^{3\mu\pi/2} (\mu+1)^2.$$

Taking log of both sides, we get

$$log2 = \frac{3\pi}{2}\mu + 2\log(\mu+1),$$

$$\therefore \qquad 2log\,2 = 3\mu\pi + 4\log(\mu+1).$$

EXERCISE

1. A ring which can slide on a rough circular wire in a vertical plane is projected from the lowest point with such velocity as will just take it to the horizontal diameter ; if the ring returns so the lowest point, show that its velocity on arrival is to its velocity on departure as

$$\sqrt{1-2\mu^2 - 3\mu e^{-\mu\pi}} : \sqrt{1-2\mu^2 + 3\mu e^{\mu\pi}}.$$

where μ is the co-efficient of friction.

2. A heavy particle slides from a cusp, down the arc of a rough cycloid, the axis of which is vertical. Prove that its velocity at the vertex to its velocity at the same point when the cycloid is smooth, as $\sqrt{e^{-\mu\pi} - \mu^2} : \sqrt{1+\mu^2}$, where μ is the co-efficient of friction.

3. A particle of mass m is projected in a medium offering resistance $mg\left(\frac{\omega}{\omega_0}\right)^2$ when the angular velocity is ω. If u denotes the horizontal component of velocity, s the length of the arc traversed, and ψ the inclination of the direction of motion to the vertical, show that

$$\frac{du}{ds} = \frac{-gu}{\omega_0^2}, \frac{d\psi}{ds} = \frac{g}{u^2}\sin^3\psi,$$

and determine the (s, ψ) equation of the path.

4. If a particle of mass m be acted upon by equal constant force mf in the directions of the tangent and the normal to the path, and if the resistance be $mfkv^2$, prove that intrinsic equation of the path is

$$e^{2k^2 fs} - 1 = k^2u^2(e^{2\psi} - 1),$$

Where u is the velocity of projection.

5. A particle is projected with velocity u in a direction inclined at an angle α to the horizontal. Determine the horizontal and vertical components of displacement after time t, on the assumption that gravity is the only force acting.
Any number of particles are projected from the same point at the same instant. Prove that (i) at any subsequent instant the directions of their velocities are concurrent (ii) the point of concurrence rises with constant acceleration.

6. Drops of water are thrown tangentially off the rim of a wheel of radius a which is rotating in a vertical plane about its axis which is fixed and horizontal, with angular velocity $\omega > (g/a)^{1/2}$.
Show that if a horizontal ceiling at height $h > a$ above the axis of the wheel, is not to be spotted with water,

$$a\omega < \left\{gh + g(h^2 + a^2)^{1/2}\right\}^{1/2},$$

the effects of air resistance being neglected.

7. The propulsive force on a heavy rocket moving in a vertical plane is so adjusted that its speed v is constant. Rectangular axes ox, oy are taken in the plane of motion with origin O at highest point of the path. If any change of mass is neglected, prove that equation of the path is

$$y = \frac{v^2}{g}\log\sec(gx/v^2).$$

Hence, deduce that the horizontal range cannot exceed $(\pi v^2/g)$, and that to approach this range, the direction of projection must be vertical.

8. A smooth wire in the shape of a cardioid $r = a\,(1 + \cos\theta)$ is fixed in a horizontal plane. One end of a light elastic string of natural length a and

modulus mg, is fastened to a bead of mass m which slides on the wire. The bead is slightly disturbed from rest at the vertex of the wire. Prove that the angular velocity of the string when it is inclined at an angle θ to the axis of the cardioid is $(g/a)^{1/2} \tan \frac{\theta}{2}$, and that the reaction of the wire on the rod is instantaneously zero when $\cos\theta = \frac{3}{5}$.

9. A particle P is describing a parabola freely under gravity. Show that the angular velocity of the line joining P to the focus is $2gu/v^2$, where v is the velocity of the particle at P and u is the horizontal component of v.
10. Particles are projected simultaneously from a point under gravity in various directions with velocity V. Prove that at any subsequent time t, they will lie on a sphere of radius Vt, and determine the motion of the centre of the sphere.
11. A particle is projected at time $t = 0$ in a fixed vertical plane from a given point S with given velocity $\sqrt{2ga}$ of which the upward vertical component is v. Show that at time $t = 2a/v$ the particle is on a fixed parabola (independent of v), that its path touches the parabola, and that its direction of motion is then perpendicular to its direction of projection.
12. A fort is on the edge of a cliff of height h. Show that there is an annular region in which the fort is out of range of the ship, but the ship is not out of the range of the fort, of area $8\pi kh$, where $\sqrt{2gk}$ is the velocity of shells used by both.
13. A shell fired with velocity V at an elevation θ hits at an airship at height h which is moving horizontally away from the gun with velocity v. Show that if
$$(2V\cos\theta - v)(V^2\sin^2\theta - 2gh)^{1/2} = vV\sin\theta,$$
the shell might also have hit the ship if the latter had remianed stationary in the position it occupied when the gun was actually fired.
14. An aeroplane is flying with constant velocity v and at constant height h. Show that if gun is fired point blank at the aeroplane after it has passed directly over the gun and when its angle of elevation as seen from the gun is α the shell will hit the aeroplane, provided
$$2(V\cos\alpha - v)\, v\tan^2\alpha = gh,$$
where V is the velocity of the shot, the path being assumed to be parabolic.
15. A particle moving under gravity describes a parabola of vertex A and focus S, the velocity at A being u. Prove that, when the particle is at P, where $SP = r$, the components of its velocity along and perpendicular to SP are equal to the vertical and horizontal components of its velocity.

Show that the component in the direction of SP of the acceleration of P is

$$g-\frac{u^2}{r}.$$

16. A particle P moving along a horizontal straight line has a retardation ku, where u is the velocity at time t. When $t = 0$, the particle is at O and has velocity u_0. Show that $u_0 - u$ is perpendicular to OP, and that the final value of OP is u_0 / k.
A particle subjected to gravity describes a curved path in a resisting medium which causses retardation $k \times$ velociy. Show that the resultant acceleration f has a constant value and that $f = f_0 e^{-kt}$, where f_0 is the acceleration when $t = 0$.
17. A particle slides down the surface of a smooth sphere of radius a, being slightly displaced from rest at the highest point. Find where it will leave the sphere, and show that it will afterwards describe a parabola of latus rectum $16a/27$, and that it will strike the horizontal plane through the lowest point of the sphere at a distance $5(\sqrt{5}+4\sqrt{2})a/27$ from the vertical diameter.
18. A particle starts from rest at any point P in the arc of a smooth cycloid whose axis is vertical and vertex downwards, prove that the time of descent to the vertex is $\pi\sqrt{\dfrac{a}{g}}$ where a is the radius of generating circle .
Show that if particle is projected from P downwards along the curve with velocity equal to that with which it reaches A when starting from rest at p, it will now reach A in half the time taken in the preceding case.
19. A particle is constrained to move under gravity in a parabola with axis vertical and vertex upwards. Show that pressure on the curve is numerically $m(u_0^2 - gl)\rho, u_0$, is the velocity at the vertex, $2l$ the latus rectum and P the radius of curvature at any point?
What is the meaning of special case $u_0^2 = gl$?
20. A smooth narrow tube, with open ends, in the form of one complete arc of a cycloid $s = 4\,a \sin\psi$ is fixed with the tangent at the mid-point A ($\psi = 0$) of the arc vertical. A particle of mass m is placed in the tube at A and is connected to a light inextensible string. The string whose length exceeds that of the arc, lies in the tube and after passing through the upper end of the tube is attached to a particle of mass, m' which hangs freely. If the system is released from rest and $m' > m$, show that

$$(m+m')\dot{s}^2 = 8m'ga\sin\psi - 2mga\sin\psi - 4mga\psi,$$

and that if the reaction between m and the tube vanishes when $\psi = \pi / 4$, then

$$\frac{m'}{m} = \frac{(2\sqrt{2} + 1)(\pi + 4)}{14}.$$

If $m' < m$, show that m will emerge from the lower end of the tube provided $4m' < \pi m$.

21. A particle can slide on a rough cycloidal arc fixed with its axis vertical and vertex downwards. If the resistance is proportional to the speed, show that oscillations of the particle about the lowest point are isochronous.

22. P, Q are two particles each of mass m, connected by a light inextensible string of length $2l$, which passes through a small hole O in a smooth horizontal table. P is free to slide on the table and Q hangs freely. Initially OQ is of length l, and P is projected at right angles to OP with velocity $(8gl / 3)^{1/2}$. Show that in the ensuing motion Q will just reach the table.

23. A smooth wire is bent into the form $y = \sin x$ and placed in vertical plane with the axis of x horizontal. A bead of mass m slides down the wire from rest at $x = \pi / 4$. Show that pressure on the wire as the bead passes the origin is mg/ $\sqrt{2}$ and find the pressure as it passes through $x = -\pi / 2$.

24. A particle of mass m is attached by a string to a point on the circumference of a fixed circular cylinder of radius a whose axis is vertical, the string being initially horizontal and tangential to the cylinder. The particle is projected with velocity v at right angles to the string along a smooth horizontal plane so that the string winds itself round the cylinder. Show that
   (*i*) the velocity of the particle is constant,
   (*ii*) the tension in the string is inversely proportional to the length which remains straight at any moment,
   (*iii*) if the initial length of the string is l and greastest tension that string can bear is T, the string will break when it has turned through an angle

$$\frac{l}{a} - \frac{mv^2}{aT}.$$

25. A small ring fits loosely on a rough spoke (length a) of a wheel which can turn about a horizontal axle and the ring is originally at rest in contact with the lowest point of the rim, if the wheel is now made to revolve with uniform angular velocity ω, prove that the angle θ through which the wheel will turn before the ring slides is given by the equation

$$\cos(\theta - \lambda) / \cos \lambda + \omega^2 / ag = 0,$$

where λ is the angle of friction.

5

ORBITAL MOTION

5.1 CENTRAL ORBIT

If a particle is describing an orbit under the action of a force directed towards a fixed point.

(*i*) ***Prove that orbit must be a plane curve.***

At any instant the particle is moving in the plane through the tangent to the path and the fixed point and the only force acting on lies in this plane, therefore the particle continuous to move in this plane.

Alternatively

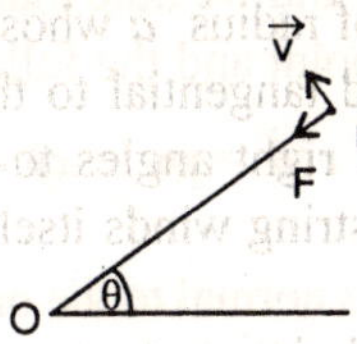

Fig. 5.1

If a particle of constant mass be moving under the action of a force $\vec{F}$ along $\overline{PO}$, where O is a fixed point, let $\vec{v}$ be the velocity of the particle P. Let $\hat{a}$ be the unit vector normal to the plane of $\vec{F}$ and $\vec{v}$. Hence, $\vec{a}.\vec{F} = O$ *i.e.*, $\hat{a} \cdot \frac{d\vec{v}}{dt} = 0$, which means the component of acceleration in the direction of $\hat{a}$ is zero. Further $\hat{a} . \vec{v} = 0$, *i.e.* the component of velocity in the direction of $\hat{a}$ is zero. Since, components of velocity and acceleration in the direction of $\hat{a}$ are zero, the particle must continue to move in the plane determined by $\vec{F}$ and $\vec{v}$, *i.e.*, the orbit produced by a central force is a plane curve.

(ii) Prove that rate of description of area by radius vector drawn from the fixed point to the particle is constant.

There is no force at right angle to the radius vector, hence, the transverse component of acceleration is zero, i.e.,

$$\frac{1}{r}\frac{d}{dt}(r^2\dot{\theta}) = 0,$$

$$\therefore \qquad r^2\,\dot{\theta} = \text{constant.}$$

Since, a sectorial element of area is $\frac{1}{2}r^2\delta\theta$, to first order hence, rate of description is

$$\frac{1}{2}r^2\frac{\delta\theta}{\delta t},$$

hence,

$$\lim_{\delta t\to 0}\frac{1}{2}r^2\frac{\delta\theta}{\delta t} = \frac{1}{2}r^2\dot{\theta} = \text{Const.}$$

It follows that $r^2\dot{\theta}$ = *twice the rate of description of the area by the radius vector.*

Usually $r^2\dot{\theta}$ is denoted by h, and we note that $r^2\dot{\theta}$ is also the moment of the velocity about the fixed point and the following forms are equivalent

$$h = r^2\dot{\theta} = pv = x\dot{y} - y\dot{x}$$

Definitions

Positions of P for which OP is normal to the orbit are called *apses* and the corresponding radius vector is called an *apse line*. The angle between two consecutive apse lines is called an *apsidal angle.* The point of nearest approach to O is called the *perihelion*, the point of farthest approach, if finite, is called the *aphelion.*

5.2 GIVEN THE ORBIT AND THE CENTRE OF FORCE, DETERMINE THE LAW OF FORCE

There are two common formulae for the law of force, one for use when the orbit is given by its polar co-ordinates *i.e.*, a relation between r and θ, the other for use when the orbit is given by its (r, p) equation, where r denotes the radius vector and p the perpendicular from the origin to the tangent or reciprocal co-ordinates (u, θ).

1. Reciprocal co-ordinates

If a particle moves under a central force given by

$$\vec{F} = mP(r)\,\hat{r}, \qquad (1)$$

where P is the magnitude of the force per unit mass. The equations of motion of the particle in a plane are given by

$$(\ddot{r} - r\dot{\theta}^2) = P(r) \tag{2}$$

$$\frac{1}{r}\frac{d}{dt}(r^2\dot{\theta}) = 0. \tag{3}$$

If P is negative, the central force is attractive and if P is positive, the central force is repulsive.

From (3)

$$r^2\dot{\theta} = \text{constant} = h \text{ (say)}. \tag{4}$$

Since, m is constant. The constant h may be evaluated from the initial conditions.

From (4),

$$\dot{\theta} = \frac{h}{r^2}. \tag{5}$$

Substituting $r = \dfrac{1}{u}$, we have $\dot{\theta} = hu^2$.

Also

$$\frac{dr}{dt} = \frac{d}{dt}\left(\frac{1}{u}\right) = \frac{d}{du}\left(\frac{1}{u}\right)\frac{du}{dt} = -\frac{1}{u^2}\frac{du}{d\theta}\frac{d\theta}{dt} = -h\frac{du}{d\theta}$$

$$\frac{d^2r}{dt^2} = \frac{d}{dt}\left(\frac{dr}{dt}\right) = \frac{d}{dt}\left(-h\frac{du}{d\theta}\right) = \frac{d}{d\theta}\left(-h\frac{du}{d\theta}\right)\frac{d\theta}{dt}$$

$$= -h\frac{d^2u}{d\theta^2}\cdot hu^2 = -h^2u^2\frac{d^2u}{d\theta^2}.$$

Substituting the values of $\dfrac{dr}{dt}$ and $\dfrac{d^2r}{dt^2}$, in (2), we get

$$-h^2u^2\frac{d^2u}{d\theta^2} - \frac{1}{u}(hu^2)^2 = P$$

$$\therefore \quad \frac{d^2u}{d\theta^2} + u = -\frac{P}{h^2u^2} \tag{6}$$

This is the differential equation of the orbit under a central force in reciprocal co-rdinates (u, θ).

Now

$$v^2 = \dot{r}^2 + r^2\dot{\theta}^2$$

$$= h^2\left(\frac{du}{d\theta}\right)^2 + h^2u^2 = h^2\left[u^2 + \left(\frac{du}{d\theta}\right)^2\right].$$

II. Pedal equation of orbit under a central force.

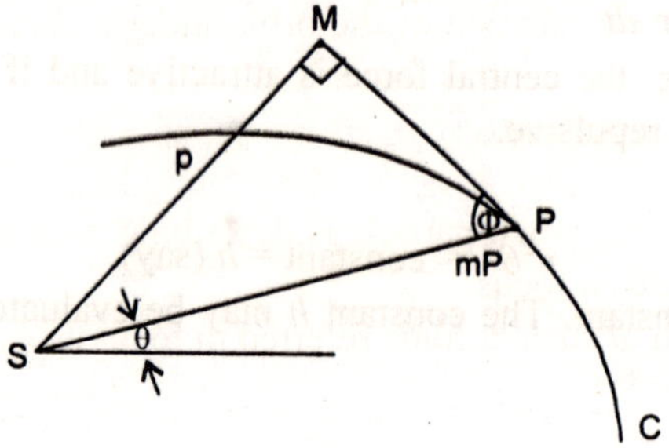

Fig 5.2

If p is the length of perpendicular from O to the tangent at P, then we have the relation

$$\frac{1}{p^2} = \frac{1}{r^2} + \frac{1}{r^4}\left(\frac{dr}{d\theta}\right)^2$$

$$= u^2 + u^4\left[\frac{d}{d\theta}\left(\frac{1}{u}\right)\right]^2,$$

since, $r = \frac{1}{u}\cdot$

$$= u^2 + u^4\left[-\frac{1}{u^2}\frac{du}{d\theta}\right]^2$$

$$= u^2 + \left(\frac{du}{d\theta}\right)^2.$$

Differentiating w.r.t. θ we get

$$\frac{-2}{p^3}\frac{dp}{d\theta} = 2u\frac{du}{d\theta} + 2\frac{du}{d\theta}\cdot\frac{d^2u}{d\theta^2},$$

$$\therefore \quad -\frac{1}{p^3}dp = udu + du\,.\frac{d^2u}{d\theta^2} = \left(u + \frac{d^2u}{d\theta^2}\right)du$$

$$= -\frac{P}{h^2u^2}du = -\frac{Pr^2}{h^2}d\left(\frac{1}{r}\right) = \frac{P}{h^2}dr$$

$$\therefore \quad \frac{h^2}{p^3}\frac{dp}{dr} = -P.$$

This is the required equation of the orbit under a central force in pedal co-ordinates.

Example

Find the law of force, when particle is describing a central conic under a central force towards a focus.

The polar equation of such a conic referred to focus as pole and the axis as the initial line is

$$\frac{l}{r} = 1 + e\cos\theta$$

$$\therefore \quad u = \frac{1+e\cos\theta}{l}$$

$$\therefore \quad \frac{du}{d\theta} = \frac{-e\sin\theta}{l}$$

$$\frac{d^2u}{d\theta^2} = \frac{-e\cos\theta}{l}.$$

Equation of the orbit is given by

$$u + \frac{d^2u}{d\theta^2} = \frac{-P}{h^2u^2}.$$

Substituting value of u and $\frac{d^2u}{d\theta^2}$ in L.H.S. only, we get

$$\frac{1+e\cos\theta - e\cos\theta}{l} = -\frac{P}{h^2u^2}$$

$$\therefore \quad P = \frac{-h^2u^2}{l} = -\frac{\mu}{r^2},$$

where

$$\mu = \frac{h^2}{l} = const.$$

Thus law of force is inverse square.

5.3 CLASSIFICATION OF ORBITS FOR AN ATTRACTIVE INVERSE SQUARE LAW

The differential equation of the particle moving under a central force P towards the origin is given by

$$\frac{h^2}{p^3}\frac{dp}{dr} = -P = \frac{\mu}{r^2}. \tag{1}$$

Integrating, we get

$$\frac{-h^2}{2p^2} = \frac{-\mu}{r} + A,$$

where A is a constant of integration.

$$\therefore \quad \frac{h^2}{p^2} = v^2 = \frac{2\mu}{r} + A' \tag{2}$$

where $A' = -2A = const.$

We know that the pedal equations of the orbits referred to focus as pole are given by

$$\frac{b^2}{p^2} = \frac{2a}{r} - 1 \text{ (ellipse)} \tag{3}$$

$$\frac{r}{p^2} = const. \text{ (parabola)} \tag{4}$$

$$\frac{b^2}{p^2} = \frac{2a}{r} + 1. \text{ (near branch of hyperbola)} \tag{5}$$

where $2a$ and $2b$ are major and minor axes respectively.

(*i*) Comapring (2) and (3), we see that
The path of the particle will be an ellipse if

$$\frac{h^2}{b^2} = \frac{2\mu}{2a} = \frac{-A'}{1}.$$

$$\therefore \quad \frac{h^2}{b^2} = \frac{\mu}{a} = -A',$$

$$\therefore \quad h^2 = \frac{\mu b^2}{a} = \mu l,\ A' = -\frac{\mu}{a}$$

The equation (2) becomes

$$v^2 = \frac{2\mu}{r} - \frac{\mu}{a} = \mu\left(\frac{2}{r} - \frac{1}{a}\right).$$

It is clear that $v^2 < \frac{2\mu}{r}$.

(*ii*) Similarly, comparing (2) and (4), we see that the path is parabola, if

$$\frac{h^2}{1} = 2\mu.consnt. = \frac{A'}{0},$$

$$\therefore \qquad h^2 = 2\mu.const., A' = 0.$$

Equation (2) becomes

$$v^2 = \frac{2\mu}{r}.$$

(*iii*) Similarly, from (2) and (5), we can prove that the path will be near branch of hyperbola, if

$$h^2 = \mu l$$

$$\therefore \qquad v^2 = \mu\left(\frac{2}{r} + \frac{1}{a}\right).$$

It is clear that $v^2 > \frac{2\mu}{r}$.

Thus, we can say that the path of the particle under inverse square law will be ellipse, parabola, or near branch of hyperbola according as

$$v^2 < \frac{2\mu}{r},\ v^2 = \frac{2\mu}{r},\ v^2 > \frac{2\mu}{r}$$

respectively,

or $A < 0$, $A = 0$, $A > 0$ respectively.

Suppose, the particle is projected with velocity V from a point at a distance c from the centre of force, then from (2) we get

$$V^2 = \frac{2\mu}{c} + A'.$$

$$\therefore \qquad A' = V^2 - \frac{2\mu}{c}.$$

A' is negative, zero, or positive according as

$$V^2 < \frac{2\mu}{c},\ V^2 = \frac{2\mu}{c},\ V^2 > \frac{2\mu}{c},$$

Therefore, the semi-major axis of the ellipse or hyperbola is given in terms of the initial velocity and distance by the equations

$$V^2 = \frac{2\mu}{c} \mp \frac{2\mu}{a}.$$

Also, when the initial circumstances are given, h, the moment of the velocity about the centre of force is known, then in each case the semi-latus rectum $l = \frac{h^2}{\mu}$, so that dimensions of the orbit are completely determined.

When the particle describes an ellipse of the semi-axes a, b $(a > b)$ under inverse square law of attraction $P = \frac{m\mu}{r^2}$, $(\mu > 0)$, now the area of the ellipse is πab and rate of description of the area by the radius vector SP is $h/2$, hence, the periodic time T is given by

$$T = \frac{\pi ab}{h/2} = \frac{2\pi ab}{\sqrt{\mu l}} = \frac{2\pi ab}{\sqrt{\mu}} \cdot \frac{a^{1/2}}{b} = \frac{2\pi a^{3/2}}{\sqrt{\mu}}.$$

Thus, square of periodic time is proportional to a^3 *i.e.*, cube of semi-major axis of the ellipse.

5.4 CIRCULAR ORBITS

Let a particle describe a circle of radius a with uniform velocity v, hence, it has acceleration v^2/a towards the centre. It follows that a particle can describe a circle under any constant force per unit mass f tending to the centre provided it is projected at right angles to the radius with velocity $\sqrt{af}$.

Circle with any point O as centre of force.

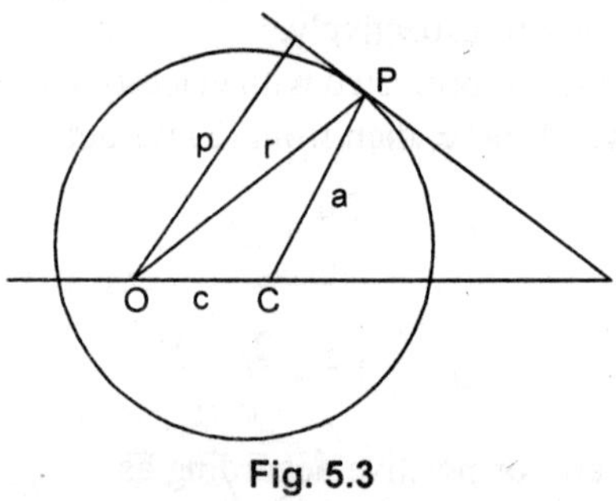

Fig. 5.3

Let a be radius and c the distance of O from the centre C. Then if P be any point on the circle and p the perpendicular from O on the tangent at P and $OP = r$, it is clear from ΔOCP that

$$c^2 = a^2 + r^2 - 2ap.$$

$$\therefore \qquad \frac{dp}{dr} = \frac{r}{a}.$$

Therefore, the force towards O under which the particle would describe the circle is

$$P = \frac{h^2}{p^3}\frac{dp}{dr} = \frac{h^3}{p^3}\cdot\frac{r}{a},$$

the velocity at P is given by

$$vp = h,$$

$$v = \frac{h}{p}.$$

In the special case in which O is on the circumference of the circle $c = a$ and $r^2 = 2ap$, so that

$$P = \frac{8a^2h^2}{r^5}, v = \frac{2ah}{r^2},$$

$$\therefore P = \frac{\mu}{r^5}, v = \frac{\sqrt{\mu/2}}{r^2}.$$

Conversely. A particle projected from a point P at a distance r from O with velocity $\frac{\sqrt{\mu/2}}{r^2}$ under the action of a force $\frac{\mu}{r^5}$ per unit mass toward O will describe a circle passing through O, and the position of the centre of the circle and therefore its radius also depends on the direction of projection; for we construct the position of C by drawing PC at right angles to the direction of projection and making the angle POC equal to OPC.

5.5 LAW VARYING AS DISTANCE

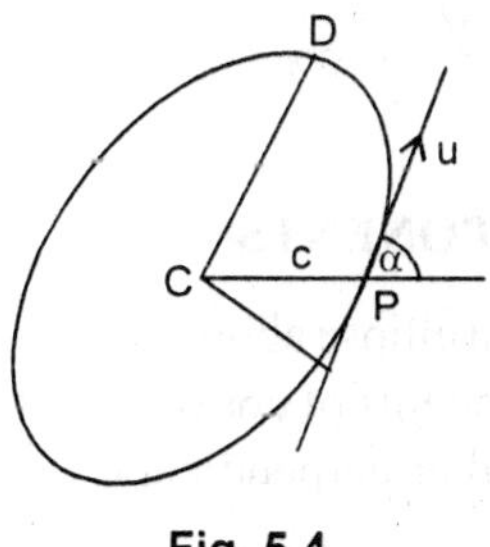

Fig. 5.4

Pedal equation of the orbit is

$$\frac{h^2}{p^3}\frac{dp}{dr} = P = \mu r.$$

Integration gives

$$\frac{h^2}{p^2} = A - \mu r^2 \tag{1}$$

But (r, p) equation of an ellipse with centre as origin is

$$\frac{a^2b^2}{p^2} = a^2 + b^2 - r^2. \tag{2}$$

From (1) and (2) we see that the orbit is in general an ellipse.

The constants involved are determined by the initial conditions. Thus, if the particle is projected with velocity u from a point at distance c from the centre of force C in a direction making an angle α with CP, we have

$$h = \text{moment of velocity about } C = \mu c \cos\alpha,$$

and since, h/p always measures the velocity, therefore substituting in (1), we get

$$u^2 = A - \mu c^2,$$

$$\therefore \quad A = u^2 + \mu c^2.$$

By comparing (1) and (2), we get

$$a^2 + b^2 = \frac{A}{\mu} = \frac{u^2}{\mu} + c^2,$$

$$a^2b^2 = \frac{h^2}{\mu} = \frac{u^2c^2\sin^2\alpha}{\mu}.$$

These equations determine the lengths of the semi-axes of the ellipse in terms of μ, u, c, α. The inclination θ of the major axis to CP is then calculated from the polar equation of the ellipse *viz.*

$$\frac{\cos^2\theta}{a^2} + \frac{\sin^2\theta}{b^2} = \frac{1}{r^2}.$$

5.6 VELOCITY COMPONENTS

When a particle describes an ellipse about a centre of force in the focus, resolve the velocity into two components of constant magnitude, one perpendicular to the radius vector and the other perpendicular to the major axis.

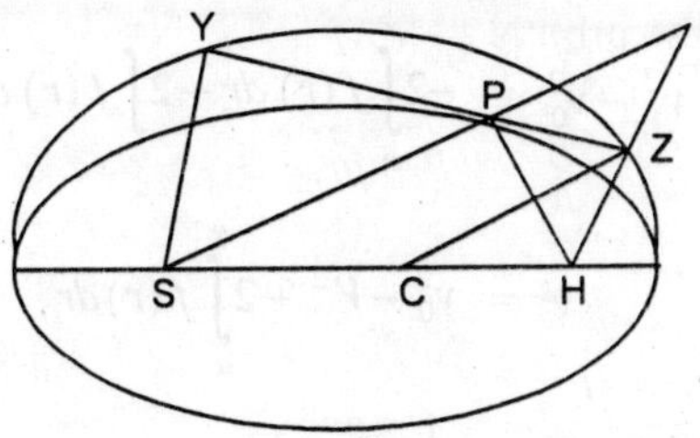

Fig. 5.5

Let S, H be the focii of an ellipse and SY, HZ perpendicular to the tangent at P. Let SP meet HZ in H', then since, the tangent is equally inclined to SP, PH, it can be easily proved that

$$HZ = ZH'$$

$\therefore$ CZ is parallel to SP.

Again the velocity at P in the ellipse is $\dfrac{h}{SP} = \dfrac{h.\,HZ}{b^2}$, and its direction is perpendicular to HZ.

But vector HZ is equivalent to $HC + CZ$; therefore velocity at P may be represented by components:

$h.\ HC/b^2$ perpendicular to HC,

and hCZ/b^2 perpendicular to CZ.

Since, $HC = ae$, where e is the eccentricity and $CZ = a$ and is perpendicular to SP, therfoer velocity components are hae/b^2 or $e\mu/h$ perpendicular to the major axis and ha/b^2 or μ/h perpendicular to SP.

5.7 VELOCITY FROM INFINITY

In connection with the central orbits the phase 'velocity from infinity' at any point of the orbit means the velocity that a particle would acquire if it moved from infinity to that particular point under the action of an attractive force in accordance with law associated with the orbit.

Let f be the acceleration towards the origin, equation of motion along tangential direction is

$$v\frac{dv}{ds} = -f\frac{dr}{ds},$$

the velocity from infinity is given by

$$\left|v^2\right|_{r=a}^{r=r} = -2\int_a^r f(r)\,dr = -2\int_a^\infty f(r)\,dr - 2\int_\infty^r f(r)\,dr\,.$$

Let v_0 be the speed at $r = a$, then

$$v^2 - v_0^2 = -2\int_a^\infty f(r)\,dr + 2\int_r^\infty f(r)\,dr$$

$$\therefore \qquad v^2 = v_0^2 - V^2 + 2\int_r^\infty f(r)\,dr$$

where

$$V^2 = 2\int_a^\infty f(r)\,dr\,.$$

This shows that when $v_0 \le V$, v is imaginary for $r = \infty$, but v is real for $r = \infty$, when $v_0 \ge V$. V is called the *escape velocity*, because for $v_0 < V$, the particle can escape to infinity.

For example, if $f = \dfrac{\mu}{r^2}$,

$$V^2 = 2\int_a^\infty f(r)dr = 2\int_a^\infty \frac{\mu}{r^2}\,dr = 2\frac{\mu}{a}\cdot$$

$$\therefore \qquad V = \sqrt{\frac{2\mu}{a}}\,.$$

Examples

1. Find the path when a particle is moving under inverse cube law.
   Let the force to the centre be μu^3, then we have

$$\frac{d^2u}{d\theta^2} + u = \frac{\mu u}{h^2},$$

$$\therefore \qquad \frac{d^2u}{d\theta^2} = \left(\frac{\mu}{h^2} - 1\right)u\,.$$

The solution of this equation is different in form according as

$$\frac{\mu}{h^2} >, =, < 1.$$

Case I. When $\dfrac{\mu}{h^2} > 1$, let $\dfrac{\mu}{h^2} - 1 = n^2$, then the equation is

$$\frac{d^2u}{d\theta^2} - n^2u = 0,$$

The solution is

$$u = Ae^{n\theta} + Be^{-n\theta},$$

where the constants A and B depend on the intial conditions, if these are such that either A or B is zero, the path is an equiangular spiral.

Case II. When $\frac{\mu}{h^2} = 1$, then equation is

$$\frac{d^2u}{d\theta^2} = 0.$$

The solution is

$$u = A\theta + B.$$

The curve is called reciprocal spiral.

Case III. When $\frac{\mu}{h^2} < 1$, then equation is

$$\frac{d^2u}{d\theta^2} + n^2u = 0,$$

The solution is

$$u = A\cos n\theta + B\sin n\theta,$$

This is a curve of infinite branches.

2. A particle of mass m describes the curve $r = a/(1 + 2\cos^2\theta)$ under a force to the origin. If the speed of the particle at the farther apse is v, find the law of force.

   When passing through A one end of the near apse, the particle receives an impulse mkv along the outward normal. Show that, if $k < 2\sqrt{3}$, the particle continuous to describe a close curve about the origin, but that if $k > 2\sqrt{3}$, the new orbit has an asymptote making an angle θ with OA given by $\cos(\alpha + 2\theta) + 2\cos\alpha = 0$, where $\tan\alpha = \frac{k}{2}$.

   The given curve

$$r = \frac{a}{1 + 2\cos^2\theta}$$

can be written as

$$u = \frac{1}{r} = \frac{1 + 2\cos^2\theta}{a} = \frac{2 + \cos 2\theta}{a}.$$

$$\therefore \quad \frac{du}{d\theta} = \frac{-2\sin 2\theta}{a}, \quad \frac{d^2u}{d\theta^2} = \frac{-4\cos 2\theta}{a}.$$

Differential equation of the path is

$$P = h^2u^2\left[u + \frac{d^2u}{d\theta^2}\right] = h^2u^2[u - 4(au-2)/a] = \frac{h^2u^2(8-3au)}{a}. \quad (1)$$

Now

$$v^2 = h^2\left[u^2 + \left(\frac{du}{d\theta}\right)^2\right].$$

At apse $\frac{du}{d\theta} = 0$, $\therefore$ $v^2 = h^2 u^2$.

Now $\frac{du}{d\theta} = 0$, gives $\sin 2\theta = 0$,

Therefore, either $\theta = 0$, or $\theta = \pi/2$, hence, either $u = 3/a$ or $1/a$.
Hence, at farther apse $u = 1/a$ and velocity is v.

$$v^2 = \frac{h^2}{p^2}, \text{ gives for } p = a.$$

$$h^2 = a^2 v^2.$$

Law of force is, from (1)

$$P = \frac{h^2u^2(8-3au)}{a}.$$

Substituting value of h^2 we get

$$P = av^2 u^2 (8 - 3au).$$

Substituting value of P in (1), we get

$$av^2u^2 (8 - 3au) = h^2u^2\left[u + \frac{d^2u}{d\theta^2}\right].$$

Integrating, we get

$$v^2 = h^2\left[u^2 + \left(\frac{du}{d\theta}\right)^2\right] = av^2(16u - 3au^2) + C. \quad (2)$$

where C a constant of integration.

At apse $u = 3/a$, $\frac{du}{d\theta} = 0$, and let $v = v_1$.

Therefore

$$v_1^2 = \frac{9h^2}{a^2} = \frac{9a^2v^2}{a^2} = 9v^2.$$

When the particle receives an impulse mkv along the normal at A, *i.e.*, when $u = 1/a$, resultant velocity v_2 becomes $\sqrt{9v^2 + kv^2} = v\sqrt{9+k^2}$.
Hence, from (2), we get

$$v^2(9+k^2) = av^2\left(\frac{48}{a} - \frac{27}{a}\right) + C.$$

$$\therefore \quad C = v^2(9 + k^2 - 21) = v^2(k^2 - 12).$$

$$\therefore \quad h^2\left[u^2 + \left(\frac{du}{d\theta}\right)^2\right] = av^3(16u - 3au^2) + (k^2 - 12)v^2,$$

$$a^2v^2\left[u^2 + \left(\frac{du}{d\theta}\right)^2\right] = av^2(16u - 3au^2) + (k - 12)v^2,$$

$$\therefore \quad u^2 + \left(\frac{du}{d\theta}\right)^2 = \frac{16u - 3au^2}{a} + \frac{k^2 - 12}{a^2}$$

$$\therefore \quad \left(\frac{du}{d\theta}\right)^2 = \frac{16au - 3a^2u^2 + k^2 - 12 - a^2u^2}{a^2}$$

$$= \frac{16au - 4a^2u^2 + k^2 - 12}{a^2} \qquad (2)$$

When particle moves to infinity then $r \to \infty$ *i.e.*, $u \to 0$, hence,

$$\left(\frac{du}{d\theta}\right)^2 \to \frac{k^2 - 12}{a^2} = -\frac{12 - k^2}{a^2},$$

which is not possible, if $k < 2\sqrt{3}$, since $\frac{du}{d\theta} \geq 0$.

Hence, distance of the particle from the origin remains finite, if $k < 2\sqrt{3}$ and C becomes negative, *i.e.*, the particle will describe a close curve.

New from (2) $\quad \left(\frac{du}{d\theta}\right)^2 = \frac{[k^2 + 4 - (2au - 4)^2]}{a^2}.$

$$= \frac{[\beta^2 - (2au - 4)^2]}{a^2},$$

where $\beta^2 = k^2 + 4 = 4\tan^2\alpha + 4 = 4\sec^2\alpha$.

$$\therefore \qquad \beta = 2\sec\alpha, \text{ or } \cos\alpha = \frac{2}{\beta}.$$

$$\frac{du}{d\theta} = \pm\frac{1}{a}\sqrt{\beta^2 - (2au-4)^2}.$$

$$\therefore \qquad \frac{a\,du}{\sqrt{\beta^2 - (2au-4)^2}} = \pm\, d\theta.$$

Integration gives

$$\frac{1}{2}\cos^{-1}\frac{2au-4}{\beta} = C_1 + \theta,$$

when $\theta = 0$, $u = \dfrac{3}{a}$,

$$\therefore \qquad C_1 = \frac{1}{2}\cos^{-1}\frac{2}{\beta} = \frac{\alpha}{2},$$

$$\therefore \qquad \frac{1}{2}\cos^{-1}\frac{2au-4}{\beta} = \frac{\alpha}{2} + \theta.$$

$$\therefore \qquad 2(au-2) = \beta\cos(\alpha + 2\theta) = \frac{2\cos(\alpha+2\theta)}{\cos\alpha}$$

$\therefore\ 2\cos(\alpha + 2\theta) - 2(au-2)\cos\alpha = 0.$

To find asymptote we take $u \to 0$, therefore value of θ is given by

$$2\cos(\alpha + 2\theta) + 4\cos\alpha = 0,$$

$$\therefore \cos(\alpha + 2\theta) + 2\cos\alpha = 0.$$

3. A particle is describing an orbit under the central force $5\mu u^3 + 8\mu c^2 u^5$ per unit mass, and is projected from an apse at the distance c with velocity $3\sqrt{\dfrac{\mu}{c}}$, prove that the orbit is

$$r = c\cos\frac{2}{3}\theta.$$

The differential equation of the path is

$$h^2\left[u + \frac{d^2u}{d\theta^2}\right] = 5\mu u + 8\mu c^2 u^3$$

Multiplying both sides by $2\dfrac{du}{d\theta}$ and integrating, we get

$$h^2\left[u^2+\left(\frac{du}{d\theta}\right)^2\right] = v^2 = 5\mu u^2 + 4\mu h^2 u^4 + C \qquad (1)$$

Initially,

$$\frac{du}{d\theta} = 0,\ u = \frac{1}{c},\ v = 3\sqrt{\frac{\mu}{c}}.$$

$$\therefore \qquad h^2 = 9\mu,\ C = 0.$$

The equation (1) reduces to

$$9\mu\left[u^2+\left(\frac{du}{d\theta}\right)^2\right] = 5\mu u^2 + 4\mu c^2 u^4,$$

$$\therefore \qquad \left(\frac{du}{d\theta}\right)^2 = \frac{5}{9}u^2 - u^2 + \frac{4}{9}c^2u^4 = \frac{4\,(c^2u^4-u^2)}{9},$$

$$\therefore \qquad \frac{du}{\sqrt{c^2u^4-u^2}} = \frac{2}{3}d\theta. \qquad (2)$$

Putting $u = 1/r$, we get

$$\frac{-dr}{\sqrt{c^2-r^2}} = \frac{2}{3}d\theta.$$

$$\therefore \qquad \cos^{-1}\frac{r}{c} = \frac{2\theta}{3}+\alpha,$$

$$\therefore \qquad r = c\cos\left(\frac{2\theta}{3}+\alpha\right),$$

and if we measure θ from the apse line, $r = c$, when $\theta = 0$, therefore $\alpha = 0$ and $r = c\cos\dfrac{2\theta}{3}$.

N.B. It should be remarked that on taking square root in (2), there a choice of signs, and whatever sign be chosen, we get the same equation of the path.

4. A particle of mass m, is moving in a circle of radius a and centre O, under a force $\mu m\left(r+\dfrac{2a^3}{r^2}\right)$ directed towards O. If P is acted on by an impulse

tangential to the path of magnitude $(3\mu)^{1/2}\, ma$, show that velocity of P is immediately doubled, and that the greatest and least distances from O in ensuring motion are a and $3a$.

Differential equation of the path is

$$h^2 u^2 \left[u + \frac{d^2u}{d\theta^2}\right] = P = \mu\left(\frac{1}{u} + 2a^3u^2\right)$$

$$\therefore \quad h^2\left[u + \frac{d^2u}{d\theta^2}\right] = \mu\left(\frac{1}{u^3} + 2a^3\right).$$

Integrating, we get

$$h^2\left[u^2 + \left(\frac{du}{d\theta}\right)^2\right] = v^2 = \mu\left(-\frac{1}{u^2} + 4a^3u\right) + C. \quad (1)$$

When the particle is moving in a circle, let v_1 be its velocity, then

$$\frac{v_1^2}{a} = 3\mu a,$$

$$\therefore \quad v_1^2 = 3\mu a^2.$$

When the particle is acted on by an impulse $(3\mu)^{1/2}\, ma$, then velocity is increased by v_2 where

$$v_2^2 = 3\mu a^2,$$

$$\therefore \quad v_1 = v_2.$$

Hence, new velocity is $v_1 + v_2 = 2v_1$.

So, in ensuing motion the initial velocity is $2a\,(3\mu)^{1/2}$ and $u = 1/a$.

From (1), we get

$$12\,\mu a^2 = \mu\,(-a^2 + 4a^2) + C,$$

$$\therefore \quad C = 9\mu a^2$$

$$h^2 = 12\mu a^4.$$

For maximum and minimum distances $\frac{du}{d\theta} = 0$.

Hence, from (1), we have

$$h^2u^2 = \mu\left(-\frac{1}{u^2} + 4a^3u\right) + 9\mu a^2$$

Putting the value of h^2, we get

$$12\mu a^4u^4 = \mu\,(-1 + 4a^3u^3 + 9a^2u^2).$$

$$\therefore \quad 12a^4u^4 - 4a^3u^3 - 9a^2u^2 + 1 = 0$$

$\therefore (au-1)(3au-1)(2au+1)^2 = 0.$

$$\therefore \quad u = \frac{1}{a}, u = \frac{1}{3a}, u = \frac{-1}{2a}.$$

Rejecting negative value, we have

$$r = a, r = 3a.$$

5. A particle is moving under the force $\frac{\mu}{r^2}$ directed towards the origin and is projected with speed v_0 in a line whose perpendicular distance from O is p, show that then the closest approach of the particle to O is $\left[\mu+\sqrt{\mu^2+p^2v_0^4}\right]/v_0^2$.

See Figure 5.2 replace p by p_1.

Let $f(r)$ be the force per unit mass along OP and the tangent at P makes an angle ϕ with OP, and the velocity at P at time t be v.

Equation of motion along the normal is

$$m\frac{v^2}{\rho} = mf\sin\phi.$$

$$\therefore \quad \frac{v^2}{\rho} = f\sin\phi. \tag{1}$$

Let p_1 be the length of perpendicular from O on the tangent at P, then

$$h = p_1 v\sin\phi = \frac{p_1}{r}, \rho = \frac{rdr}{dp_1}.$$

Substituting these values in (1), we get

$$\frac{h^2}{p_1^3}\frac{dp_1}{dr} = f(r) = \frac{\mu}{r^2}.$$

Integrating, we get

$$\frac{h^2}{2p_1^2} = -\frac{\mu}{r}+C$$

Initially, $v = v_0$ $r = \infty$, $p_1 = p$.

$$\therefore \quad C = \frac{v_0^2}{2}.$$

$$\therefore \quad \frac{h^2}{2p_1^2} = -\frac{\mu}{r}+\frac{v_0^2}{2}.$$

$$\therefore \qquad \frac{h^2}{p_1^2} = -\frac{2\mu}{r} + v_0^2 \,. \qquad (2)$$

Now,

$$h = p_1 v = p v_0.$$

$$\therefore \qquad \frac{p^2 v_0^2}{p_1^2} = -\frac{2\mu}{r} + v_0^2 \,.$$

But

$$v^2 = h^2 \left[u^2 + \left(\frac{du}{d\theta} \right)^2 \right].$$

For closest approach $\frac{du}{d\theta} = 0$,

$$\therefore \qquad v^2 = h^2 u^2.$$

$$\therefore \qquad \frac{h^2}{p_1^2} = \frac{h^2}{r^2}.$$

$$\therefore \qquad p_1 = r.$$

Hence, from (2), we get

$$\frac{p^2 v_0^2}{2r^2} = -\frac{\mu}{r} + \frac{v_0^2}{2}.$$

$$\therefore \qquad r^2 v_0^2 - 2\mu r - p^2 v_0^2 = 0.$$

Hence,

$$r = \frac{\mu \pm \sqrt{\mu^2 + p^2 v_0^4}}{v_0^2}.$$

Hence, for closest approach

$$r = \frac{\mu + \sqrt{\mu^2 + p^2 v_0^4}}{v_0^2}.$$

5.8 PRINCIPLE OF ENERGY AND MOMENTUM APPLIED TO CENTRAL ORBITS

Problems on central orbits can be solved by applying principles of energy and momentum. If f denotes the force per unit mass towards the origin, the work done in the small displacement ds is

$$-f\frac{dr}{ds}\cdot ds = -fdr.$$

Components of velocity are $\dot{r}, r\dot{\theta}$, we have

$$\frac{1}{2}(\dot{r}^2 + r^2\dot{\theta}^2) = -\int fdr + C \qquad (1)$$

where C is a constant.

Also, the moment about O of the momentum is constant, therefore

$$r^2\dot{\theta} = h. \qquad (2)$$

Equation (1) and (2) are sufficient for the determination of the orbits, if f is given function of r.

Example

A particle moves in an orbit under a central acceleration $\frac{\mu}{r^2}$ along the radius vector r; obtain the equations of energy and angular momentum. If the particle is projected with velocity u at right angles to the radius, at distance c from the origin, prove that

$$\left(\frac{dr}{dt}\right)^2 = [2\mu/c - u^2(1 + c/r)][c/r - 1]$$

The equation of energy is

$$\frac{1}{2}(\dot{r}^2 + r^2\dot{\theta}^2) = -\int\frac{\mu}{r^2}dr + C,$$

$$\therefore \quad \frac{1}{2}(\dot{r}^2 + r^2\dot{\theta}^2) = \frac{\mu}{r} + C. \qquad (1)$$

and the equation of angular momentum is

$$r^2\dot{\theta} = h. \qquad (2)$$

Initially,

$$r = c,\ \dot{r} = 0,\ r\dot{\theta} = u.$$

Hence, from (1) and (2), we get

$$cu = h.$$

$$\therefore \quad \frac{1}{2}(\dot{r}^2 + r^2\dot{\theta}^2) = \frac{\mu}{r} + \frac{u^2}{2} - \frac{\mu}{c}.$$

$$r^2\dot{\theta} = cu.$$

Eliminating $\dot{\theta}$ we get

$$\dot{r}^2 = 2\mu\left(\frac{1}{r}-\frac{1}{c}\right)+u^2-\frac{c^2u^2}{r^2}$$
$$= [2\mu/c - u^2(1+c/r)]\,[c/r-1]$$

5.9 REPULSIVE FORCE

The motion of a particle under the action of a central repulsive force can be discussed in like manner.

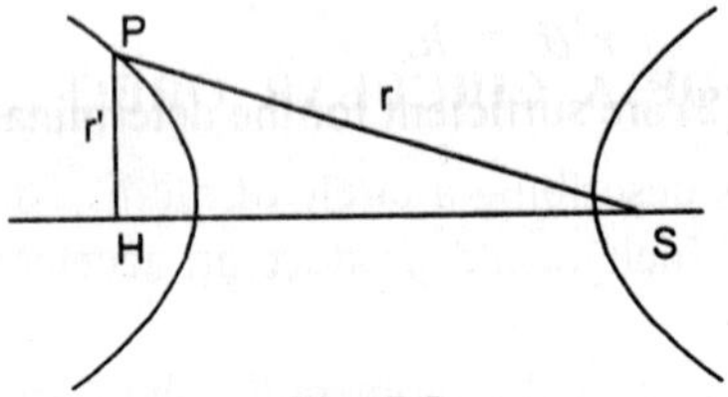

Fig. 5.6

For example, when the repulsive force is varying inversely as the square of the distance, we may write

$$\frac{h^2}{p^3}\frac{dp}{dr} = -\frac{\mu}{r^2},$$

$$\therefore \quad \frac{h^2}{p^2} = C-\frac{2\mu}{r}. \tag{1}$$

We know that for hyperbola

$$\frac{b^2}{p^2} = \frac{r'}{r}.$$

so that for points on the branch remote from the focus S, we have

$$\frac{b^2}{p^2} = \frac{r-2a}{r} = 1-\frac{2a}{r}. \tag{2}$$

Comparing (1) and (2), we get

$$C = \frac{\mu}{a},\ h^2 = \frac{\mu b^2}{a}.$$

Hence, the particle will describe the farther branch of the hyperbola under a repulsive force $\frac{\mu}{r^2}$, from (1) the velocity would be given by

$$v^2 = \frac{\mu}{a}-\frac{2\mu}{r} = \mu\left(\frac{1}{a}-\frac{2}{r}\right).$$

5.10 STABILITY OF AN ORBIT

If, when a particle is slightly disturbed from its orbit, it continuous to describe a path never deviating from its original path, we may call the originl motion a *stable* one.

In order to determine whether a given orbital motion is stable or not, we assume that the inverse radius $u = u_0 + x$, where u_0 is the value in the undisturbed orbit and x a small quantity. If the solution of the equation of motion, neglecting higher powers of x, gives a small periodic value, the assumption is justified and the motion is stable, otherwise, it is unstable.

5.11 STABILITY OF A CIRCULAR ORBIT

Suppose a particle is describing a circle of radius $1/u_0$ under the action of a force to the centre which would produce an acceleration $f(u)$ at a distance $1/u$.

Let h be the moment of the momentum about the centre in the circular orbit, then hu_0 is the velocity, and the acceleration $f(u_0)$ is (velocity)^2/radius *i.e.*, $h^2u_0^{\ 3}$.

$$\therefore \qquad h^2 = \frac{f(u_0)}{u_0^3}.$$

Now, suppose that without altering h, so that $1/u_0$ remains apsidal distance, the particle is slightly disturbed from the circular orbit. The equation for the path with the law of force $f(u)$ is

$$u + \frac{d^2u}{d\theta^2} = \frac{f(u)}{h^2u^2}, = \frac{u_0^3 f(u)}{u^2 f(u_0)}.$$

Putting $u = u_0 + x$, where x is a small, and retaining only the first powers of x, we get

$$\frac{d^2x}{d\theta^2} + u_0 + x = \frac{u_0^3 f(u_0 + x)}{f(u_0)(u_0 + x)^2}$$

$$= \frac{u_0 \{f(u_0) + xf'(u_0)\}}{f(u_0)(1 + 2x / u_0)}$$

$$= \frac{u_0 [f(u_0) + xf'(u_0)](1 - 2x / u_0)}{f(u_0)}$$

$$\therefore \quad \frac{d^2x}{d\theta^2} + x\left[3 - \frac{u_0 f'(u_0)}{f(u_0)}\right] = 0.$$

If the coefficient of x is positive, *i.e.*, when $\frac{u_0 f'(u_0)}{f(u_0)} < 3$, and we denote it by n^2, then we have

$$\frac{d^2x}{d\theta^2} + n^2 x = 0.$$

Its solution is

$$x = A\cos(n\theta + \alpha).$$

This shows that x is periodic in θ and oscillates in its values, so that motion is stable. Also, if it vanishes for a given value of θ, it will next vanish when θ is increased by π/n, so that the apsidal angle or angle between two consecutive apses in π/n, *i.e.*, $\dfrac{\pi}{\sqrt{3 - \dfrac{u_0 f'(u_0)}{f(u_0)}}}$.

The condition for stability is that the expression under the square root must be positive. If the force varies as the kth power of the distance,

i.e., $f(u) \propto \frac{1}{u^k}$.

$$\frac{u_0 f'(u_0)}{f(u_0)} = -k,$$

so that $3 + k$ must be positive and k must not be less than -3.

When the force varies as the distance, $k = 1$, and the apsidal angle $= \pi/2$. The disturbed circular orbit becomes in fact an ellipse with the centre of force in the centre and there are four apses at the ends of the axes.

When the force varies inversely as the square of the distance, $k = -2$ and the apsidal angle $= \pi$. In this case, the disturbed orbit is an ellipse with the centre of force in a focus and there are two apses at the ends of the major axis.

Example

If $f(r) = \frac{\mu}{r^2} e^{-\lambda r}$, where μ and λ are positive constants, prove that a circular orbit of radius a is stable, provided $a\lambda < 1$. Show that if $a\lambda$ is small, the apse line is nearly a circle of mean radius a and will advance in each revolution through an angle $\pi a\lambda$ approximatley.

Differential equation of the orbit is

$$h^2u^2\left[u+\frac{d^2u}{d\theta^2}\right] = \mu u^2 e^{-\lambda/u}. \quad (1)$$

When orbit is a circle of radius $r = a$ *i.e.*, $u = 1/a$, then

$$h^2 = \mu a e^{-\lambda a}.$$

When the orbit is slightly disturbed, let

$$u = x+\frac{1}{a},$$

where x is small.

$$\frac{du}{d\theta} = \frac{dx}{d\theta}, \frac{d^2u}{d\theta^2} = \frac{d^2x}{d\theta^2}.$$

Putting these values in (1), we get

$$\mu a e^{-\lambda a}\left[x+\frac{1}{a}+\frac{d^2x}{d\theta^2}\right] = \mu e^{-\lambda a/(1+ax)}$$

$$= \mu e^{-\lambda a(1-ax)}$$

$$= \mu e^{-\lambda a+\lambda a^2x}.$$

$$\therefore \qquad x+\frac{1}{a}+\frac{d^2x}{d\theta^2} = \frac{e^{\lambda a^2 x}}{a} = \frac{1+\lambda a^2x}{a}.$$

$$\therefore \qquad \frac{d^2x}{d\theta^2} = -(1-a\lambda)\,x.$$

The motion will be stable, if

$$1-a\lambda > 0, \text{ i.e., } a\lambda < 1.$$

The solution will be

$$x = A\cos\{\theta\sqrt{1-a\lambda}+a\}$$

$$\therefore \qquad \frac{dx}{d\theta} = 0,$$

gives

$$\sin\{\theta\sqrt{1-a\lambda}+\alpha\} = 0.$$

The roots differ by $\dfrac{\pi}{\sqrt{1-a\lambda}} = \pi\left(1+\dfrac{a\lambda}{2}\right).$

Neglecting multiple of 2π, we get $\lambda\pi a$ approximately as an angle through which apse line will advance in each revolution.

5.12 ORBIT IN RESISTING MEDIUM

(*i*) Central orbit with resistance varying as the velocity

Taking the mass unity, the resistance as $k\dot{s}$.

Equation of motion along the radial direction is

$$\frac{1}{r}\frac{d}{dt}(r^2\dot{\theta}) = -k\dot{s}r\frac{d\theta}{ds},$$

$$\therefore \qquad r^2\dot{\theta} = he^{-kt}.$$

$$\therefore \qquad p\dot{s} = he^{-kt}. \qquad (1)$$

where h is a constant, and p denotes the central perpendicular on the tangent to the path.

Let P be the central force per unit mass, by resolving along the normal,

$$\frac{\dot{s}^2}{\rho} = P\frac{p}{r},$$

so that from (1), we have

$$\frac{h^2}{p^2}.e^{-2kt} = Pp\frac{dr}{dp}. \qquad (2)$$

But

$$\frac{1}{p^2} = u^2 + \left(\frac{du}{d\theta}\right)^2.$$

$$\therefore \qquad u + \frac{d^2u}{d\theta^2} = -\frac{dp}{d\theta}/p^3\frac{du}{d\theta} = -\frac{dp}{du}/p^3 = r^2\frac{dp}{dr}/p^3.$$

and from (2), we get

$$u + \frac{d^2u}{d\theta^2} = \frac{Pe^{2kt}}{h^2u^2}. \qquad (3)$$

(*ii*) Central orbit with resistance varying as the square of velocity.

With the same notations, resolving along the transverse direction, we have

$$\frac{1}{r}\frac{d}{dt}(r^2\dot{\theta}) = -k\dot{s}^2r\frac{d\theta}{ds} = -k\dot{s}r\dot{\theta},$$

$$\therefore \qquad \frac{1}{r^2\dot{\theta}}\frac{d}{dt}(r^2\dot{\theta}) = -k\dot{s}$$

$$\therefore \qquad r^2\dot{\theta} = he^{-ks}. \qquad (4)$$

where h is a constant,

$$p\dot{s} = he^{-ks} \qquad (5)$$

Then resolving along the normal, we get

$$\frac{\dot{s}^2}{\rho} = \frac{Pp}{r}$$

$$\therefore \quad \frac{h^2 e^{-2ks}}{p^2} = Pp.\frac{dr}{dp} \quad (6)$$

But as in above

$$u + \frac{d^2u}{d\theta^2} = r^2 \frac{dp}{dr} / p^3,$$

so that

$$u + \frac{d^2u}{d\theta^2} = \frac{pe^{2ks}}{h^2u^2}. \quad (7)$$

***(iii)* To find the resistance necessary for the description of a given orbit with a given central force.**

Taking R as the resistance per unit mass, and with the foregoing notation and resolutions, we have

$$\frac{1}{r}\frac{d}{dt}(r^2\dot{\theta}) = -\frac{Rp}{r}$$

$$\frac{\dot{s}^2}{\rho} = \frac{Pp}{r}$$

$$\therefore \quad \frac{d}{dt}(p\dot{s}) = -Rp$$

$$\therefore \quad \frac{d}{dt}(p^2\dot{s}^2) = -2Rp^2\dot{s}.$$

$$\therefore \quad R = \frac{-d}{ds}(p^2\dot{s}^2)/2p^2$$

$$\therefore \quad R = -\frac{d}{ds}\left[Pp^3\frac{dr}{dp}\right]/2p^2.$$

5.13 ORBIT DESCRIBED UNDER THE ACTION OF A FORCE WITH GIVEN RADIAL AND TRANSVERSE COMPONENTS

If F, T denote the force to the origin and the transverse force, we have

$$\ddot{r} - r\dot{\theta}^2 = -F, \quad \frac{1}{r}\frac{d}{dt}(r^2\dot{\theta}) = T. \quad (1)$$

or writing $r^2\dot{\theta} = h$ (not constant), we get

$$\dot{h} = Tr. \tag{2}$$

Hence,

$$2h\dot{h} = 2Tr^3\dot{\theta}.$$

$$\therefore \quad \frac{dh^2}{d\theta} = \frac{2T}{u^3} \tag{3}$$

Then since,

$$\frac{d}{dt} = hu^2 \frac{d}{d\theta},$$

$$\therefore \quad \frac{h^2}{r^3} - F = \ddot{r} = hu^2 \frac{d}{d\theta}\left[hu^2 \frac{d}{d\theta}\left(\frac{1}{u}\right)\right]$$

$$= -hu^2 \frac{d}{d\theta}\left(h\frac{du}{d\theta}\right)$$

$$= -h^2u^2 \frac{d^2u}{d\theta^2} - \frac{T}{u}\frac{du}{d\theta} \text{ from (3),}$$

$$\therefore \quad h^2u^2\left[u + \frac{d^2u}{d\theta^2}\right] = F - \frac{T}{u}\frac{du}{d\theta} \tag{4}$$

where,

$$h^2 = 2\int \frac{T}{u^3}\, d\theta.$$

In terms of a field of force of potential energy V per unit mass at (r, θ), we may put

$$F = -\frac{\partial V}{\partial r} = u^2 \frac{\partial V}{\partial u}$$

$$T = -\frac{1}{r}\frac{\partial V}{\partial \theta} = -u\frac{\partial V}{\partial \theta}$$

Hence, (4) becomes

$$h^2u^2\left[u + \frac{d^2u}{d\theta^2}\right] = u^2 \frac{\partial V}{\partial u} + \frac{\partial u}{\partial \theta}\cdot\frac{\partial V}{\partial \theta}.$$

5.14 KEPLER'S LAWS OF PLANETARY MOTION AND NEWTON3'S LAW OF GRAVITATION

The following laws were published by Kepler (1571–1630) as a result of observations on the motion of planets of solar system:

1. Relative to Sun the planets describe ellipses with the Sun as one of focus.
2. The radius vector drawn from the Sun to a planet sweeps out areas at a constant rate.
3. The squares of the periodic times are proportional to the cubes of the semi-major axis of the elliptic orbits, and hence also to the cubes of the mean distances from the Sun.

The discovery of these laws was followed some sixty years later by Newton's enunciation of *Law of Gravitation.*

Newton's Law of Gravitation: It states that every particle in the universe attracts every other particle with a force which is directly proportional to the product of their masses and inversely proportional to the square of the distance between them. Thus if m_1, m_2 denote the masses of two particles and r distance between them, then force of attraction between them in $\gamma m_1 \cdot m_2/r^2$, where γ is a constant and is known as *constant of Gravitation.* This constant represents the attraction between two particles of unit mass at unit distance apart.

In his principia Newton proved a series of propositions in Mechanics which enable us to follow a process of reasoning from Kepler's Law of gravitation.

Necessary Modification of Kepler's Laws: If we assume the sun to be at rest, the truth of Kepler's Laws, based on Newton's Law of gravitation has already been established. Actually, first law is quite correct, we shell show that it is the centroid of the Sun and planet which is at rest, when there are no other bodies present. However, since the mass of the Sun is very very large in comparison of any planet in the solar system, the centroid very nearly coincides with the Sun.

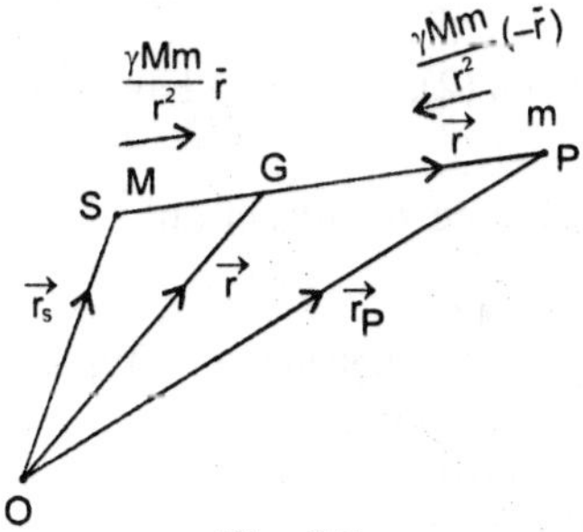

Fig. 5.7

Let S be the Sun and P a planet of the solar system, no other bodies being present. Let M be the mass of the Sun and m the mass of planet and let

$$\overrightarrow{OS} = \vec{r}_s, \overrightarrow{OP} = \vec{r}_p, \overrightarrow{SP} = \vec{r}$$

G is the centroid of the system, so that

$$\overline{OG} = \bar{r} = \frac{(M\vec{r}_s + m\vec{r}_p)}{M+m}.$$

By Newton's Law of gravitation, s and P are subjected to forces $\frac{\gamma\, Mm}{r^2}\hat{r}$ and $\frac{\gamma\, Mm}{r^2}(-\hat{r})$, respectively. Hence, their equations of motion are

$$\frac{\gamma\, Mm}{r^2}\hat{r} = M\ddot{\vec{r}}_s \tag{1}$$

$$\frac{\gamma\, Mm}{r^2}(-\hat{r}) = m\ddot{\vec{r}}_p \tag{2}$$

Adding (1) and (2), we get

$$M\ddot{\vec{r}}_s + m\ddot{\vec{r}}_p = \vec{0} \tag{3}$$

$$\ddot{\bar{r}} = \vec{0}.$$

i.e., the centroid of the two body system has no acceleration.

From (1) and (2), we have

$$\ddot{\vec{r}} = \ddot{\vec{r}}_p - \ddot{\vec{r}}_s = \frac{-\gamma\,(M+m)\,\hat{r}}{r^2}.$$

Thus planet is subjected to a force $\frac{-\mu\hat{r}}{r^2}$, where $\mu = \gamma\,(M + m)$ and so relative to Sun it describes a conic with Sun as focus Were the Sun at rest, we should have $\ddot{\vec{r}}_p = \ddot{\vec{r}}$ and from (2), we would have $\ddot{\vec{r}} = \frac{-\gamma\, M\hat{r}}{r^2}$, *i.e.*, the planet would be subjected to a force of attraction $\frac{-\mu'\hat{r}}{r^2}$ per unit mass, where $\mu' = \gamma M$. The ratio $\frac{\mu'}{\mu} = \frac{M}{M+m}$ is near to unity, since $M >> m$, so the error involved in treating S as fixed is negligible.

Examples

1. If the orbit of m relative to M is a circle of radius a described with velocity v, show that

$$v = \sqrt{\frac{\gamma (M+m)}{a}}.$$

If the relative velocity of particle is suddenly changed to $v + u$, without change of direction, show that, if $(u/v)^2$ is negligible, the particle will describe relative to M an ellipse of eccentricity $2u/v$.

(*i*) The orbit of m relative to M is a circle of radius a, and the speed is v, normal to the circle at a point in the radial direction.

$$\ddot{\vec{r}} = \frac{-v^2}{a}\hat{r} + \dot{v}\hat{t},\ \vec{r} = a\hat{r}.$$

where $\hat{t}$ is the unit vector in the direction of tangent to the circle. From previous article

$$\ddot{\vec{r}} = \frac{-\gamma (M+m)}{r^2}\hat{r}.$$

For circle $r = a$, therefore, we have

$$\frac{v^2}{a} = \frac{\gamma (M+m)}{a^2},\ \dot{v} = 0$$

$$\therefore \quad v = \sqrt{\frac{\gamma (M+m)}{a}}.$$

(*ii*) For the motion of m relative to M,

$$\frac{h^2}{p^3}\frac{dp}{dr} = \frac{\mu}{r^2},\ \mu = \gamma(M+m) = av^2$$

$$\therefore \quad \frac{h^2}{p^2} = \frac{2\mu}{r} + C.$$

Further $h = a\,(u + v)$ and when $r = a$, $p = a$

$$\therefore \quad C = (v+u)^2 - \frac{2\gamma (M+m)}{a} = (v+u)^2 - 2v^2 = u^2 + 2uv - v^2.$$

Since, C is negative, the orbit of m relative to M is an ellipse. The pedal equation of the ellipse may be written as

$$\frac{a^2(v+u)^2}{p^2} = \frac{2av^2}{r} - v^2 + 2uv + u^2,$$

$$\therefore \quad \frac{a^2\,(v+u)^2}{v^2 - 2vu - u^2} / p^2 = \frac{2av^2}{v^2 - 2uv - u^2} / r - 1.$$

Let

$$\alpha = \frac{av^2}{v^2 - 2uv - u^2}, \beta^2 = \frac{a^2(v+u)^2}{v^2 - 2uv - u^2}.$$

$$\therefore \quad \frac{\beta^2}{p^2} = \frac{2\alpha}{r} - 1.$$

This is an ellipse with semi-major axes α and β. Hence, if e is eccentricity, then

$$\beta^2 = \alpha^2(1 - e^2)$$

$$\therefore \quad e^2 = 1 - \frac{\beta^2}{\alpha^2} = 1 - \left(1 + \frac{u}{v}\right)^2 \left(1 - \frac{2u}{v} - \frac{u^2}{v^2}\right)$$

$$= 1 - \left(1 + \frac{2u}{v}\right)\left(1 - \frac{2u}{v}\right),$$

retaining only first oder terms of u/v.

$$= 1 - \left(1 - \frac{4u^2}{v^2}\right) = \frac{4u^2}{v^2}$$

$$\therefore e \cong \frac{2u}{v}.$$

2. A particle P is describing a circle with acceleration μ/SP^2 directed towards a fixed point S, collides and coalesces with an equal particle which is at rest. Show that, if the composite particle moves under the same law of force, it will describe an ellipse of eccentricity 3/4. Show also that the periods of description of the circle and the ellipse are in the ratio $7\sqrt{7} : 8$.

Let V be the velocity after the collision and V' before the collisions. By principle of conservation of linear momentum, we have

$$mV' = 2mV,$$

$$\therefore \quad V = \frac{V'}{2}. \tag{1}$$

When the particle is describing a circle of radius c, then

$$\frac{V'^2}{c} = \text{normal acceleration} = \frac{\mu}{c^2}.$$

$$V'^2 = \frac{\mu}{c}. \tag{2}$$

From (1) and (2), we get

$$V^2 = \frac{V'^2}{4} = \frac{\mu}{4c},$$

as $V^2 < \frac{2\mu}{c}$, so that the path is an ellipse.

Differential equation of the path is

$$h^2u^2\left[u + \frac{d^2u}{d\theta^2}\right] = \mu u^2,$$

$$\therefore \qquad h^2\left[u + \frac{d^2u}{d\theta^2}\right] = \mu.$$

Multiplying by $\frac{2du}{d\theta}$ and integrating, we get

$$v^2 = h^2\left[u^2 + \left(\frac{du}{d\theta}\right)^2\right] = 2\mu u + A.$$

Initially, $v = V = \sqrt{\frac{\mu}{4c}}$, $u = \frac{1}{c}$, $\frac{du}{d\theta} = 0$,

$$\therefore \qquad A = -\frac{7\mu}{4c},\ h^2 = \frac{\mu c}{4}.$$

$$\therefore \qquad v^2 = 2\mu u - \frac{7\mu}{4c} = \mu\left(\frac{2}{r} - \frac{7}{4c}\right).$$

Let $a = \frac{4c}{7}$.

$$\therefore \qquad v^2 = \mu\left(\frac{2}{r} - \frac{1}{a}\right).$$

Hence, path is an ellipse with semi-major axis $4c/7$

If l is semi-latus rectum, then

$$\therefore \qquad h = \sqrt{\mu l}$$

$$\therefore \qquad \sqrt{\mu c/4} = \sqrt{\mu a(1-e^2)}$$

where e is the eccentricity

$$\therefore \quad c = 4a(1-e^2) = \frac{16c}{7}(1-e^2).$$

$$\therefore \quad 1-e^2 = \frac{7}{16}$$

$$\therefore \quad e = \frac{3}{4}.$$

3. A particle P of unit mass, describes an ellipse under the action of force $\frac{\mu}{SP^2}$, directed towards a fixed point S. If ω_1, ω_2 are the greatest and least angular velocities of SP, show that the mean angular velocity of SP, is

$$\frac{2(\omega_1^{1/2}.\omega_2^{1/2})}{\omega_1^{1/2}+\omega_2^{1/2}}.$$

We know that

$$r^2\dot{\theta} = h,$$

$$\therefore \quad \dot{\theta} = \frac{h}{r^2}.$$

Angular velocity is greatest when distance is minimum, *i.e.*, when $r = a(1-e)$

Angular velocity is least, when distance is maximum, *i.e.*, when $r = a(1+e)$

$$\omega_1 = \frac{h}{a^2(1-e)^2}, \quad \omega_2 = \frac{h}{a^2(1+e)^2}$$

$$\therefore \quad \omega_1\omega_2 = \frac{h^2}{a^4(1-e^2)^2}.$$

$$\therefore \quad (\omega_1\omega_2)^{3/4} = \frac{h^{3/2}}{a^3(1-e^2)^{3/2}}.$$

$$\omega_1^{1/2}+\omega_2^{1/2} = \frac{\sqrt{h}}{a}\left[\frac{1}{1+e}+\frac{1}{1-e}\right] = \frac{2\sqrt{h}}{a(1-e^2)}$$

$$\therefore \quad \frac{2(\omega_1\omega_2)^{3/4}}{\omega_1^{1/2}+\omega_2^{1/2}} = \frac{h}{a^2(1-e^2)^{1/2}} = \frac{\sqrt{\mu}}{a^{3/2}}. \tag{1}$$

$$\text{Hence, angular velocity} = \frac{2\pi}{T} = \frac{2\pi}{2\pi a^{3/2}/\sqrt{\mu}} = \frac{\sqrt{\mu}}{a^{3/2}}.$$

$$= \frac{2(\omega_1\omega_2)^{3/4}}{\omega_1^{1/2}+\omega_2^{1/2}}.$$

4. A particle is describing a circle of radius a under an inverse square law of force $\frac{\mu}{r^2}$ per unit mass directed towards the centre of the circle. An outward radial velocity $\left(\frac{\mu}{5a}\right)^{1/2}$ is suddenly impressed on it in such a way as to leave the transverse velocity unaltered; prove that it will describe an ellipse and find the new periodic time.

Let V be the velocity when the particle is moving in a circle of radius a.

$$\therefore \quad \frac{V^2}{a} = \frac{\mu}{a^2},$$

$$\therefore \quad V^2 = \frac{\mu}{a}.$$

$$\text{New (velocity)}^2 = \frac{\mu}{a} + \frac{\mu}{5a} = \frac{6\mu}{5a} = V_1^2 \text{ (say)}.$$

Since, $V_1^2 < \frac{2\mu}{a}$, the new path will be an ellipse.

Equation of motion is

$$v\frac{dv}{dr} = -\frac{\mu}{r^2},$$

Integration gives

$$v^2 = \frac{2\mu}{r} + C\cdot$$

When $r = a$, $v^2 = V_1^2 = \frac{6\mu}{5a}$.

$$\therefore \quad C = -\frac{4\mu}{5a}.$$

$$\therefore \quad v^2 = \frac{2\mu}{r} - \frac{4\mu}{5a} = \mu\left(\frac{2}{r} - \frac{4}{5a}\right).$$

Hence, if a' is the semi-major axis of the ellipse, then

$$a' = \frac{5a}{4}.$$

$$\text{Periodic time } T = \frac{2\pi a'^{3/2}}{\sqrt{\mu}} = \frac{2\pi\,(5a/4)^{3/2}}{\sqrt{\mu}} = \frac{5\pi a^{3/2}\sqrt{5}}{4\sqrt{\mu}}.$$

5. A particle of mass m moving in a circle of radius c under an attractive force $\frac{\mu}{r^2}$ per unit mass towards the centre, collides and co-alesces with a particle of mass λm which is at rest. Show that the orbit of the combined mass is an ellipse with major axis c cosec$^2\alpha$, latus rectum $4c\cos^2\alpha$, and eccentricity $-\cos 2\alpha$, where $\sec^2\alpha = 2(1+\lambda)^2$.

If V is the velocity when particle is moving in a circle of radius c, then

$$V^2 = \frac{\mu}{c}.$$

Let V' be the velocity after collisions, then from the principle of conservation of momentum, we have

$$m(1+\lambda)V' = mV.$$

$$\therefore \quad V' = \frac{V}{1+\lambda},$$

$$\therefore \quad V'^2 = \frac{V^2}{(1+\lambda)^2} = 2V^2\cos^2\alpha = \frac{2\mu\cos^2\alpha}{c}$$

After collision differential equation of the path is

$$h^2u^2\left[u + \frac{d^2u}{d\theta^2}\right] = \mu u^2.$$

$$\therefore \quad h^2\left[u + \frac{d^2u}{d\theta^2}\right] = \mu.$$

Multiplying by $\frac{2du}{d\theta}$ and integrating, we get

$$v^2 = h^2\left[u^2 + \left(\frac{du}{d\theta}\right)^2\right] = 2\mu u + A,$$

When $\quad u = \frac{1}{c},\ u^2 = V'^2 = \frac{2\mu\cos^2\alpha}{c},\ \frac{du}{d\theta} = 0.$

$$h^2 = 2\mu c\cos^2\alpha,\ A = \frac{2\mu\cos^2\alpha}{c} - \frac{2\mu}{c}$$

$$= -\frac{2\mu\sin^2\alpha}{c}.$$

$$\therefore \quad v^2 = \mu\left(2u - \frac{2\sin^2\alpha}{c}\right) = \mu\left(\frac{2}{r} - \frac{2\sin^2\alpha}{c}\right).$$

Let
$$a' = \frac{c}{2\sin^2\alpha} = \frac{c\operatorname{cosec}^2\alpha}{2}.$$

Hence, we get

$$v^2 = \mu\left(\frac{2}{r} - \frac{1}{a'}\right).$$

Therefore new path is an ellipse of semi-major axis $a' = c \operatorname{cosec}^2 \alpha$. If l' is the semi-latus rectum, then

$$h^2 = \mu l' = 2\mu c \cos^2 \alpha.$$

$$\therefore \quad l' = 2c \cos^2 \alpha.$$

Hence, latus rectum $= 2l' = 4c \cos^2 \alpha$.

We know that

$$l' = a'(1 - e'^2),$$

$$\therefore \quad 2c\cos^2\alpha = \frac{c\operatorname{cosec}^2\alpha}{2}.(1 - e'^2)$$

$$\therefore \quad 1 - e'^2 = 4\cos^2\alpha \sin^2\alpha = \sin^2 2\alpha.$$

$$\therefore \quad e'^2 = \cos^2 2\alpha,$$

$$\therefore \quad e' = \pm \cos 2\alpha,$$

But

$$\sec^2\alpha = 2(1 + \lambda)^2,$$

$$\therefore \quad (1 + \lambda)^2 = \frac{1}{2\cos^2\alpha} = \frac{1}{1 + \cos 2\alpha}.$$

$(1 + \lambda)$ will be real if $1 + \cos 2\alpha > 0$,

$$\cos 2\alpha > -1.$$

$$\therefore \quad e' = -\cos 2\alpha.$$

so that $e' < 1$, which is the necessary condition for path to be an ellipse.

5.15 DISTURBED ELLIPTIC MOTION

When a body moving in a central orbit is subjected to the action of small impulsive forces, the effect is to produce small instantaneous changes in the elements of the orbit. In the case of elliptic motion about a focus, the elements are the length of the major axis, the eccentricity and the longitude of the apse. When the body is subjected to small finite and continuous forces, such as the action of resisting medium, the orbit is not an ellipse, but the motion can be represented by supposing the body moves in an ellipse, the elements of which are changing continuously. The ellipse is then called the *instantaneous orbit* and it may be defined as the orbit which would be described after any instant of time, if the action of the disturbing forces were suspended at the instant.

I. Small Tangential Impulse

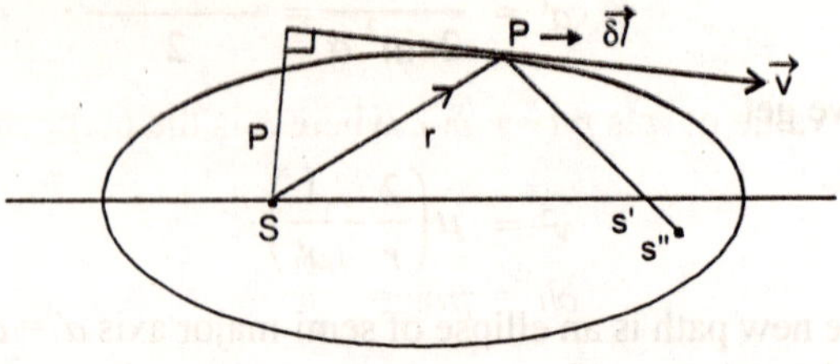

Fig. 5.8

Let the particle P of mass be describing an ellipse under inverse square law $\frac{\mu}{r^2}$ towards focus, S and S' be the focii. The velocity of P is $\vec{v}$ and the particle receives an impulse of $\delta\vec{l}$ in the direction of $\vec{v}$.

The velocity at P before the impulse is given by

$$v^2 = \mu\left(\frac{2}{r} - \frac{1}{a}\right),$$

$$\therefore \qquad 2v\delta v = \frac{\mu}{a^2}\delta a,$$

since, r does not change instantaneously.

$$\therefore \qquad \delta a = \frac{2a^2 v}{\mu}\delta v,$$

where

$$\delta_v = \frac{|\delta\vec{l}|}{m}.$$

Hence, length of new semi-major axis $\cong a + \dfrac{2a^2 v\delta v}{\mu}$.

Since, the direction of motion and the radius vector are unchanged by the impulse, therefore the increment of the major axis means an increment in the second focal distance PS', *i.e.*, the second focus is displaced to S'', so that the apse line turned through an angle $S'SS''$.

Now

$$SP + PS' = 2a,$$

$$SP + PS'' = 2\left(a + \frac{2a^2 v\delta v}{\mu}\right)$$

$$\therefore \quad S'S'' = \frac{4a^2 v\delta v}{\mu}.$$

The new value of h is $p\,(v + \delta v)$, where p is the perpendicular from S on $\vec{v}$. Hence,

$$\delta h = p\delta v = \frac{h\delta v}{v}.$$

Further, since,

$$h^2 = \mu l = \frac{\mu b^2}{a}$$

$$\therefore \quad 2 \log h = \log \mu + 2 \log b - \log a,$$

$$\therefore \quad \frac{2\delta h}{h} = \frac{2\delta b}{b} - \frac{\delta a}{a}.$$

$$\therefore \quad \frac{\delta b}{b} = \frac{\delta v}{v} + \frac{av\delta v}{\mu}.$$

This gives the increment in the minor axis.
Since, the periodic time in the undisturbed orbit is

$$T = \frac{2\pi ab}{h},$$

$$\therefore \quad \log T = \log (2\pi) + \log a + \log b - \log h$$

$$\therefore \quad \frac{\delta T}{T} = \frac{\delta a}{a} + \frac{\delta b}{b} - \frac{\delta h}{h},$$

$$\therefore \quad \frac{\delta T}{T} = \left[\frac{2av}{\mu} + \frac{1}{v} + \frac{av}{\mu} - \frac{1}{v}\right]\delta v = \frac{3av}{\mu}\delta v$$

This gives change in periodic time.
Further, since, $\angle S''SS'$ is small.

$$\therefore \quad \angle S''SS' = \frac{S'S''}{SS'} \sin PS'S$$

$$= \frac{S'S''}{SS'} \cdot \frac{PS}{PS'} \sin PSS'$$

$$= \frac{2av\delta v}{\mu e}\left[\frac{r}{2a-r}\right]\sin PSS'$$

$$= \frac{2arv\delta v}{\mu e\,(2a-r)}\sin PSS'$$

$$= \frac{2\delta v}{e}\left[\frac{ar}{\mu\,(2a-r)}\right]^{1/2}\sin PSS'$$

$$= \frac{2\delta v}{e}\left[\frac{ar}{\mu\,(2a-r)}\right]^{1/2}\sin(\theta-\overline{\omega}),$$

where $\angle PSS' = \theta - \overline{\omega}$.

This gives the rotation of the apse line.

$$h^2 = \mu a\,(1-e^2).$$

$$\therefore \quad 2h\delta h = \mu\,(1-e^2)\,\delta a - 2\mu ae\,\delta e,$$

$$\therefore \quad \mu ae\,\delta e = va^2\,(1-e^2)\,\delta v - \frac{h^2}{v}\delta v$$

$$= \frac{b^2}{v}\left(v^2 - \frac{\mu}{a}\right)\delta v.$$

$$\therefore \quad \delta e = \frac{2b^2\delta v}{\mu aev}\left(\frac{\mu}{r} - \frac{\mu}{a}\right)$$

$$= \frac{2a(1-e^2)}{ev}\left(\frac{1}{r} - \frac{1}{a}\right)\delta v.$$

This gives change in the eccentricity.

Continuous tangential force: Let the mass of the body be m and let it be acted upon by a continuous small tangential disturbing force mf. We can represent this force as the limit of succession of small impulses producing a change of velocity δv at intervals of δt, so that $\dot{v} = f$. Then the small rates of the elements of the orbit are given by, just replacing δv by f, *i.e.*,

$$\dot{a} = \frac{2va^2 f}{\mu}, \quad \dot{\overline{\omega}} = \frac{2f\sqrt{\dfrac{ar}{\mu\,(2a-r)}}}{e}\sin(\theta-\overline{\omega}),$$

$$\dot{e} = \frac{2a\,(1-e^2)\left(\dfrac{1}{r} - \dfrac{1}{a}\right)f}{ve}.$$

II. Small Normal Impulse

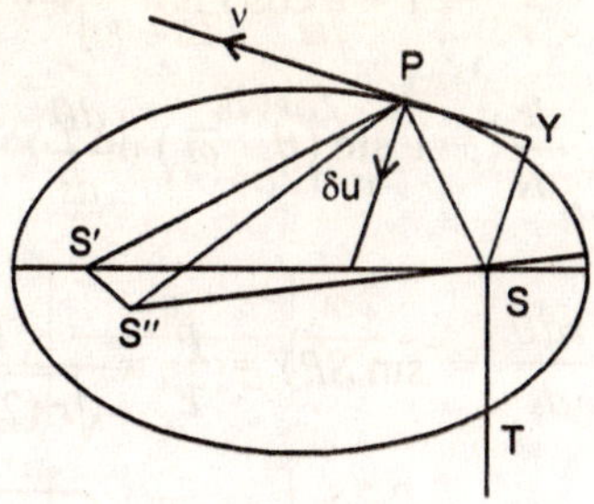

Fig. 5.9

Let an impulse give the body a small velocity δu along the inward normal to the ellipse. To the first order of δu this leaves the square of velocity unchanged, since

$$v^2 = \frac{2\mu}{r} - \frac{\mu}{a} \tag{1}$$

$$\therefore \quad \delta a = 0. \tag{2}$$

But the direction of velocity, *i.e.*, the tangent to the path, is turned through an angle $\frac{\delta u}{v}$ and the tangent is equally inclined to the focal distances PS, PS', so if S'' is the new second focus, we have

$$S'PS'' = \frac{2\delta u}{v}, \text{ and since, } \delta a = 0,$$

$$\therefore \quad S''P = S'P, \; S'S'' = \frac{2S'P\delta u}{v}.$$

Hence,

$$\angle S'SS'' = \frac{S'S''}{S'S} \cos SS'P = \frac{S'P\delta u}{aev} \cos SS'P$$

$$= \{SS' + SP \cos PSS'\}\, \delta u/aev,$$

$$= \frac{\{2ae + r\cos PSS'\}\delta u}{e} \sqrt{\frac{r}{\mu a\,(2a - r)}}. \tag{3}$$

Again

$$h^2 = \mu a\,(1 - e^2),$$

$$\therefore \quad h\delta h = -\,\mu ae\;\delta e. \tag{4}$$

But δh is increment in the moment of the velocity about S, so that

$$\delta h = PY.\,\delta u = \frac{rdr}{ds}\,\delta u,$$

where $$\frac{l}{r} = 1 + e\cos PSS' = 1 + e\cos(\theta - \overline{\omega}), \text{ say.}$$

$$\therefore \quad \frac{l}{r^2}\cdot\frac{dr}{ds} = e\sin(\theta - \overline{\omega})\cdot\frac{d\theta}{ds}.$$

But

$$\frac{rd\theta}{ds} = \sin SPY = \frac{p}{r} = \frac{b}{\sqrt{r(2a-r)}}.$$

$$\therefore \quad \frac{dr}{ds} = \frac{eb}{l}\sin(\theta - \overline{\omega})\sqrt{\frac{r}{2a-r}}.$$

$$\therefore \quad \delta h = \frac{be\sin(\theta - \overline{\omega})\,\delta u}{l}\sqrt{\frac{r^3}{2a-r}}.$$

From (4) it follows that

$$\delta e = -\delta u\sin(\theta - \overline{\omega})\sqrt{\frac{r^3}{\mu a(2a-r)}}. \qquad (5)$$

Continuous normal force: If f' denotes the acceleration effect of the normal force, then

$$\dot{a} = 0,$$

$$\dot{\overline{\omega}} = \frac{f'\{2ae + r\cos(\theta - \overline{\omega})\}}{e}\sqrt{\frac{r}{\mu a(2a-r)}},$$

$$\dot{e} = -f'\sin(\theta - \overline{\omega})\sqrt{\frac{r^3}{\mu a(2a-r)}}$$

III. Small Radial Impulse

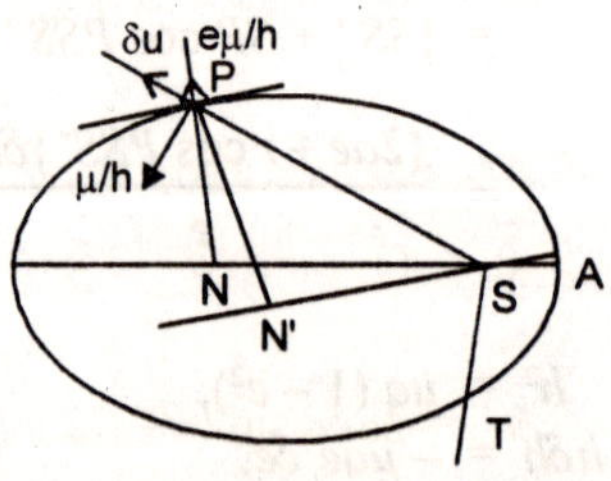

Fig. 5.10

Let an impulse in the direction SP give the body a small velocity δu. We know that velocity in the undisturbed orbit can be represented by components $\frac{\mu}{h}$ at right angles to SP and $\frac{\mu e}{h}$ along the ordinate NP. Let PN' be the perpendicular to the new position of the apse line. The moment of velocity about S is unaltered, therefore the velocity in the new orbit is component of $\frac{\mu}{h}$ at right angles to SP and $\frac{e'\mu}{h}$ along $N'P$, where $e' = e + \delta e$. But these two components are equivalent to $\frac{\mu}{h}$ at right angles to SP, $\frac{e\mu}{h}$ along NP and δu along SP, so that $\frac{e'\mu}{h}$ along N$'P$ is the resultant of $\frac{e\mu}{h}$ along NP and δu along SP.

Then we have,

$$\delta\overline{\omega} = \angle NSN' = \angle NPN' = \frac{\delta u \sin SPN}{\frac{e\mu}{h}}$$

$$= \frac{-\delta u}{e}\cos(\theta-\overline{\omega})\sqrt{\frac{a(1-e^2)}{\mu}} \tag{1}$$

Again by resolving in the direction NP, we get

$$\frac{e'\mu\cos\delta\overline{\omega}}{h} = \frac{e\mu}{h} + \delta u\cos SPN,$$

whence

$$\frac{\mu\delta e}{h} = \delta u\sin(\theta-\overline{\omega}).$$

$$\therefore \qquad \delta e = \delta u\sin(\theta-\overline{\omega})\sqrt{\frac{a(1-e^2)}{\mu}} \tag{2}$$

Again

$$h^2 = \mu e(1-e^2)$$

$$\therefore \qquad 0 = (1-e^2)\,\delta a - 2ae$$

$$\therefore \qquad \delta a = 2e\delta u\sin(\theta-\overline{\omega})\sqrt{\frac{a^3}{\mu(1-e^2)}} \tag{3}$$

Continuous radial force : If f_1 denotes the accelerating effect of the radial force, then we deduce that

$$\dot{\overline{\omega}} = -\frac{f_1 \cos(\theta - \overline{\omega})}{e}\sqrt{\frac{a(1-e^2)}{\mu}},$$

$$\dot{e} = f_1 \sin(\theta - \overline{\omega})\sqrt{\frac{a(1-e^2)}{\mu}},$$

and

$$\dot{a} = 2ef_1 \sin(\theta - \overline{\omega})\sqrt{\frac{a^3}{\mu(1-e^2)}}.$$

IV. Small Transverse Impulse

Let an impulse at right angle to SP, increases the velocity by a small amount δu. Then the components of the velocity in the new orbit are $\frac{\mu}{h'}$, at right angles to SP and $\frac{e'\mu}{h'}$ along $N'P$, where $h' = h + \delta h$, $e' = e + \delta e$, $\delta h = r\delta u$.

Resolving along and at right angles to SP, we get

$$\frac{e'\mu}{h'}\cos(SPN - \delta\overline{\omega}) = \frac{e\mu}{h}\cos SPN.$$

$$\frac{\mu}{h'} - \frac{e'\mu}{h}\sin(SPN - \delta\overline{\omega}) = \frac{\mu}{h} + \delta u - \frac{e\mu}{h}\sin SPN$$

These equations reduce to

$$\delta e \cos SPN + e\delta\overline{\omega} \sin SPN - \frac{er}{h}\delta u \cos SPN = 0,$$

$$-\delta e \sin SPN + e\delta\overline{\omega} \cos SPN - \delta u\left(\frac{h}{\mu} + \frac{r}{h} - \frac{er \sin SPN}{h}\right) = 0.$$

From these equations, we get

$$\delta e = \left[\frac{r}{h}\{e + \cos(\theta - \overline{\omega})\} + \frac{h\cos(\theta - \overline{\omega})}{\mu}\right]\delta u \qquad (1)$$

$$\delta\overline{\omega} = \frac{\sin(\theta - \overline{\omega})\left(h + \frac{\mu r}{h}\right)\delta u}{\mu e} \qquad (2)$$

Since,

$$h^2 = \mu a(1 - e^2),$$

$$\therefore \quad 2hr\,\delta u = \mu(1 - e^2)\,\delta a - 2\mu ae\,\delta e,$$

$$\therefore \qquad \delta a = \frac{2a^2 \delta u \{1 + e\cos(\theta - \overline{\omega})\}}{h}. \tag{3}$$

The effects of a continuous transverse force may be written as above.

V. Small Change in the Magnitude of the Absolute Force

Let μ receives an increment $\delta\mu$, the velocity remaining unchanged. Then since,

$$v^2 = \mu\left(\frac{2}{r} - \frac{1}{a}\right),$$

$$\therefore \qquad \delta a = \frac{-a(2a - r)\,\delta\mu}{r\mu} \tag{1}$$

Since, the direction of motion is unaltered, the new second focus S'' lies on PS' such that

$$S'S'' = 2\delta a.$$

Then using Fig. 5.10 We have

$$\delta\overline{\omega} = \frac{S'S''}{SS'}\sin PS'S = \frac{\sin PS'S\,\delta a}{ae}.$$

$$= -\frac{\delta\mu}{e\mu}\frac{PS'}{PS}\sin PS'S = -\frac{\delta\mu}{e\mu}\sin(\theta - \overline{\omega}) \tag{2}$$

Again

$$h^2 = \mu a(1 - e^2) \text{ and } h \text{ is unchanged; therefore}$$

$$2\mu ae\,\delta e = (\mu\,\delta a + a\,\delta\mu)(1 - e^2),$$

which with the help of (1) gives

$$\delta e = \left\{\frac{1 - e^2}{e}\right\}\left\{\left(1 - \frac{a}{r}\right)\frac{\delta\mu}{\mu}\right\} \tag{3}$$

If the change in μ is continuous at rate $n\mu$ per unit time, we get

$$\dot{a} = -\frac{na(2a - r)}{r}, \tag{4}$$

$$e\dot{\overline{\omega}} = -n\sin(\theta - \omega), \tag{5}$$

$$\dot{e} = \frac{n(1 - e^2)}{e}\left(1 - \frac{a}{r}\right). \tag{6}$$

5.16 ELLIPTIC HARMONIC MOTION

Let a particle P moves under a central force μr per unit mass along PO, where O is the centre of the force and $\overrightarrow{OP} = \vec{r}$.

The equation of motion is

$$\ddot{\vec{r}} = -\mu\vec{r}. \tag{1}$$

Let Ox, Oy through O in the plane of motion be co-ordinate axes and $\hat{i}, \hat{j}$ the unit vectors along Ox, Oy, then from (1), we obtain

$$\ddot{x} = -\omega^2 x,\ \ddot{y} = -\omega^2 y. \tag{2}$$

where $\omega^2 = \mu$. these equations are a pair of superposed simple harmonic motions. then general solutions of (2) are

$$x = A_1 \cos \omega t + A_2 \sin \omega t,$$
$$y = B_1 \cos \omega t + B_2 \sin \omega t.$$

Solving for cos ωt and sin ωt, we get

$$\cos \omega t = \frac{B_2 x - A_2 y}{\Delta},\ \sin \omega t = \frac{A_1 y - B_1 x}{\Delta},\ \Delta = A_1 B_2 - A_2 B_1.$$

Now, we have

$$1 = \cos^2 \omega t + \sin^2 \omega t, \text{ for locus of } P$$

$$(B_2 x - A_2 y)^2 + (A_1 y - B_1 x)^2 = (A_1 B_2 - A_2 B_1)^2.$$

This locus being of second degree in x, y and having no first order terms in x, y, represents a conic having O as centre. This equation can be written as

$$(B_2^2 + B_1^2)x^2 + (A_2^2 + A_1^2)y^2 - (A_2 B_2 + A_1 B_1)xy = (A_1 B_2 - A_2 B_1)^2.$$

The discriminate of second degree terms is

$$4(A_2 B_2 + A_1 B_1)^2 - 4(B_2^2 + B_1^2)(A_2^2 + A_1^2)$$
$$= 4[2A_1 A_2 B_1 B_2 - B_1^2 A_2^2 - A_1^2 B_2^2]$$
$$= -4(A_1 B_2 - A_2 B_1)^2 < 0, \text{ in general.}$$

Thus the conic is an ellipse, with centre O. This type of motion is called *elliptic harmonic motion* with period $2\pi/\omega$.

Examples

1. A particle of unit mass is describing an ellipse under a central force $\frac{\mu}{r^2}$ directed towards a fixed point S, when the particle is at P, its radial component of velocity is suddenly destroyed. Show that the focal chord PSP' of the original elliptical orbit becomes the major axis of the new elliptical orbit.

   Equation of the elliptic path is

$$\frac{l}{r} = 1 + e \cos \theta,$$

$$\therefore \quad \frac{dr}{dt} = \frac{e \sin\theta}{l} r^2 \dot{\theta} = e \sin \theta \frac{h}{l} = u_0 \text{ (say)}$$

$$\frac{rd\theta}{dt} = \frac{h}{r} = \frac{h(1+e\cos\theta)}{l} = v_0 \text{ (say)}.$$

Let $\theta = \alpha$, when $\frac{dr}{dt}$ is destroyed, so that only velocity is v_0. Originally the particle was describing an ellipse and when u_0 has been destroyed, the velocity is decreased, so the new orbit also will be an ellipse. We can prove that new orbit will be an ellipse, by finding the value of constant of intergration C.

Differential equation of the path is

$$\frac{vdv}{dr} = \frac{-\mu}{r^2}.$$

Integration gives

$$v^2 = \frac{2\mu}{r} + C,$$

when

$$v = (v_0)_{\theta=\alpha} = \frac{h(1+e\cos\alpha)}{l}, r = \frac{l}{1+e\cos\alpha}.$$

$$\therefore \quad \frac{h^2(1+e\cos\alpha)^2}{l^2} = \frac{2\mu(1+e\cos\alpha)}{l} + C$$

$$\therefore \quad C = -\frac{\mu}{l}(1-e^2\cos^2\alpha).$$

As C is negative, hence now orbit is an ellipse.

Now

$$v^2 = \frac{2\mu}{r} - \frac{\mu(1-e^2\cos^2\alpha)}{l} = \mu\left(\frac{2}{r} - \frac{1-e^2\cos^2\alpha}{l}\right).$$

If a' is the semi-major axis of the new ellipse, then

$$a' = \frac{l}{1-e^2\cos^2\alpha},$$

$$\therefore \quad 2a' = \frac{2l}{1-e^2\cos^2\alpha}.$$

For original ellipse

$$r_1 = \frac{1}{1-e\cos\alpha}, r_2 = \frac{1}{1+e\cos\alpha}.$$

Since, if $\theta = \alpha$, at P, then $\theta = \pi + \alpha$ at P', the other end of the focal chord PSP'.

$$\therefore \qquad r_1 + r_2 = \frac{2l}{1 - e^2 \cos^2 \alpha} = 2a'.$$

2. Two particles of equal masses are moving in a coplanar ellipse under a central attractive force $\frac{\mu}{r^2}$ per unit mass. The axis of each orbit is 2a and the latus recta are $2l_1$, $2l_2$ respectively. The particles collide at a point at a distance c from the centre of the force and coalesce. Prove that the latus rectum of the orbit of the combined particles is $\frac{\left(\sqrt{l_1} + \sqrt{l_2}\right)^2}{4}$, and that the major axis of the new ellipse is $2\left[\frac{2}{c}\sin^2\frac{\alpha}{2} + \frac{1}{a}\cos^2\frac{\alpha}{2}\right]^{-1}$, α being the angle between the velocities before the collision.

Let mass of each particle be m and v_1, v_2 be their velocities before collision. Now for ellipse

$$v^2 = \mu\left(\frac{2}{r} - \frac{1}{a}\right).$$

$$\therefore \qquad v_1^2 = \mu\left(\frac{2}{c} - \frac{1}{a}\right) = v_2^2 .$$

By principle of linear momentum

$$2m \text{ (resultant velocity)} = m\sqrt{v_1^2 + v_2^2 + 2v_1v_2\cos\alpha}\ .$$

$$\therefore \text{ resultant velocity} = \frac{1}{2}\sqrt{2v_1^2\,(1 + \cos\alpha)}$$

$$= v_1\cos\frac{\alpha}{2}.$$

As this velocity is $\leq v_1$, hence, now orbit is also ellipse. Let V be the velocity of the new ellipse, then

$$V^2 = \mu\left(\frac{2}{c} - \frac{1}{a}\right)\cos^2\frac{\alpha}{2}$$

If a' be the semi-major axis of the ellipse, then

$$\frac{2}{c} - \frac{1}{a'} = \left(\frac{2}{c} - \frac{1}{a}\right)\cos^2\frac{\alpha}{2},$$

$$\therefore \quad \frac{1}{a'} = \frac{2}{c} - \left(\frac{2}{c} - \frac{1}{a}\right)\cos^2\frac{\alpha}{2}$$

$$= \frac{2}{c}\sin^2\frac{\alpha}{2} + \frac{1}{a}\cos^2\frac{\alpha}{2}.$$

$$\therefore \quad 2a' = 2\left[\frac{2}{c}\sin^2\frac{\alpha}{2} + \frac{1}{a}\cos^2\frac{\alpha}{2}\right]^{-1}.$$

Now $\quad h_1^2 = \mu l_1 \; h_2^2 = \mu l_2.$

If h be the angular momentum of the combined particles, the law of conservation of angular momentum, gives

$$2mh = m(h_1 + h_2),$$

$$\therefore \quad h = \frac{1}{2}(h_1 + h_2) = \frac{\sqrt{\mu}}{2}(\sqrt{l_2} + \sqrt{l_2})$$

$$\therefore \quad \frac{h^2}{\mu} = \frac{(\sqrt{l_1} + \sqrt{l_2})^2}{4}.$$

Let l be the latus rectum of the new ellipse, then

$$l = \frac{(\sqrt{l_1} + \sqrt{l_2})^2}{4}.$$

3. A missile P is projected from the Earth's surface at A with speed $(2gd)^{1/2}$, the earth being supposed spherical, of radius c and centre O, and $d < c$. The missile is supposed to be attracted by a force $\frac{gc^2}{r^2}$ per unit mass towards O, where $OP = r$. Show that the path of P is an ellipse having O as focus, and that the speed v of P is given by $v^2 = 2g\left(d - c + \frac{c^2}{r}\right)$. Show that the locus of the second focus H for all paths having A in vertical plane with the same speed, is a circle of centre A and radius $\frac{cd}{c-d}$.

Equation of motion of P is given by

$$\frac{h^2}{p^3}\frac{dp}{dr} = \frac{gc^2}{r^2}.$$

Integration gives,

$$v^2 = \frac{h^2}{p^2} = \frac{2gc^2}{r} + C.$$

Initially,

$r = c,\ v^2 = 2gd$

$$\therefore \quad C = 2g\,(d - c).$$

$$\therefore \quad v^2 = 2g\left(d - c + \frac{c^2}{r}\right).$$

$$= gc^2\left(\frac{2}{r} + \frac{2d - 2c}{c^2}\right).$$

Let $\mu = gc^2$, then

$$v^2 = \frac{h^2}{p^2} = \mu\left(\frac{2}{r} - \frac{2c - 2d}{c^2}\right).$$

This is the pedal equation of the ellipse having O as focus whose semi-major axis and semi-minor axis are given by

$$a = \frac{c^2}{2(c - d)},\ b = \frac{hc}{\sqrt{2\mu\,(d - c)}}.$$

Now

$$h = Vc,\ \therefore\ h^2 = V^2 c^2.$$

But

$$h^2 = \mu l = gc^2\, a\,(1 - e^2),$$

$$\therefore \quad V^2 c^2 = gc^2\, a\,(1 - e^2),$$

$$\therefore \quad V^2 = 2gd = ga\,(1 - e)^2,$$

$$\therefore \quad 1 - e^2 = \frac{2d}{a} = \frac{4d\,(c - d)}{c^2},$$

$$\therefore \quad e^2 = 1 - \frac{4dc - 4d^2}{c^2} = \frac{(c - 2d)^2}{c^2},$$

$$\therefore \quad e = \pm\,\frac{c - 2d}{c}.$$

Since, for ellipse $e < 1$, and $d < c$ (given). Hence,

$$e = \frac{c - 2d}{c}$$

The first focus O lies on the circle, Σ_1, having A as centre and radius $a\,(1 + e)$, while the other focus H lies on a circle Σ_2 having A as centre and radius $a\,(1 - e)$.

Now

$$a(1-e) = \frac{c^2}{2(c-d)}\left[1-\frac{c-2d}{c}\right] = \frac{cd}{c-d}.$$

4. **A particle of mass *m*, is attracted towards the origin *O* of rectangular co-ordinates *x*, *y* by a force mn²*OP*. Initially, the particle is projected from the point (*a*, 0) with velocity *u* parallel to the axis of *y*. Prove that the particle describes the ellipse**

$$\left(\frac{x}{a}\right)^2 + \left(\frac{y}{b}\right)^2 = 1,$$

and the eccentric angle of *P* on the ellipse increases at a constant rate. Show that the hodograph of the motion is an ellipse which is similar to the orbit.

Equation of motion is

$$m\ddot{\vec{r}} = -mn^2\vec{r},$$

$$\therefore \quad \ddot{x} = -n^2x, \ \ddot{y} = -n^2y.$$

The solution of these equations are

$$x = A_1 \cos nt + A_2 \sin nt, y = A_3 \cos nt + A_4 \sin nt.$$

$$\therefore \quad \dot{x} = -A_1 n \sin nt + A_2 n \cos nt, \ \dot{y} = -A_3 n \sin nt + A_4 n \cos nt.$$

Initially, $t = 0, x = a, \dot{x} = 0, y = 0, \dot{y} = u$

$$\therefore \quad A_1 = a, A_2 = 0, A_3 = 0, A_4 = \frac{u}{n}.$$

$$\therefore \quad x = a\cos nt, \ \dot{x} = -an\sin nt, \ y = \frac{u}{n}\sin nt, \ \dot{y} = u\cos nt.$$

Hence, *P* describes an ellipse given by

$$\left(\frac{x}{a}\right)^2 + \left(\frac{ny}{u}\right)^2 = 1,$$

and the hodograph of the motion is also an ellipse given by

$$\left(\frac{x}{an}\right)^2 + \left(\frac{y}{u}\right)^2 = 1.$$

Let $\phi = nt$, so that the equation of the path of *P* is given by

$$x = a\cos\phi, y = b\sin\phi,$$

where $b = u/n$.

Thus the path is an ellipse whose eccentric angle is ϕ. Hence, its rate of change $= \dfrac{d\phi}{dt} = n = \cos nt$.

5. A particle of unit mass is describing an elliptic orbit of axes $2a$ and $2b$ under an attraction μr to the centre when it is at a distance, r from C; prove that its speed is given by $v^2 = \mu(a^2 + b^2 - r^2)$, and that the moment of its velocity about C is $\sqrt{\mu\, ab}$.

If, in the above orbit, $a = 2b$ and the intensity of attraction is suddenly increased in the ratio 1 : 4, when the particle is at a distance $3b/2$ from C, prove that axes of the new orbit are

$$\frac{(\sqrt{79} \pm \sqrt{15})\, b}{4}.$$

Equation of the path of the particle is

$$\frac{h^2}{p^3}\frac{dp}{dr} = \mu r.$$

Integration gives

$$\frac{h^2}{p^2} = v^2 = -\mu r^2 + A.$$

When $r = b$, $p = b$,

$$\therefore \qquad A = \frac{h^2}{b^2} + \mu b^2,$$

when $r = a$, $p = a$

$$\therefore \qquad A = \frac{h^2}{a^2} + \mu a^2,$$

$$\therefore \qquad \frac{h^2}{a^2} + \mu a^2 = \frac{h^2}{b^2} + \mu b^2,$$

$$\therefore \qquad h^2\left(\frac{1}{a^2} - \frac{1}{b^2}\right) = \mu(b^2 - a^2),$$

$$\therefore \qquad h^2 = \mu a^2 b^2,$$

$$\therefore \qquad h = \sqrt{\mu}\, ab.$$

This is the moment of velocity about C. Substituting the value of h in the value of A, we get

$$A = \mu(a^2 + b^2),$$

$$\therefore \quad v^2 = \frac{h^2}{p^2} = \mu(a^2 + b^2 - r^2).$$

When $a = 2b$, $r = 3b/2$, then

$$v^2 = \mu\left(4b^2 + b^2 - \frac{9b^2}{4}\right) = \frac{11b^2}{4}.$$

When the intensity of attraction is suddenly increased in the ratio 1 : 4, then the equation of the path is given by

$$\frac{h^2}{p^3}\frac{dp}{dr} = 4\mu r = \mu' r,$$

where $\mu' = 4\mu$.
Integration gives

$$\frac{h^2}{p^2} = v^2 = -\mu' r^2 + A',$$

When $r = 3b/2$, $v^2 = \dfrac{11\mu b^2}{4} = \dfrac{11\mu' b^2}{16}$,

$$\therefore \quad A' = \frac{11\mu' b^2}{16} + \frac{9\mu' b^2}{4} = \frac{47\mu' b^2}{16},$$

$$\therefore \quad v^2 = \mu'\left(\frac{47b^2}{16} - r^2\right).$$

If the semi-axes of the new ellipse are a', b', then

$$a'^2 + b'^2 = \frac{47b^2}{16}. \qquad (1)$$

Now h is constant

$$\therefore \quad h^2 = \mu a^2 b^2.$$

Putting $a = 2b$, we get

$$h^2 = 4\mu b^4 = \mu' b^4$$

In second case

$$h^2 = \mu' a'^2 b'^2.$$

Equation both values of h^2 we get

$$a'b' = b^2 \qquad (2)$$

From (1) and (2), we get

$$a' + b' = \frac{\sqrt{79}b}{4}, \; a' - b' = \frac{\sqrt{15}b}{4}$$

$$\therefore \qquad 2a' = \frac{(\sqrt{79}+\sqrt{15})\,b}{4},\; 2b' = \frac{(\sqrt{79}-\sqrt{15})\,b}{4}.$$

6. Two particles P and Q each of unit mass, attract one another with a force $n^2\,PQ$. The two particles move under their mutual attraction and also under the influence of a fixed centre of force O which attracts with a force n^2 times the distance from O. Show that G, the centre of mass of P and Q describes an ellipse about O as centre with period $2\pi/n$ and that the orbit of P or Q relative to G is an ellipse with G as centre described in the period $2\pi / n\sqrt{3}$.

   (*i*) If initially OPQ is a straight line with $OP = PQ = a$, and Q is at rest, while P is projected at right angles to OPQ with speed v, show that the axes of the ellipse described by G are of lengths $3a$ and v/n. Find the lengths of the axes described by P relative to G.

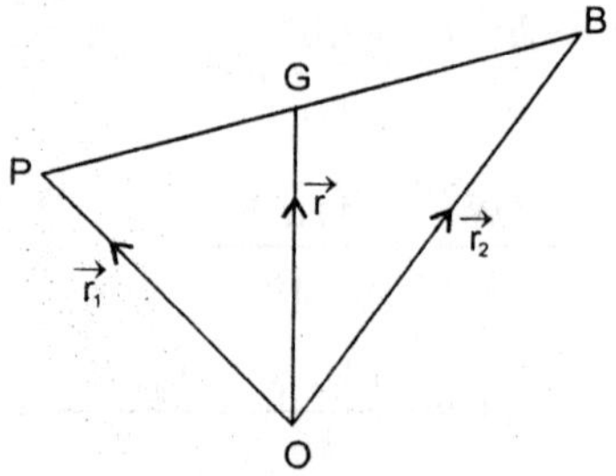

Fig. 5.11

Let $\qquad \overrightarrow{OP} = \vec{r}_1, \overrightarrow{OQ} = \vec{r}_2, \vec{R} = \vec{r}_1 + \vec{r}_2.$

Position vector of $G = \overrightarrow{OG} = \dfrac{\vec{r}_1 + \vec{r}_2}{2}.$

Hence, equation of motion of G is given by

$$2.\frac{1}{2}(\ddot{\vec{r}}_1 + \ddot{\vec{r}}_2) = -n^2(\vec{r}_1 + \vec{r}_2),$$

$$\therefore \qquad \vec{R}_1 = -n^2\vec{R}. \qquad (1)$$

This is an elliptic harmonic motion with period $2\pi/n$.

(*ii*) Force at P and $Q = n^2\overrightarrow{PQ} = 2n^2\overrightarrow{PG}$.

G will be attracted by P or Q with a force $n^2\overrightarrow{GP}$. In order to reduce it to rest, we apply additional force $n^2\overrightarrow{PG}$ on P and G both. Then G is

reduced to rest and force at P will be $3n^2\overrightarrow{PG} = -3n^2\vec{\alpha}$, if $\overrightarrow{GP} = \vec{\alpha}$.
Hence, equation of motion of P relative to G is

$$\ddot{\vec{\alpha}} = -3n^2\vec{\alpha}. \qquad (2)$$

This represents, elliptic harmonic motion with period $2\pi / n\sqrt{3}$,

(*iii*) Equation of motion of G is

$$\ddot{\vec{R}} = -n^2\vec{R}$$

Taking

$$\vec{R} = x\hat{i} + y\hat{j},$$

we get

$$\ddot{x} = -n^2x, \ \ddot{y} = -n^2y.$$

Solution of these equations are given by

$$x = A\cos(nt + r), y = B\cos(nt + \beta).$$

Initially, $t = 0, x = 3a/2, y = 0, \ \dot{x} = 0, \ \dot{y} = v/2.$

$$\therefore A = 3a/2, B = -v/2n, r = 0, \beta = \pi/2.$$

$$\therefore x = \frac{3a}{2}\cos nt, y = \frac{v}{2n}\sin nt.$$

These are parametric equations of the ellipse with axes $3a$ and v/n.
From (2), we have

$$\ddot{\vec{\alpha}} = -3n^2\vec{a}.$$

Let

$$\vec{\alpha} = x_1\hat{i} + y_1\hat{j}.$$

Hence, we get equations of motion of P relative to G as

$$\ddot{x}_1 = -3n^2x_1, \ \ddot{y}_1 = -3n^2y_1$$

Solutions of these equations are

$$x_1 = A_1\cos(\sqrt{3}nt + \alpha_1), y_1 = B_1\cos(\sqrt{3}nt + \beta_1).$$

Initially, $x_1 = a/2, \ \dot{x}_1 = 0, y_1 = 0, \ \dot{y}_1 = v/2.$

$$\therefore \quad \alpha_1 = 0, A_1 = a/2, \beta_1 = \pi/2, B_1 = -v/2n\sqrt{3}.$$

$$\therefore \quad x_1 = \frac{a}{2}\cos\sqrt{3}\,nt, y_1 = \frac{v}{2n\sqrt{3}}\sin\sqrt{3}\,nt.$$

This represents an ellipse with axes a and $v/\sqrt{3}n$.

7. A comet travelling in an elliptic orbit under an attraction $\frac{\mu}{r^2}$ per unit mas has its tangential velocity increased by a small amount δv. Taking $2a$ t be axis and e the eccentricity of the former orbit, show that the comet' least distance from the Sun is increased by $4\delta v\left[\frac{a^3(1-e)}{\mu(1+e)}\right]^{1/2}$

At the point A, the comet is at greatest distance from Sun S. The velocit at A is given by

$$v^2 = \mu\left(\frac{2}{SA}-\frac{1}{a}\right) = \mu\left[\frac{2}{a(1+e)}-\frac{1}{a}\right] = \frac{\mu(1-e)}{a(1+e)}. \quad (1)$$

We know that

$$\delta a = \frac{2a^2\,\delta v}{\mu},\ \delta e = \frac{(1-e^2)(av^2-\mu)\,\delta v}{\mu e v}.$$

Substituting the value of v^2 from (1), we get

$$\delta e = \frac{(1-e^2)\,\delta v}{ev}\left[\frac{1-e}{1+e}-1\right] = -\frac{2(1-e)\,\delta v}{v}.$$

Now the comet is at shortest distance from Sun S at A'.

Let $\quad r' = SA' = a(1-e).$

$\therefore \quad \delta r' = \delta a\,(1-e) - a\,\delta e$

$$= \frac{(1-e)\,2a^2\,v\,\delta v}{\mu} + \frac{2a(1-e)\,\delta v}{v}$$

$$= 2a(1-e)\left(\frac{av}{\mu}+\frac{1}{v}\right)\delta v$$

$$= 2a(1-e)\left[\frac{\sqrt{a(1-e)}}{\sqrt{\mu(1+e)}}+\frac{\sqrt{a(1+e)}}{\sqrt{\mu(1-e)}}\right]\delta v$$

$$= \frac{4a^{3/2}(1-e)\,\delta v}{\sqrt{\mu(1-e^2)}}$$

$$= 4\left[a^3(1-e)/\mu(1+e)\right]^{1/2}\delta v.$$

EXERCISE

1. A particle is describing an ellipse of eccentricity e about the centre of force at focus. Prove with usual notation.

$$v^2 = \mu\left(\frac{2}{r} - \frac{1}{a}\right), h^2 = \mu a (1 - e^2).$$

When the particle is at one end of the minor axis, its velocity is doubled. Prove that the new path is a hyperbola of eccentricity $(9 - 8e^2)^{1/2}$.

2. If two particles P, Q describe the same ellipse under the same force to the centre C, prove that the area of the triangle CPQ is invariable.

3. A fixed nucleus S, having positive charge Ze, repels a particle P having mass m and positive charge e'. P is projected from a great distance at infinity with initial speed v_0 in the direction whose perpendicular distance from s is d, the medium being a vacuum. Show that its ultimate direction makes an angle ϕ with an initial direction where $\cot\frac{\phi}{2} = \frac{dmv_0^2}{Zee'}$.

4. A particle is projected from a given point P under the action of forces whose accelerating effects are μPS, $\mu PS'$ directed towards fixed points S, S'. Find the magnitude and direction of the velocity of projection in order that the orbit may be ellipse with S, S' focii.

5. If $P = \mu(u^2 - au^3)$, where $a > 0$, and the particle is projected from an apse at a distance a from the centre of force with velocity $(\mu c/a^2)^{1/2}$, where $a > c$, prove that the other apsidal distance of the orbit is $a(a + c)/(a - c)$ and find the apsidal angle.

6. Two particles are describing the same ellipse about a centre of force in the centre in opposite directions, the mass of one being double that of the other. If the particles meet and coalesce at the end of the minor axis, show that the new orbit trisects the major axis of the old.

7. A particle describes an ellipse with major axis $2a$ and eccentricity e under a force directed to the focus S. When the particle is at the end of the latus rectum through S its velocity is suddenly turned through a right angle without change of magnitude. Prove that the particle will now describe an ellipse whose major axis is still of length $2a$ but is inclined at an angle

$\tan^{-1}\frac{1-e^2}{e}$ to the major axis of the original orbit. Find the eccentricity of the new orbit.

8. A particle of unit mass describes an ellipse under the action of central force μc. Show that the normal component of the acceleration at any instant is $ab\mu^{3\,2}/v$, where v is the velocity at that instant and a, b are semi-axes of the ellipse.

9. Show that speed v of a particle moving under a central force μ/r^2 per unit mass towards the pole (origin) is given by $v^2 = C + 2\mu/r$, and show that the path is one branch of hyperbola when $C > 0$. When at a great distance from the Sun, a comet moves towards it with a speed V in the direction of the straight line whose perpendicular distance from the Sun's centre is p. Show that the angle through which comet's path is ultimately deflected is $2 \cot^{-1} (pv^2/M\gamma)$, where M is the mass of the Sun and γ is a constant of gravitation.
10. A particle is repelled from a centre of force O with a force μr per unit mass, where r is the distance of the particle from O. Show that, if the particle is projected from a piont P in any direction with velocity $OP\sqrt{\mu}$ its path is a rectangular hyperbola with O as centre.
11. A particle is projected from a point A at right angles to SA, and is acted on by a force varying inversely as the square of the distance towards S. If the intensity of the force is unity at unit distance from SA, and the veloicty of projection is 1/2, prove that the eccentricity of the orbit is 3/4, and find the periodic time.
12. A star Q of mass M, initially at rest, is approached by a star of mass m whose velocity V at a great distance from Q is along a line at a perpendicular distance from Q. Show that the orbit of P relative to Q is a hyperbola with semi-major axis of length $a = \dfrac{\gamma (M+m)}{V^2}$ and that the velocity of P realtive to Q is displaced through an angle 2α, where $\tan \alpha = \dfrac{a}{p}$, γ being constant of gravitation.
Show further that Q acquires a velocity of magnitude $2\gamma mV [\gamma^2 (M+m)^2 + p^2V^4]^{-1/2}$ and determine its direction.
13. A particle P moves under a force $\omega^2 \overrightarrow{PO}$ per unit towards the fixed point O. If its initial position is A and its initial velocity is represented by $\omega \overrightarrow{OC}$, show that its position at time t is given by

$$\overrightarrow{OP} = \overrightarrow{OA} \cos \omega t + \overrightarrow{OC} \sin \omega t.$$

and that the orbit is an ellipse with centre O.
14. A particle describes an ellipse under the action of a force to focus, if the particle arriving at any point in the orbit is diverted by a blow in the direction of the normal so that the tangent to the new orbit at that point passes through the second focus, prove that the latus rectum of the new path is four times that of the old.
15. A particle is projected from infinity with velocity V so as to pass through a fixed point at a distance c of undisturbed orbit. If it is attracted to the

fixed point with acceleration μu^2, where u is reciprocal of radius vector, show that the equation of the orbit is

$$u = \frac{\mu}{V^2c^2} + \frac{\cos\theta}{c}\sqrt{1+\frac{\mu^2}{V^4c^2}},$$

θ being measured from the apse.

16. Show that, if a particle describes an ellipse under a force to focus, the velocity at the mean distance from the centre of force is a mean proportional between the velocities at the ends of any diameter.

17. A planet of small mass is describing an elliptical orbit round Sun of large mass. When the planet is at perihelion, the mass of Sun is suddenly doubled. Show that the planet will continue to describe an elliptic orbit, but that its velocity at perihelion to its former velocity at aphelion as twice the major axis of the former orbit is to the major axis of the new one.

18. A particle acted on by a central attractive force μu^3 is projected with a velocity $\frac{\sqrt{\mu}}{a}$ at an angle $\pi/4$ with its initial distance a from the centre of force, show that its orbit is equiangular spiral $r = ae^{-\theta}$.

19. If a particle is projected from an apse at a distance a with velocity from infinity under the action of a central force μr^{-2n-3}, prove that the path is $r = a^n \cos n\theta$.

20. Find the velocity necessary for the description of a circular orbit of radius a under a central force $2\mu a^2u^5 - \mu u^3$. Show that orbit is unstable and that if a slight disturbance takes place inwards the path may be represented by $r = a \tan h\ \theta$.

21. A particle of mass m moves under a central force $m\mu\ \{3au^4 - 2\,(a^2 - b^2)\,u^5\}$, $a > b$, and is projected from an apse at a distance $a + b$, with velocity $\frac{\sqrt{\mu}}{a+b}$. Show that its orbit is $r = a + b\cos\theta$.

22. Prove that, if the law of force is $3\mu u^3 + 2\mu a^2u^5$ and a particle is projected in direction making an angle $\cot^{-1} 2$ with the initial distance a, and with a velocity equal to that in a circle at the same distance, the orbit is

$$au = \tan(\theta + \pi/4).$$

23. If the law of force be $2\mu(u^3 - a^2u^5)$ and the particle be projected from an apse at a distance a with velocity $\frac{\sqrt{\mu}}{a}$ show that it will be at a distance r from the centre after a time

$$\frac{r\sqrt{r^2 - a^2} + a^2\cos h^{-1} r/a}{2\sqrt{\mu}}.$$

24. A particle describes a circle of radius a so that its hodograph is a second circle of radius b. If the pole of the hodograph is at a distance c from its centre, where c/b is small, show that the time of a complete revolution is approximately

$$\frac{2\pi a\,(1 + 3c^2 / 4b^2)}{b}$$

25. Two gravitating particles of mass m and M, starting from rest at an infinite distance apart, are allowed to fall freely towards one another. Show that when their distance apart is a, their relative velocity of approach is $\sqrt{\frac{2\gamma\,(M+m)}{a}}$, γ being the constant of gravitation.

Impulses of magnitude I are applied to the particles in the opposite directions perpendicular to their line of motion at the instant when their distance apart is a. Show that p, their distance of closest approach during the subsequent motion, is a positive root of a certain quadratic equation, and that if I is a small, p is given approximately by

$$\frac{p}{a^2} = \frac{I^2 (M+m)}{2\gamma\, M^2 m^2}.$$

26. When a body moving in an ellipse about the focus is at an end of the latus rectum, a small velocity u is communicated to it parallel to the axis. Prove that the change of eccentricity is ehu/μ and that the apse line is turned through an angle $2hu/e\mu$.
27. In an elliptic orbit about the focus, if the direction of motion at any point is turned through a small angle $\delta\alpha$, show that the apse line is turned through an angle $\frac{\delta\alpha\,(1 + e^2 - r/a)}{e^2}$.
28. A particle is describing an ellipse under a force to the focus when at a certain point the intensity (μ) of the force is increased by ($1/n$)th. Show that the semi-major axis is increased by $\frac{a^2 v^2}{n\mu}$, where v is the velocity, and find the position of the point (1) if the eccentricity is unaltered, (2) if the position of the apse line is unaltered. Can these conditions be fulfilled at one point?
29. If the velocity at any point in an ellipse described about centre of force in the focus be suddenly increased by a small given quantity, prove that alteration in the position of the transverse axis is proportional to the cosine of the angle which the direction of the motion at the point makes with the radius vector.

30. Find the la:v of force to the pole when the path is cardiod $r = a(1 - \cos\theta)$; and prove that, if F be the force at the apse, and v the velocity,

$$3v^2 = 4aF.$$

31. A body is describing an ellipse under the action of the force to a focus. When the body is at one extremity of the minor axis the law of force is changed without instantaneous change of the magnitude of the force or of the velocity; if the force now varies as the distance, prove that the periodic time is the same as before.

32. A curve described under the action of a central force is such that, if at any moment the component velocity along the radius vector be destroyed by an impulse along the radius vector; the particle will proceed to describe a circle. Prove that the curve is a reciprocal spiral.

33. A particle of mass m, is acted upon by a central force which is some function of the distance r and by a transverse force $m\mu r$. Prove that, if the particle moves completely round any close curve, the square of the velocity is increased by 4μ (area of the curve).

34. A particle describes the curve $r^2 = a^2 \cos 2\theta$ under the action of a force to the pole. Prove that the square of the force varies as the seventh power of the angular velocity of the radius vector of the particle.

35. A particle m is projected from an apse with the velocity from infinity under the action of the force $\frac{m\mu}{r^3}.\log\frac{r}{a}$ directed to a centre, at a distance a; find the equation of the orbit.

36. A particle under the action of a central force $\mu(r + a)/r^3$ is projected from an apse at a distance a with velocity which is to the that in a circle at the same distance $1:\sqrt{2}$. Show that the equation of the central orbit is $r(2 + \theta^2) = 2a$, and that the particle will arrive at origin in time $\pi\sqrt{\frac{a^3}{8\mu}}$.

37. A particle is projected at a distance a from a centre of force with a velocity $2\sqrt{\mu/a}$, and at an angle $\pi/4$ with the radius vector, the force being $\mu\left(\frac{3}{r^3}+\frac{a^2}{r^5}\right)$; determine the orbit described, and show that the time to the centre of the force is $\frac{a^2(2-\pi/2)}{\sqrt{2\mu}}$.

38. A body is describing an ellipse of eccentricity e under the action of a force tending to focus and when at the nearer apse the centre of force is transferred to the other focus. Prove that the eccentricity of the new orbit is $\dfrac{e(3+e)}{1-e}$.

39. A particle describes an ellipse of major axis $2a$ under the action of a force to focus. When the particle is at end of minor axis the centre of force is suddenly moved a small distance $a\alpha$ towards the particle. Show that the eccentricity of the orbit is unaltered, and that the apse line is turned through a small angle $\alpha\sqrt{e^2-1}$.

40. The co-ordinates (x, y), (x_1, y_1) of the simultaneous position of two equal particles are given by the equations

$$x = a(\theta - 2\sin\theta),\ x_1 = a\theta,$$
$$y = a(1-\cos\theta),\ y_1 = -a(1 - a\cos\theta);$$

prove that if they move under their mutual attractions, the law of force will be the inverse fifth power of the distance.

41. Two bodies of masses m_1 and m_2 move under their mutual gravitation; prove that if when m_1 is at rest the velocity of m_2 is at right angles to the line joining the masses and equal to $\sqrt{\dfrac{\gamma(m_1+m_2)}{d}}$ where γ is a constant of integration and d the distance of bodies, then the path of m_1 in space is a succession of cycloids, the body coming to rest at a cusp at equal intervals of time.

42. A particle is projected from an apse at a distance c with velocity $\dfrac{\sqrt{2\mu c^3}}{\sqrt{3}}$. The force to the centre being $m\mu(r^5 - c^4 r)$, prove that orbit is $x^4 + y^4 = c^4$.

43. Show that the time average of the kinetic energy of a planet is equal to the value at the end of the minor axis of its orbit. Show aslo that the time mean of the distance is $a(1 + e^2/2)$.

44. A particle describes an ellipse under a force to the focus S, and SY is a perpendicular upon the tangent to the orbit; show that the angular velocity of SY is minimum or maximum when the particle is at the farther apse according to the eccentricity is less than or greater than 1/3.

6

MOTION OF SYSTEM OF PARTICLES

6.1 CONSERVATION THEOREMS FOR SYSTEM OF PARTICLES

I. Conservation theorems for linear momentum

Let a system consists of n particles moving generally in space, and m be the constant mass of a typical particle P, such that $\vec{OP} = \vec{r}$, O being a fixed origin. Let G be the C.G. of the system and $\vec{OG} = \bar{\vec{r}}$, M the total mass of the system, hence,

$$M = \sum m, \ M\bar{\vec{r}} = \sum m\vec{r},$$

the summation here and in the following being taken over the entire system.

Differentiating the last result with respect to t, we get

$$M\dot{\bar{\vec{r}}} = \sum m\dot{\vec{r}}.$$

$M\dot{\bar{\vec{r}}}$ is the linear momentum of a single particle of mass M concentrated at G, and $\sum m\dot{\vec{r}}$ is the linear momentum of the system of particles. Thus we see that at *any instant the total linear momentum of the system of particles moving generally in space is equal to that of a single particle concentrated at the centroid of the system, having a velocity equal to that of the centroid and mass equal to the total mass of the system.*

Denoting the total linear momentum of the system by $\vec{p}$, we have

$$\dot{\vec{p}} = \sum m\ddot{\vec{r}} = M\ddot{\bar{\vec{r}}},$$

Thus we see at *any instant the rate of change of linear momentum of the system is equal to the product of total mass of the system and the acceleration of the centroid.*

Now, the particle P experiences two types of forces:

(1) The external forces due to agents outside the system.

(2) The internal forces experienced by given particle due to all other particles of the system.

Suppose the particle is acted upon by an external force $\vec{F}$ and internal force $\vec{F}'$. Since, total force on P is $\vec{F}+\vec{F}'$ hence, its equation of motion is

$$m\ddot{\vec{r}} = \vec{F}+\vec{F}'$$

Taking summation over the entire system, we get

$$\sum m\ddot{\vec{r}} = \sum \vec{F} + \sum \vec{F}'.$$

From Newton's third law action and reaction are equal and opposite, the internal forces occur in equal and opposite pairs, and hence they contribute nothing to the total system. In other words, the forces exerted by any two particles of the sustem on each other are equal and opposite; and lie along the line joining them, therefore sum of the internal forces is zero. Hence, we have

$$\sum \vec{F} = \sum m\ddot{\vec{r}} = \dot{\vec{p}}$$

This shows that *at any instant the sum of external forces acting on the system is equal to the rate of change of linear momentum of the system.*
In particular case, when there is no external force *i.e.*, when

$$\sum \vec{F} = 0, \dot{\vec{p}} = 0 \therefore \vec{p} = const.$$

Thus we get principle of conservation of linear momentum for the system which states *in the absence of external force the linear momentum of the system is constant.*

II. Conservation theorem for angular momentum

Now angular momentum or moment of mementum of the particle P about the fixed point O is given by

$$\vec{r} \times m\dot{\vec{r}}.$$

Hence, the total angular momentum about O is

$$\vec{H} = \sum (\vec{r} \times m\dot{\vec{r}}) \tag{1}$$

The physical angular momentum about a line specified by a unit vector $\vec{a}$ is then $\vec{a}.\vec{H}$.

Differentiating (1) with respect to t,

$$\therefore \quad \dot{\vec{H}} = \sum (\dot{\vec{r}} \times m\dot{\vec{r}}) + \sum (\vec{r} \times m\ddot{\vec{r}}),$$

$$\therefore \quad \dot{\vec{H}} = \sum (\vec{r} \times m\ddot{\vec{r}}). \tag{2}$$

(2) *shows that the rate of change of angular momentum about the fixed point O is equal to the total moment of rate of change of linear momentum about O.*

$$m\ddot{\vec{r}} = \vec{F} + \vec{F}',$$

$$\therefore \quad \dot{\vec{H}} = \sum(\vec{r} \times \vec{F}) + \sum(\vec{r} \times \vec{F}').$$

Let suffixes i, j refer to different particles of the system and let $\vec{F}'_{ij}$ denote the force which particle P_j exerts on particle P_i then $\vec{F}'_{ij} = -\vec{F}'_{ji}$, in virtue of Newton's third Law of motion ($i \neq j$). Further

$$\sum(\vec{r} \times -\vec{F}') = \sum_i\sum_j \vec{r}_i \times \vec{F}'_{ji} \qquad (i \neq j)$$

$$= \sum_i\sum_j \left(\vec{r}_i - \vec{r}_j\right) \times \vec{F}'_{ji} + \sum_i\sum_j \left(\vec{r}_j \times \vec{F}'_{ji}\right)$$

Now, $\vec{F}'_{ji}$ is along the line joining P_i, P_j, hence, $(\vec{r}_i - \vec{r}_j)$ and $\vec{F}'_{ji}$ are parallel, therefore

$$\left(\vec{r}_i - \vec{r}_j\right) \times \vec{F}'_{ji} = \vec{0}.$$

Since,

$$\vec{F}'_{ji} = -\vec{F}'_{ij}$$

$$\therefore \quad \sum_i\sum_j \left(\vec{r}_j \times \vec{F}'_{ji}\right) = -\sum_i\sum_j \left(\vec{r}_j \times \vec{F}'_{ij}\right) = -\sum_j\sum_i \left(\vec{r}_j \times \vec{F}'_{ij}\right)$$

$$= -\sum(\vec{r} \times \vec{F}').$$

Thus

$$\sum(\vec{r} \times \vec{F}') = \vec{O}.$$

Hence,

$$\dot{\vec{H}} = \sum(\vec{r} \times \vec{F}). \tag{3}$$

From equation (3) we see that *the rate of change of vector angular momentum about a fixed point for a system of particles moving generally in space is equal to the sum of moments of the external forces acting on that system about the point.*

In the particular case, if external forces are absent, then

$$\dot{\vec{H}} = 0,$$

$$\therefore \quad \vec{H} = \text{const.} \tag{4}$$

This is the principle of conservation of angular momentum, which states that *in case of absence of external forces the angular momentum of the system of particles is conserved.*

Let $\vec{a}$ specify the unit vector along an axis passing through the fixed point *O*.

Taking scalar product of equation (3) with $\vec{a}$, we get

$$\vec{a}.\dot{\vec{H}} = \vec{a}.\sum(\vec{r} \times \vec{F}) = \sum\left\{\vec{r}.\left(\vec{a} \times \vec{F}\right)\right\}. \tag{5}$$

Equation (5) states that *the rate of angular momentum about any axis through a fixed point for such a system is equal to physical moments of the external forces acting on that system about the line.*

Examples

1. Three interacting particles of mass m, $2m$, $4m$ have each a velocity v along the positive directions of x, y, z respectively. Find the velocity of the first particle if due to forces of interaction the third particle stops moving and the velocity of the second particle is $(2\vec{j} + \vec{k})\, v_2$, where $\vec{i}, \vec{j}, \vec{k}$ are unit vectors along exes of co-ordinates.

Initial linear momentum $= mv\vec{i} + 2mv\vec{j} + 4mv\vec{k} = m(\vec{i} + 2\vec{j} + 4\vec{k})v.$

Let $\vec{v}_1$ be the velocity of the first particle, when third particle stops moving. Now Angular momentum $= m\vec{v}_1 + 2m(2\vec{j} + \vec{k})v_2$.

Using principle of conservation of linear momentum, we get

$$m(\vec{i} + 2\vec{j} + 4\vec{k})v = m\vec{v}_1 + 2m(2\vec{j} + \vec{k})v_2.$$

$$\therefore \quad \vec{v}_1 = v\vec{i} + 2(v - 2v_2)\vec{j} + 2(2v - v_2)\vec{k}.$$

2. A particie moving with velocity 30m / sec. collides with another particle of the same mass which is initially at rest. After collision the first particle moves along a line making an angle 30° to its initial position. Find the velocity of each particle after collision and the angle made by the second particle with the initial direction of the first.

Let m be the mass of each particle $v_1 = 30$ m/sec. the velocity of the first particle before collision and v'_1, v'_2, be the velocities of the first and second particle after collision, $\theta_1 = 30°$, θ_2 the angles which v'_1, v'_2, make with the initial direction of the motion of the first particle.

Using the principle of conservation of linear momentum along and perpendicular to the initial direction of the first particle, we have

$$mv_1 = mv'_1\cos\theta_1 + mv'_2\cos\theta_2, \tag{1}$$

$$0 = mv'_1\sin\theta_1 - mv'_2\sin\theta_2, \tag{2}$$

Using principle of conservation of energy

$$\frac{1}{2}mv_1^2 = \frac{1}{2}mv_1'^2 + \frac{1}{2}mv_2'^2,$$

$$\therefore \quad v_1^2 = v_1'^2 + v_2'^2. \tag{3}$$

Squaring (1) and (2) and adding, we get

$$(v_1 - v_1'\cos\theta_1)^2 + v_1'^2 \sin^2\theta_1 = v_2'^2$$

$$\therefore \quad v_1^2 + v_1'^2 - 2v_1v_1'\cos\theta_1 = v_1^2 - v_1'^2, \text{ using (3)}$$

$$\therefore v_1' = v_1\cos\theta_1.$$

Putting $v_1 = 30, \theta_1 = 30^\circ$, we get

$$v_1' = 30\cos 30^\circ = 15\sqrt{3} \cong 26m / \text{sec}.$$

From (3), we get

$$v_2'^2 \cong (30)^2 - (26)^2 \cong 56 \times 4 \cong 15m / \text{sec}.$$

From (2)

$$\sin\theta_2 = \frac{v_1'\sin\theta_1}{v_2'} \cong \frac{15\sqrt{3}\sin 30^\circ}{15} \cong \frac{\sqrt{3}}{2}$$

$$\therefore \theta_2 \cong 60^\circ.$$

3. Two particle of masses m_1, m_2 collide and stick together on collision. If m_2 is at rest at origin and m_1 is moving with velocity v_1 along +ve direction of x-axis before collision, then find the ratio of K.E. after collision to the initial K.E.

Momentum along x-axis before collision = m_1v_1.
Let V be the velocity of the system after collision, then
Momentum after collision = $(m_1 + m_2)V$.
Using principle of conservation of linear momentum, we get

$$m_1v_1 = (m_1 + m_2)V,$$

$$\therefore \quad V = \frac{m_1v_1}{(m_1 + m_2)}. \tag{1}$$

K.E. before collision is

$$T_1 = \frac{1}{2}(m_1v_1^2).$$

K.E. after the collision is

$$T_2 = \frac{1}{2}(m_1 + m_2)V^2.$$

Using (1)

$$T_2 = \frac{1}{2}\frac{m_1^2 v_1^2}{(m_1 + m_2)}.$$

$$\therefore \quad \frac{T_2}{T_1} = \frac{m_1}{(m_1 + m_2)}.$$

4. A hemisphere of mass M is free to slide on a smooth horizontal table. A particle of mass m is placed on the hemisphere at an angular distance α from the vertex, determine the motion, the surface of the hemisphere being smooth.

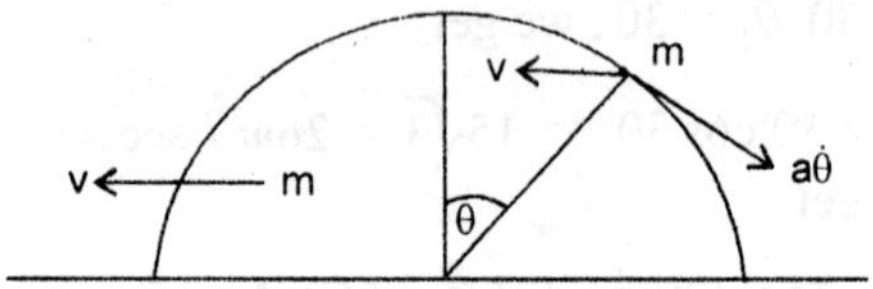

Fig. 6.1

Suppose at time t the particle is at an angular distance θ from the vertex, and that hemisphere has acquired velocity v, the velocity of the particle being $a\dot{\theta}$ relative to the hemisphere, if a is the radius of the hemisphere. Since, there is no external force acting on the system the total horizontal momentum remains zero throughout the motion,

$$\therefore \quad Mv + m\,(v - a\dot{\theta}\cos\theta) = 0. \tag{1}$$

Gain in K.E. = Work done.

$$\therefore \frac{1}{2}Mv^2 + \frac{m}{2}(v^2 + a^2\dot{\theta}^2\cos^2\theta - 2av\dot{\theta}\cos\theta) = mga\,(\cos\alpha - \cos\theta). \tag{2}$$

Eliminating v from (1) and (2), we get

$$a\dot{\theta}^2 = \frac{2g(M + m)(\cos\alpha - \cos\theta)}{(M + m\sin^2\theta)}, \tag{3}$$

which gives the velocity of the particle relative to the hemisphere, and then from (1). we get

$$v = \frac{ma\dot{\theta}\cos\theta}{(M + m)} \tag{4}$$

$$= m\cos\theta\sqrt{\frac{2ag(\cos\alpha - \cos\theta)}{(M + m)(M + m\sin^2\theta)}}.$$

Let R be the mutual reaction between hemisphere and the particle.

Resolving horizontally for the hemisphere, we get

$$M\dot{v} = R\sin\theta. \tag{5}$$

From (4) and (5) we get

$$\frac{ma(\ddot{\theta}\cos\theta - \dot{\theta}^2\sin\theta)}{(M+m)} = R\sin\theta.$$

The particle will leave the hemisphere, when $R = 0$.

$$\therefore\ ma(\ddot{\theta}\cos\theta - \dot{\theta}^2\sin\theta) = 0,$$

$$\therefore\quad \ddot{\theta}\cos\theta = \dot{\theta}^2\sin\theta \tag{6}$$

Differentiating (3) with respect to t, we get

$$2a\dot{\theta}\ddot{\theta} = \frac{2(M+m)g\left[(M+m\sin^2\theta)\sin\theta\dot{\theta} - (\cos\alpha - \cos\theta)2m\sin\theta\cos\theta\dot{\theta}\right]}{(M+m\sin^2\theta)^2}.$$

$$\ddot{\theta} = \frac{(M+m)g\left[(M+m\sin^2\theta)\sin\theta - (\cos\alpha - \cos\theta)2m\sin\theta\cos\theta\right]}{a(M+m\sin^2\theta)^2}.$$

Putting this value of $\ddot{\theta}$ and $\dot{\theta}^2$ in (6), we get

$$m\cos^3\theta - (M+m)(3\cos\theta - 2\cos\alpha) = 0.$$

5. A bead of Mass M can slide on a smooth straight horizontal wire and a particle of mass m is attached to the bead by a light string of length l. The particle is held in contact with the wire with the string taut and is then let fall. Prove that when the string is inclined at an angle θ the bead will have slipped the distance

$\dfrac{ml(1-\cos\theta)}{(M+m)}$ along the wire, and that the angular velocity ω of the string will be given by $(M + m\cos^2\theta)\, l\omega^2 = 2(M+m)g\sin\theta$.

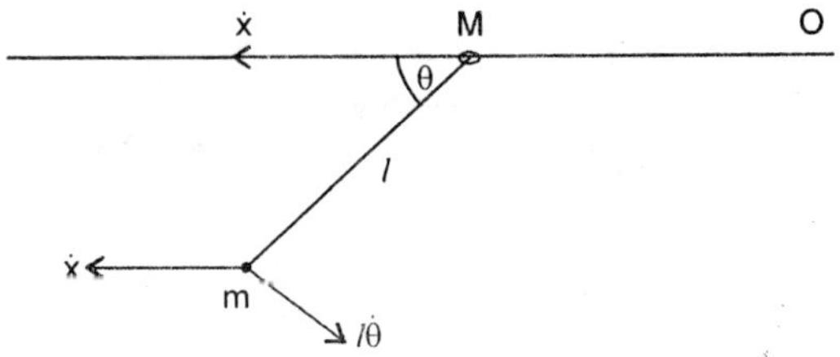

Fig. 6.2

Suppose that the bead moves a distance x while the string turns throught an angle θ. Then velocity of the bead is $\dot{x}$ and the particle m has a velocity $l\dot{\theta}$ relative to the bead in addition to the velocity of the bead.

Since, there is no external horizontal force, the total horizontal momentum remains zero, therefore

$$M\dot{x} + m(\dot{x} - l\dot{\theta}\sin\theta) = 0. \qquad (1)$$

Integrating, we get

$$Mx + m\,(x + l\cos\theta) = const.$$

Initially, $x = \theta = 0$,

$$\therefore \quad \text{const.} = ml.$$

$$\therefore Mx + m\,(x + l\cos\theta) = ml,$$

$$\therefore x = \frac{ml(1-\cos\theta)}{(M+m)}.$$

Now gain in K.E. = Work done by gravity

$$\therefore \frac{1}{2}M\dot{x}^2 + \frac{m}{2}(\dot{x}^2 + l^2\dot{\theta}^2 - 2\dot{x}l\dot{\theta}\sin\theta) = mgl\sin\theta. \qquad (2)$$

Eleminating $\dot{x}$ from (1) and (2), we get

$$\frac{(M+m)m^2l^2\sin^2\theta\dot{\theta}^2}{(M+m)^2} + ml^2\dot{\theta}^2 - \frac{2m^2l^2\dot{\theta}^2\sin^2\theta}{M+m} = 2mgl\sin\theta$$

$$\therefore \quad (M + m\cos^2\theta)\,l\,\dot{\theta}^2 = 2\,(M+m)g\sin\theta.$$

6. A fine smooth wire of mass m forms an equilateral triangle ABC. The triangle can move horizontally in a vertical plane, the uppermost side BC passing through smooth fixed rings in a horizontal line of masses m and m' are free to slide on the wires BC, CA. The system begins to move with beads at B and C respectively. Prove that the velocity of the wire at any instant, while both the beads are moving on it, is equal to the difference of speeds of the beads relative to the wire, and the acceleration of the wire is

$$\frac{\sqrt{3}(m'-m)g}{(4M+3m+3m')}.$$

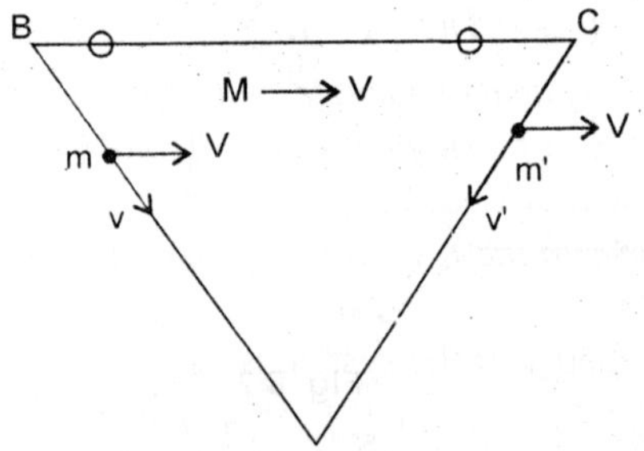

Fig. 6.3

At time t from the start, let V be the velocity of the wire and v, v' velocities of m and m' relative to the wire.

Since there is no horizontal force acting on the system as a whole, the horizontal momentum remains zero.

$\therefore MV + m(V + v\cos 60°) + m'(V - v'\cos 60°) = 0$ (1)

Resolving along BA for m and along CA for m', we get

$$\begin{aligned} m(\dot{v} + \dot{V}\cos 60^\circ) &= mg\cos 30^\circ \\ m'(\dot{v}' - \dot{V}\cos 60\ ) &= m' g\cos 30^\circ . \end{aligned} \qquad (2)$$

Subtracting these, we get

$$(\dot{v} - \dot{v}') = -\dot{V}.$$

Integrating, we get

$$(v - v') = -V - \text{const.}$$

Initially, $v = v' = V = 0$.

$$\therefore \quad \text{const.} = 0.$$

$$\therefore \quad V = v' - v.$$

Differentiating (1), we get

$$(M + m + m')\dot{V} = \frac{(m'\dot{v}' - m\dot{v})}{2}.$$

With the help of (2) it becomes

$$(M + m + m')\dot{V} = \frac{\left[m'\left(\dot{V} + \sqrt{3}g\right) + m\left(\dot{V} - \sqrt{3}g\right)\right]}{4}$$

$$\therefore \quad (4M + 3m + 3m')\dot{V} = \sqrt{3}\,(m' - m)\,g.$$

6.2 USE OF CENTROID

Let G be the centroid of the system such that

$$\overrightarrow{OP} \equiv \vec{r}, \overrightarrow{OG} = \bar{\vec{r}}, \overrightarrow{GP} = \vec{r}\,' = \vec{r} - \bar{\vec{r}}.$$

(*i*) *Show that K.E. of a system of particles moving generally in space is equal to the sum of the K.E. of a single particle of total mass equal to that of the system, concentrated at C.G. of the system and moving with velocity of centroid, together with the K.E. of the system in its motion relative to the C. G. of the system.*

The total kinetic energy of the system is

$$T = \frac{1}{2}\sum m\dot{\vec{r}}^2$$

$$= \frac{1}{2}\sum m\left(\dot{\vec{r}}' + \dot{\bar{\vec{r}}}\right)^2$$

$$= \frac{1}{2}\sum m\dot{\vec{r}}'^2 + \sum m\dot{\vec{r}}'\dot{\bar{\vec{r}}} + \frac{1}{2}\sum m\dot{\bar{\vec{r}}}^2$$

$$= \frac{1}{2}\sum m\dot{\vec{r}}'^2 + \dot{\bar{\vec{r}}}.\sum m\dot{\vec{r}}' + \frac{1}{2}\dot{\bar{\vec{r}}}^2\sum m.$$

Now

$$\sum m\vec{r}' = \vec{0}, \therefore \sum m\dot{\vec{r}}' = \frac{d}{dt}\left(\sum m\vec{r}'\right) = \vec{0}. \quad (A)$$

$$\therefore T = \frac{1}{2}M\dot{\bar{\vec{r}}}^2 + \frac{1}{2}\sum m\dot{\vec{r}}'^2 \quad (1)$$

Here, first term is the K.E. of a particle of mass M, moving with velocity $\dot{\bar{\vec{r}}}$ and second term is the K. E. of the system in its motion relative to the centroid. Hence, the required result is proved.

(*ii*) *Prove that the angular momentum of the system about a fixed point O is equal to the sum of angular momentum of a single particle of total mass equal to that of the entire system, concentrated at the C.G. of the system and moving with velocity of its centroid together with the angular mementum of the system in its motion relative to the centroid.*

The angular momentum of the system about O is

$$\vec{H} = \sum(\vec{r} \times m\dot{\vec{r}})$$

$$= \sum\left\{(\bar{\vec{r}} + \vec{r}') \times m(\dot{\bar{\vec{r}}} + \dot{\vec{r}}')\right\}$$

$$= \bar{\vec{r}} \times (M\dot{\bar{\vec{r}}}) + \bar{\vec{r}} \times (\Sigma m\dot{\vec{r}}') + (\Sigma m\vec{r}') \times \dot{\bar{\vec{r}}} + \Sigma(\vec{r}' \times m\dot{\vec{r}}').$$

Using (A), we get

$$\vec{H} = \bar{\vec{r}} \times (M\dot{\bar{\vec{r}}}) + \sum(\vec{r}' \times m\dot{\vec{r}}'). \quad (2)$$

Here, first term represents the angular momentum about O of the particle of mass M, concentrated at its centroid and moving with velocity of the centroid and second term represents angular momentum about the centroid in its motion relative to the centroid.

Hence, the required result is proved.

(*iii*) *Prove that the rate of change of angular momentum of the system about O is equal to the moment of the rate of change of linear momentum about O of a single particle of total mass equal to that of the entire system, concetrated at its C. G. and having velocity of C. G., together with the moment of the rate of change of momentum about the C.G. of the system in its motion relative to the centroid.*

Differentiating (2) with respect to t, we get

$$\dot{\vec{H}} = \vec{\bar{r}} \times (M\ddot{\vec{\bar{r}}}) + \sum (\vec{r}' \times m\ddot{\vec{r}}'). \tag{3}$$

As in the above we can interpret both the terms and thus required result is proved.

6.3 MOVING ORIGINS

If the origin O is moving with velocity $\vec{v}_0$, then also we can prove that all the principles proved earlier will be true.

6.4 IMPULSIVE FORCES

(*i*) *Prove that the total external impulse applied to the system of particles is equal to the total change in linear momentum produced.*

Let the system of particles be subjected to impulsive forces. Suppose $\vec{I}, \vec{I}'$ are the resultant external and internal impulses acting on the particle P of mass m. Let $\vec{v}, \vec{v}_0$ be the instantaneous velocities after and before the impulse. We know that

Total impulse = Change in momentum.

$$\therefore \quad \vec{I} + \vec{I}' = m(\vec{v} - \vec{v}_0).$$

Summing over all the particles,

$$\sum \vec{I} + \sum \vec{I}' = \vec{p} - \vec{p}_0.$$

From Newton's Third Law of motion and the nature of the impulsive force,

$$\sum \vec{I}' = \vec{0}.$$

$$\therefore \quad \sum \vec{I} = \vec{p} - \vec{p}_0.$$

Hence, the result.

(*ii*) *Prove that irrespective of whether the origin is stationary or moving, the total vector sum of moments of the external impulses about it is equal to the instantaneous increase in angular momentum produced about the origin.*

Let $\overrightarrow{OP} = \vec{r}$, we have

$$\sum \{\vec{r} \times (\vec{I} + \vec{I}')\} = \sum \{\vec{r} \times m(\vec{v} - \vec{v}_0)\}.$$

By arguments similar to those previously invoked,

$$\sum \vec{r} \times \vec{I}' = \vec{0}.$$

$$\therefore \qquad \sum \vec{r} \times \vec{I} = \vec{H} - \vec{H}_0 .$$

Hence, the result.

Examples

1. Two particles each of mass m are connected by a light rigid rod, of length $2a$, and the system rests on a smooth horizontal table. A blow in the plane of the table acts on the particle A and gives it a velocity u perpendicular to AB. Describe the resulting motion and find the displacements of the particles from their original positions after time t.

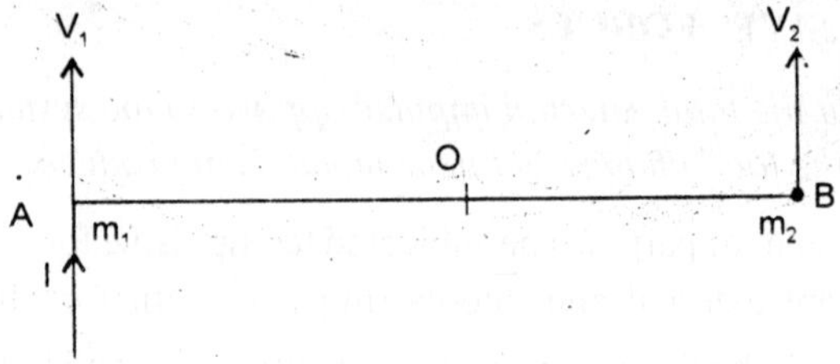

Fig. 6.4

The centroid G of the rod moves in the direction of impulse and so A and B must move in the same direction.

Let v_1, v_2 be the velocities of particles at time t.

Gain in linear momentum $= m\,(v_1 + v_2)$.

By principle of linear momentum

$$I = m\,(v_1 + v_2).$$

$$\therefore \qquad v_1 + v_2 = \frac{I}{m}. \qquad (1)$$

The moment of impuse I about B is equal to the gain of angular momentum

$$I \,.\, 2a = mv_1 2a,$$

$$\therefore \qquad v_1 = \frac{I}{m}. \qquad (2)$$

Hence, from (1) and (2), $v_2 = 0$.

Since

$$I = mu,$$

$$\therefore \qquad u = \frac{I}{m} = v_1 .$$

Velocity of the centroid

$$\frac{1}{2}(v_1 + v_2) = \frac{v_1}{2} = \frac{u}{2}.$$

Hence, centroid moves a distance $\frac{ut}{2}$.

The rod rotates through an angle θ = distance/a = $\frac{ut}{2a}$.

2. Particles A, B, C, of masses m_1, m_2, m_3 lie on a smooth horizontal table and are connected together by taut inextensible strings AB, BC. The angle ABC is $\pi - \alpha$, α bring acute. If an impulse is applied to C along BC, show that B starts moving in direction making with AB an angle

$\tan^{-1}\left\{\left(1 + \frac{m_1}{m_2}\right)\tan\alpha\right\}$. Find the initial velocity of A.

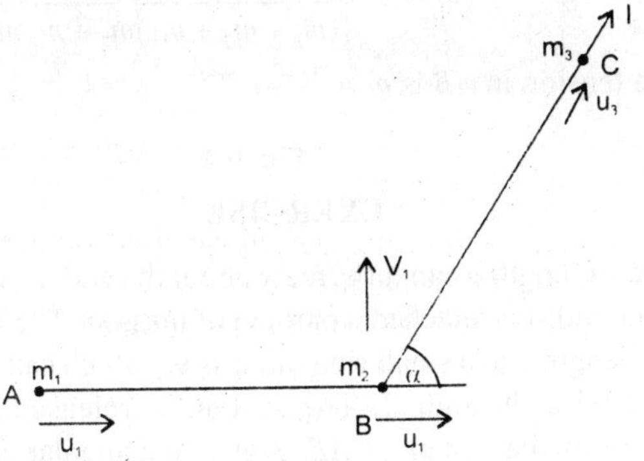

Fig. 6.5

Since, an impulsive tension is produced in AB, the motion of A is along AB. Let u_1 be its velocity in this direction. On C an impulse is acting along BC and an impulsive tension along CB, let its velocity be u_3. Since, A and B are connected by a taut string, B's velovity along AB will be u_1 and its velocity perpendicular to AB will be v_1 (say).

Since, BC is taut, the component of B's velocity along BC is u_3.

$$\therefore \quad u_1 \cos\alpha + v_1 \sin\alpha = u_3. \tag{1}$$

Considering linear momentum changes along and perpendicular BC, we get

$$m_3 u_3 + m_2 (u_1 \cos\alpha + v_1 \sin\alpha) + m_1 u_1 \cos\alpha = I. \tag{2}$$

$$m_2 v_1 \cos\alpha - m_2 u_1 \sin\alpha - m_1 u_1 \sin\alpha = 0. \tag{3}$$

From (3)

$$\frac{v_1}{u_1} = \left(1 + \frac{m_1}{m_2}\right) \tan\alpha \cdot \qquad (4)$$

From (1) and (2)

$$m_3(u_1 \cos\alpha + v_1 \sin\alpha) + m_2(u_1 \cos\alpha + v_1 \sin\alpha) + m_1 u_1 \cos\alpha = I,$$

$$\therefore \qquad (m_1 + m_2 + m_3)u_1 \cos\alpha + (m_2 + m_3)v_1 \sin\alpha = I,$$

Substituting value of v_1 from (4) we get

$$(m_1 + m_2 + m_3)u_1 \cos\alpha + (m_2 + m_3)\left(1 + \frac{m_1}{m_2}\right) u_1 \sin\alpha \tan\alpha = I.$$

$$\therefore (m_1 + m_2 + m_3)m_2 u_1 \cos^2\alpha + (m_2 + m_3)(m_1 + m_2)u_1 \sin^2\alpha = Im_2 \cos\alpha,$$

$$\therefore u_1 = \frac{Im_2 \cos\alpha}{\{(m_1 + m_2 + m_3)m_2 + m_1 m_3 \sin^2\alpha\}}.$$

The impulsive tension in AB is $m_1 u_1$.

EXERCISE

1. A light rod of length a can turn freely about the end A, which is fixed, and to the other end B is attached a particle of mass m. The end B is joined by a string of length a to a small ring of mass m, which can slide on a smooth horizontal wire through A. The system is released from rest, when $AC = 2a$. Show that when $\angle CAB = 60°$, the angular acceleration of the rod is $-g/16a$, and find the tension in the string at this instant.
2. A smooth wedge of mass M and angle α is free to move on a smooth horizontal plane in a direction perpendicular to its edge. A particle of mass m is projected vertically up the face of the wedge with velocity V. Prove that it returns to the point on the wedge from which it was projected after a time

$$\frac{2V(M + m\sin^2\alpha)}{(M + m)g\sin\alpha}.$$

3. A gun of weight W is mounted so that it is free to recoil along a horizontal railway track. When pointed at an elevation α and in a vertical plane making an angle ϕ with the track, it fires a shell of weight εW with muzzle velocity V. If ε is so small that its square and higher powers can be neglected, prove that the shell leaves the gun with an absolute velocity

$V(1 - \varepsilon \cos^2 \phi \cos^2 \alpha)$ at an elevation $\alpha + \varepsilon \sin \alpha \cos \alpha \cos^2 \phi$. Prove that the maximum range is

$$R(1 - \varepsilon \cos^2 \phi),$$

where R is the range when there is no recoil.

4. A particle of mass m is placed on a smooth slant surface of a wedge of mass M and angle α. The wedge moves (edge foremost) along a smooth horizontal plane under the action of a constant horizontal force P perpendicular to its edge, all the motion being parallel to the same plane. Prove that the acceleration is

$$\frac{(P - mg \sin \alpha \cos \alpha)}{(M + m \sin^2 \alpha)},$$

and find the pressure between the particle and the wedge.

5. A wedge of mass M and angle $\pi/6$ rests on a smooth horizontal table. One end of light elastic string of natural length l and modulus $2mg$ is fastened to a point O on a smooth inclined face of the wedge. A particle of mass m is tied to the other end of the string and is held at O. The system is released form rest. Write down the energy and horizontal momentum equations for the system and show that when it is next instantaneously at rest, the particle is at a distance $2l$ from O. Find the velocity of the particle relative to the wedge when the string is just stretched, and show that the velocity of the wedge is greatest when the particle is at a distance $5l/4$ from O.

6. A smooth wedge of inclination α is placed on a horizontal table; a string, to the extremities of which masses m, m' attached, passes round a smooth peg which projects from the upper surface of the wedge, the particles being in contact with the same face of the wedge. Prove that, if M be the mass of the wedge its acceleration is

$$\frac{(m - m')^2 g \sin \alpha \cos \alpha}{M(m + m') + (m - m')^2 \sin^2 \alpha + 4mm'},$$

assuming that the motion is in the plane of greatest slope.

7. Two particles A and B, each of mass m are attached to the ends of a light inextensible string of length l, and B moves in a smooth straight groove in a smooth horizontal table. Initially A is at rest on the table and B is at rest in a groove, AB being perpendicular to the groove and of length $l/2$. If A is projected with velocity v along the table and parallel to the groove, show that when B begins to move its velocity is $3v/7$.
Find the impulsive tension in the string and the inpulsive reaction of the groove.

8. Two particles of mass M and m ($M > m$) are placed on two smooth faces of a light wedge which rests on a smooth horizontal plane. The faces of the wedge are inclined to the horizontal at angles α and β respectively. If

the system starts from rest, show that the smaller particle will move up the face on which it is placed, if

$$\tan\beta < \frac{M\sin\alpha\cos\alpha}{(M\sin^2\alpha + m)}.$$

9. Four particles of equal mass are attached to a light inextensible string at points A, B, C, D where $AB = BC = CD$, and placed with the three parts of the string taut and forming three sides of a regular hexagon. An impulse is given to the particle at A, so that it moves in the direction with speed u. Prove that the particle at D begins to move with speed $u/13$.
10. A particle of mass m is placed on the inclined face of a wedge of mass M which rests on a rough horizontal table. Prove that, if the particle slides down, the wedge will begin to move, provided that

$$\frac{m}{M} > \frac{\cos\lambda\sin\lambda'}{\cos\alpha\sin(\alpha-\lambda-\lambda')},$$

where α is the inclination of the face of the wedge to the horizontal, λ is the angle of friction for the particle and the wedge, and λ' is the angle of friction for the wedge and the table and $\alpha > \lambda + \lambda'$.
11. Two equal particles A, B lie on a smooth horizontal table, joined by a taut inextensible string. A third equal particle C moving on the table at right angles to the string hits at its middle point. Prove that in subsequent motion C does not leave the string before A, B collide.
Show also that at the time of this collision C has only one-ninth of its original K.E.
12. Three particle A, B, C of masses $m, 2m, 3m$, respectively lie on a smooth horizontal table at the vertices of an equilateral triangle. A and B, B and C, are connected by light inextensible string. C is given a velocity V parallel to AB and in a sense from A to B. Show that A eventually begins to move with velocity $2V/19$.
13. A smooth wedge of mass M resting on a horizontal plane is subject to smooth constraints so that it can move only along the plane in a direction at right angles to the intersection of slant faces with the plane. A particle of mass m is moving along a face of the wedge which is inclined to the horizontal at an angle α so that the component of velocity of the particle perpendicular to the line of greatest slope is u. Show that the wedge moves with constant acceleration and that the path of the particle on the surface of the wedge is a parabola of latus rectum

$$\frac{2u^2(M + m\sin^2\alpha)}{g(M+m)\sin\alpha}.$$

Verify that the principle of conservation of energy holds good in this case.

14. A ring of mass m slides on a smooth vertical rod; attached to the ring is a light string passing over a smooth peg distant a from the rod, and at the other end of the string is a mass M ($M > m$). The ring is held on a level with the peg and released; show that it first comes to rest after falling a distance

$$\frac{2mMa}{(M^2+m^2)}.$$

15. Two equal particles A, B attached to the ends of a light string of length a are placed on a smooth horizontal table with the string AB perpendicular to the edge of the table and B just over the edge. The system is released from rest in this position. Prove that when first the string is horizontal, the distance of B from the vertical through the edge of the table is a $(\pi-2)/4$, and find the tension in the string.

16. A particle of mass m' is attached by a light inextensible string of length l to a ring of mass m free to slide on a smooth horizontal rod. Initially the two masses are held with the string taut along the rod and they are then set free. Prove that the greatest velocity of the string is

$$\sqrt{\frac{2g\,(m+m')}{lm}}.$$

Also show that the time of small oscillations about the vertical is

$$2\pi\sqrt{\frac{lm}{g\,(m+m')}}.$$

17. Two particles, each of mass m, are joined by a rod of negligible mass and of length a. One particle rests on a smooth horizontal plane and the other end is vertically above it. The upper particle is given a small displacement, so that the rod beings to fall. Show that when its inclination to the vertical is 60° the velocity of the lower particle is $\sqrt{\frac{ga}{14}}$.

18. Particles A, B, each of mass m are joined by an inelastic string of length $2a$. Initially AB is vertical with B at a distance a below A. A is given a horizontal velocity V, while B is simultaneously released from rest. Find the inclination of the string to the vertical when it becomes taut and the time which elapses before the string is horizontal.

19. Two masses m_1 and m_2 at A and B are connected by a rigid rod AB of negligible mass. Their velocities are suddenly changed by the application of impulses $\vec{I}_1$ and $\vec{I}_2$. Find the magnitude of the impulsive reaction of the rod on m_1 and change in K.E. of the system.

20. An inelastic sphere of mass m falls with velocity V on a perfectly rough inelastic inclined plane of mass M and angle α, which rests on a smooth horizontal plane. Show that the vertical velocity of the centre of the sphere immediately after the impact is.

$$\frac{5(M+m)V\sin^2\alpha}{7M+2m+5m\sin^2\alpha}.$$

21. A perfectly rough circular disc of mass M and radius c impinges upon a rod of mass m and length $2a$ capable of turning freely about a pivot at its centre. If the point of contact is distant b from the centre of the rod, and the direction of the motion of the centre of the disc makes angles α, β with the rod before and after the collision, show that

$$2(3Mb^2 + ma^2)\tan\beta = 3(cma^2 - 3Mb^2)\tan\alpha.$$

7

MOMENT AND PRODUCTS OF INERTIA

In previous chapters we dealt with dynamics of particles, now we will deal with dynamics of rigid body. A rigid body is considered as a system of paticles rigidly connected such that during its motion, the distance between any two particles of the body remains invariable. The particles of the body are held together by cohesive forces such that by Newton's third law, the action and reaction between any pair of particles are equal and opposite. A rigid body has invariable size and shape. In this book we will be dealing with motion of rigid bodies.

7.1(a) MOMENT OF INERTIA OF A BODY ABOUT A LINE

If m is the mass of an elemennt of a rigid body and r the distance of the element from the given line, then $\sum mr^2$ is called the moment of inertia of the body about the straight line. Hence, to determine the moment of inertia of a body of mass M, take an element of the body, multiply by square of its perpendicular distance from the given line. The sum of all such quantities is the moment of inertia of the body about the line.

If this sum be denoted by Mk^2, where M is the mass of the body, then k is called *radius of gyration* of the body about the given line.

(b) PRODUCT OF INERTIA

If (x, y) be the co-ordinates of an element of the mass m referred to two mutually perpendicular lines Ox, Oy then $\sum mxy$ is called the product of inertia of the body with respect of lines Ox and Oy.

If mutually perpendicular axes Ox, Oy, Oz be taken in the space and (x, y, z) are the coordinates of the element m of the body, then the quantities $\sum myz$, $\sum mzx$, $\sum mxy$ are the products of inetia of the body with respect to the pairs of axes y and z, z and x, x and y respectively. Similarly, the quantities $\sum m(y^2 + z^2)$, $\sum m(z^2 + x^2)$, and $\sum m(x^2 + y^2)$, are called moments of inertia about axes x, y and z respectively. Generally we use the following notations.

$$A = \sum m(y^2 + z^2),\ B = \sum m(z^2 + x^2),\ C = \sum m(x^2 + y^2),$$
$$D = \sum myz,\ E = \sum mzx,\ F = \sum mxy.$$

These quantities are also called second moments of mass distribution about the x-, y-, z-coordinates axes respectively. Here, summations are taken throughout the system of particles. In the case, when the distribution is continuous, the above expressions are replaced by $A = \int(y^2 + z^2)$, etc. the integration now being taken throughout the continuous system. In general discussion the summation signs will be used but for computing A, B, ... in particular cases, the precise form of course, have to be adopted.

In next article we will find the moments of inertia in particular cases.

Example

Find the M.I. of the distribution about an axis through O having direction cosines $\lambda\ \mu\ v$, in terms of D.C.' s and A, B,......., F.

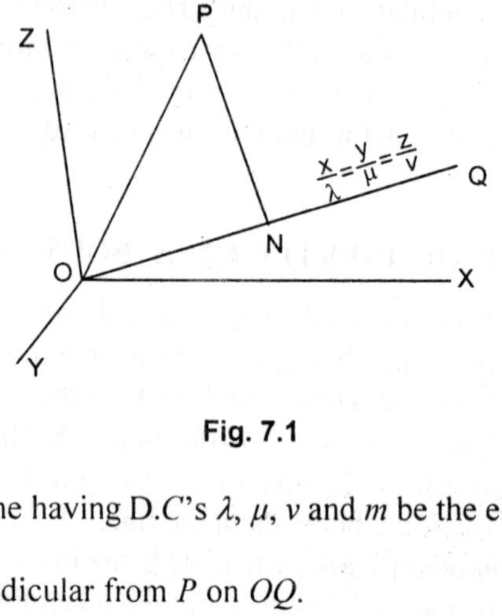

Fig. 7.1

Let OQ be the line having D.C's λ, μ, v and m be the element of mass m at point P (x, y, z).

Draw PN perpendicular from P on OQ.

Now,

$$OP^2 = x^2 + y^2 + z^2,\ ON = \lambda x + \mu y + vz.$$

$$\therefore \quad PN^2 = OP^2 - ON^2 = (x^2 + y^2 + z^2) - (\lambda x + \mu y + vz)^2$$

$$= x^2(1 - \lambda^2) + y^2(1 - \mu^2) + z^2(1 - v^2) - 2\mu vyz - 2v\lambda zx - 2\lambda\mu xy$$

$$= x^2(\mu^2 + v^2) + y^2(v^2 + \lambda^2) + z^2(\lambda^2 + \mu^2) - 2\mu vyz - 2v\lambda zx - 2\lambda\mu xy.$$

$$\because \quad \lambda^2 + \mu^2 + z^2 = 1.$$

$$= \lambda^2(y^2 + z^2) + \mu^2(z^2 + x^2) + v^2(x^2 + y^2) - 2\mu vyz - 2v\lambda zx - 2\lambda\mu xy$$

If I denote the M.I. about OQ, then

$$I = \sum mPN^2 = \sum m[\lambda^2(y^2 + z^2) + \mu^2(z^2 + x^2) + v^2(x^2 + y^2) - 2\mu vyz - 2v\lambda zx - 2\lambda\mu xy]$$

$$= \lambda^2 \Sigma m(y^2 + z^2) + \mu^2 \Sigma m(z^2 + x^2) + v^2 \Sigma m(x^2 + y^2)$$
$$- 2\mu v \Sigma myz - 2v\lambda \Sigma mzx - 2\lambda\mu \Sigma mxy$$
$$= A\lambda^2 + B\mu^2 + Cv^2 - 2D\mu v - 2Ev\lambda - 2F\lambda\mu.$$

7.2 THE THEOREMS OF PARALLEL AND PERPENDICULAR AXES

(*i*) Referred to axes *OX*, *OY*, *OZ*, let the co-ordinates of the centroid *G* be $\bar{x}, \bar{y}, \bar{z}$ and let *GX'*, *GY'*, *GZ'* be the axes through *G* parallel to *OX*, *OY*, *OZ* through *O*. Let an element of mass *m* be situated at the point *P* which has co-ordinates *x*, *y*, *z* referred to *OX*, *OY*, *OZ* and *x'*, *y'*, *z'* referred to *GX'*, *GY'*, *GZ'* so that

$$x = \bar{x} + x',\ y = \bar{y} + y',\ z = \bar{z} + z'.$$

$$A = \text{M.I. about } OX = \Sigma m(y^2 + z^2)$$
$$= \Sigma m\,[(\bar{y} + y')^2 + (\bar{z} + z')^2]$$
$$= \Sigma m[(\bar{y}^2 + \bar{z}^2) + (y'^2 + z'^2) + 2\bar{y}y' + 2\bar{z}z']$$
$$= M(\bar{y}^2 + \bar{z}^2) + \Sigma m(y'^2 + z'^2) + 2\bar{y}\Sigma my' + 2\bar{z}\Sigma mz'$$
$$= M(\bar{y}^2 + \bar{z}^2) + A' + 0 + 0 = M(\bar{y}^2 + \bar{z}^2) + A',$$

because from the centroids property $\Sigma my' = 0 = \Sigma mz'$ and where $M = \Sigma m$, the total mass of the system and $A' = \Sigma m(y'^2 + z'^2)$, the M.I. about *GX'*

Hence,

$$A = A' + Mh_1^2, \tag{1}$$

where $h_1 = (\bar{y}^2 + \bar{z}^2)^{\frac{1}{2}}$, the distance of the centroids from *OX*.

Equation (1) is the *parallel axis theorem for M.I.*

Further, we have

$$D = \Sigma myz = \Sigma m(\bar{y} + y')(\bar{z} + z')$$
$$= \Sigma m(\bar{y}\bar{z} + y'\bar{z} + \bar{y}z' + y'z')$$
$$= M\bar{y}\bar{z} + \bar{z}\Sigma my' + \bar{y}\Sigma mz' + \Sigma my'z'$$
$$= M\bar{y}\bar{z} + 0 + 0 + D' = M\bar{y}\bar{z} + D'. \tag{2}$$

where, $D' = \Sigma my'z'$ = P. I. with respect to *GY'* and *GZ'*, here we have used centroids property $\Sigma my' = \Sigma mz' = 0$.

Equation (2) is the *parallel axis theorem for P.I.*

(*ii*) Now suppose the distribution is such that particles lie in the plane $z = 0$, then we have

$$A = \Sigma my^2,\ B = \Sigma mx^2,\ C = \Sigma m\,(x^2 + y^2),\ D = 0,\ E = 0,\ F = \Sigma mxy.$$

Thus

$$C = A + B. \tag{3}$$

Equation (3) is the *perpendicular theorem for a laminar distribution.* It states that *the M.I. of a plane distribution with respect to any normal axis is equal to the sum of the moments of inertia about any two perpendicular axes in the plane of distribution and passing through the intersection of the normal with the plane of distribution.*

Cor. If there are two parallel axes A and B at a distance a and b from the C.G. of the body of mass M, then

M.I. about $A - Ma^2$ = M.I. about $B - Mb^2$

7.3 PARTICULAR CASES OF M.I.

1. A thin rod of mass M and length $2a$.

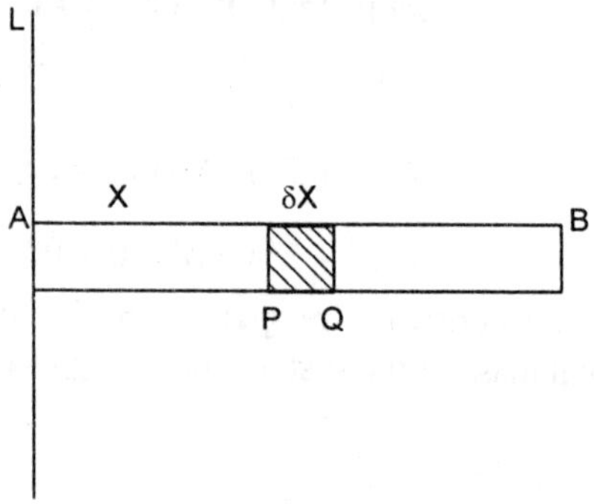

Fig. 7.2

Case I. About an axis through an end of the rod and perpendicular to the rod.

Let AB be the rod and PQ and element of the rod such that $AP = x$ and $PQ = \delta x$.

The mass of $PQ = M\dfrac{\delta x}{2a}$.

Hence, M.I. of the rod about an axis AL through A perpendicular to the rod

$$= \Sigma M\frac{\delta x}{2a}x^2 = \frac{M}{2a}\int_0^{2a} x^2 dx = M\frac{4a^2}{3}.$$

Case II. About an axis through the middle point O of the rod.

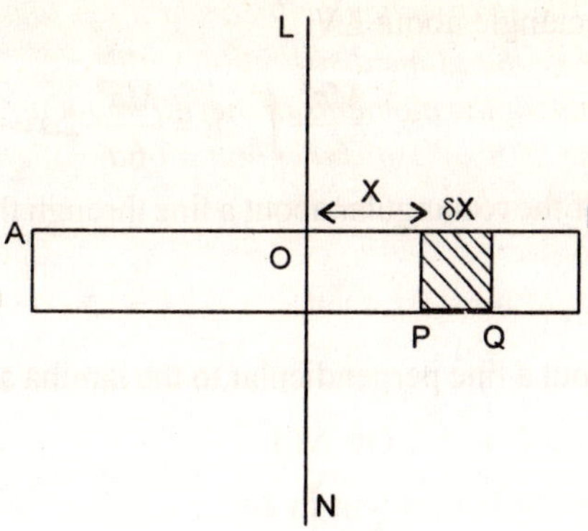

Fig. 7.3

Let us consider an element $PQ = \delta x$ at a distance x from O and let OL be the axis perpendicular to the rod.
M.I. of the rod about OL

$$= \sum M \frac{\delta x}{2a} x^2 = \frac{M}{2a}\int_{-a}^{a} x^2 dx = M\frac{a^2}{3}.$$

2. Rectangular Lamina

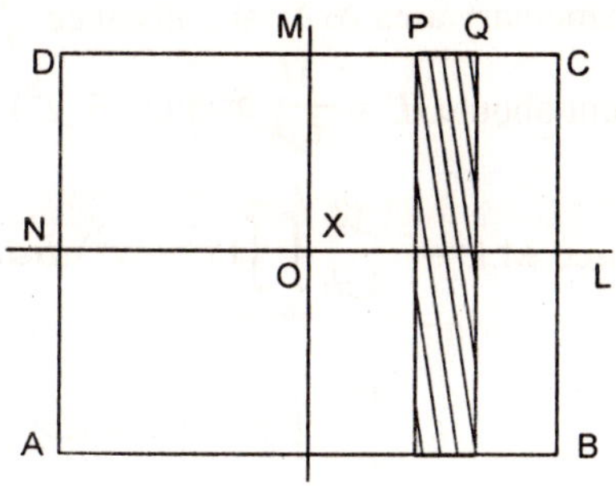

Fig. 7.4

Let $ABCD$ be a rectangular lamina having side $AB = 2a$ and $AD = 2b$, with centre O and mass M. Let NL be the axis through O parallel to AB. Let us consider an elementary strip of breadth δx at a distance x from O, parallel to AD.

Mass of this strip $$= 2b\,\delta x \frac{M}{4ab} = \frac{M}{2a}\delta x\,.$$

Case I. M. I. of this strip about $LN = \frac{M}{2a}\delta x \frac{b^2}{3}$.

$\therefore$ M.I. of the rectangle about LN

$$= \frac{Mb^2}{6a}\int_{-a}^{a}\delta x = \frac{Mb^2}{6a}2a = \frac{Mb^2}{3}.$$

Similarly, M.I. of the rectangular about a line through the centre parallel to $AD = \frac{Ma^2}{3}$.

Case II. M.I. about a line perpendicular to the lamina and passing through the centre O.

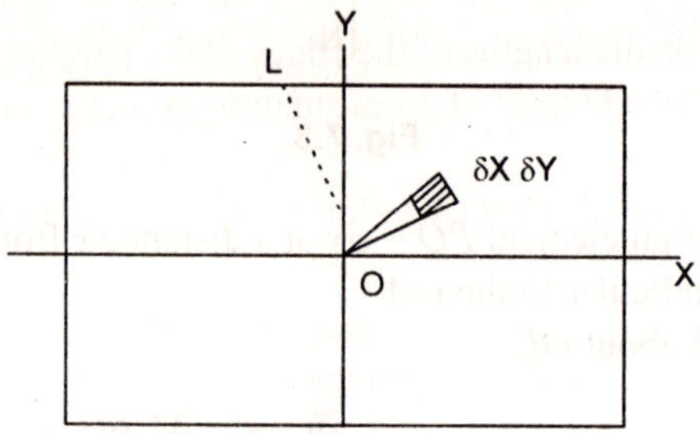

Fig. 7.5

Let OL be a line perpendicular to the lamina and passing through O. Let us consider an elementary area $\delta x \delta y$ at a distance $\sqrt{x^2+y^2}$ from O.

M.I. of this element about $OL = \frac{M}{4ab}\delta x \delta y (x^2+y^2)$.

$$\therefore \quad \text{required M.I.} = \frac{M}{4ab}\int_{-b}^{b}\int_{-a}^{a}(x^2+y^2)\,dx\,dy$$

$$= \frac{4M}{4ab}\int_{0}^{b}\int_{0}^{a}(x^2+y^2)\,dx\,dy$$

$$= \frac{M}{ab}\left[\frac{x^3}{3}y + \frac{y^3}{3}x\right]_{x=0,\,y=0}^{x=a,\,y=b}$$

$$= \frac{M}{ab}\cdot\frac{1}{3}(a^3b+b^3a) = \frac{M}{3}(a^2+b^2).$$

Alternatively,

By theorem of perpendicular axes, we have

M.I. about OL = M.I. about OX + M.I. about OY

$$= \frac{Mb^2}{3} + \frac{Ma^2}{3} = \frac{M}{3}(a^2 + b^2).$$

3. Rectangular Parallelopiped

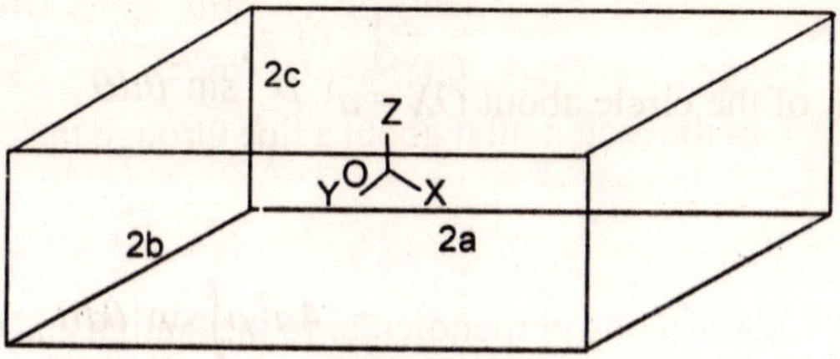

Fig. 7.6

Let $2a$, $2b$, $2c$ be the lengths of the edges of the parallelopiped with centre O. Take OX, OY, OZ axes of co-ordinates parallel to the edges $2a$, $2b$, $2c$ respectively.

Let us consider an elementary volume $\delta x \delta y \delta z$ at a point $P(x, y, z)$.

M.I. of this element about $OX = \dfrac{M}{8abc}\, \delta x \delta y \delta z\, (y^2 + z^2)$

Hence, M.I. of the parallelopiped about OX

$$= \frac{M}{8abc} \int_{-a}^{a} \int_{-b}^{b} \int_{-c}^{c} (y^2 + z^2)\, dx\, dy\, dz$$

$$= \frac{M}{8abc} \frac{8abc}{3}(b^2 + c^2) = \frac{M}{3}(b^2 + c^2).$$

4. Circumference of the Circle

Case I : M.I. about a radius.

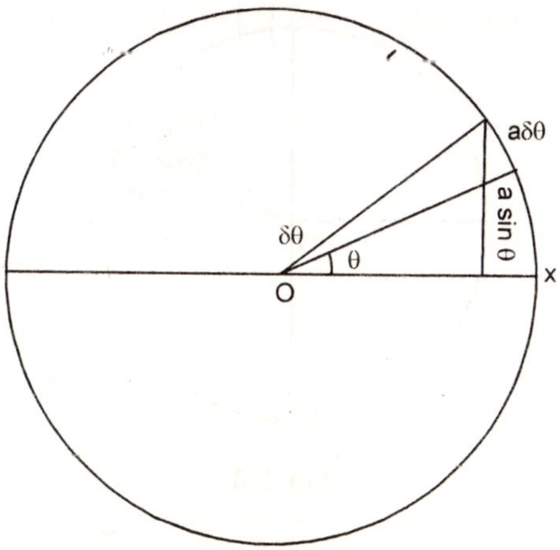

Fig. 7.7

Let O be the centre and OX a radius of the circle. Let us consider an elementary are $a\delta\theta$. Its mass = $a\delta\theta\rho$ where ρ is the density of the circle and its distance from the radius $OX = a\sin\theta$.

$\therefore$ Its M.I. about $OX = a\delta\theta\rho\,(a\sin\theta)^2 = a^3\rho\sin^2\theta\delta\theta$.

Hence, M.I. of the circle about $OX = a^3\rho\int_0^{2\pi}\sin^2\theta d\theta$

$$= 4a^3\rho\int_0^{\frac{\pi}{2}}\sin^2\theta d\theta$$

$$= 4a^3\rho\frac{1}{2}\frac{\pi}{2} = a^3\rho\pi.$$

where $\rho = \dfrac{M}{2\pi a}$,

$$\therefore \text{M.I.} = a^{\delta}\pi\frac{M}{2\pi a} = M\frac{a^2}{2}.$$

Case II. M.I. about an axis through O.

M.I. of the elementary mass about an axis through $O = \rho a\delta\theta.\, a^2$

$\therefore$ M.I. of the circle about an axis through O.

$$= \int_0^{2\pi}\rho a^3 d\theta = 2\pi a^3\rho = 2\pi a^3\frac{M}{2\pi a} = Ma^2.$$

5. Circular Disc

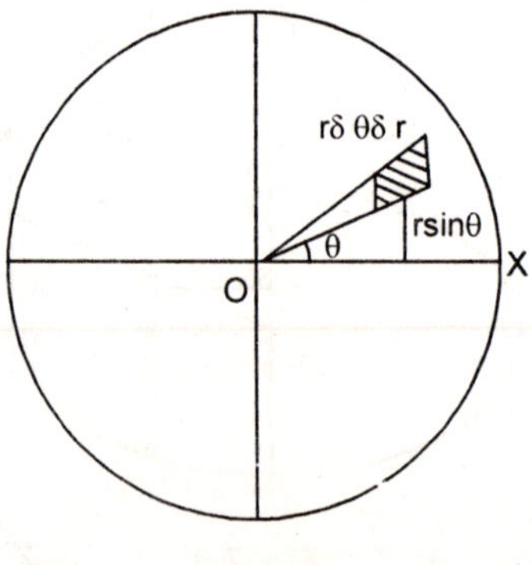

Fig. 7.8

Case I : About its radius OX.

Let us consider an elemantary area $r\delta\theta\delta r$. If ρ is the density and M the mass of the disc then,

$$\rho = \frac{M}{\pi a^2}.$$

Mass of the elementary area is $\rho r\delta\theta\delta r$ and its distance from the radius OX is $r\sin\theta$.

$\therefore$ M.I. of the element about $OX = r\delta\theta\delta r\rho(r\sin\theta)^2 = r^3\rho\sin^2\theta\delta\theta\delta r$.

$$\therefore \text{ M.I. of the disc about } OX = \rho\int_0^{2\pi}\int_0^{a} r^3 \sin^2 \theta d\theta dr$$

$$= 4\rho\int_0^{\pi/2}\int_0^{a} r^3 \sin^2 \theta d\theta dr$$

$$= 4\rho\left[\frac{r^4}{4}\right]_0^a \frac{1}{2}\frac{\pi}{2} = \frac{\pi\rho a^4}{4}$$

$$= \frac{Ma^2}{4}.$$

This is the M.I. of the disc about any diameter through O.

Case II : About a normal through O.

Let OZ be a line through O normal to the disc. By perpendicular axis theorem, M.I. about OZ = M.I. about OX + M.I. about OY.

$$= \frac{Ma^2}{4} + \frac{Ma^2}{4} + \frac{Ma^2}{2}.$$

6. An Elliptic Disc

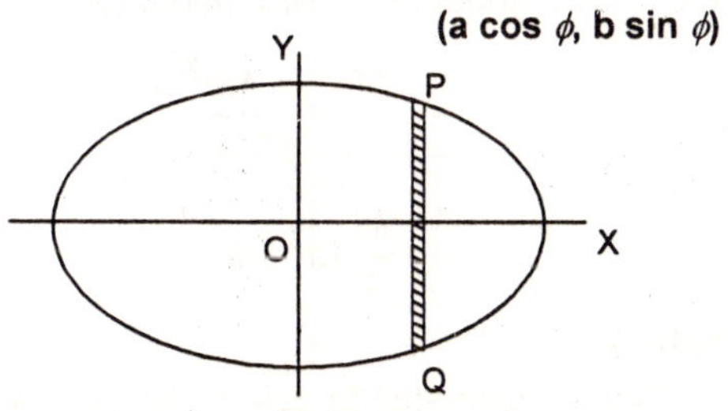

Fig. 7.9

Case I : M.I. about major and minor axis. Let us consider an elementary strip PQ such that P has co-ordinates $(a\cos\phi,\ b\sin\phi)$. Length and breadth

of the strip PQ are $2b\sin\phi$ and $\delta x = a\sin\phi\delta\phi$ respectively. Let OA and OB be semi major and minor axes, if ρ is the density of the disc,
Mass of the strip $= 2b\sin\phi.\ a\ \sin\phi.\ \delta\phi.\rho = 2ab\sin^2\phi.\delta\phi.\rho$

$\therefore$ M.I. of the strip about $OA = 2ab\sin^2\phi.\delta\phi\rho.\ \dfrac{b^2\sin^2\phi}{3} = \dfrac{2ab^3\sin^4\phi\delta\phi.\rho}{3}.$

$\therefore$ M.I. of the elliptic disc about $OA = \displaystyle\int_0^{\pi}\frac{2ab^3\rho\sin^4\phi}{3}d\phi$

$$= \frac{4ab^3\rho}{3}\int_0^{\frac{\pi}{2}}\sin^4\phi d\phi$$

$$= \frac{4ab^3\rho}{3}.\frac{3}{4}.\frac{1}{2}.\frac{\pi}{2}$$

$$= \frac{\pi ab^3\rho}{4},$$

but $\rho = \dfrac{M}{\pi ab}.$

$\therefore$ M.I. of the elliptic disc about $OA = \dfrac{Mb^2}{4}.$

Similarly, M.I. of the elliptic disc about $OB = \dfrac{Ma^2}{4}.$

Case II : M.I. of the elliptic disc about an axis through O perpendicular to the disc.

Let OC be the axis through O perpendicular to the disc. By perpendicular axis theorem.

M.I. about OC = M.I. about OA + M.I. about OB

$$= \frac{Mb^2}{4} + \frac{Mb^2}{4}$$

$$= \frac{M}{4}(a^2 + b^2).$$

7. A Hollow Sphere

The hollow sphere is generated by revolving a semicircular arc about its bounding diameter.

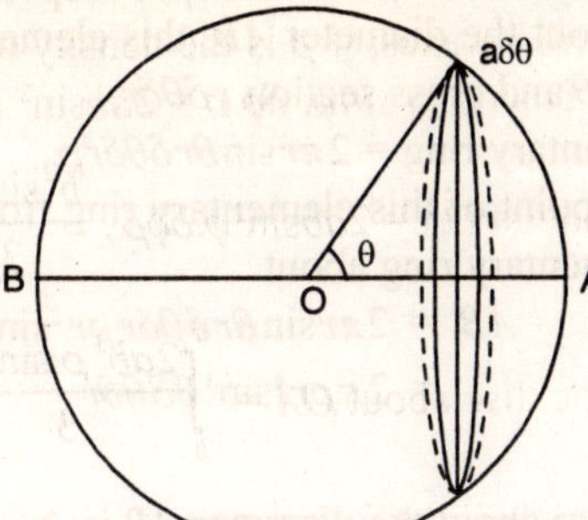

Fig. 7.10

Let us consider an element $a\delta\theta$ of the arc. When revolved about the diameter AB, this arc generates a circular ring of radius $a\sin\theta$ and width $a\delta\theta$.

Mass of the elementary ring = $2\pi\, a\sin\theta a\delta\theta\rho$.

Distance of every point of this ring from $AB = a\sin\theta$.

$\therefore$ M.I. of the elementary ring about

$$OA = 2\pi a\sin\theta a\delta\theta\rho a^2\sin^2\theta = 2\pi a^4\rho\sin^3\theta\delta\theta.$$

M.I. of the hollow sphere about the diameter AB

$$= 2\pi a^4\rho\int_0^{\pi}\sin^3\theta d\theta$$

$$= 4\pi a^4\rho\int_0^{\frac{\pi}{2}}\sin^3\theta d\theta$$

$$= 4\pi a^4\rho.\frac{2}{3} = \frac{8\pi a^4\rho}{3} = \frac{2}{3}Ma^2,$$

Since, $M = 4\pi a^2\rho$.

8. A Solid Sphere

A solid sphere is generated by revolving a semicircular area about its bounding diameter.

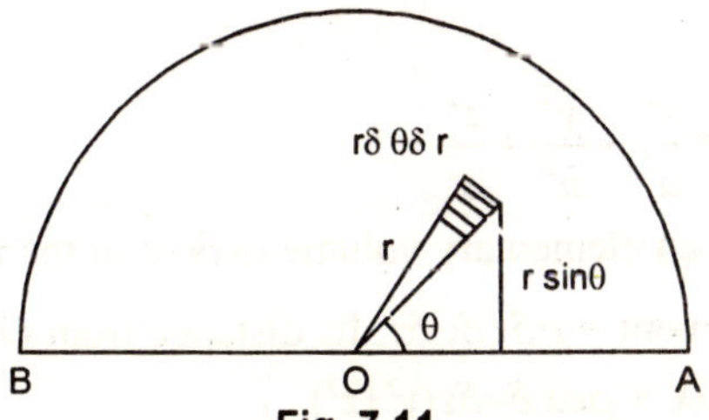

Fig. 7.11

Let us consider an elementary area $r\delta\theta\delta r$ at a distance r from the centre. When revolved about the diameter AB, this element generates a circular ring of radius $r\sin\theta$ and cross section $r\delta\theta\delta r$.

Mass of this elementary ring = $2\pi r\sin\theta r\delta\theta\delta r\rho$.

Distance of every point of this elementary ring from $AB = r\sin\theta$.

$\therefore$ M.I. of the elementary ring about

$$AB = 2\pi r\sin\theta r\delta\theta\delta r\rho r^2\sin^2\theta$$
$$= 2\pi\rho r^4\sin^3\theta\delta\theta\delta r.$$

$\therefore$ M.I. of the sphere about the diameter AB is $2\pi\rho\int_0^a\int_0^\pi r^4\sin^3\theta d\theta dr$

$$= 4\pi\rho\int_0^a\int_0^{\frac{\pi}{2}} r^4\sin^3\theta d\,\theta dr$$

$$= 4\pi\rho\left[\frac{r^5}{5}\right]_0^a\frac{2}{3} = \frac{8\pi\rho a^5}{3.5} = \frac{2}{5}Ma^2.$$

Since,

$$M = \frac{4}{3}\pi a^3\rho.$$

9. An Ellipsoid

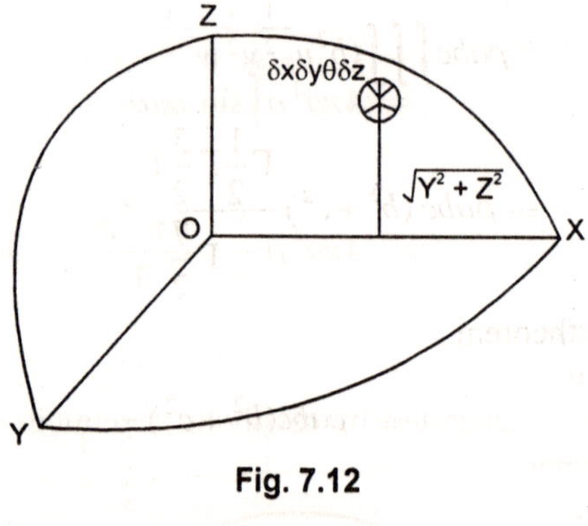

Fig. 7.12

Let ellipsoid be $\frac{x^2}{a^2}+\frac{y^2}{b^2}+\frac{z^2}{c^2}=1$.

Let us consider an elementary volume $\delta x\delta y\delta z$ in the positive octant.

Mass of the element = $\rho\delta x\,\delta y\,\delta z$. Its distance from $OX = \sqrt{y^2+z^2}$.

$\therefore$ Its M.I. about $OX = \rho\delta x\,\delta y\,\delta z(y^2+z^2)$.

Hence, M.I. of the ellipsoid about OX

$$= 8\int_0^a\int_0^b\int_0^a \rho(y^2+z^2)\,dx\,dy\,dz.$$

Where $\frac{x^2}{a^2}+\frac{y^2}{b^2}+\frac{z^2}{c^2}\le 1$,

the integration being extended over positive octant of the ellipsoid. Put

$$u = \frac{x^2}{a^2} \text{ i.e., } x = au^{\frac{1}{2}}, \quad \therefore dx = \frac{a}{2}u^{-\frac{1}{2}}du,$$

$$v = \frac{y^2}{b^2} \text{ i.e., } y = bv^{\frac{1}{2}}, \quad \therefore dy = \frac{b}{2}v^{-\frac{1}{2}}dv,$$

$$w = \frac{z^2}{c^2} \text{ i.e., } z = cw^{\frac{1}{2}}, \quad \therefore dz = \frac{c}{2}w^{-\frac{1}{2}}dw.$$

Hence,

$$u+v+w \le 1.$$

$\therefore$ M.I. of the ellipsoid about OX

$$= 8\rho\iiint \frac{1}{8}abc(b^2v+c^2w)u^{-\frac{1}{2}}v^{-\frac{1}{2}}w^{-\frac{1}{2}}\,du\,dv\,dw$$

$$= \rho abc\iiint (b^2u^{-\frac{1}{2}}v^{\frac{1}{2}}w^{-\frac{1}{2}}+c^2u^{-\frac{1}{2}}v^{-\frac{1}{2}}w^{\frac{1}{2}})\,du\,dv\,dw.$$

$$= \rho abc\,(b^2+c^2)\frac{\Gamma\frac{1}{2}\Gamma\frac{3}{2}\Gamma\frac{1}{2}}{\Gamma\frac{7}{2}},$$

by Dirichlet's theorem.

$$= \rho abc(b^2+c^2)\frac{\pi}{\frac{5}{2}\cdot\frac{3}{2}}$$

$$= \frac{4abc\rho\pi}{3}\cdot\frac{b^2+c^2}{5} = \frac{1}{5}M(b^2+c^2)\cdot$$

Since,

$$M = \frac{4}{3}\pi abc\rho.$$

Examples

1. Show that M.I. of a paraboloid of revolution about its axis is $\frac{M}{3}\times$ square of the radius of its base.

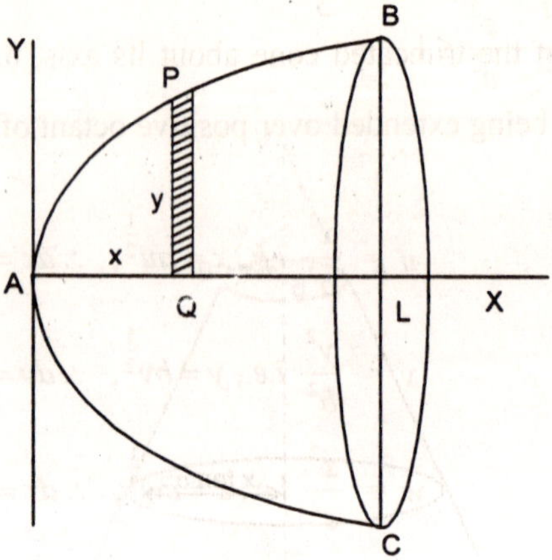

Fig. 7.13

Let the paraboloid of revolution be generated by revolving arc APB of the parabola $y^2 = 4ax$ about axis Ax and let b be the radius of the base of the parabolid of revolution.

$\therefore$ Co-ordinates of B are $(b^2/4a, b)$

Let us consider a strip of breadth δx at a distance x from the vertex A. This strip when revolved about x-axis generates a circular disc of radius y.

Mass of the elementary disc $= \pi y^2 \delta x \rho$.

Hence M = mass of the paraboloid of revolution

$$= \pi\rho \int_0^{b^2/4a} y^2 dx = \pi\rho \int_0^{b^2/4a} 4ax\, dx = \frac{\pi\rho b^4}{8a}.$$

M.I. of the strip about $AX = \pi y^2 \delta x \rho \frac{y^2}{2}$.

Hence, M.I. of the paraboloid of revolution about AX

$$= \frac{\pi}{2}\rho \int_0^{b^2/4a} y^4 dx = \frac{\pi}{2}\rho \int_0^{b^2/4a} (4ax)^2 dx$$

$$= 8\pi\rho a^2 \int_0^{b^2/4a} x^2 dx = 8\pi a^2 \rho \left[\frac{x^3}{3}\right]_0^{b^2/4a}$$

$$= \frac{\pi\rho b^6}{24a} = \frac{Mb^2}{3}$$

$$= \frac{M}{3} \text{ square of the radius of the base.}$$

2. Find the M.I. of the truncated cone about its axis, the radii of its ends being a and b.

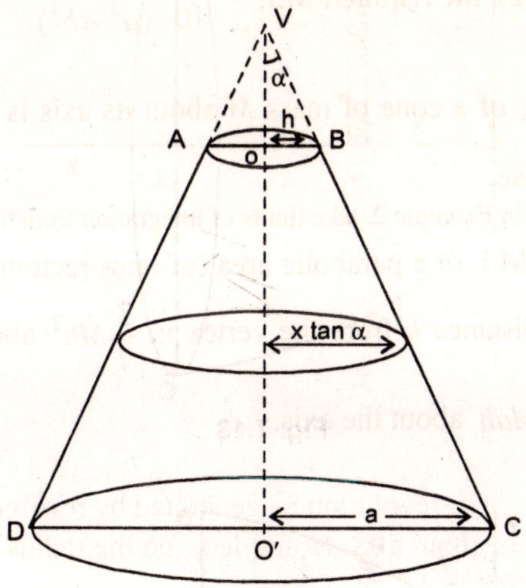

Fig. 7.14

Let $ABCD$ be the truncated cone of semi vertical angle α.
Let us consider a circular disc of breadth δx at a distance x from the vertex V.
Radius of the disc $= x\tan\alpha$ and its mass $= \pi x^2 \tan^2 \alpha \delta x \rho$

M.I. of the disc about $OO' = \pi x^2 \tan^2 \alpha \delta x \rho \cdot \frac{x^2 \tan^2 \alpha}{2}$.

Hence, M.I. of the truncated cone about OO'

$$= \frac{1}{2}\pi\rho\tan^4\alpha \int_{b\cot\alpha}^{a\cot\alpha} x^4 dx = \frac{\pi\rho\tan^4\alpha}{2} \cdot \frac{1}{5}(a^5 - b^5)\cot^5\alpha$$

$$= \frac{\pi\rho\cot\alpha}{10}(a^5 - b^5). \tag{1}$$

Mass

$$M = \int_{b\cot\alpha}^{a\cot\alpha} \pi x^2 \tan^2 \alpha \rho dx$$

$$= \pi\rho\tan^2\alpha\frac{\cot^3\alpha}{3}(a^3-b^3) = \frac{\pi\rho\cot\alpha}{3}(a^3-b^3).$$

$$\therefore \quad \pi\rho\cot\alpha = \frac{3M}{a^3-b^3}. \tag{2}$$

From (1) and (2), the required M.I. $= \dfrac{3M}{10}\dfrac{(a^5-b^5)}{(a^3-b^3)}$.

3. Show that M.I. of a cone of mass M about its axis is $\frac{3}{10}Ma^2$, a being radius of its base.

Hint. Proceed as in Example 2, take limits of integration from 0 to $a\cot\alpha$.

4. Show that the M.I. of a parabolic area (of latus rectum $4a$) cut off by an ordinate at a distance h from the vertex is $\frac{3}{7}Mh^2$ about tangent at the vertex and $\frac{4}{5}Mah$ about the axis.

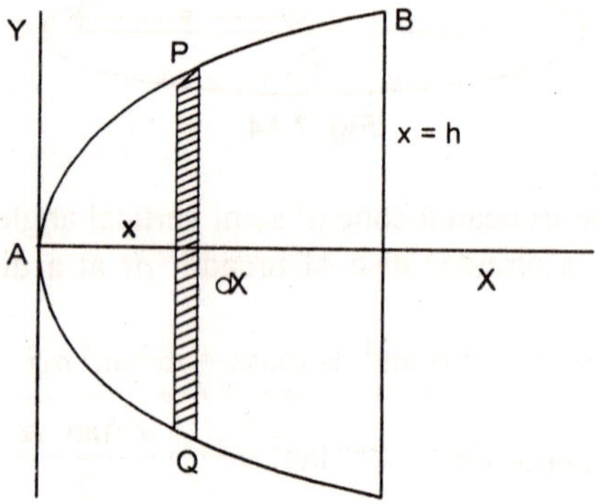

Fig. 7.15

Let equation of the bounding parabola by $y^2 = 4ax$.
Consider a strip PQ of breadth δx at a distance x from the vertex A.
M = Mass of the parabolic area is

$$\int_0^h 2y\rho dx = 2\rho\int_0^h \sqrt{4ax}\,dx = 4\rho\sqrt{a}\frac{2}{3}h^{\frac{3}{2}} = \frac{8}{3}\rho\sqrt{a}\cdot h^{\frac{3}{2}} \tag{1}$$

AY is the tangent at the vertex A (0, 0).
M.I. of the strip about $AY = 2y\rho\delta x.x^2$.

$\therefore$ M.I. of the whole area about $AY = \int_0^h 2y\rho x^2 dx = 2\rho \int_0^h 2\sqrt{ax}\, x^2 dx$

$$= 4\rho a^{\frac{1}{2}} \int_0^h x^{\frac{5}{2}} dx = \frac{8}{7}\rho a^{\frac{1}{2}} h^{\frac{7}{2}}. \qquad (2)$$

From (1) and (2) M.I. of the parabolic area about $AY = \frac{3}{7}Mh^2$.

Again M.I. of the strip about axis $AX = 2y\delta x \rho \frac{y^2}{3}$.

$\therefore$ M.I. of the whole area about AX

$$= \frac{2}{3}\rho \int_0^h y^3 dx = \frac{2}{3}\rho \int_0^h (4ax)^{\frac{3}{2}} dx$$

$$= \frac{2}{3}\rho . 8a^{\frac{3}{2}} . \frac{2}{5}h^{\frac{5}{2}} = \frac{32}{15}\rho a^{\frac{3}{2}} h^{\frac{5}{2}}. \qquad (3)$$

From (1) and (3), M.I. of the parabolic area about $AX = \frac{4}{5}Mah$.

5. Find the M.I. about the x-axis of the portion of the ellipsoid $\frac{x^2}{a^2} + \frac{y^2}{b^2} + \frac{z^2}{c^2} = 1$, which lies in the positive octant, supposing the law of volume density to be $\rho = \mu xyz$.

Consider an elementary volume $\delta x \delta y \delta z$ at a point x, y, z.

Distance of this element from x-axis $= \sqrt{y^2 + z^2}$.

M.I. of the element about x-axis $= \rho(y^2 + z^2)\, \delta x\, \delta y\, \delta z$.

$$= \mu xyz\,(y^2 + z^2)\, \delta x\, \delta y\, \delta z.$$

Hence, M.I. of the octant of the ellipsoid

$$= \iiint \mu xyz(y^2 + z^2) dx\, dy\, dz.$$

where

$$\frac{x^2}{a^2} + \frac{y^2}{b^2} + \frac{z^2}{x^2} = \leq 1.$$

Put

$$u = \frac{x^2}{a^2}, v = \frac{y^2}{b^2}, w = \frac{z^2}{c^2}.$$

Hence,

$$u + v + w \le 1.$$

$$\therefore \text{ required M.I.} = \mu \iiint \frac{1}{8} a^2 b^2 c^2 (b^2 v + c^2 w) du\, dv\, dw$$

$$= \frac{1}{8} \mu a^2 b^2 c^2 \iiint (b^2 u^{1-1} v^{2-1} w^{1-1} + c^2 u^{1-1} v^{1-1} w^{2-1}) du\, dv\, dw.$$

$$= \frac{1}{8} \mu a^2 b^2 c^2 \left[b^2 \frac{\Gamma 1 \Gamma 2 \Gamma 1}{\Gamma(1+2+1+1)} + c^2 \frac{\Gamma 1 \Gamma 1 \Gamma 2}{\Gamma(1+1+2+1)} \right]$$

$$= \frac{1}{8} \mu a^2 b^2 c^2 (b^2 + c^2) \frac{\Gamma 2}{\Gamma 5} = \frac{1}{192} \mu a^2 b^2 c^2 (b^2 + c^2).$$

M = Mass of the ellipsoid

$$= \iiint \mu x\, yz\, dx\, dy\, dz,$$

where

$$\frac{x^2}{a^2} + \frac{y^2}{b^2} + \frac{z^2}{c^2} \le 1.$$

$$= \frac{1}{8} \mu a^2 b^2 c^2 \iint du\, dv\, dw,$$

where $u + v + w \le 1.$

$$= \frac{1}{8} \mu a^2 b^2 c^2 \frac{\Gamma 1 \Gamma 1 \Gamma 1}{\Gamma 4} = \frac{1}{8} \mu a^2 b^2 c^2 . \frac{1}{6} \quad (2)$$

From (1) and (2) required M.I. $= \frac{1}{4} M (b^2 + c^2)$.

7.4 SIMPLE PROPOSITIONS

I. *Sum of M.I. About any two axes is greater than third.*

We have $A + B - C = \sum m (y^2 + z^2) + \sum m (x^2 + z^2) - \sum m (x^2 + y^2)$

$= 2 \sum mz^2 =$ positive.

Hence, proved.

II. *The sum of M.I. about any three rectangular axes meeting at a given point is always constant and is equal to twice the M.I. about that point.*

We have $A + B + C = 2\sum m(x^2 + y^2 + z^2) = 2\sum mr^2$.

This shows that $A + B + C$ is independent of the direction of axes.

III. *The sum of M.I. of a body with reference to any plane through a given point and its normal at that point is constant and is equal to the M.I. of the body with reference to that point.*

Let the given point be taken as origin and plane as xy plane.

If C' is the M.I. w.r.t. xy plane and C the M.I. about its normal at origin (z-axis), then $C + C' = \frac{1}{2}(A + B + C)$, from prop. II, which is independent of the direction of the axes.

$$\therefore C' = \frac{1}{2}(A + B - C),$$

Thus if A', B', C' are M.I. with reference to the planes of $y, z; z, x$ and x, y then $A' = \frac{1}{2}(B + C - A)$, $B' = \frac{1}{2}(C + A - B)$, $C' = \frac{1}{2}(A + B - C)$.

IV. $A > 2D$, $B > 2E$, $C > 2F$

We know that A.M. > G.M.

A.M. of $y^2, z^2 = \frac{1}{2}(y^2 + z^2)$

G.M. of $y^2, z^2 = yz$

$$\therefore \quad \frac{1}{2}(y^2 + z^2) > yz$$

$$\therefore \quad \Sigma m(y^2 + z^2) > 2\Sigma myz$$

$$\therefore \quad A > 2D.$$

Similarly, we can prove $B > 2E$, $C > 2F$.

7.5 ANGULAR MOMENTUM OF A RIGID BODY AND ABOUT FIXED AXIS

Let a rigid body be rotating about an axis through O with angular velocity $\vec{\omega}$, P be a piont having position vector $\vec{r}$ and velocity $\vec{v}$.

Now

$$\vec{v} = \vec{\omega} \times \vec{r},$$

vector angular momentum about O is

$$\vec{H} = \sum \vec{r} \times m\vec{v} = \sum m[\vec{r} \times (\vec{\omega} \times \vec{r})]$$

$$= \sum mr^2 \vec{\omega} - \sum m(\vec{r}\,.\vec{\omega})\,\vec{r} \qquad (1)$$

This gives vector angular velocity about O.
Writing

$$\vec{r} = x\vec{i} + y\vec{j} + z\vec{k},\ \vec{\omega} = \omega_1\vec{i} + \omega_2\vec{j} + \omega_3\vec{k},\ \vec{H} = h_1\vec{i} + h_2\vec{j} + h_3\vec{k}$$

in (1) and equating coefficients of $\vec{i}, \vec{j}, \vec{k}$ we get

$$h_1 = \sum mr^2\omega_1 - \sum m(\omega_1 x + \omega_2 y + \omega_3 z)\,x$$

$$= \sum m(r^2 - x^2)\,\omega_1 - \left(\sum mxy\right)\omega_2 - \left(\sum mxz\right)\omega_3$$

$$= \sum m\,(y^2 + z^2)\,\omega_1 - \left(\sum mxy\right)\omega_2 - \left(\sum mxz\right)\omega_3$$

$$= A\omega_1 - F\omega_2 - E\omega_3$$

Similarly

$$h_2 = B\omega_2 - D\omega_3 - F\omega_1.$$
$$h_3 = C\omega_3 - E\omega_1 - D\omega_2.$$

Hence, we get matrix form $\begin{bmatrix} h_1 \\ h_2 \\ h_3 \end{bmatrix} = \begin{bmatrix} A & -F & -E \\ -F & B & -D \\ -E & -D & C \end{bmatrix} \begin{bmatrix} \omega_1 \\ \omega_2 \\ \omega_3 \end{bmatrix}$

The symmetrical 3 × 3 matrix on the R.H.S. is called *inertia matrix*.

7.6 COPLANAR DISTRIBUTION

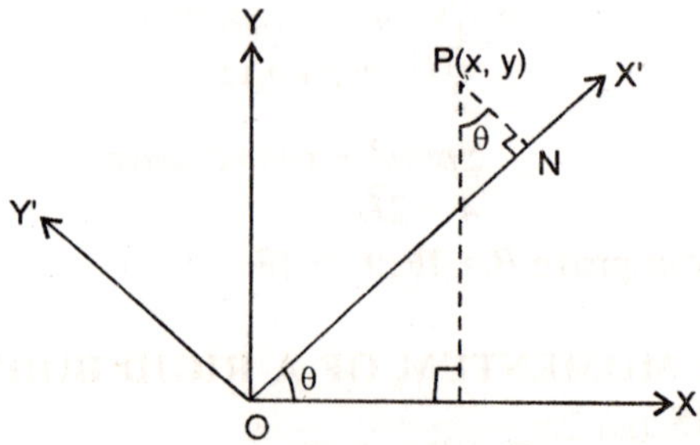

Fig. 7.16

Let A and B be the M.I. and F the P.I. of a plane lamina about two axes OX and OY in the plane of lamina.

If m be the element of mass at the point P (x, y), then

$$A = \sum my^2,\ B = \sum mx^2,\ F = \sum mxy.$$

Through O, take axes OX', OY' inclined at angle θ, $\theta + \frac{\pi}{2}$ to OX. If (x', y') be the coordinates of P referred to OX' and OY', then

$$x' = x\cos\theta + y\sin\theta,\ y' = y\cos\theta, - x\sin\theta$$

M.I. of the plane lamina about OX'

$$A' = \sum my'^2 = \sum m(y\cos\theta - x\sin\theta)^2$$

$$= \sum m(y^2\cos^2\theta - 2xy\cos\theta\sin\theta + x^2\sin^2\theta)$$

$$= \cos^2\theta\sum my^2 - \sin 2\theta\sum mxy + \sin^2\theta\sum mx^2$$

$$= A\cos^2\theta + B\sin^2\theta - F\sin 2\theta. \tag{1}$$

To find M.I. about OY' replace θ by $\theta + \frac{\pi}{2}$, we get M.I. about OY'

$$B' = A\sin^2\theta + B\cos^2\theta + F\sin 2\theta. \tag{2}$$

P.I. about OX', OY'

$$F' = \sum mx'y' = \sum m(x\cos\theta + y\sin\theta)(y\cos\theta - x\sin\theta)$$

$$= \sum m[xy(\cos^2\theta - \sin^2\theta) + \sin\theta\cos\theta(y^2 - x^2)]$$

$$= \cos 2\theta\sum mxy + \frac{1}{2}\sin 2\theta\sum m(y^2 - x^2)$$

$$= \cos 2\theta\sum mxy + \frac{1}{2}\sin 2\theta\left(\sum my^2 - \sum mx^2\right)$$

$$= F\cos 2\theta + \frac{1}{2}(A - B)\sin 2\theta \tag{3}$$

If OX', OY' are principal axes, then $F' = 0$. Let $\theta = \alpha$ denote a solution for which $F' = 0$.

$$\therefore \qquad \tan 2\alpha = \frac{2F}{B - A}. \tag{4}$$

But

$\tan 2\alpha = \tan 2\left(\alpha + \frac{\pi}{2}\right)$, so that $\theta = \alpha + \frac{\pi}{2}$ is also a solution.

Suppose that $B > A$, then (4) shows that 2α and hence α are acute angles. Also from (1) and (2)

$$A' = \frac{1}{2}(A+B) + \frac{1}{2}[(B-A)\cos 2\theta + 2F\sin 2\theta],$$

$$B' = \frac{1}{2}(A+B) + \frac{1}{2}[(B-A)\cos 2\theta + 2F\sin 2\theta].$$

Thus

$$\text{min. } A' = \frac{1}{2}\left[(A+B) - \sqrt{(B-A)^2 + 4F^2}\right]$$

$$\text{max. } B' = \frac{1}{2}\left[(A+B) + \sqrt{(B-A)^2 + 4F^2}\right].$$

Thus extreme values being attained for as defined by (4). Thus the greatest and least M.I. for lines through O in the plane of distribution are attained along the principal axes.

7.7 TO FIND WHETHER A GIVEN STRAIGHT LINE IS, AT ANY POINT OF ITS LENGTH, A PRINCIPAL AXIS OF A MATERIAL SYSTEM. IF THE LINE IS PRINCIPAL AXIS, THEN TO DETERMINE THE OTHER TWO PRINCIPAL AXES

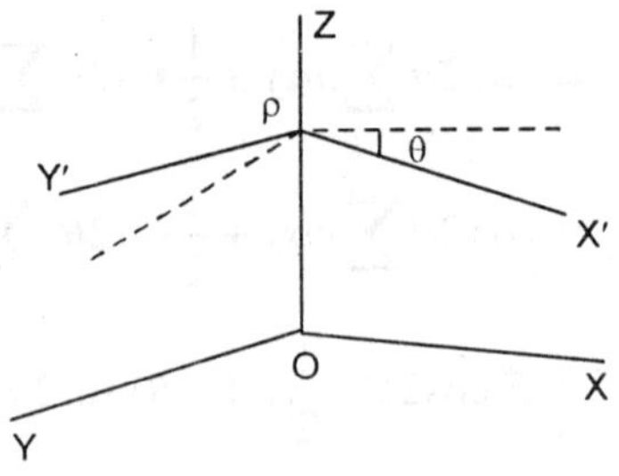

Fig. 7.17

Take the straight line as axis of z, some point O on it as origin and let P be a point at which it is principal axis and $OP = h$. Let PX', PY' be the other principal axes at P and θ be the angle between PX' and OX.

The necessary and sufficient conditions that OZ may be principal axis at P are

$$\sum my'z' = \sum mz'x' = \sum mx'y' = 0, \tag{1}$$

where

$$x' = x\cos\theta + y\sin\theta,\ y' = y\cos\theta - x\sin\theta,\ z' = z - h.$$

Proceeding as in 7.6 we get three equations

$$D\cos\theta - E\sin\theta - Mh\,(\bar{y}\cos\theta - \bar{x}\sin\theta) = 0, \tag{2}$$

$$E\cos\theta + D\sin\theta - Mh\,(\bar{x}\cos\theta + \bar{y}\sin\theta) = 0, \tag{3}$$

$$\frac{1}{2}\sin 2\theta\,(A - B) + F\cos 2\theta = 0. \tag{4}$$

where $\bar{x}$, $\bar{y}$ are the coordinates of C.G. of the system. Eliminating h from (2) and (3), we get

$$E\bar{y} = D\bar{x}. \tag{5}$$

This is the condition for OZ to be principal axis at some points of its length and then from (2) and (5) we get

$$h = \frac{D}{M\bar{y}} = \frac{E}{M\bar{x}}. \tag{6}$$

This determines the position of the pint required. From (4) we get

$$\tan 2\theta = \frac{2F}{B - A}.$$

This determines the position of the other principal axes at P.

Examples

1. A uniform rectangular plate whose sides are of lengths $2a$, $2b$ has a portion cut in the form of a square whose centre is the centre of the rectangle and whose mass is half the mass of the rectangle. Show that the axes of greatest and least M.I. at a corner of the rectangle make angles θ, $\frac{\pi}{2} + \theta$ with a side where

$$\tan 2\theta = \frac{6ab}{5(a^2 - b^2)}.$$

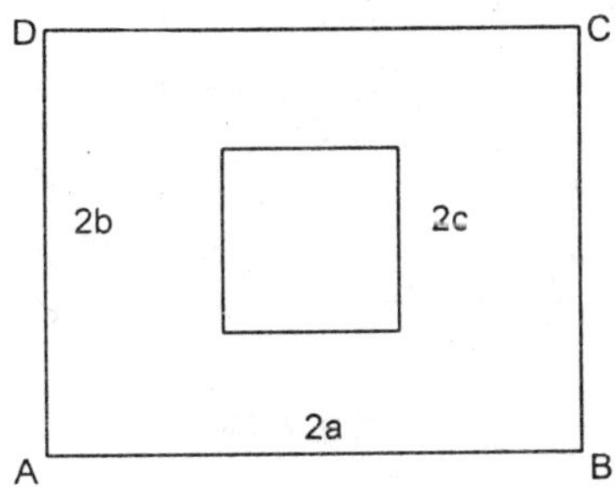

Fig. 7.18

Let $2c$ be the length of the side of the square. If M is the mass of the rectangle, then mass of the square is $\frac{1}{2}M$.

A = M.I. of remaining portion about AB
= M.I. of the rectangle – M.I. of the square

$$= \frac{4}{3}Mb^2 - \left\{\frac{1}{3}\left(\frac{M}{2}\right)c^2 + \frac{M}{2}b^2\right\}$$

$$= \frac{M}{6}(5b^2 - c^2),$$

B = M.I. of the remaining portion about AD

$$= \frac{4}{3}Ma^2 - \left\{\frac{1}{3}\left(\frac{M}{2}\right)c^2 + \frac{M}{2}a^2\right\} = \frac{M}{6}(5a^2 - c^2),$$

F = P.I. of the remaining portion about AB, AD

$$= Mab - \frac{M}{2}ab = \frac{M}{2}ab$$

Hence,

$$\tan 2\theta = \frac{2F}{B-A} = \frac{Mab}{\frac{5}{6}(a^2 - b^2)} = \frac{6ab}{5(a^2 - b^2)}.$$

2. Find the principal axes at the corner of the cube.

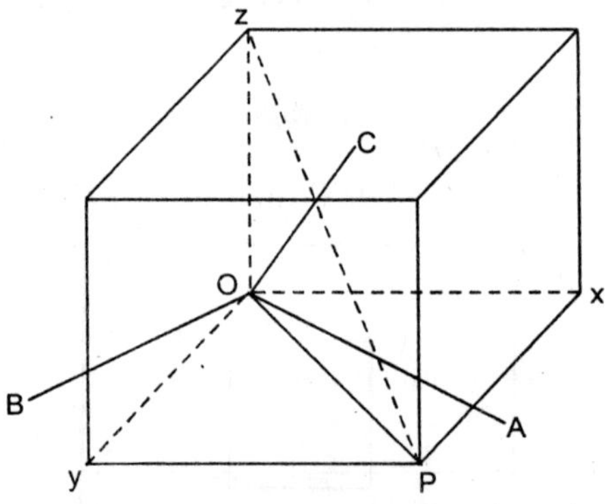

Fig. 7.19

Let O be the corner of the cube of edges $2a$ and C the C.G. Take O as origin and three co-terminus edges OX, OY, OZ as the fixed axes. The coordinates of C are $(0, 0, a\sqrt{3})$.

M.I. of the cube about any line C through is $\frac{2}{3}Ma^2$ and P.I. about any two perpendicular lines through C is zero. Hence by parallel axes theorem P. I about OC and any two perpendicular lines through O are zero. Hence, OC and two perpendicular lines through O are the principal axes.
Draw a line OA perpendicular to the plane containing OZ, OC, OP, where OP is the diagonal of the base face of the cube, then OA lies in the XOY plane and is perpendicular to OC and OP both. Thus OA is another principal axis, the third principal axis OB is then perpendicular to OA and OC. Thus OA, OB, OC are the principal at O. Let A, B, C denote M.I. about them.
Then

$$C = \frac{2}{3}Ma^2.$$

$$A = B = \frac{2}{3}Ma^2 + 3Ma^2 = \frac{11}{3}Ma^2,$$

using parallel axes theorem.

D.C.'s of OP are $\frac{1}{\sqrt{2}}, \frac{1}{\sqrt{2}}, O,$

D.C.'s of OZ are 0, 0, 1,

and D.C.'s of OC are $\frac{1}{\sqrt{3}}, \frac{1}{\sqrt{3}}, \frac{1}{\sqrt{3}}$.

D.C.'s of OA which is perpendicular to OP and OZ both are $\frac{1}{\sqrt{2}}, -\frac{1}{\sqrt{2}}, 0$.

Again D.C's of OB which is perpendicular to OC and OA are $\frac{1}{\sqrt{6}}, \frac{1}{\sqrt{6}}, -\frac{2}{\sqrt{6}}$.

Thus D.C.'s of a set of principal axes at O are

$$\left(\frac{1}{\sqrt{2}}, -\frac{1}{\sqrt{2}}, 0\right), \left(\frac{1}{\sqrt{6}}, \frac{1}{\sqrt{6}}, -\frac{2}{\sqrt{6}}\right), \left(\frac{1}{\sqrt{3}}, \frac{1}{\sqrt{3}}, \frac{1}{\sqrt{3}}\right).$$

3. A square lamina of side a has particles of masses m, $2m$, $3m$, $4m$ at its vertices. Show that the principal moments of inertia at the centre are $2ma^2$, $3ma^2$, $5ma^2$, and find the directions of principal axes.

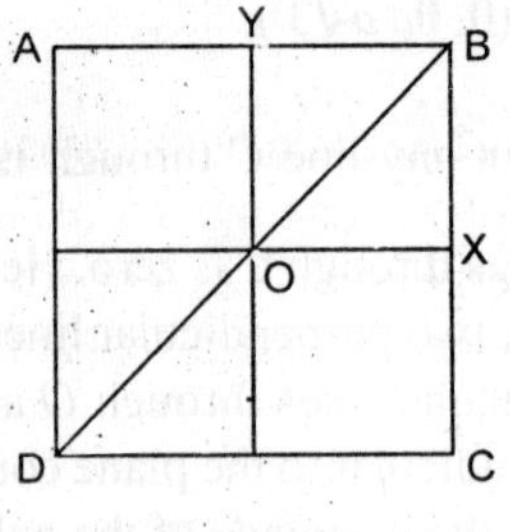

Fig. 7.20

Let O be the centre of the square, OX, OY two perpendicular axes in the plane of the square, OZ normal to the square through O. The coordinates of the vertices are $A\left(-\frac{a}{2}, \frac{a}{2}, 0\right)$, $B\left(\frac{a}{2}, \frac{a}{2}, 0\right)$, $C\left(\frac{a}{2}, -\frac{a}{2}, 0\right)$, $D\left(-\frac{a}{2}, -\frac{a}{2}, 0\right)$.

$$A = \text{M.I. about } OX = m\frac{a^2}{4} + 2m\frac{a^2}{4} + 3m\frac{a^2}{4} + 4m\frac{a^2}{4} = \frac{5}{2}ma^2.$$

Similarly, M.I. about $OY = \frac{5}{2}ma^2$.

$\therefore A = B$

$$F = \text{P.I. about } OX, OY$$

$$= m\left(\frac{-a}{2}\right)\left(\frac{a}{2}\right) + 2m\left(\frac{a}{2}\right)\left(\frac{a}{2}\right) + 3m\left(\frac{a}{2}\right)\left(\frac{-a}{2}\right) + 4m\left(\frac{-a}{2}\right)\left(\frac{-a}{2}\right) = \frac{1}{2}ma^2.$$

If A^*, B^* are principal moments of inertia, then

$$A^*, B^* = \frac{1}{2}\left[A + B \pm \sqrt{(A-B)^2 + 4F^2}\right]$$

$$= \frac{1}{2}\left[\frac{5}{2}ma^2 + \frac{5}{2}ma^2 \pm \sqrt{0 + m^2a^4}\right]$$

$$= \frac{1}{2}[5ma^2 \pm ma^2] = 3ma^2, 2ma^2.$$

But

$$C^* = A^* + B^* = 3ma^2 + 2ma^2 = 5ma^2.$$

Directions of principal axes are given by

$$\tan 2\theta = \frac{2F}{B-A} = \frac{2F}{0} = \infty.$$

$$\therefore \qquad \theta = \frac{\pi}{4}, \frac{3\pi}{4}.$$

Hence, principal axes are given by diagonals of the square.

4. The centre of mass of a plane lamina of mass M is G and GX, GY are principal axes of inertia at G. Ox, Oy are axes in the plane of lamina parallel to GX, GY respectively. Prove that P.I. of the lamina with respect to Ox, Oy is $M\alpha\beta$, where α, β are coordinates of G with respect to Ox, Oy.

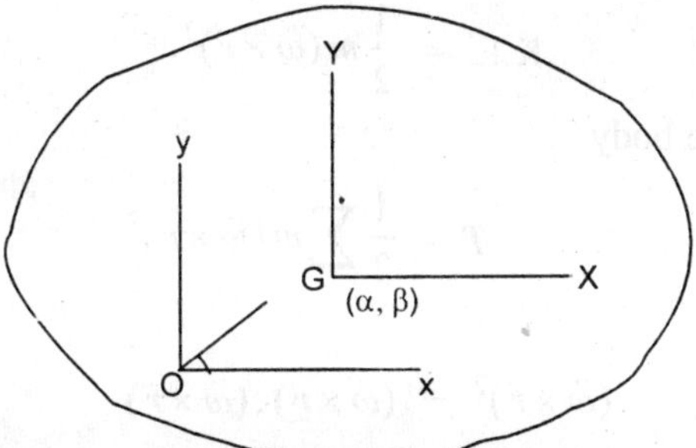

Fig 7.21

If the principal moments of inertia are $Mn\alpha^2$, about GX and $Mm\beta^2$ about GY, and one of the axes of inertia at O is inclined at an angle of 45° to Ox, prove that

$$(n-1)/(m-1) = \beta^2/\alpha^2.$$

By parallel axes theorem

$$A = A' + M\beta^2 = Mn\alpha^2 + M\beta^2 = M(n\alpha^2 + \beta^2),$$

$$B = B' + M\alpha^2 = Mm\beta^2 + M\alpha^2 = M(m\beta^2 + \alpha^2),$$

$$F = F' + M\alpha\beta = 0 + M\alpha\beta = M\alpha\beta.$$

Now

$$\tan 2\theta = \frac{2F}{B-A},$$

$$\therefore \qquad \tan 90° = \frac{2F}{B-A} = \infty.$$

$$\therefore \qquad B = A,$$

$$\therefore \qquad Mn\alpha^2 + M\beta^2 = Mm\beta^2 + M\alpha^2,$$

$$\therefore \qquad (n-1)\alpha^2 = (m-1)\beta^2,$$

$$\therefore \qquad \frac{n-1}{m-1} = \frac{\beta^2}{\alpha^2}.$$

7.8 KINETIC ENERGY OF THE BODY ROTATING ABOUT A FIXED POINT

Let a rigid body rotate instantaneously with angular velocity $\vec{\omega}$ about an axis through a fixed point O of the body. Let m be the mass of the particle at the point P of the body and $\vec{OP} = \vec{r}$.

Velocity of $P = \vec{\omega} \times \vec{r}$

Its $\quad$ K.E. $= \dfrac{1}{2} m (\vec{\omega} \times \vec{r})^2$.

$\therefore$ K.E. of the body

$$T = \frac{1}{2} \sum m (\vec{\omega} \times \vec{r})^2 \tag{1}$$

Now

$$\begin{aligned}(\vec{\omega} \times \vec{r})^2 &= (\vec{\omega} \times \vec{r}).(\vec{\omega} \times \vec{r}) \\ &= \vec{\omega}.[\vec{r} \times (\vec{\omega} \times \vec{r})]\end{aligned} \tag{2}$$

From (1) and (2) we get

$$\begin{aligned}T &= \frac{1}{2} \sum m [\vec{\omega}.\{\vec{r} \times (\vec{\omega} \times \vec{r})\}] \\ &= \frac{1}{2} \vec{\omega}.[m \{\vec{r} \times (\vec{\omega} \times \vec{r})\}]\end{aligned}$$

But angular momentum $\vec{H}$ of the body about O is given by

$$\vec{H} = \sum m\vec{r} \times (\vec{\omega} \times \vec{r}).$$

$$\therefore \quad T = \frac{1}{2} \vec{\omega}.\vec{H}. \tag{3}$$

Let $\quad \vec{\omega} = \omega_1 \vec{a}_1 + \omega_2 \vec{a}_2 + \omega_3 \vec{a}_3$,

where $\vec{a}_1, \vec{a}_2, \vec{a}_3$ denote unit vectors along three principal axes through O.

Now

$$\vec{H} = A^* \omega_1 \vec{a}_1 + B^* \omega_2 \vec{a}_2 + C^* \omega_3 \vec{a}_3,$$

where A^*, B^*, C^* are the principal M.I.

Substituting these values of $\vec{\omega}$ and $\vec{H}$ in (3), we get

$$\mathrm{T} = \frac{1}{2} [A^* \omega_1^2 + B^* \omega_2^2 + C^* \omega_3^2] \tag{4}$$

Let $\vec{i}, \vec{j}, \vec{k}$ be the unit vectors along any other set of mutually perpendicular axes through O and let

$$\vec{r} = x\vec{i} + y\vec{j} + z\vec{k}, \omega = (\lambda\vec{i} + \mu\vec{j} + \nu\vec{k}).$$

where λ, μ, ν are the D.C.'s of the instantaneous axis of rotation.

Now

$$\vec{\omega} \times \vec{r} = \omega\,[(\mu z - \nu y)\vec{i} + (\nu x - \lambda z)\vec{j} + (\lambda y - \mu x)\vec{k}]$$

$$\therefore \quad (\vec{\omega} \times \vec{r})^2 = \omega^2\,[(\mu z - \nu y)^2 + (\nu x - \lambda z)^2 + (\lambda y - \mu x)^2]$$

$$= \omega^2\,[\lambda^2 (y^2 + z^2) + \mu^2 (z^2 + x^2) + \nu^2 (x^2 + y^2) - 2\lambda\mu xy - 2\mu\nu yz - 2\nu\lambda zx]]$$

Putting this value of $(\vec{\omega} \times \vec{r})^2$, we get

$$T = \frac{1}{2}\omega^2\,[\lambda^2 \sum m(y^2 + z^2) + \mu^2 \sum m(z^2 + x^2) + \nu^2 \sum m(x^2 + y^2) - 2\lambda\mu \sum mxy - 2\mu\nu \sum myz - 2\nu\lambda \sum mzx]$$

$$= \frac{1}{2}\omega^2\,[A\lambda^2 + B\mu^2 + C\nu^2 - 2D\mu\nu - 2E\nu\lambda - 2F\lambda\mu].$$

$$= \frac{1}{2}\omega^2 I. \tag{5}$$

7.9 CAUCHY'S MOMENTAL ELLIPSOID AND ELLIPSOID OF GYRATION

If we consider the quadratic surface whose equation is

$$Ax^2 + By^2 + Cz^2 - 2Dyz - 2Ezx - 2Fxy = 1,$$

where A, B, C, D, E, F are the M.I. and P.I. of a given body with respect to the axes of reference $OXYZ$, it follows from the equation

$$I = A\lambda^2 + B\mu^2 + C\nu^2 - 2D\mu\nu - 2E\nu\lambda - 2F\lambda\mu,$$

that the reciprocal of the square of any radius vector of the quadratic is equal to the M.I. of the body about this radius. The quadric therefore is the same whatever be the axes of reference provided origin is unchanged, and consequently its equation referred to any other axes $OX'Y'Z'$ having the same orogin is

$$A'x'^2 + B'y'^2 + C'z'^2 - 2D'y'z' - 2E'z'x' - 2F'x'y' = 1,$$

where A', B', C', D', E$'$, F$'$, are the M.I. and P.I. with respect to these axes.

This quadratic is called the *momental ellipsoid* of the body at the point O; its principal axes are called the *principal axes of inertia* of the body at O. The equation of quadratic referred to these axes contains no product terms, and therefore the P.I. with respect to them are zero. the M.I. with respect to these axes are called the *principal moments of inertia* of the body at the the point O.

The momental ellipsoid is also called the *ellipsoid of inertia;* its polar reciprocal with regard to its centre is another ellipsoid, which is called the *ellipsoid of gyration*. Equation of the ellipsoid of gyration is given as

$$\frac{x^2}{A}+\frac{y^2}{B}+\frac{z^2}{C}=\frac{1}{M}. \tag{1}$$

Let l', m', n' be the $D.C.$'s of perpendicular from O to the tangent plane to the ellipsoid of gyration and p' its length, then from (1),

$$Mp'^2 = Al'^2 + Bm'^2 + Cn'^2 \tag{2}$$

so that the M.I. about this perpendicular is Mp'^2. This is the characteristics property of an ellipsoid of gyration, *viz.*, that the M.I. about the central perpendicular on a tangent plane is equal to the mass times square of the length of this perpendicular.

7.10 EQUIMOMENTAL SYSTEMS

Definition : Sytems are said to be equimomental, if they have equal M.I. about every line in space.

Necessary and sufficient conditions for two systems to be equimomental are :

(*i*) The same total mass;

(*ii*) The same Centroid;

(*iii*) The same principal axes and same principal M.I. at the centroid.

Proof :

(*a*) Suppose that the stated conditions hold. Denote by M the total mass of each system, G the centroid, A^*, B^*, C^* M.I. at G.

In both cases, M.I. about the line through G, having $D.C.$'s λ, μ, ν *w.r.t.* principal axes is

$A^*\lambda^2 + B^*\mu^2 + C^*\nu^2$. Hence, by parallel axes theorem, the M.I. about a parallel line distant h from this line is

$$A^*\lambda^2 + B^*\mu^2 + C^*\nu^2 + Mh^2,$$

in both cases. Thus the two systems have equal M.I. about any line in space, and the systems are equimomental.

(*b*) Suppose the systems are equimomental, now we will derive conditions (*i*), (*ii*), (*iii*).

Let G_1, G_2 be the centroids of two systems, and M_1, M_2 be their masses. Since the systems are equimomental, they have same M.I. about line G_1G_2. The M.I.S. about any parallel line distant d from G_1G_2 are respectively $I + M_1d^2$, $I + M_2d^2$, by using parallel axes theorem.

Since, the systems are equimomental,

$$\therefore \quad I + M_1d^2 = I + M_2d^2$$

$\therefore \quad M_1 = M_2 = M$, say

Let J be the M.I. of either system about the line through G_1 perpendicular to G_1G_2. Then the M.I. of the first system about the parallel line through G_2 is $J + M(G_1G_2)^2$, and that of the second system is $J - M(G_1G_2)^2$.

$$\therefore \quad J + M(G_1G_2)^2 = J - M(G_1G_2)^2,$$

$$\therefore \quad G_1 \equiv G_2.$$

The systems have equal moments of inertia about any line through their common centroid G. Thus their momental ellipsoids at G are equal and so are the principal axes and principal moments of inertia at G.

7.11 EQUIMOMENTAL SYSTEM OF A UNIFORM PLANE TRIANGULAR LAMINA.

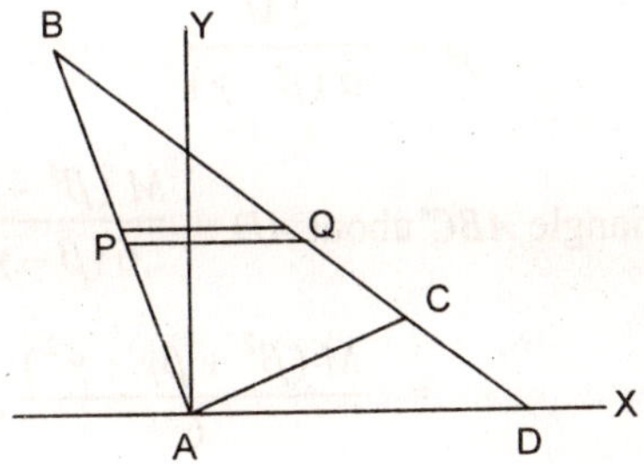

Fig. 7.22

Let AX, AY be rectangular axes through the corner A of the triangle ABC; BC meet AX at D. Let β, γ be the distance of B, C from AX and $AD = \alpha$.

Let us consider a strip PQ of thickness δy at a distance y from AD. Now triangle BAD and BPQ are similar

$$\therefore \quad \frac{\beta - y}{\beta} = \frac{PQ}{\alpha}$$

$$\therefore \quad PQ = \alpha\left(1 - \frac{y}{\beta}\right).$$

$$\text{M.I. of triangle } BAD \text{ about } AX = \int_0^\beta \rho\alpha\left(1 - \frac{y}{\beta}\right) y^2\, dy$$

$$= \rho\alpha \int_0^\beta \left(y^2 - \frac{y^3}{\beta}\right) dy$$

$$= \rho\alpha \frac{\beta^3}{12}.$$

Similarly, M.I. of triangle CAD about $AD = \rho\alpha \frac{\gamma^3}{12}$.

Hence, M.I. of triangle ABC about AD = M.I. of the triangle BAD about AD – M.I. of the triangle CAD about AD

$$= \frac{\rho\alpha}{12}(\beta^3 - \gamma^3).$$

$$\text{Mass of } \Delta ABC = \frac{1}{2}\rho\alpha(\beta - \gamma)$$

$$\therefore \qquad \rho = \frac{2M}{\alpha(\beta-\gamma)}.$$

Hence, M.I. of triangle ABC about $AD = \dfrac{M(\beta^3 - \gamma^3)}{6(\beta - \gamma)}$

$$= \frac{M(\beta^2 + \beta\gamma + \gamma^2)}{6}.$$

This may be written as $\dfrac{M}{3}\left[\left(\dfrac{\beta}{2}\right)^2 + \left(\dfrac{\gamma}{2}\right)^2 + \left(\dfrac{\beta+\gamma}{2}\right)^2\right]$

This is the M.I. about AD of three particles each of mass $M/3$ placed at the middle points of the sides of the triangle.

Total mass of these three particles $= \dfrac{M}{3} + \dfrac{M}{3} + \dfrac{M}{3} = M$

Distance of C.G. of these masses from AD

$$= \frac{M}{3}\left(\frac{\beta+\alpha}{2} + \frac{\gamma+\beta}{2} + \frac{\alpha+\gamma}{2}\right)$$

$$= \frac{M}{3}(\alpha + \beta + \gamma).$$

This is the distance of C.G. of triangle ABC from AD.

Hence, both systems have same M.I. about an axis GX' parallel to AD. But direction of this line is arbitrary, therefore the systems have same equal M.I. about all lines through G in the plane; and by parallel axes theorem, about all lines in the plane. Since, for plane body $C = A + B$, therefore for all axes at right angles to the plane. Hence, the systems are equimomenal.

Hence, *equimomental for a triangular lamina is system of three particles each of mass one third of that of triangle and placed at the middle points of the sides of the triangle.*

Examples

1. Show that a uniform triangular lamina of mass m is equimomental with three particles each of mass $m/12$ placed at the angular points and a particle of mass $3m/4$ placed at the C.G. of the triangle.

   Let α, β, γ be the distances of the vertices of the triangle from a line l in its plane.

   $$\text{Total mass of the four masses} = 3 \times \frac{m}{12} + \frac{3}{4}m = m. \quad (1)$$

   Distance of C.G. of four masses from the line l is

   $$= \left[\frac{m}{12}\alpha + \frac{m}{12}\beta + \frac{m}{12}\gamma + \frac{3m}{4} \times \frac{1}{3}(\alpha + \beta + \gamma)\right] / m = \frac{1}{3}(\alpha + \beta + \gamma). \quad (2)$$

   M.I. of the four particles about line l is

   $$= \frac{m}{12}\alpha^2 + \frac{m}{12}\beta^2 + \frac{m}{12}\gamma^2 + \frac{m}{12}(\alpha + \beta + \gamma)^2.$$

   $$= \frac{m}{6}[\alpha^2 + \beta^2 + \gamma^2 + \alpha\beta + \beta\gamma + \gamma\alpha] \quad (3)$$

   From (1) – (3) we see that triangular lamina and system of given particles, satisfy the conditions for the two systems to be equimomental, hence, they are equimomental.

2. *ABCD* is a uniform parallelogram of mass M, at the middle points of the four sides are placed masses each equal to $M/6$ and at the intersection of the diagonals, particle of mass $M/3$. Show that these five particles and the parallelogram are equimomental system.

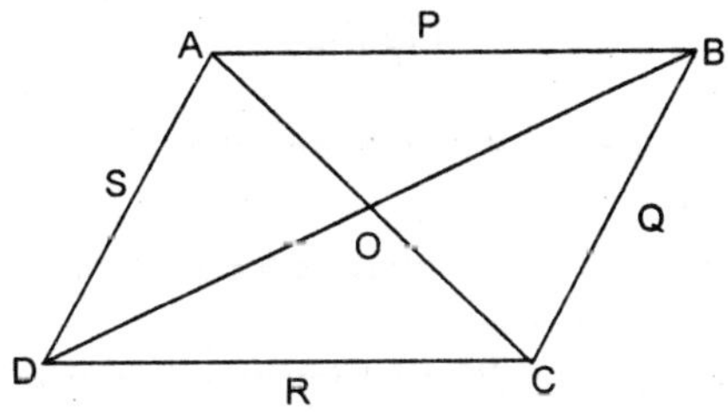

Fig. 7.23

Parallelogram *ABCD* = Triangle *ABC* + triangle *ADC* (each of mass $M/2$)

Let P, Q, R, S be the middle points of the sides of parallelogram.
Now triangle ADC is equimomental to three masses each of mass $M/6$ placed at the points R, S, O. Similarly triangle ABC is equimomental to three particles each of mass $M/6$ placed at the points P, Q, O.
Thus two triangles (or parallelogram) are equimomental to a system of masses each of mass $M/6$ paced at the points P, Q, R, S and a mass $\frac{M}{6}+\frac{M}{6}=\frac{M}{3}$ at O.

3. Show that the M.I. of a regular polygon of n sides about any straight line through its centre is

$$\frac{Mc^2}{24}\cdot\frac{2+\cos(2\pi/n)}{1-\cos(2\pi/n)}.$$

where n is the number of sides and c is the length of each side.

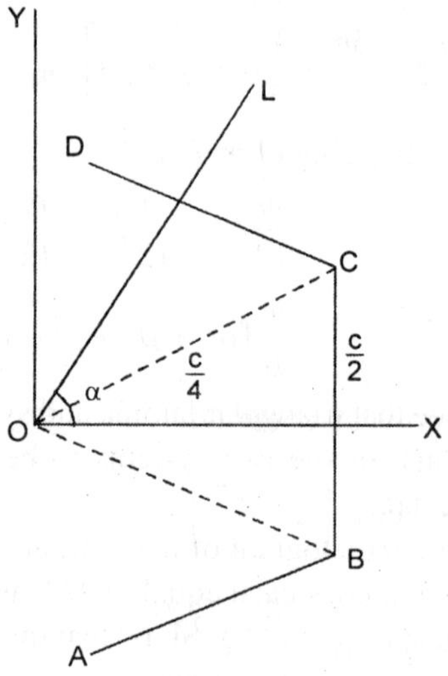

Fig. 7.24

Let $ABCD$ .... be the regular polygon of n sides each of length c and having O as centre.
The polygon can be divided into n isosceles triangles each of mass M/n.
Let BOC be one such triangle.
Take OX a line bisecting BC, and a perpendicular line Oy as the axes of reference.

$$\angle BOC = 2\pi/n,$$

$$\therefore \quad \angle COX = \pi/n.$$

ΔBOC can be replaced by three particles each of mass $\frac{M}{3n}$, placed at the middle points of its sides.

M.I. of triangle BOC about OX

$$= \frac{M}{3n}\left[\left(\frac{c}{4}\right)^2 + \left(\frac{c}{4}\right)^2\right] = \frac{Mc^2}{24n}.$$

M.I. of triangle BOC about OY

$$= \frac{M}{3n}\left[\left(\frac{c}{2}\times\cot\frac{\pi}{n}\right)^2 + 2\left(\frac{c}{4}\cot\frac{\pi}{n}\right)^2\right]$$

$$= \frac{Mc^2}{8n}\cot^2\frac{\pi}{n}.$$

Due to symmetry P.I. about OX, $OY = 0$

M.I. of triangle about any line OL, which makes an angle α with OX

$$= \frac{Mc^2}{24n}\cos^2\alpha + \frac{Mc^2}{8n}\cot^2\frac{\pi}{n}\sin^2\alpha.$$

Replacing α by $\alpha + \frac{2\pi}{n}$, $\alpha + \frac{4\pi}{n}$, ..., we get M.I. of other triangles about OL

Hence, M.I. of the polygon about OL

$$= \frac{Mc^2}{24n}\left[\cos^2\alpha + \cos^2\left(\alpha + \frac{2\pi}{n}\right) + \cos^2\left(\alpha + \frac{4\pi}{n}\right) + \text{.......} + n \text{ terms}\right]$$

$$+ \frac{Mc^2}{8n}\cot^2\frac{\pi}{n}\left[\sin^2\alpha + \sin^2\left(\alpha + \frac{2\pi}{n}\right) + \sin^2\left(\alpha + \frac{4\pi}{n}\right) + \text{......} + n \text{ terms}\right]$$

$$= \frac{Mc^2}{48n}\left[1 + \cos 2\alpha + 1 + \cos\left(2\alpha + \frac{4\pi}{n}\right) + \text{......}\right]$$

$$+ \frac{Mc^2}{16n}\cot^2\frac{\pi}{n}\left[1 - \cos 2\alpha + 1 - \cos\left(2\alpha + \frac{4\pi}{n}\right) + \text{......}\right]$$

Now

$$\cos 2\alpha + \cos\left(2\alpha + \frac{4\pi}{n}\right) + \text{.....} + n \text{ terms}$$

$$= \frac{\cos[2\alpha + (n-1)(2\pi/n)]}{\sin(2\pi/n)} = 0.$$

Hence, M.I. of the polygon about OL

$$= \frac{Mc^2}{48n}.n + \frac{Mc^2}{16n}.n.\cot^2\frac{\pi}{n}$$

$$= \frac{Mc^2}{48}\left[1 + 3\cot^2\frac{\pi}{n}\right]$$

$$= \frac{Mc^2}{48}\left[\frac{1 + 2\cos^2\pi/n}{\sin^2\pi/n}\right]$$

$$= \frac{Mc^2}{24}.\frac{2+\cos 2\pi/n}{1-\cos 2\pi/n}.$$

4. The parameter of a uniform lamina of mass M, is a regular hexagon $ABCDEF$ of side $2a$. Find a system particles equimomental with the lamina and hence or otherwise, show that the M.I. of the lamina about AD is $5Ma^2/6$. Show also that the M.I. of the lamina about any line in its plane passing through the intersection of AD and BE has the same constant value.

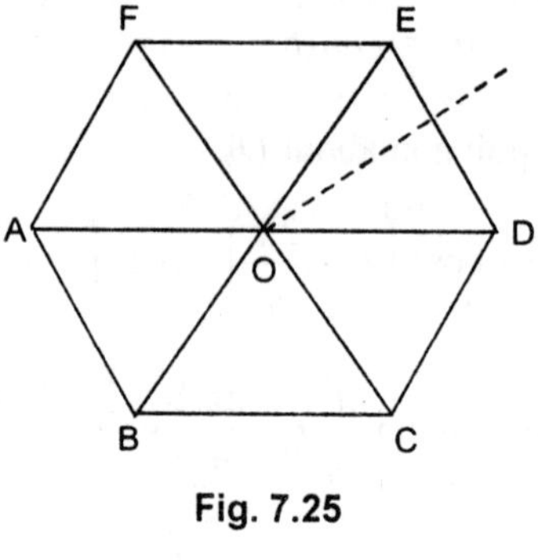

Fig. 7.25

Put $n = 6$ and $c = 2a$ in Example 3. We get
M.I. of the pane lamina about any in its plane passing through its centre O

$$= \frac{Ma^2}{6}\left[\frac{2+\cos\pi/3}{1-\cos\pi/3}\right]$$

$$= \frac{Ma^2}{6}\left[\frac{2+1/2}{1-1/2}\right] = \frac{5ma^2}{6} = \text{const.}$$

Since, this is M.I. about any line in the plane of lamina passing through *O*, hence it is also M.I. about *A.D.*

5. **If *ABCDE* .... is a regular polygon enclosing of uniform surface density ($n \geq 3$). If *O* be the centre of the circumscribing circle, prove that the momental ellipsoid at *O* has an equation of the form $x^2 + y^2 + 2z^2 = const.$**

From example 3. M.I. of the polygon about any line in the plane of lamina passing through the centre is

$$= \frac{Mc^2}{24} \cdot \frac{2 + \cos 2\pi / n}{1 - \cos 2\pi / n} = const.$$

Let OX, OY be axes of reference in the plane of the lamina, then M.I. about OX, OY are same.

$$\therefore \qquad A = B = \frac{Mc^2}{24} \cdot \frac{2 + \cos 2\pi / n}{1 - \cos 2\pi / n}.$$

But $C = A + B = 2a$, and Products of inertia about pairs of axes are zero.

Momental ellipsoid at *O* is given by

$$Ax^2 + By^2 + Cz^2 = const.$$

$$\therefore \quad Ax^2 + Ay^2 + 2Az^2 = const.$$

$$\therefore \quad x^2 + y^2 + 2z^2 = const.$$

We observe that section of this ellipsoid with the plane of the lamina $z = 0$ is a circle

$$x^2 + y^2 = const.$$

6. Show that a uniform solid cuboid of mass *M* is equimomental with
   (*i*) masses *M*/24 at the middle points of its sides and *M*/2 at the centre.
   (*ii*) masses *M*/24 at its centres and 2*M*/3 at its centre.

   Let 2*a* be the length of each side of the cuboid, *G* its centre, and coterminous edges *OA*, *OB*, *OC* be the axes of reference.

   (*i*) Coordinates of the middle points of its edges are
   $(a, 0, 0), (a, a, 0), (0, a, a), (0, a, 0), (a, 0, a), (0, 0, a), (a, 2a, a), (a, a, 2a). (2a, a, 2a), (2a, a, a), (a, 2a, 2a), (2a, 2a, a)$.

Total mass of 12 masses each of mass *M*/24 placed at the middle points of the sides and *M*/2 placed at the centre $= 12.\frac{M}{24} + \frac{M}{2} = M =$ Total mass of the cuboid.

If $\bar{x}, \bar{y}, \bar{z}$ are the coordinates of C.G. of the system of particles, then

$$\bar{x} = \frac{\frac{M}{24}(a + a + a + a + a + 2a + 2a + a + 2a) + \frac{M}{2}a}{M} = a$$

Similarly, we can prove that $\bar{y} = a = \bar{z}$.

M.I. of the system of particles about OA, is

$$\frac{M}{24}(a^2 + 2a^2 + a^2 + a^2 + a^2 + a^2 + 5a^2 + 5a^2 + 5a^2 + 5a^2 + 8a^2 + 5a^2) + \frac{M}{2}.2a^2$$

$$= \frac{8}{3}Ma^2.$$

Similarly, M.I. about OB, OC are also $\frac{8}{3}Ma^2$. These are moments of inertia for cuboid also.

P.I. about OA, OB is

$$\frac{M}{24}(a^2 + a^2 + a^2 + 2a^2 + a^2 + 2a^2 + 4a^2) + \frac{1}{2}Ma^2 = Ma^2.$$

This is also the P.I. of the cuboid about OA, OB.

Similarly, we can prove that product of inertia about OB, OC and OC, OA are also Ma^2, these results hold for cuboid also.

Thus, the system of particles has same mass, same C.G. same M.I. about OA, OB, OC and same P. I about pairs of these lines, hence, the system of particles is equimomental with cuboid.

(*ii*) Total mass of the system of particles $= 8.\frac{M}{24} + \frac{2}{3}M = M.$

Coordinates of corners are (0, 0, 0), (2*a*, 0, 0), (0, 2*a*, 0), (0, 0, 2*a*), (2*a*, 2*a*, 0), (0, 2*a*, 2*a*), (2*a*, 0, 2*a*), (2*a*, 2*a*, 2*a*).

If $\bar{x}, \bar{y}, \bar{z}$ are coordinates of the system of particles,

$$\bar{x} = \frac{\frac{M}{24}(2a + 2a + 2a + 2a) + \frac{2}{3}Ma}{M} = a$$

Similarly, we can prove that $\bar{y} = \bar{z} = a$.

M.I. about OA is

$$\frac{M}{24}(4a^2 + 4a^2 + 4a^2 + 4a^2 + 8a^2 + 8a^2) + \frac{2}{3}M2a^2 = \frac{8}{3}Ma^2.$$

Similarly, we can prove that moments of inertia about OB, OC are also $\frac{8}{3}Ma^2$.

P.I. about OA, OB is

$$\frac{M}{24}(4a^2 + 4a^2) + \frac{2}{3}Ma^2 = Ma^2.$$

Similarly, we can prove that products of inertia about OB, OC and OC, OA are also Ma^2.

Thus, arguing as in case (i), we can say that this system of particles is also equimomental with cuboid.

7.12 GENERAL MOTION OF RIGID BODY

Since, a rigid body a is a collection of particles such that the distances between all pairs of them remain invariant, hence, results of system of particles will be applicable to rigid body also.

1. **Kinetic Energy :** Suppose O is a fixed point in space, P a point fixed in the body and G is the centroid of the body, so that $\vec{OP} = \vec{r}$, $\vec{OG} = \vec{\bar{r}}$, $\vec{GP} = \vec{r}\,' = \vec{r} - \vec{\bar{r}}$.

From Chapter VI, we have K.E.

$$T = \frac{1}{2}M\dot{\vec{\bar{r}}}^2 + \frac{1}{2}\sum m\dot{\vec{r}}\,'^2 ,$$

where M is the total mass of the body, m the mass of the particle at P.

Suppose $\vec{\omega}$ is the angular velocity of the body. Then

$$\dot{\vec{r}}\,' = (\vec{\omega} \times \vec{r}\,')$$

Now

$$(\vec{\omega} \times \vec{r}\,')^2 = \vec{\omega} \,.\, \{\vec{r}\,' \times (\vec{\omega} \times \vec{r}\,')\},\ \dot{\vec{\bar{r}}} = \vec{\bar{v}}\,.$$

$$\therefore \quad T = \frac{1}{2}M\vec{\bar{v}}^2 + \frac{1}{2}\sum m\vec{\omega} \,.\, \{\vec{r}\,' \times (\vec{\omega} \times \vec{r}\,')\}$$

$$= \frac{1}{2}M\,\vec{\bar{v}}^2 + \frac{1}{2}\,\vec{\omega} \,.\, \sum m\,\{\vec{r}\,' \times (\vec{\omega} \times \vec{r}\,')\}$$

$$= \frac{1}{2}\,M\vec{\bar{v}}^2 + \frac{1}{2}\vec{\omega} \,.\, \vec{H}' \qquad (1)$$

since,

$$\sum \{\vec{r}\,' \times m\,(\vec{\omega} \times \vec{r}\,')\} = \vec{H}'.$$

If $\vec{\alpha}_1$, $\vec{\alpha}_2$, $\vec{\alpha}_3$ are unit vectors along the principal axes at G; A^*, B^*, C^* the corresponding principal M.I.

$$\vec{\omega} = \omega_1\vec{a}_1 + \omega_2\vec{a}_2 + \omega_3\vec{a}_3,$$

then

$$\vec{H}' = A^*\omega_1\vec{a}_1 + B^*\omega_2\vec{a}_2 + C^*\omega_3\vec{a}_3.$$

Substituting values of $\vec{\omega}$ and $\vec{H}'$ in (1), we get

$$T = \frac{1}{2} M \bar{\vec{v}}^2 + \frac{1}{2}(A^* \omega_1^2 + B^* \omega_2^2 + C^* \omega_3^2).$$

First term is K.E. due to translation of a single particle of mass M at G, and moving with velocity of centroids and second term is K.E. of the rigid body about the point G, now considered fixed, with angular velocity $\vec{\omega}$. *Thus K.E. is the sum of two energies one due to translation and the other due to rotation, which are independent of each other.*

2. **Angular momentum :** From Chapter VI, we know that angular momentum about a moving point O is given by

$$\vec{H} = \bar{\vec{r}} \times M\bar{\vec{v}} + \vec{H}',$$

where

$$\vec{H}' = \sum \vec{r}' \times (m\dot{\vec{r}}') = A.M. \text{ about } G.$$

But

$$\vec{H}' = A^* \omega_1 \vec{a}_1 + B^* \omega_2 \vec{a}_2 + C^* \omega_3 \vec{a}_3,$$

$$\therefore \quad \vec{H} = \bar{\vec{r}} \times M\bar{\vec{v}} + A^* \omega_1 \vec{a}_1 + B^* \omega_2 \vec{a}_2 + C^* \omega_3 \vec{a}_3.$$

Again, it is seen that angular momentum consists of two terms one due to translation and second due to roation, which are independent of each other.

3. **Moment of rate of change of momentum :** We know that moment of the rate of change of momentum about O is

$$\sum(\vec{r} \times m\dot{\vec{v}}) = \sum(\bar{\vec{r}} + \vec{r}') \times (m\dot{\vec{v}}) = \bar{\vec{r}} \times \sum m\dot{\vec{v}} + \sum \vec{r}' \times m\dot{\vec{v}} \qquad (1)$$

Now

$$\vec{v} = \bar{\vec{v}} + \dot{\vec{r}}'$$

$$\therefore \quad \sum (\vec{r}' \times m\dot{\vec{v}}) = (\sum m\vec{r}') \times \dot{\bar{\vec{v}}} + \sum \vec{r}' \times (m\ddot{\vec{r}}')$$

Now

$$\vec{r}' = \vec{r} - \bar{\vec{r}}$$

$$\therefore \quad \sum m\vec{r}' = \sum m\vec{r} - \sum m\bar{\vec{r}}$$

$$= M\bar{\vec{r}} - \bar{\vec{r}} \sum m$$

$$= M\bar{\vec{r}} - M\bar{\vec{r}} = 0.$$

$$\therefore \quad \sum (\vec{r}' \times m\dot{\vec{v}}) = \sum \vec{r}' \times m\ddot{\vec{r}}'$$

Substituting value of $\sum \vec{r}' \times m\dot{\vec{v}}, \sum m\vec{v} = M\vec{\bar{v}}$

$$\vec{H}' = \sum (\vec{r}' \times m\dot{\vec{r}}')$$

in (1), we get

$$\sum \vec{r} \times m\dot{\vec{v}} = \vec{\bar{r}} \times M\dot{\vec{\bar{v}}} + \dot{\vec{H}}'$$

where

$$\vec{H}' = A^* \omega_1 \vec{a}_1 + B^* \omega_2 \vec{a}_2 + C^* \omega_3 \vec{a}_3 .$$

Example

A uniform rigid rod AB of mass M moves so that A and B have velocities $\vec{u}_1$ and $\vec{u}_2$ at any instant. Show that the K.E. is then

$$T = \frac{1}{6} M (\vec{u}_1^2 + \vec{u}_1 . \vec{u}_2 + \vec{u}_2^2) .$$

Let G be the C.G. of the rod. The velocity of G is $\frac{1}{2}(\vec{u}_1 . \vec{u}_2)$.

The rod can be replaced by three particles $M/6$, $M/6$, $2M/3$ at A, B, G moving with velocities $\vec{u}_1, \vec{u}_2, \frac{1}{2}(\vec{u}_1 + \vec{u}_2)$.

K.E.

$$\text{T} = \frac{1}{2}\frac{M}{6}\vec{u}_1^2 + \frac{1}{2}\frac{M}{6}\vec{u}_2^2 + \frac{1}{2}\frac{2M}{3}\frac{1}{4}(\vec{u}_1 + \vec{u}_2)^2$$

$$= \frac{M}{12}(\vec{u}_1^2 + \vec{u}_2^2 + \vec{u}_1^2 + 2\vec{u}_1 . \vec{u}_2 + \vec{u}_2^2)$$

$$= \frac{M}{6}(\vec{u}_1^2 + \vec{u}_2^2 + 2\vec{u}_1 . \vec{u}_2).$$

EXERCISE

1. A uniform square lamina is bounded by the lines $x = 2c$, $y = 2c$, and a corner is cut off by the line $\frac{x}{a} + \frac{y}{b} = 2$. Show that if, $c > a$ and $c > b$, the two principal axes at the centre of the square which are in its own plane are inclined to the axis of x at angles given by

$$\tan 2\theta = \frac{ab - 2c(a+b) + 3c^2}{(a-b)(a+b-2c)}.$$

2. A homogeneous right circular cone is of mass M; its semi-vertical angle α, and the length of a slant side is l. Show that its M.I. about its axis is

$$\frac{3}{10} Ml^2 \sin^2 \alpha,$$

and its M.I. about a line through its vertex perpendicular to its axis is

$$\frac{3}{4} Ml^2 \sin^2 \alpha \left(\cos^2 \alpha + \frac{1}{5}\right).$$

3. Find the equation of momental ellipsoid at the vertex of uniform solid right circular cone of mass M, height h and semi-vertical angle α. Find also four particles, one at the vertex and three in plane perpendicular to the axis, which form a system equimomental with the cone.
4. Show that M.I. of the area enclosed by two loops of the lemniscate

$$r^2 = a^2 \cos 2\theta.$$

about the axis of the curve is

$$\frac{(3\pi - 8)\, a^2}{48} \times \text{mass of the area}.$$

5. Prove that a uniform tetrahedron is equimomental with particles each one-twentieth of its mass at its four corners and a particle of four-fifth of its mass at its C.G.
6. Find the M.I. of a hollow cubical box about its axis through the C.G. of the box and perpendicular to one of the faces.
7. Each of three equal uniform rods AB, BC, CD is of mass m and of length $2a$, and each is perpendicular to other two. Prove that the principal M.I. at the centre of inertia of the system are ma^2, $\frac{11}{3} ma^2$, $4ma^2$.
8. Show that the M.I. of an anchor-ring about its axis is $2\pi\rho a^2 c\left(c^2 + \frac{3}{4}a^2\right)$, where a is the radius of the generating circle, c is the distance of its centre from the axis of the anchor-ring, and ρ the density.
9. From any point a tangent cone is drawn to the ellipsoid of gyration at the C.G. of a body. Prove that the M.I. about every generator of the reciprocal cone is the same.
10. Show that the envelope of lines in the plane of an area about which that area has a constant M.I. is a set of confocal ellipses and hyperbolas. Hence, find the directions of principal axes at any point.

11. Show that the momental ellipsoid at the centre of an elliptic plate is

$$\frac{x^2}{a^2}+\frac{y^2}{b^2}+z^2\left(\frac{1}{a^2}+\frac{1}{b^2}\right)=const.$$

12. Find how the principal axes of inertia are arranged in the plane body. Write down the conditions that particles m_i at (x_i, y_i), where $i = 1, 2, ...$ may be equimomental to a given plate. Show that six quantities m_1, m_2, x_1, x_2, y_1, y_2 can be eliminated from these conditions.

If three equal particles are equimomental to a given plate, prove that the area of the triangle formed by them is $3\sqrt{3/2}$ times the product of the principal radii of gyration at the C.G.

13. Show that momental ellipsoid at a point on the rim of hemisphere is

$$2x^2+7(y^2+z^2)-\frac{15}{4}xz = const.$$

14. Find the momental ellipsoid at the centre of a solid ellipsoid.

15. Show that any lamina is dynamically equivalent to three particles, each one third of the mass of the lamina, placed at the corners of a maximum triangle inscribed in the ellipse, whose equation referred to the principal axes at the centre of inertia is $\frac{x^2}{A}+\frac{y^2}{B}=2$, where MA, MB are the principal moments of inertia about OX, OY.

16. A smoothly jointed framework is in the form of a parallelogram formed by attaching the ends of a pair of rods of mass m and length $2a$ to those of a pair of rods of m as m' and length $2b$. Masses M are attached to each of the four corners. Express the angular momentum of the system about the origin of coordinates, in terms of the coordinates (x, y) of the centre of gravity and the angle θ, and φ between the two pairs of sides and the axis of x.

17. Find the principal axes of a right circular cone at a point on the circumference of the base; and show that one of them will pass through its centre of gravity, if the vertical angle of the cone is given by $\tan\frac{\alpha}{2}-\frac{1}{2}$.

18. The lengths AB and AD of the sides of a rectangle $ABCD$ are $2a$, $2b$; show that inclination to AB of the principal axes at A is $\frac{1}{2}\tan^{-1}\frac{3ab}{2(a^2-b^2)}$.

19. Show that at the centre of the quadrant of an ellipse the principal axes in the plane are inclined at an angle $\frac{1}{2}\tan^{-1}\left(\frac{4}{\pi}\frac{ab}{a^2-b^2}\right)$ to the axes.

20. A wire is in the form of a semicircle of radius a; show that at one end of the diameter, the principal axes in its plane are inclined to the diameter at angles $\frac{1}{2}\tan^{-1}\frac{4}{5}$ and $\frac{\pi}{2}+\frac{1}{2}\tan^{-1}\frac{4}{5}$.

21. At the vertex C of the triangle ABC, which is right angled at C, prove that the principal axes are perpendicular to the plane and two others are inclined to the sides at an angle

$$\frac{1}{2}\tan^{-1}\frac{ab}{a^3-b^2}.$$

22. Show that the M.I. of elliptic area of mass M and semi-axes a and b about a diameter of length r is

$$\frac{1}{4}M\frac{a^2b^2}{r^2}.$$

23. Show that there is momental ellipse at an angular point of a triangular area which touches the opposite side at its middle point and bisects the adjacent sides.

24. Show that at an extremity of the bounding diameter of a semi-circular lamina, the principal axes make an angle $\frac{1}{2}\tan^{-1}\left(\frac{8}{3\pi}\right)$ to the diameter.

25. A uniform lamina is bounded by a parabolic arc, of latus rectum $4a$, and a double ordinate at a distance b from the vertex. If $b=\frac{a}{3}(6+4\sqrt{7})$, show that two of the principal axes at the end of a latus rectum are the tangent and normal there.

26. Prove that the principal radii of gyration at the vertex C of the triangle ABC are the roots of the equation

$$x^4-\frac{a^2+b^2+c^2}{36}x^2+\frac{\Delta^2}{108}=0$$

where Δ is the area of the triangle.

8

TWO DIMENSIONAL MOTION

8.1 INTRODUCTION

Two dimensional or plane motion

If all the particles of a rigid body move in parallel planes, the body is said to have two *dimensional or plane motion.* A rigid lamina moving in its own plane is an important case of two dimensional motion. In general the motion of a rigid lamina consists of translation and rotation and that there is a point in its plane which is instantaneously at rest. Moreover, with the knowledge of instantaneous angular velocity of the lamina and the linear velocity of its centroid, the velocity of any other particle of it at the considered instant can be determined.

D'Alemberet's principle. It states that reversed effective forces acting on each particle of the rigid body and the external forces acting on the body are in equilibrium.

Note:

1. This principle reduces the solution of the dynamical problem concerning the motion of the rigid body to that of statical problem.
2. The motion of the rigid body can be considered as made up of two motions: namely the motion of the centre of gravity and the motion of rotation about the centre of gravity.
   (*i*) The motion of the centre of gravity G of the body is same as if all the mass of the body were collected into a particle at G and if all the external forces were acting at G in directions parallel to that in which they act.
   (*ii*) The motion of a rigid body about its C.G. is the same as if the G, were fixed and the same forces acted on the body.

For the motion of rigid planar distribution, to fix the position of the centroid two co-ordinates are required, to fix the orientation of the lamina, angular velocity is required. Thus three independent variables are involved in the instantaneous motion of the lamina, two linear velocity components of its centroid and angular velocity of the distribution.

8.2 ROLLING AND SLIDING

In case of rough surfaces, the frictional force comes into play. Hence, usual law of friction *i.e.*, friction is a self adjusting force, which tends to stop the relative motion of the point at which it acts. But it cannot exceed the limiting friction μR where R is corresponding normal reaction.

Rolling

In this case, the point of contact is instantaneously at rest. We assume a frictional force F in a direction opposite to what would be the direction of relative motion and assume that the point of contact is at relative rest. We also have a geometrical equation corresponding to rolling. If after obtaining the solution, F has a positive value, then it means the direction of frictional force is correct, otherwise the direction of friction is opposite.

In this case $F<\mu R$ and comes into play only when the body begins to slide.

Sliding

The point of contact has velocity ($\neq 0$) and the friction $F = \mu R$. Directon of sliding friction is opposite to that of the motion of the point of contact. In this case, there is no geometrical equation.

Perfectly rough surface

A surface is said to be perfectly rough when sufficient friction to prevent sliding can be exerted. **Thus only rolling can take place between two perfectly rough surfaces.**

Note :

1. -In applying the principle of work and energy, the rolling friction may be omitted, because the point of contact is at rest, but sliding friction should not be omitted.
2. If one body be rolling on another, the energy equation of the two bodies is written by treating them as one system, omitting the reaction between them.

8.3 MOTION OF A BODY ON AN INCLINED PLANE

A body moves from rest down an inclined plane having inclination α to the horizontal plan. Discuss the motion.

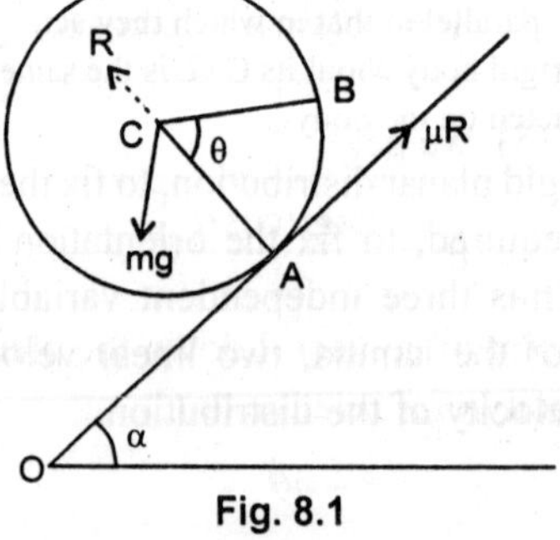

Fig. 8.1

Let C be the centre of the body and at distance a from the plane. Let CB be a line fixed in the body which was initially normal to the plane and makes an angle θ with the normal at time t.

Let us assume that the fiction is not sufficient to produce pure rolling, therefore the body slides as well as turns. So the maximum normal friction acts up the plane. Let x be the distance moved by the centre of gravity C parallel to the inclined plane in time t and θ be the angle through which the body turns.

Now equations of motion are

$$m\ddot{x} = mg\sin\alpha - \mu R \tag{1}$$

$$0 = R - mg\cos\alpha \tag{2}$$

$$mk^2\ddot{\theta} = \mu Ra \tag{3}$$

From (1) and (2)

$$\ddot{x} = g(\sin\alpha - \mu\cos\alpha) \tag{4}$$

From (2) and (3)

$$\ddot{\theta} = \frac{\mu ga\cos\alpha}{k^2} \tag{5}$$

Integrating (4), we get

$$\dot{x} = g(\sin\alpha - \mu\cos\alpha)t + A.$$

Initially, $\dot{x} = 0$, $t = 0$,

$$\therefore \quad A = 0.$$

$$\therefore \quad \dot{x} = g(\sin\alpha - \mu\cos\alpha)t.$$

Integrating again, we have

$$x = g(\sin\alpha - \mu\cos\alpha)\frac{t^2}{2} + B.$$

Initially, t = 0, $x = 0$

$$\therefore \quad B = 0.$$

$$\therefore \quad x = g(\sin\alpha - \mu\cos\alpha)\frac{t^2}{2}. \tag{6}$$

Similarly integrating (5), we get

$$\theta = \frac{\mu ga\cos\alpha}{2k^2}t^2. \tag{7}$$

The velocity of the point of contact A down the plane =
Velocity of C + velocity of A relative to C

$$= \dot{x} - a\dot{\theta}.$$

$$= g(\sin\alpha - \mu\cos\alpha)t - \frac{\mu g a^2 \cos\alpha\, t}{k^2} \qquad (8)$$

Let us consider equation (8).

$$= \frac{g}{t}(t\sin\alpha - \mu\cos\alpha - \mu\frac{a^2}{k^2}\cos\alpha)$$

Case I.

If $\sin\alpha > \mu\cos\alpha\left(1+\frac{a^2}{k^2}\right)$, *i.e.*, if $\mu < \frac{k^2}{k^2+a^2}\tan\alpha$, the velocity of the point of contact is positive for all values of t *i.e.*, the point of contact always slides down and maximum friction μR acts. Hence, above equations hold through out the motion.

Case II

If $\sin\alpha = \mu\cos\alpha\left(1+\frac{a^2}{k^2}\right)$ *i.e.*, if $\mu = \frac{k^2}{k^2+a^2}\tan\alpha$.

Then the velocity of the point of contact is zero for all values of t *i.e.*, the motion of the body is pure rolling throughout and the maximum friction μR acts. Hence, above equation of motion hold.

Case III

If $\sin\alpha < \mu\cos\alpha\left(1+\frac{a^2}{k^2}\right)$, *i.e.* if $\mu > \tan\alpha\frac{k^2}{k^2+a^2}$.

The velocity of the point of contact appears to be negative, when maximum friction μR were allowed to act, the point of contact will slide upwards which is impossible because that amount of friction (not more) will only act which is sufficient to keep the point of contact at rest. Hence, the motion is of pure rolling from the very start and remains same throughout and the maximum friction μR is not exerted. Therefore in this case the above equations of motion will not hold good.

If F be frictional force now acting, then equations of motion become

$$m\ddot{x} = mg\sin\alpha - F \qquad (1)'$$

$$0 = R - mg\cos\alpha \qquad (2)'$$

$$mk^2\ddot{\theta} = Fa \qquad (3)'$$

Since point of contact is at rest

$$\therefore \quad \dot{x} - a\dot{\theta} = 0,$$

$$\therefore \quad \dot{x} = a\dot{\theta} \qquad (4)'$$

$$\therefore \qquad F = m\frac{k^2}{a^2}\ddot{x}.$$

Putting this value of F in (1)′, we get

$$\ddot{x} = \frac{a^2}{a^2+b^2} g \sin\alpha.$$

Integrating, we get

$$\dot{x} = a\dot{\theta} = \frac{a^2}{a^2+k^2} gt \sin\alpha + A',$$

Initially $t = 0$, $\dot{x} = 0, = a\dot{\theta}$

$$\therefore \qquad A' = 0.$$

$$\therefore \qquad \dot{x} = a\dot{\theta} = \frac{a^2}{a^2+k^2} gt \sin\alpha \tag{5}$$

Integrating again, we have

$$x = a\theta = \frac{a^2}{2(a^2+k^2)} gt^2 \sin\alpha + B'$$

Initially, $t = 0, x = 0, a\theta = 0.$

$$\therefore \qquad B' = 0.$$

$$\therefore \qquad x = a\theta = \frac{a^2}{2(k^2+a^2)} gt^2 \sin\alpha.$$

Particular cases,

(i) For solid sphere $\qquad k^2 = \frac{2}{5}a^2, \therefore \ddot{x} = \frac{5}{7} g \sin\alpha.$

(ii) For hollow sphere $\qquad k^2 = \frac{2}{3}a^2, \therefore \ddot{x} = \frac{3}{5} g \sin\alpha.$

(iii) For circular disc $\qquad k^2 = \frac{a^2}{2}, \therefore \ddot{x} = \frac{2}{3} g \sin\alpha,$

(iv) For circular ring, $\qquad k^2 = a^2, \therefore \ddot{x} = \frac{1}{2} g \sin\alpha.$

8.4 MOTION OF A CIRCULAR DISC ON A HORIZONTAL PLANE

A uniform, circular disc is projected with its plane vertical along a rough horizontal plane with a velocity v of translation and angular velocity ω about the centre.

Discuss the motion.

Case I

when $v \to v > a\omega$ and ω is clockwise.

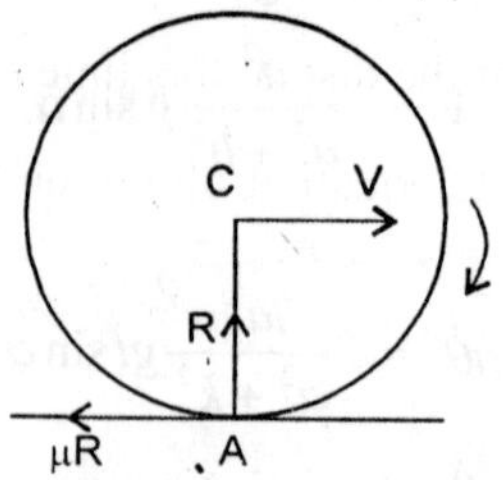

Fig. 8.2 (i)

In this case the initial velocity of the point of contact is $v - a\omega$. Since, $v > a\omega$ hence, the velocity has the direction of v. Hence, μR acts in the direction $\leftarrow$.

Equations of motion are

$$m\ddot{x} = -\mu R,$$

$$0 = R - mg$$

$$\therefore \quad \ddot{x} = -\mu g \tag{1}$$

$$m\frac{a^2}{2}\ddot{\theta} = \mu R a$$

$$\therefore \quad a\ddot{\theta} = 2\mu g \tag{2}$$

Integrating (1), we get

$$\dot{x} = -\mu g t + A,$$

Initially, $t = 0$, $\dot{x} = v$,

$$\therefore \quad A = v.$$

$$\therefore \quad \dot{x} = -\mu g t + v. \tag{3}$$

Integrating, (2), we get

$$a\dot{\theta} = 2\mu g t + B,$$

Initially $t = 0$, $\dot{\theta} = \omega$

$$\therefore \quad B = a\omega$$

$$\therefore \quad a\dot{\theta} = 2\mu g t + a\omega \tag{4}$$

Rolling commences when $\dot{x} = a\dot{\theta}$. If it happens after time t_1, from (3) and (4) we have

$$-\mu g t_1 + v = 2\mu g t_1 + a\omega$$

$$\therefore \qquad t_1 = \frac{v - a\omega}{3\mu g}.$$

Substituting this value of t_1 in (3), we get

Velocity of the centre of the disc at this time $t_1 = \dfrac{2v + a\omega}{3}$ (5)

When rolling commences equation of motion become

$$m\ddot{x} = -F, m\frac{a^2}{2}\ddot{\phi} = Fa \qquad (6)$$

Since, there is no sliding,

$$\therefore \qquad \dot{x} = a\dot{\phi} \quad \ddot{x} = a\ddot{\phi}$$

From these equations, we get

$$F = 0, \ \ddot{x} = 0, \quad a\ddot{\phi} = 0.$$

Since, $F = 0$, so no friction is required.

Integrating $\ddot{x} = 0$, we get

$$\dot{x} = const. = \frac{2a + a\omega}{3}, \text{ from (5)}$$

Therefore the disc continues to roll with constant angular velocity $\dfrac{2a + a\omega}{3}$.

Case II

When $v \to v < a\omega$ and is ω clockwise.

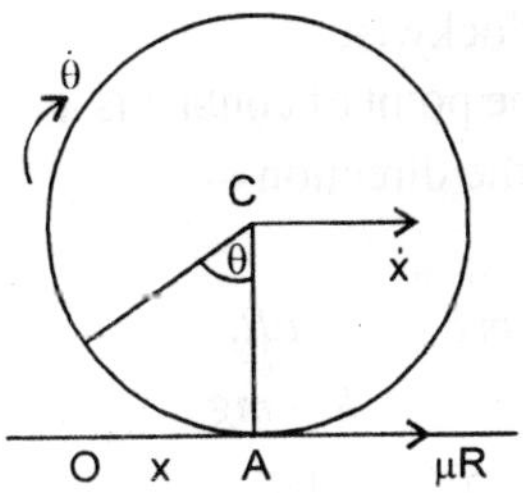

Fig. 8.2 (ii)

Now initial velocity is $v - a\omega$ in the directon $\leftarrow$ since , $v < a\omega$, where uR acts in the directon $\rightarrow$

Equations of motion are

$$m\ddot{x} = \mu R$$

$$0 = R - mg.$$

$$\therefore \qquad \ddot{x} = \mu g \qquad (1)$$

$$m\frac{a^2}{2}\ddot{\theta} = -\mu Ra = -\mu mga$$

$$\therefore \qquad a\ddot{\theta} = -2\mu R \qquad (2)$$

Integrating these equations and using the initial conditions, $t = 0, \dot{x} = v, \dot{\theta} = \omega$ we get

$$\dot{x} = \mu gt + v, \qquad (3)$$

$$a\dot{\theta} = -2\mu gt + a\omega. \qquad (4)$$

Pure rolling commences when $\dot{x} = a\dot{\theta}$. If this happens after time t_1, then from (3) and (4), we get

$$t_1 = \frac{a\omega - v}{3\mu g}.$$

Substituting this value in (9), we have

$$\dot{x} = \frac{2v - a\omega}{3}.$$

When pure rolling commences equations of motion are same as equation (6) of case I, which gives $F = 0$. Hence, rolling continues with constant angular velocity $\frac{2v - a\omega}{3}$.

Case III

When $v \rightarrow$ and ω is anti-clockwise.

Now inital velocity of the point of contact is $v + a.\omega$. So its direction is $\rightarrow$, hence, friction μR acts in the direction $\leftarrow$.

Equation of motion are

$$m\ddot{x} = -\mu R,$$

$$0 = R - mg.$$

$$\therefore \qquad \ddot{x} = -\mu g \qquad (1)$$

$$m\frac{a^2}{2}\ddot{\theta} = -\mu Ra$$

$$\therefore \qquad a\ddot{\theta} = -2\mu g \qquad (2)$$

Integrating (1) and (2), and using initial conditions, $t = 0, \dot{x} = v, a\dot{\theta} = \omega$ we get

$$\dot{x} = -\mu gt + v, \qquad (3)$$

$$a\dot{\theta} = -2\mu gt + a\omega \qquad (4)$$

Pure rolling commences with (*i.e.*, the point of contact is at rest), when

$$\dot{x} + a\dot{\theta} = 0.$$

If this happens after time t_1, from (3) and (4) we get

$$-\mu g t_1 + v + (-2\mu g t_1 + a\omega) = 0.$$

$$t_1 = \frac{v + a\omega}{3\mu g}.$$

Substituting this value of t_1 in (3) and (4), we have

$$\dot{x} = a\dot{\theta} = \frac{2v - a\omega}{3}.$$

It is evident that when $2v > a\omega$, then $\dot{x} = a\dot{\theta}$ is positive, so pure rolling continues with uniform velocity $\frac{2v - a\omega}{3}$.

If $2v < a\omega$, then the velocity is negative (backward).

In this case from equation (3), we see that the velocity of the centre becomes zero, when $t = \frac{v}{\mu g}$ and at that time we see from equation (4), that

$$a\dot{\theta} = -2v + a\omega.$$

This is positive because $2v < a\omega$.

Hence, when $2v < a\omega$, then the disc begins to move backward before pure rolling begins.

8.5 TWO SPHERES IN CONTACT

Two unequal smooth spheres one on the top of the other are in unstable equilibrium, the lower resting on a smooth table. The system is slightly disturbed, show that the spheres will separate when the line joining their centres make angle θ with the vertical given by the equation

$$m \cos^3 \theta = (M + m)(3 \cos\theta - 2),$$

where M is the mass of the lower and m that of the upper sphere.

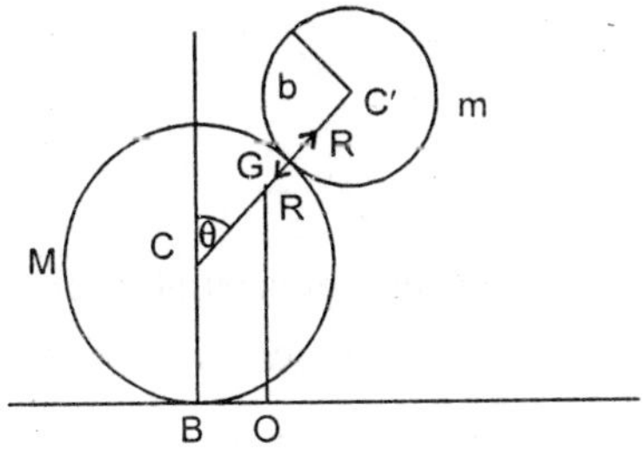

Fig. 8.3

Let C and C' be the centres and a and b the radii of the lower and the upper spheres whose masses are M and m respectively. Let R be the normal reaction between the spheres.

If G be their common centre of gravity, then

$$M.CG = m.C'G$$

$$\therefore \quad \frac{CG}{m} = \frac{C'G}{M} = \frac{CG + C'G}{M+m} = \frac{a+b}{M+m}.$$

$$\therefore \quad CG = \frac{m(a+b)}{M+m}, C'G = \frac{M(a+b)}{M+m}.$$

Since, There is no horizontal force on the system, therefore common centre of gravity G moves in a vertical line; let this vertical through G meet the horizontal line in a point O; then O is a fixed point. Let O be the origin and horizontal and vertical lines through O the co-ordinate axes.

For the horizontal motion of the lower sphere, we have

$$R\sin\theta = M\frac{d^2}{dt^2}(CG\sin\theta) = \frac{Mm(a+b)}{M+m}\left[\cos\theta\ddot{\theta} - \sin\theta\dot{\theta}^2\right] \quad (1)$$

For the vertical motion of the upper sphere,

$$R\cos\theta - mg = m\frac{d^2}{dt^2}[a + (a+b)\cos\theta] = m(a+b)\left[-\sin\theta\ddot{\theta} - \cos\theta\dot{\theta}^2\right] \quad (2)$$

Eliminating R, we get

$$\ddot{\theta}(M + m\sin^2\theta) + m\sin\theta\cos\theta\dot{\theta}^2 = \frac{(M+m)g\sin\theta}{a+b}. \quad (3)$$

Integrating, we get

$$\dot{\theta}^2(M + m\sin^2\theta) = -\frac{2g(M+m)\cos\theta}{a+b} + const.$$

Since, motion starts from rest at the highest point,

$$\therefore \quad \theta = 0, \text{ when } \dot{\theta} = 0,$$

$$\therefore \quad Const. = \frac{2g(M+m)}{a+b}$$

Hence, we get

$$\dot{\theta}^2(M + m\sin^2\theta) = \frac{2g(M+m)}{a+b}(1 - \cos\theta). \quad (4)$$

By (1), R vanishes, *i.e.*, the spheres separate, when

$$\cos\theta\ddot{\theta} = \sin\theta\dot{\theta}^2 \quad (5)$$

from (3) and (5), we get

$$\dot{\theta}^2 = \frac{g\cos\theta}{a+b}$$

at this instant, substituting this value of $\dot{\theta}^2$ in (4), we have

$$(M + m\sin^2\theta)\frac{g\cos\theta}{a+b} = \frac{2g}{a+b}(M+m)(1-\cos\theta).$$

$$\therefore \quad (M + m - m\cos^2\theta)\cos\theta = 2(M+m)(1-\cos\theta).$$

$$\therefore \quad m\cos^3\theta = (M+m)(3\cos\theta - 2).$$

There are no forces acting so as to turn either sphere about its centre, so that neither of them has any rotary motion.

Examples

1. A cylinder; whose centre of inertia is at a distance c from axis, rolls on a horizontal plane. If it be just started from a position of unstable equilibrium, show that the normal reaction of the plane when the centre of mass is in its lowest position is $\left[1 + \frac{4c^2}{(a+c)^2 + k^2}\right]$ times its weight, when k is the radius of gyration about an axis through the centre of mass.

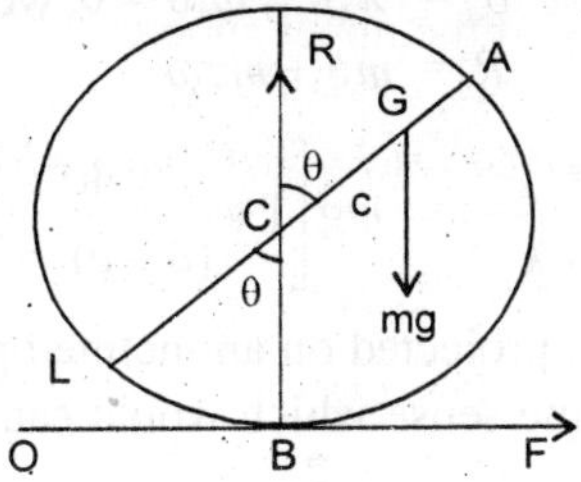

Fig. 8.4

Let O be the initial position of the point of contact when centre of gravity was vertically above the centre of the figure. Let G be the centre of gravity, R the normal reaction and F the frictional force and at any time t the radius through G turn through an angle θ.

Referred to O as origin and horizontal and vertical lines as axes, the co-ordinates (x, y) of G are given by

$$x = a\theta + c\sin\theta, \; y = a + c\cos\theta, \; \because CG = c.$$

Equations of motion of G are given by

$$m\ddot{x} = m\frac{d^2}{dt^2}(a\theta + c\sin\theta) = F \quad (1)$$

$$m\ddot{y} = m\frac{d^2}{dt^2}(a + c\cos\theta) = R - mg. \quad (2)$$

Energy equation is

$$\frac{1}{2}m\left[\dot{x}^2+\dot{y}^2+k^2\dot{\theta}^2\right] = \text{Work done by forces}$$

$$\therefore\ \frac{1}{2}m\left[(a\dot{\theta}+c\cos\theta\dot{\theta})^2+(-c\sin\theta\dot{\theta})^2+k^2\dot{\theta}^2\right]=mg\,(c-c\cos\theta).$$

Let ω be the angular velocity, when G is in its lowest position, *i.e.*, $\dot{\theta}=\omega$, when $\theta = \pi$, then we have

$$\frac{1}{2}m\left[(a-c)^2+k^2\right]\omega^2 = 2\,mgc.$$

$$\therefore\qquad \omega^2 = \frac{4gc}{(a-c)^2+k^2}.$$

From (2), we have

$$R = mg - mc\,(\sin\theta\ddot{\theta}+\cos\theta\dot{\theta}^2)$$

Putting

$$\theta = \pi, \dot{\theta}=\omega, \ddot{\theta}=0, \text{ we have}$$

$$R = mg + mc\omega^2$$

$$= mg + mc\frac{4gc}{(a-c)^2+k^2} = mg\left[1+\frac{4c^2}{(a-c)^2+k^2}\right].$$

2. A sphere of radius a is projected on an inclined plane with velocity V and angular velocity Ω in the sense which would cause it to roll up, if $V > a\Omega$ and the co-efficient of friction $> \frac{2}{7}\tan\alpha$, show that the sphere will cease to ascend at the end of time

$$\frac{5V+2a\Omega}{5g\sin\alpha}.$$

where α is the inclination of the plane.

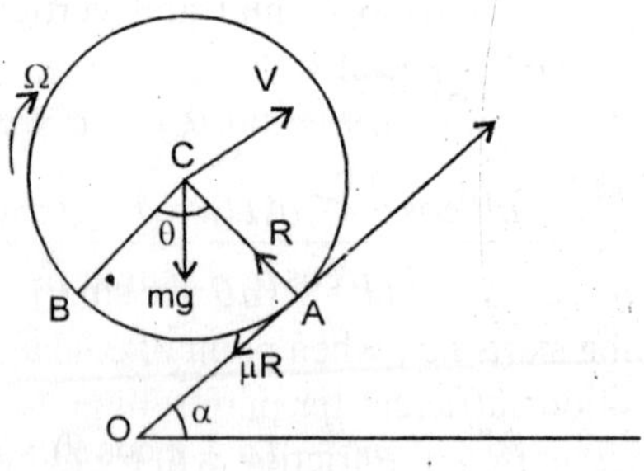

Fig. 8.5

Let C be the centre and CB radius (fixed in the body) of the sphere. Initaially CB was normal to the plane. Let at time t, CB makes an angle θ with normal to the plane (line fixed in space).
Initial velocity of the point of contact A is up the plane = velocity of centre C + velocity of A relative to $C = V - a\Omega$ = +ve, since $V > a\Omega$
Hence, friction μR acts down the plane. Hence, sphere slides as well as turns.
Equations of motion are

$$m\ddot{x} = -mg\sin\alpha - \mu R \tag{1}$$

$$O = R - mg\cos\alpha \tag{2}$$

$$m\frac{2}{5}a^2\ddot{\theta} = \mu Ra \tag{3}$$

From these equations, we get

$$\ddot{x} = -g(\sin\alpha + \mu\cos\alpha)$$

$$a\ddot{\theta} = \frac{5}{2}\mu g\cos\alpha$$

Integrating, and using initial conditions, $\dot{x} = V, \dot{\theta} = \Omega, t = 0$, we get

$$\dot{x} = -g(\sin\alpha + \mu\cos\alpha)t + V \tag{4}$$

$$a\dot{\theta} = \frac{5}{2}\mu gt\cos\alpha + a\Omega \tag{5}$$

The velocity of the point of contact is zero, when $\dot{x} - a\dot{\theta} = 0$.
Hence, from (4) and (5) we obtain

$$-g(\sin\alpha + \mu\cos\alpha)t_1 + V - \frac{5}{2}\mu gt_1\cos\alpha - a\Omega = 0,$$

$$\therefore \qquad t_1 = \frac{2(V - a\Omega)}{g(7\mu\cos\alpha + 2\sin\alpha)}.$$

Substituting the value of t_1 in (4), we get

$$\dot{x} = V - g(\sin\alpha + \mu\cos\alpha).\frac{2(V - a\Omega)}{g(7\mu\cos\alpha + 2\sin\alpha)}$$

$$= \frac{5\mu V\cos\alpha + 2a\Omega(\sin\alpha + \mu\cos\alpha)}{(7\mu\cos\alpha + 2\sin\alpha)} = U \text{ (say)}$$

Now when the rolling starts *i.e.*, when point of contact has been brought to rest let F be the friction sufficient for pure rolling. Because now the point of contact is at rest, therefore the friction will try to keep it at rest if it can; hence, friction acts upwards.

Now equations of motions are

$$m\ddot{y} = -mg\sin\alpha + F \quad (i)$$

$$O = R - mg\cos\alpha \quad (ii)$$

$$m\frac{2}{5}a^2\ddot{\phi} = -Fa. \quad (iii)$$

Since, throughout this motion, point of contact is at rest, therefore its velocity

$$\dot{y} - a\dot{\phi} = 0,$$

$$\therefore \quad \ddot{y} = a\ddot{\phi} \quad (iv)$$

From (*iii*) and (*iv*)

$$m\ddot{y} = -\frac{5}{2}F.$$

Substituting value of $m\ddot{y}$ in (1), we get

$$F = \frac{2}{7}mg\sin\alpha.$$

Also we note that

$$\mu R = \mu mg\cos\alpha > \frac{2}{7}\tan\alpha . mg\cos\alpha > \frac{2}{7}mg\sin\alpha.$$

Hence, condition

$$F < \mu R$$

is satisfied. Therefore the motion of the sphere is of pure rolling and above equations are true.

Eliminating F from these equations, we obtain

$$\ddot{y} = -g\sin\alpha - \frac{2}{5}a\ddot{\phi}$$

$$= -g\sin\alpha - \frac{2}{5}\ddot{y}$$

$$\therefore \quad \ddot{y} = -\frac{5}{7}g\sin\alpha.$$

Integrating, and using initial condition $t = 0,\ \dot{y} = U$, we get

$$\dot{y} = -\frac{5}{7}gt\sin\alpha + U.$$

The rolling will stop when $\dot{y} = 0$.

$$\therefore \quad O = -\frac{5}{7} g t_2 \sin\alpha + U,$$

$$\therefore \quad t_2 = \frac{7U}{5g\sin\alpha}.$$

Hence, total time $= t_1 + t_2$

$$= \frac{2(V - a\Omega)}{g(7\mu\cos\alpha + 2\sin\alpha)} + \frac{7}{5g\sin\alpha} \cdot \frac{5\mu V\cos\alpha + 2a\Omega(\sin\alpha + \mu\cos\alpha)}{(7\mu\cos\alpha + 2\sin\alpha)}$$

$$= \frac{10(V - a\Omega)\sin\alpha + 35\mu V\cos\alpha + 14a\Omega(\sin\alpha + \mu\cos\alpha)}{5g\sin\alpha(7\mu\cos\alpha + 2\sin\alpha)}$$

$$= \frac{5V + 2a\Omega}{5g\sin\alpha}.$$

3. A uniform sphere of radius a, is rotating about a horizontal diameter with angular velocity Ω and is gently placed on a rough plane, which is inclined at an angle α to the horizontal, the sense of rotation being such as to cause the sphere to move up the plane along the line of greatest slope. Show that, if co-efficient of friction be tan α, the centre of sphere will remain at rest for a time $\frac{2a\Omega}{5g\sin\alpha}$ and will then move downwards with acceleration $\frac{5}{7}\sin\alpha$.

If the body be a thin circular hoop instead of sphere, show that the time is $\frac{a\Omega}{g\sin\alpha}$ and acceleration $\frac{1}{2} g\sin\alpha$.

Before being placed on the inclined plane the sphere was rotating with angular velocity Ω about a horizontal diameter.

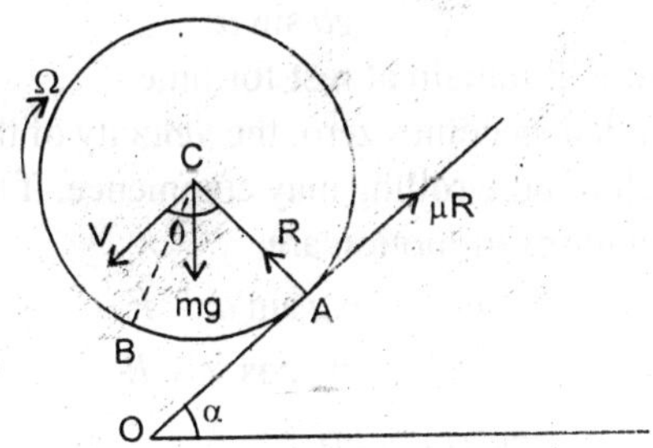

Fig. 8.6

Since, it is gently placed on the inclined plane, it means that at this time velocity of centre is zero. Now the sense of rotation at the time of placing is such as to cause the sphere to move up the plane, it means the sense of Ω is clock-wise. Hence, initial velocity of the point of contact A is down the plane, which is given by
Velocity of centre C + Velocity of A relative to $C = 0 + a\Omega$ which is positive. Hence, initial velocity of the point of contact is down the plane, so friction μR acts up the plane.
Equations of motion are

$$m\ddot{x} = mg\sin\alpha - \mu R \tag{1}$$
$$O = R - mg\cos\alpha \tag{2}$$
$$mk^2\ddot{\theta} = -\mu Ra \tag{3}$$

where $\mu = \tan\alpha$.
Substituting $\mu = \tan\alpha$ in (1) and using (2), we get

$$m\ddot{x} = mg\sin\alpha - \tan\alpha . mg\cos\alpha = 0,$$

$$\therefore \quad \ddot{x} = 0. \tag{4}$$

and putting value of μ in (3) and using (2), we get

$$mk^2\ddot{\theta} = -\tan\alpha . mg\cos\alpha a = -mga\sin\alpha \tag{5}$$

Integrating (4) and (5) and using initial conditions

$t = 0,\ \dot{x} = 0, \dot{\theta} = \Omega$ we get

$$\dot{x} = 0. \tag{6}$$
$$k^2\dot{\theta} = -gat\sin\alpha + k^2\Omega \tag{7}$$

From (6) and (7) it is evident that the centre does not move at all but the sphere goes on revoling.

The rotation will cease when $\dot{\theta} = 0$,
Hence, from (7), we get

$$O = -gat_1\sin\alpha + k^2\Omega$$

$$\therefore \quad t_1 = \frac{k^2\Omega}{ga\sin\alpha} \tag{8}$$

Therefore centre will remain at rest for time t_1.
Now when $\dot{x}$ and $\dot{\theta}$ becomes zero, the velocity of the point of contact *i.e.*, $\dot{x} + a\dot{\theta} = 0$ therefore pure rolling may commence, if friction is sufficient for rolling, then equations of motion are

$$m\ddot{y} = mg\sin\alpha - F \tag{i}$$
$$O = mg\cos\alpha - R \tag{ii}$$
$$mk^2\ddot{\phi} = Fa \tag{iii}$$
$$\dot{y} - a\dot{\phi} = 0. \tag{iv}$$

From (i), (iii), (iv), we get

$$ma\ddot{\phi} = mg\sin\alpha - F$$

$$\frac{a^2}{k^2}F = mg\sin\alpha - F$$

$$\therefore \qquad F = \frac{mg\sin\alpha}{1+\frac{a^2}{k^2}}$$

Now $F < mg\sin\alpha$.

From (*ii*)

$\mu R = \mu mg\cos\alpha = \tan\alpha\, mg\cos\alpha = mg\sin\alpha$

$$\therefore \qquad F < \mu R.$$

Hence, body rolls and equations (*i*) – (*iii*) hold.

Putting value of F in (*i*), we get

$$m\ddot{y} = mg\sin\alpha - \frac{mg\sin\alpha}{1+\frac{a^2}{k^2}}$$

$$\therefore \qquad \ddot{y} = \frac{a^2 g\sin\alpha}{a^2+k^2} \qquad \text{(v)}$$

If the body is sphere $k^2 = \frac{2}{5}a^2$, then from (8)

$$t_1 = \frac{2a\Omega}{5g\sin\alpha}$$

From (v),

$$\ddot{y} = \frac{5}{7}g\sin\alpha.$$

If the body be circular hoop, then

$$k^2 = a^2$$

then

$$t_1 = \frac{a\Omega}{g\sin\alpha}, \ddot{y} = \frac{1}{2}g\sin\alpha.$$

4. An inclined plane of mass M is capable of moving on a smooth horizontal table. A perfectly rough sphere of mass m is placed on its inclined face and rolls down under the action of gravity. If y be the horizontal distance advanced by the inclined plane and x the part of the plane rolled over by the sphere, prove that

$$(M+m)y = mx\cos\alpha, \frac{7}{5}x - y\cos\alpha = \frac{1}{2}gt^2\sin\alpha,$$

where α is the inclination of the plane to the horizon.

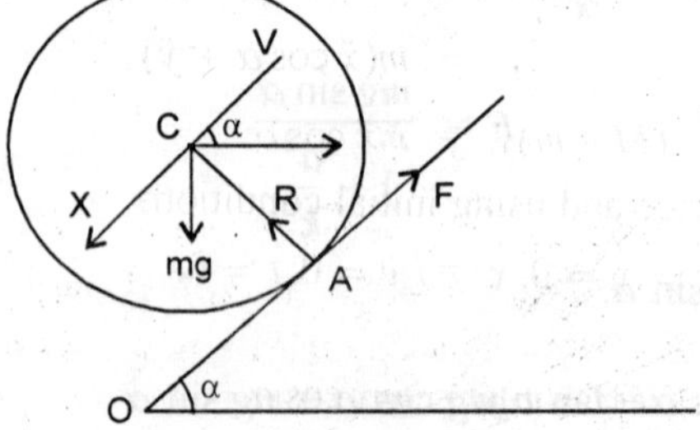

Fig. 8.7

Centre C of the sphere has two accelerations one $\ddot{x}$ parallel to the and other $\ddot{y}$ in the horizontal direction.

Equations of motion are

$$m(\ddot{x} - \ddot{y}\cos\alpha) = mg\sin\alpha - F \tag{1}$$

$$m\ddot{y}\sin\alpha = mg\cos\alpha - R \tag{2}$$

$$m\frac{2}{5}a^2\ddot{\theta} = Fa \tag{3}$$

Also

$$\dot{x} = a\dot{\theta} \tag{4}$$

Equation of the motion of the plane is

$$M\ddot{y} = R\sin\alpha - F\cos\alpha \tag{5}$$

From (3) and (4), we have

$$m\frac{2}{5}\ddot{x} = F$$

Putting this value of F in (1), we get

$$\frac{7}{5}\ddot{x} - \ddot{y}\cos\alpha = g\sin\alpha$$

Integrating and using the initial conditions $t = 0, \dot{x} = \dot{y} = 0$, we get

$$\frac{7}{5}\dot{x} - \dot{y}\cos\alpha = gt\sin\alpha$$

Inegrating again and using initial conditions $t = 0, x = y = 0$, we get

$$\frac{7}{5}x - y\cos\alpha = \frac{1}{2}gt^2 \sin\alpha.$$

From (1), and (2), (5) we get

$$M\ddot{y} = m(g\cos\alpha - \ddot{y}\sin\alpha)\sin\alpha + m(\ddot{x} - \ddot{y}\cos\alpha - g\sin\alpha)\cos\alpha$$

$$= m(\ddot{x}\cos\alpha + \ddot{y})$$

$$\therefore \qquad (M+m)\ddot{y} = m\ddot{x}\cos\alpha.$$

Integrating twice and using initial conditions

$$x = y = 0, \dot{x} = \dot{y} = 0, t = 0,$$

we get

$$(M+m)y = mx\cos\alpha.$$

8.6 MOTION OF A UNIFORM ROD WITH ONE POINT IN CONTACT WITH HORIZONTAL PLANE

A uniform rod is held in a vertical position with one end resting upon a rough table, and when released rotates about the end in contact with the table. Discuss the motion.

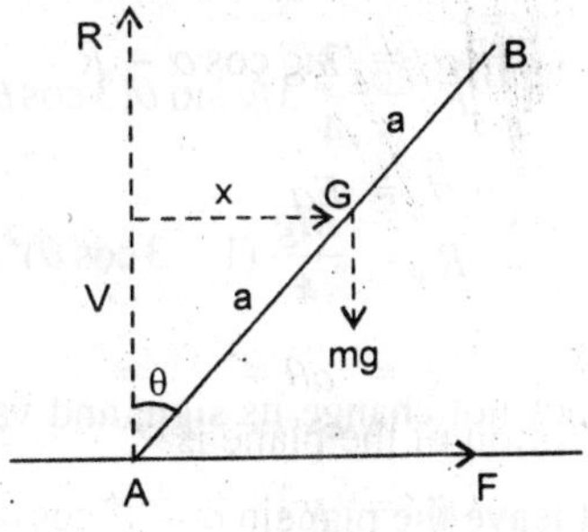

Fig. 8.8

Let AB be the rod of length $2a$ and mass M, which can rotate about the end A.

Let at time t the rod make an angle θ with the vertical.

Referred to A as origin and horizontal and vertical lines as axes, the co-ordinates (x, y) of the centre of gravity G of the rod are given by

$$x = a\sin\theta, y = a\cos\theta.$$

Let F be the frictional force and R the normal reaction at A, the equations of motion of G are given by

$$M\ddot{x} = M\left[a\cos\theta\ddot{\theta} - a\sin\theta\dot{\theta}^2\right] = F \qquad (1)$$

$$M\ddot{y} = M\left[-a\sin\theta\ddot{\theta} - a\cos\theta\dot{\theta}^2\right] = R - Mg \qquad (2)$$

Energy of the rod is

$$\frac{1}{2}M\left[(a\dot{\theta})^2 + \frac{1}{3}a^2\dot{\theta}^2\right] = \frac{2}{3}Ma^2\dot{\theta}^2.$$

Work done by forces = $Mga(1 - \cos\theta)$.
Hence, energy equation gives

$$\frac{2}{3}Ma^2\dot{\theta}^2 = Mga(1 - \cos\theta)$$

$$\therefore \qquad \dot{\theta}^2 = \frac{3g}{2a}(1 - \cos\theta) \qquad (3)$$

Differentiating *w.r.t.* and dividing by $2\dot{\theta}$, we get

$$\ddot{\theta} = \frac{3g}{4a}\sin\theta. \qquad (4)$$

Substituting the values of $\dot{\theta}^2$ and $\ddot{\theta}$ from (3) and (4) in (1) and (2), we have

$$F = \frac{3}{4}Mg\sin\theta(3\cos\theta - 2)$$

$$R = \frac{Mg}{4}(1 - 3\cos\theta)^2.$$

We note that R does not change its sign, and vanishes when $\cos\theta = \frac{1}{3}$. Thus the end does not leave the plane.

From expression of F, it is clear the F changes its sign as θ passes through the angle $\cos^{-1}\left(\frac{2}{3}\right)$, and hence, its direction is reversed.

When $\cos\theta = \frac{1}{3}$, $R = 0$, Hence, ratio $\frac{F}{R}$ becomes infinite when $\cos\theta = \frac{1}{3}$; hence, unless the plane be infinitely rough, there will be sliding at least when $\cos\theta = \frac{1}{3}$. In practice, the end A of the rod begins to slip before $\cos\theta = \frac{1}{3}$. The end A will slip backward or forward as the slipping occurs before or after the inclination of the rod is $\cos^{-1}\left(\frac{2}{3}\right)$.

8.7 MOTION OF A UNIFORM ROD IN CONTACT WITH TWO PLANES.

A uniform rod slides down in a vertical plane, its ends being in contact with two smooth planes, one horizontal and other vertical. If it started from rest at angle α with the horizontal, discuss the motion.

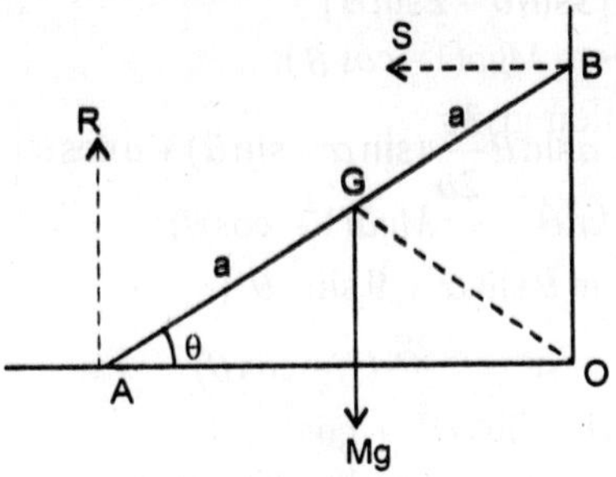

Fig. 8.9

Let R and S be the reactions at the ends A and B of the rod AB of length $2a$ and mass M. Let the rod make an angle θ with the horizontal.

The co-ordinates of the centre of gravity G referred to O as origin are given by

$$x = a\cos\theta,\ y = a\sin\theta.$$

$$\therefore\ \ddot{x} = -a\cos\theta\dot{\theta}^2 - a\sin\theta\ddot{\theta},\ \ddot{y} = -a\sin\theta\dot{\theta}^2 + a\cos\theta\ddot{\theta}$$

The equations of motion of G are

$$M\ddot{x} = S$$

$$\therefore\quad M\left[-a\cos\theta\dot{\theta}^2 - a\sin\theta\ddot{\theta}\right] = S \tag{1}$$

$$M\ddot{y} = R - Mg$$

$$\therefore\quad M\left[-a\sin\theta\dot{\theta}^2 + a\cos\theta\ddot{\theta}\right] = R - Mg \tag{2}$$

Energy equation gives

$$\frac{1}{2}M[\dot{x}^2 + \dot{y}^2 + k^2\dot{\theta}^2] = Mga(\sin\alpha - \sin\theta)$$

$$\therefore\quad \frac{1}{2}M\left[a^2\theta^2 + \frac{1}{3}a^2\dot{\theta}^2\right] = Mga(\sin\alpha - \sin\theta)$$

$$\therefore\quad \dot{\theta}^2 = \frac{3g}{2a}(\sin\alpha - \sin\theta) \tag{3}$$

Differentiating (3), w.r.t. t, we get

$$\ddot{\theta} = -\frac{3g}{4a}\cos\theta \tag{4}$$

Substituting values of $\dot{\theta}^2$ and $\ddot{\theta}$ in (1) and (2), we have

$$S = M\left[-a\cos\theta.\frac{3g}{2a}(\sin\alpha - \sin\theta) + a\sin\theta.\frac{3g}{4a}\cos\theta\right]$$

$$= \frac{3}{4}Mg\cos\theta[3\sin\theta - 2\sin\alpha] \tag{5}$$

$$R = Mg + M\left[-a\sin\theta\frac{3g}{2a}(\sin\alpha - \sin\theta) + a\cos\theta\left(-\frac{3g}{4a}\cos\theta\right)\right]$$

$$= \frac{Mg}{4}[1 - 6\sin\theta\sin\alpha + 9\sin^2\theta]$$

$$= \frac{Mg}{4}[(3\sin\theta - \sin\alpha)^2 + \cos^2\alpha] \tag{6}$$

From (5) it is evident that S = O when $\sin\theta = \frac{2}{3}\sin\alpha$ and when this value of θ is attained the value of S becomes negative. Hence, the end B never leaves wall when $\sin\theta = \frac{2}{3}\sin\alpha$.

From (6), we see that R is always +ve, *i.e.*, the end never leaves the plane.

When end B leaves the wall $\sin\theta = \frac{2}{3}\sin\alpha$ and S = O, hence, equations (1)–(4) do not hold good for further motion.

Putting $\sin\theta = \frac{2}{3}\sin\alpha$ in (3), the angular velocity becomes $\sqrt{\frac{g}{2a}\sin\alpha}$.

Second part of the motion

When the end B has left the wall, let R_1 be the normal reaction at A and the rod is inclined at an angle ϕ to the horizontal.

Now the equations of motion of G are

$$M\ddot{x} = O \tag{1}$$

$$M\ddot{y} = R_1 - Mg \tag{2}$$

Taking moments about G, we have

$$m\frac{a^2}{3}\ddot{\phi} = -R_1 a\cos\phi \tag{3}$$

Now

$$y = a\sin\phi, \quad \therefore \ddot{y} = a(-\sin\phi\dot{\phi}^2 + \cos\phi\ddot{\phi}) \tag{4}$$

From (2) and (4),

$$Ma\left[-\sin\phi\dot{\phi}^2+\cos\phi\ddot{\phi}\right] = R_1 - Mg$$

$$\therefore \qquad R_1 = Mg + Ma\left[-\sin\phi\dot{\phi}^2+\cos\phi\ddot{\phi}\right].$$

Substituting value of R_1 in (3), we get

$$M\frac{a^2}{3}\ddot{\phi} = -a\cos\phi\, Mg - Ma^2\cos\phi\left[-\sin\phi\dot{\phi}^2+\cos\phi\ddot{\phi}\right]$$

$$\therefore \qquad a^2\left(\frac{1}{3}+\cos^2\phi\right)\ddot{\phi} - a^2\cos\phi\sin\phi\dot{\phi}^2 = -ag\cos\phi$$

$$\therefore \qquad \left(\frac{1}{3}+\cos^2\phi\right)\ddot{\phi} - \cos\phi\sin\phi\dot{\phi}^2 = -\frac{g}{a}\cos\phi \qquad (5)$$

Integrating, we have

$$\left(\frac{1}{3}+\cos^2\phi\right)\dot{\phi}^2 = -\frac{2g}{a}\sin\phi + C' \qquad (6)$$

When

$$\sin\phi = \frac{2}{3}\sin\alpha,\ \dot{\phi} = \sqrt{\frac{g}{2a}\sin\alpha}$$

$$\therefore \qquad C' = \frac{g}{2a}\sin\alpha\left[\frac{1}{3}+1-\frac{4}{9}\sin^2\alpha\right]+\frac{4g}{3a}\sin\alpha$$

$$= \frac{2g}{a}\sin\alpha\left(1-\frac{1}{9}\sin^2\alpha\right)$$

$$\therefore \left(\frac{1}{3}+\cos^2\phi\right)\dot{\phi}^2 = -\frac{2g}{a}\left(\sin\phi - \sin\alpha + \frac{1}{9}\sin^3\alpha\right).$$

When rod reaches the horizontal $\phi = 0, \dot{\phi} = \Omega$, then

$$\frac{4}{3}\Omega^2 = \frac{2g\sin\alpha}{a}\left(1-\frac{1}{9}\sin^2\alpha\right).$$

$$\therefore \qquad \Omega^2 = \frac{3g\sin\alpha}{2a}\left(1-\frac{1}{9}\sin^2\alpha\right).$$

Note: from (1), we see that $\dot{x} = const.$ during the second part of the motion and is equal to the value of $\dot{x}$ from part I, when $\sin\theta = \frac{2}{3}\sin\alpha$, *i.e.*, $\dot{x} = -a\sin\theta\dot{\theta} = \frac{1}{3}\sqrt{2ag\sin^3\alpha}.$

Examples

1. A uniform rod is held at an inclination α to the horizon with one end in contact with a horizontal table whose co-efficient of friction is μ. If it be released; show that it will commence to slide if

$$\mu < \frac{3\sin\alpha\cos\alpha}{1+3\sin^2\alpha}.$$

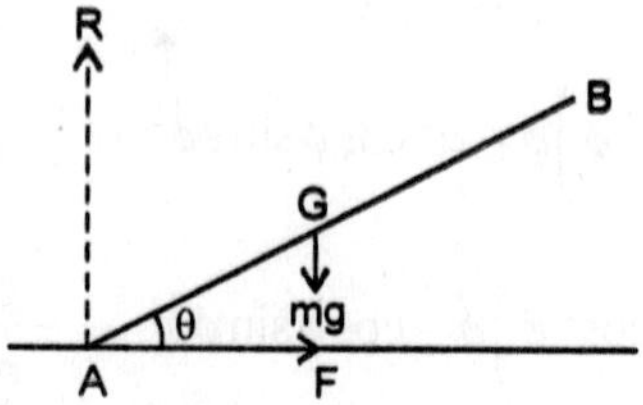

Fig. 8.10

Let AB be the rod of length $2a$ and mass m, and at time t it makes an angle θ with the horizontal. Let F be the frictional force sufficient to prevent sliding. The co-ordinates of G of the C.G. are given by

$$x = a\cos\theta,\quad y = a\sin\theta.$$

Hence, equations of motion of G are given by

$$m\ddot{x} = m\left[-a\cos\theta\dot{\theta}^2 - a\sin\theta\ddot{\theta}\right] = F \qquad (1)$$

$$m\ddot{y} = m\left[-a\sin\theta\dot{\theta}^2 + a\cos\theta\ddot{\theta}\right] = R - mg \qquad (2)$$

Energy equation is

$$\frac{1}{2}m\left[(a\dot{\theta})^2 + \frac{1}{3}a^2\dot{\theta}^2\right] = mga(\sin\alpha - \sin\theta)$$

$$\therefore \quad \frac{2}{3}a^2\dot{\theta}^2 = ga(\sin\alpha - \sin\theta)$$

$$\therefore \quad \dot{\theta}^2 = \frac{3g}{2a}(\sin\alpha - \sin\theta) \qquad (3)$$

Differentiating w.r.t. t and dividing by $2\dot{\theta}$, we get

$$\ddot{\theta} = -\frac{3g}{4a}\cos\theta. \qquad (4)$$

Substituting values of $\dot{\theta}^2$ and $\ddot{\theta}$, from (3) and (4) in (1) and (2), we have

$$F = m\left[-a\cos\theta\cdot\frac{3g}{2a}(\sin\alpha - \sin\theta) - a\sin\theta\left(-\frac{3g}{4a}\cos\theta\right)\right]$$

$$= \frac{3}{4} mg \cos\theta \,[3\sin\theta - 2\sin\alpha.]$$

When $\theta = \alpha, F = \frac{3}{4} mg \cos\alpha . \sin\alpha.$

$$R = mg + m\left[-a\sin\theta \cdot \frac{3g}{2a}(\sin\alpha - \sin\theta) + a\cos\theta\left(-\frac{3g}{4a}\cos\theta\right)\right]$$

$$= \frac{mg}{4}\left[4 - 6\sin\theta\sin\alpha + 6\sin^2\theta - 3\cos^2\theta\right].$$

When $\theta = \alpha$,

$$R = \frac{mg}{4}(4 - 3\cos^2\alpha) = \frac{mg}{4}(1 + 3\sin^2\alpha).$$

The end A will commence to slide if

$$\mu < \frac{F}{R}$$

$$\therefore \qquad \mu < \frac{3\sin\alpha\cos\alpha}{1 + 3\sin^2\alpha}.$$

2. A rough uniform rod of length $2a$ is placed on a rough table at right angles to its edge; if its centre of gravity be initially at distance b beyond the edge, show that the rod will begin to slide when it has turned through an angle $\tan^{-1}\dfrac{\mu a^2}{a^2 + 9b^2}$, where μ is a co-efficient of friction.

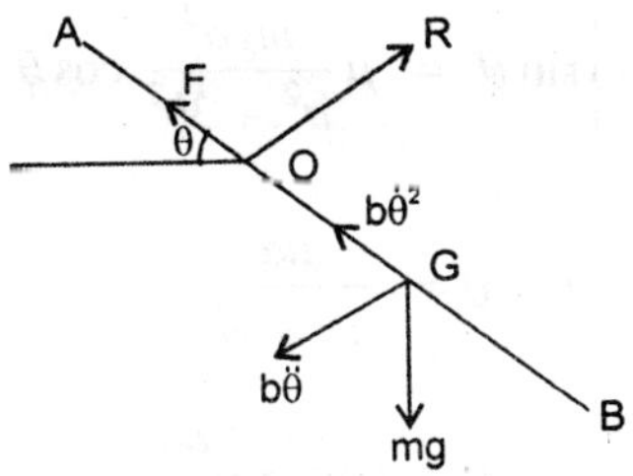

Fig. 8.11

Let there be no sliding when the rod has turned through an angle θ.
Let F be the frictional force and R the normal reaction.
Equations of motion of centre of gravity G are

$$mb\ddot{\theta} = mg\cos\theta - R \qquad (1)$$

$$mb\dot{\theta}^2 = F - mg\sin\theta \qquad (2)$$

Taking moments about the point of contact O, we have

$$mk^2\ddot{\theta} = mgb\cos\theta$$

but

$$k^2 = \frac{a^2}{3} + b^2$$

$$\therefore \quad \left(\frac{a^2}{3} + b^2\right)\ddot{\theta} = gb\cos\theta$$

$$\ddot{\theta} = \frac{3gb\cos\theta}{a^2 + 3b^2} \qquad (3)$$

Integrating and using initial conditions $\theta = 0 = \dot{\theta}$, we get

$$\dot{\theta}^2 = \frac{6gb\sin\theta}{a^2 + 3b^2} \qquad (4)$$

Substituting values of $\ddot{\theta}$ and $\dot{\theta}$ from (3) and (4) in (1) and (2), we have

$$R = mg\cos\theta - \frac{3mgb^2\cos\theta}{a^2 + 3b^2} = \frac{mga^2\cos\theta}{a^2 + 3b^2}.$$

$$F = mg\sin\theta + \frac{6mgb^2\sin\theta}{a^2 + 3b^2} = \frac{a^2 + 9b^2}{a^2 + 3b^2}mg\sin\theta.$$

The sliding commences, when

$$F = \mu R$$

$$\therefore mg\left(\frac{a^2 + 9b^2}{a^2 + 3b^2}\right)\sin\theta = \mu\frac{mga^2}{a^2 + 3b^2}\cos\theta$$

$$\therefore \quad \tan\theta = \frac{\mu a^2}{a^2 + 9b^2}$$

$$\therefore \quad \theta = \tan^{-1}\frac{\mu a^2}{a^2 + 9b^2}.$$

3. The lower end of a uniform rod, inclined at an angle α to the horizon, is placed on a smooth horizontal table. A horizontal force is applied at its lower end of such a magnitude that the rod rotates in vertical plane with constant angular velocity ω. Show that when the rod is inclined at an angle θ to the horizon, the magnitude of the force is

$$mg\cot\theta - ma\omega^2\cos\theta,$$

where m is the mass of the rod.

Let F be the force applied at the lower end A of the rod and R the normal reaction. At any time t, let θ be the inclination of the rod to the horizon. Since, the rod rotates with uniform angular velocity ω.

$$\therefore \qquad \dot{\theta} = \omega, \ddot{\theta} = 0.$$

Equation of motion of the centre of gravity G along the vertical is

$$R - mg = m\frac{d^2}{dt^2}(a\sin\theta) = ma(-\sin\theta\dot{\theta}^2 + \cos\theta\ddot{\theta})$$

$$\therefore \qquad R = mg - ma\omega^2 \sin\theta$$

(Here, the end is not fixed, so equation of horizontal motion is not required.)

Taking moments about G, we have

$$mk^2\ddot{\theta} = Fa\sin\theta - Ra\cos\theta$$

$$\therefore \qquad 0 = Fa\sin\theta - Ra\cos\theta$$

$$\therefore \qquad F = R\cot\theta$$

Substituting the value of R, we get

$$F = (mg - ma\omega^2\sin\theta)\cot\theta$$

$$= mg\cot\theta - ma\omega^2\cos\theta.$$

4. A uniform rod is held nearly vertical with one end resting on an imperfectly rough plane. It is released from rest and falls forwards. The inclination to the vertical at any instant is θ, prove that
   (i) If the co-efficient of friction is less than a certain finite amount, then lower end of the rod will slip backwards before $\sin^2\frac{\theta}{2} = \frac{1}{6}$.
   (ii) However great the co-efficient of friction may be, the lower end will begin to slip forwards at a value of $\sin^2\frac{\theta}{2}$ between $\frac{1}{6}$ and $\frac{1}{3}$.

From Art. 8.6, we have

$$F = \frac{3}{4}mg\sin\theta(3\cos\theta - 2), \quad R = \frac{1}{4}Mg(1 - 3\cos\theta)^2$$

Now F vanishes when

Either $\sin\theta = 0$, $\therefore \theta = 0$ or $3\cos\theta = 2$. $\therefore \cos\theta = \frac{2}{3}$ $\therefore 2\sin^2\frac{\theta}{2} = \frac{1}{3}$

$$\therefore \sin^2\frac{\theta}{2} = \frac{1}{6}$$

The value of F is positive for all values of θ lying between 0 and $\cos^{-1}\frac{2}{3}$ and is a continuous function of θ, hence, between these two minimum values lies the maximum value of F. Let F_1 be the maximum value of F.

We also observe that for $0 \le \theta \le \cos^{-1}\frac{2}{3}$, the value of $R \le Mg$.

Here for finite values of μ, $F_1 > \mu Mg$ and therefore for this value of μ sliding will take place, before $\cos^{-1}\frac{2}{3}$ *i.e.*, $\sin^2\frac{\theta}{2} = \frac{1}{6}$. Since, F is positive (in the forward direction) hence, slipping will start in backward direction.

Second Part

When $\theta > \cos^{-1}\frac{2}{3}$, F will change its sign. *i.e.*, the direction of F is reversed if

$$F' = -F = \frac{3}{4} Mg \sin\theta(2 - 3\cos\theta).$$

Now the slipping will start when $F' > \mu R$.
i.e., when

$3Mg \sin\theta(2 - 3\cos\theta) > \mu(1 - 3\cos\theta)^2$.

Now as θ increases from $\cos^{-1}\frac{2}{3}$ to $\cos^{-1}\frac{1}{3}$, the left hand of the expression increases while right hand side of the expression decreases from 1 to 0. Hence, for some value of θ between $\cos^{-1}\frac{2}{3}$ and $\cos^{-1}\frac{1}{3}$ *i.e.*, for $\sin^2\frac{\theta}{2}$ between $\frac{1}{6}$ and $\frac{1}{3}$ the condition will hold good and the slipping will start in the forward direction.

5. A uniform rod is placed in contact with a horizontal table, at an inclination α to the horizon, and is allowed to fall. When it becomes horizontal, show that its angular velocity is, $\sqrt{\frac{3g}{4a}\sin\alpha}$, whether the plane is perfectly smooth or perfectly rough. Show also that the end of the rod will not leave the plane in either case.

Case I

When plane is perfectly rough and end O is fixed.

The energy equation gives

$$\frac{1}{2}m\left[a^2\dot{\theta}^2 + \frac{1}{3}a^2\dot{\theta}^2\right] = mga(\sin\alpha - \sin\theta)$$

$$\therefore \qquad \dot{\theta}^2 = \frac{3g}{2a}(\sin\alpha - \sin\theta) \qquad (1)$$

When rod becomes horizontal $\theta = 0,\ \dot{\theta} = \omega$,

Then

$$\omega^2 = \frac{3g}{2a}\sin\alpha$$

$$\therefore \qquad \omega = \sqrt{\frac{3g}{2a}\sin\alpha}\,.$$

Differentiating (1), we get

$$\ddot{\theta} = -\frac{3g}{4a}\cos\theta \qquad (2)$$

The equation of motion of C.G. is

$$R - mg = m\frac{d^2}{dt^2}(a\sin\theta) = ma\,(-\sin\theta\dot{\theta}^2 + \cos\theta\ddot{\theta}),$$

Substituting the values of $\dot{\theta}^2$ and $\ddot{\theta}$, we get

$$R = mg + ma\left[-\sin\theta\cdot\frac{3g}{2a}(\sin\alpha - \sin\theta) + \cos\theta\left(-\frac{3g}{4a}\cos\theta\right)\right]$$

$$= \frac{mg}{4}[4 - 6\sin\alpha\,\sin\theta + 6\sin^2\theta - 3\cos^2\theta]$$

$$= \frac{mg}{4}[(1 - 3\sin\alpha\sin\theta)^2 - 9\sin^2\alpha\sin^2\theta + 9\sin^2\theta]$$

$$= \frac{mg}{4}[(1 - 3\sin\alpha\sin\theta)^2 + 9\sin^2\theta\cos^2\alpha]$$

This shows that R is always positive and O will never leave the plane.

Case II

When the plane is perfectly smooth

When the plane is perfectly smooth, there is no frictional force, *i.e.*, there is no force in the horizontal direction, hence, centre of gravity G descends in a vertical line, the only velocity of G is along the vertical.

Now

$$y = a\sin\theta,\ \dot{y} = a\cos\theta\dot{\theta}.$$

Energy equation is given by

$$\frac{1}{2}m\left[a^2\cos^2\theta\dot{\theta}^2 + \frac{1}{3}a^2\dot{\theta}^2\right] = mga\,(\sin\alpha - \sin\theta)$$

$$\therefore \quad \dot{\theta}^2\left(\cos^2\theta + \frac{1}{3}\right) = \frac{2g}{a}(\sin\alpha - \sin\theta) \tag{3}$$

When rod becomes horizontal

$$\theta = 0,\ \dot{\theta} = \omega \text{ (say)}$$

From (1), we get

$$\omega^2 = \frac{3g}{2a}\sin\alpha,\ \therefore \omega = \sqrt{\frac{3g}{2a}\sin\alpha}.$$

Differentiating (3), we get

$$\ddot{\theta}\left(\cos^2\theta + \frac{1}{3}\right) - \dot{\theta}^2\sin\theta\cos\theta = -\frac{g}{a}\cos\theta.$$

Substituting the value of $\dot{\theta}^2$ from (3), we get

$$\ddot{\theta}\left(\cos^2\theta + \frac{1}{3}\right) - \frac{2g\sin\theta\cos\theta(\sin\alpha - \sin\theta)}{a\left(\cos^2\theta + \frac{1}{3}\right)} = -\frac{g}{a}\cos\theta.$$

$$\therefore \quad \left(\cos^2\theta + \frac{1}{3}\right)^2\ddot{\theta} = -\frac{g}{a}\cos\theta\left[\sin^2\theta - 2\sin\alpha\sin\theta + \frac{4}{3}\right]$$

$$= -\frac{g}{a}\cos\theta\left[(\sin\alpha - \sin\theta)^2 + \frac{1}{3} + \cos^2\alpha\right] \tag{4}$$

Taking moments about G

$$m\frac{a^2}{3}\ddot{\theta} = -Ra\cos\theta$$

$$\therefore \quad R = -\frac{a}{3}\sec\theta\, m\ddot{\theta} = +ve$$

Hence, end O will never leave the plane.

8.8 MOTION OF BODIES CONNECTED WITH STRINGS

Examples

1 One end of string is fixed to a point on the rim of a uniform circular disc of radius a and mass M, and the string is wound round the rim several times. The free end of the string is attached to a fixed point and the disc is held so that the part of the string not in contact with it is vertical. If the

string is let go, show that the acceleration of its centre is $\frac{2}{3}g$ and the tension in the string is $\frac{1}{3}Mg$.

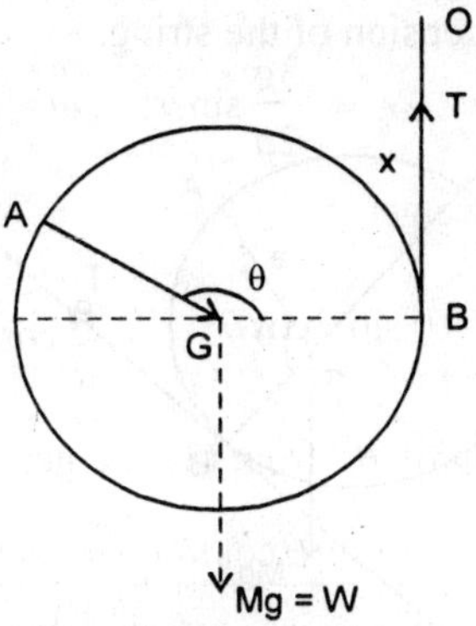

Fig. 8.12

Let A be the fixed end of the string, O the initial position of the centre G of the disc and T the tension in the string.

Since, all the forces acting on the system are vertical, therefore G will move vertically downwards. Let x be the vertical displacement of G, when the disc turns through an angle θ in time t.

The equations of the motion of G is

$$M\ddot{x} = Mg - T \tag{1}$$

Taking moment about G

$$M\frac{a^2}{2}\ddot{\theta} = Ta \tag{2}$$

Since,

$$OG = \text{arc } AB$$

$$\therefore \quad x = a\theta$$

$$\therefore \quad \ddot{x} = a\ddot{\theta} \tag{3}$$

From (2) and (3)

$$M\dot{x} = 2T \tag{4}$$

Eliminating T from (1) and (4), we get

$$M\ddot{x} = Mg - \frac{1}{2}M\ddot{x}$$

$$\therefore \ddot{x} = \frac{2}{3}g. \tag{5}$$

From (4) and (5),

$$T = \frac{1}{3}Mg.$$

2 A cylinder rolls down a smooth plane whose inclination is α, unwrapping, as it goes, a fine string fixed to the highest point O of the plane; find its acceleration and the tension of the string.

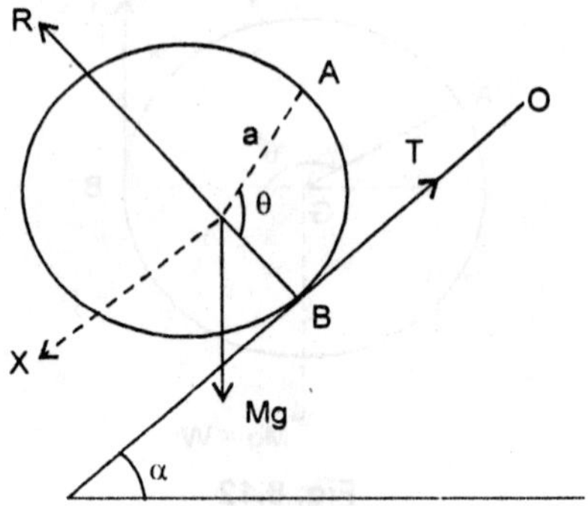

Fig. 8.13

Let T be the tension in the string when the cylinder has rolled down a distance x along the plane in time t, and θ and the angle turned by the cylinder.

Since, the string is taut, there is pure rolling.

$$\therefore \quad x = a\theta$$

$$\therefore \quad \ddot{x} = a\ddot{\theta} \tag{1}$$

The equations of motion of the centre of gravity of the cylinder are

$$M\ddot{x} = Mg\sin\alpha - T \tag{2}$$

Taking moments about the centre, we have

$$Mk^2\ddot{\theta} = T \times a \tag{3}$$

$$\therefore \quad M\frac{a^2}{2}\ddot{\theta} = T \times a$$

$$\therefore \quad \frac{1}{2}M\ddot{x} = T \tag{4}$$

Putting this value of T in (2), we get

$$\frac{3}{2}M\ddot{x} = Mg\sin\alpha$$

$$\therefore \quad \ddot{x} = \frac{2}{3}g\sin\alpha.$$

Putting this value of $\ddot{x}$ in (4), we get

$$T = \frac{1}{3} Mg \sin \alpha.$$

3. Two equal cylinders, each of mass m, are bound together by an elastic string, whose tension is T, and roll with their axes horizontal down a rough plane of inclination α. Show that their acceleration is $\frac{2}{3} g \sin \alpha \left[1 - \frac{2\mu T}{mg \sin \alpha} \right]$, where μ is the co-efficient of friction between the cylinders.

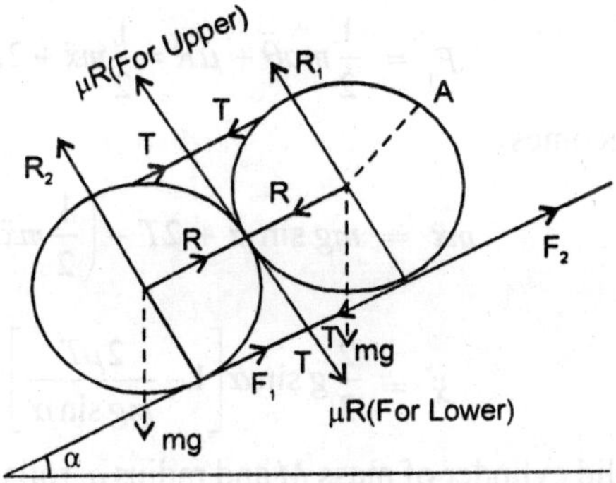

Fig. 8.14

Let F_1, F_2 be the frictional forces on two cylinders due to the plane and R_1, R_2 the normal reactions. At the point of contact of two cylinders, the frictional force μR acts away from the upper cylinder and towards the plane for lower cylinder.

Let at time t the cylinders move through a distance x along the plane, and θ be the angle turned by them.

Since, there is no slipping

$$\therefore \quad x = a\theta$$

$$\therefore \quad \ddot{x} = a\ddot{\theta} \tag{1}$$

Equations of motion of the upper sphere are

$$m\ddot{x} = mg \sin\alpha + 2T - F_1 - R \tag{2}$$

$$0 = R_1 - mg \cos\alpha + \mu R \tag{3}$$

$$mk^2\ddot{\theta} = F_1 a - \mu R a \tag{4}$$

Equations of motion of the lower cylinder are

$$m\ddot{x} = mg \sin\alpha + 2T - F_2 + R \tag{5}$$

$$0 = R_2 - mg \cos\alpha - \mu R \tag{6}$$

$$mk^2\ddot{\theta} = F_2 a - \mu Ra \tag{7}$$

From (4) and (7),

$$F_1 = F_2.$$

Subtracting (2) from (5), we get

$$R = 2T \tag{8}$$

Then from (4), we get

$$F_1 = \frac{mk^2\ddot{\theta}}{a} + \mu R.$$

Now

$$k^2 = \frac{a^2}{2}$$

$\therefore$

$$F_1 = \frac{1}{2}ma\ddot{\theta} + \mu R = \frac{1}{2}m\ddot{x} + 2\mu T.$$

Hence, (2) becomes

$$m\ddot{x} = mg\sin\alpha + 2T - \left(\frac{1}{2}m\ddot{x} + 2\mu T\right) - 2T.$$

$\therefore$

$$\ddot{x} = \frac{2}{3}g\sin\alpha\left[1 - \frac{2\mu T}{mg\sin\alpha}\right].$$

4. A uniform solid cylinder of mass M and radius a, rolls on a rough inclined plane with its axis perpendicular to the line of greatest slope. As it rolls, the cylinder winds a light string which passes over a fixed light pulley and supports a freely hanging mass m, the part of the string between the pulley and cylinder being parallel to the line of greatest slope. If α be the inclination of the plane, find the acceleration of the centre of the cylinder and the tension of the string.

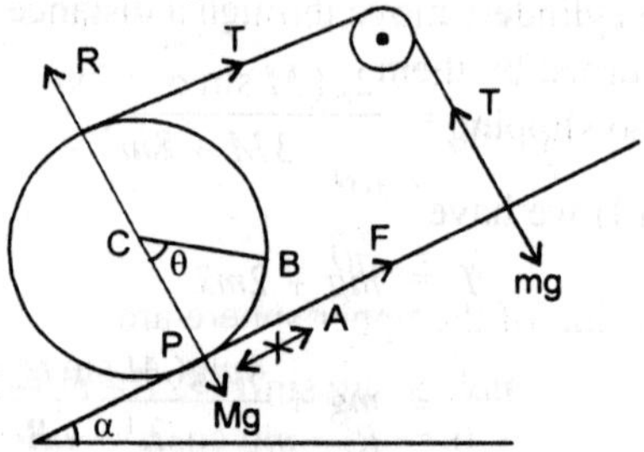

Fig. 8.15

Let A be the point of contact initially when the cylinder was at rest. At time t, when the centre of the cylinder has described the distance x, B be

the position of A and P be the new point of contact. Let $\angle PCB = \theta$, R be the normal reaction, F the friction and T the tension of the string.
Now the centre of the cylinder rolls down through a distance x, it winds up a length x of the string and an additional length x of the string moves along the cylinder. Hence, as the cylinder rolls down distance x, the mass m moves up a distance $y = 2x$.
Equations of motion of the cylinders are

$$M\ddot{x} = Mg\sin\alpha - F - T \tag{1}$$

$$M\frac{a^2}{2}\ddot{\theta} = (F - T)\,a \tag{2}$$

For pure rolling $x = a\theta$

$$\therefore \quad \ddot{x} = a\ddot{\theta} \tag{3}$$

Equation of motion for mass m is

$$m\ddot{y} = T - mg$$

But $\ddot{y} = 2\ddot{x}$

$$\therefore \quad 2m\ddot{x} = T - mg \tag{4}$$

From (2) and (3), we have

$$\frac{1}{2}M\ddot{x} = F - T \tag{5}$$

Adding (1) and (5), we get

$$\frac{3}{2}M\ddot{x} = Mg\sin\alpha - 2T \tag{6}$$

Eliminating T from (4) and (6), we get

$$(4m + \frac{3}{2}M)\ddot{x} = Mg\sin\alpha - 2mg$$

$$\therefore \quad \ddot{x} = \frac{2g(M\sin\alpha - 2m)}{3M + 8m}.$$

From equation (4) we have

$$T = mg + 2m\ddot{x}$$

$$= mg + \frac{4mg(M\sin\alpha - 2m)}{3M + 8m}$$

$$= \frac{m\,Mg(3 + 4\sin\alpha)}{3M + 8m}.$$

5. A fine string has two masses M and M' tied to its ends and passes over a rough pulley, of mass m, whose centre is fixed, if the string

does not slip over the pulley, show that M will descend with acceleration

$\dfrac{M-M'}{M+M'+m\dfrac{k^2}{a^2}}g$ where a is the radius of the pulley.

If the pulley be not sufficiently rough to prevent sliding and M be the descending mass, show that the acceleration is $\dfrac{M-M'e^{\mu\pi}}{M+M'e^{\mu\pi}}g$ and that the pulley will now spin with angular acceleration $\dfrac{2MM'ga(e^{\mu\pi}-1)}{mk^2(M+M'e^{\mu\pi})}$.

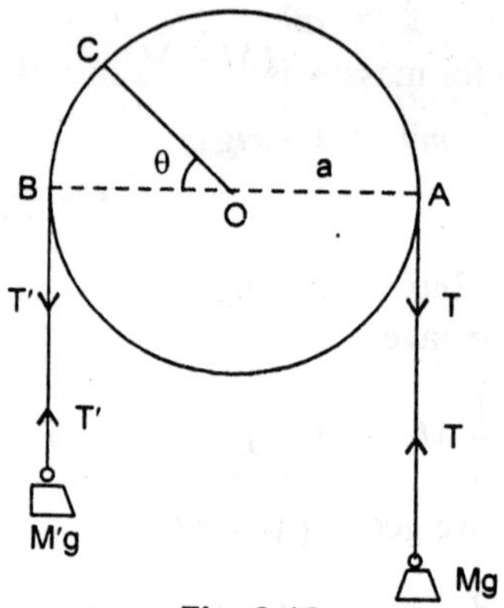

Fig. 8.16

Let T, T' be the tensions in the string when pulley has turned through an angle θ and the depths of M and M' below the centre of pulley x and y. Then equation of motion of pulley is

$$mk^2\ddot{\theta} = (T-T')a \qquad (1)$$

Equations of motion of the weights are

$$M\ddot{x} = Mg-T,\ M'\ddot{y} = M'g-T' \qquad (2)$$

Now

$$x+y = const. \text{ throughout the motion.}$$

$$\therefore \qquad \ddot{x} = -\ddot{y} \qquad (3)$$

Case I

Let pulley be rough enough to prevent sliding, the string and pulley at A are always moving with the same velocity. Then

$$\dot{x} = a\theta$$

$$\therefore \qquad \ddot{x} = a\ddot{\theta} \qquad (4)$$

From (2), we get

$$T - T' = -M'g + M\ddot{y} - M\ddot{x} + Mg$$

With the help of (3), we have

$$T - T' = g(M - M') - (M + M')\ddot{x}$$
$$= g(M - M') - (M + M')a\ddot{\theta}$$

Putting this value of T – T', in (1), we get

$$mk^2\ddot{\theta} = g(M - M')a - (M + M')a^2\ddot{\theta}$$

$$\therefore \quad \ddot{\theta} = \frac{g(M - M')a}{(M + M')a^2 + mk^2}$$

$$\therefore \quad \ddot{x} = a\ddot{\theta} = \frac{(M - M')g}{(M + M') + \frac{mk^2}{a^2}}.$$

This is the constant acceleration, with which M descends. If the pulley be uniform disc then $k^2 = \frac{a^2}{2}$, then this acceleration is

$$= \frac{(M - M')g}{(M + M') + \frac{m}{2}}.$$

Case II

When pulley is not sufficiently rough then $T = T'e^{\mu\pi}$.

From (2)

$$M\ddot{x} = Mg - T, \; M'\ddot{y}e^{\mu\pi} = (M'g - T')e^{\mu\pi}$$

$$\therefore \quad -M\ddot{x}e^{\mu\pi} = M'ge^{\mu\pi} - T'e^{\mu\pi}$$
$$= M'ge^{\mu\pi} - T$$

$$\therefore \quad (M + M'e^{\mu\pi})\ddot{x} = (M - M'e^{\mu\pi})g$$

$$\therefore \quad \ddot{x} = \frac{(M - M'e^{\mu\pi})g}{M + M'e^{\mu\pi}}$$

Putting this value of $\ddot{x}$ in (2), we get

$$T'e^{\mu\pi} = T = Mg - M\left[\frac{M - M'e^{\mu\pi}}{M + M'e^{\mu\pi}}\right]g$$

$$= \frac{2MM'e^{\mu\pi}}{M + M'e^{\mu\pi}}g.$$

From (1), we get

$$mk^2\ddot{\theta} = (e^{\mu\pi} - 1)T'a$$

$$= (e^{\mu\pi} - 1)\frac{2MM'ga}{M + M'e^{\mu\pi}}$$

$$\therefore \quad \ddot{\theta} = \frac{2ga(e^{\mu\pi} - 1)}{mk^2} \cdot \frac{MM'}{M + M'e^{\mu\pi}}$$

8.9 IMPULSIVE FORCES

If ω and ω' the angular velocities just before and just after the impulsive forces, then

Change in moment of momentum about the axis of rotation $= MK^2(\omega' - \omega)$

Let L be the moment of the external forces about this axis. Since, change in the moment of momentum of the system about the axis is equal to the moment of the impulses about it, therefore

$$Mk^2(\omega' - \omega) = L.$$

Examples

1 A rod of mass nM, is lying on a horizontal table, and has one end fixed, a particle of mass M is in contact with it. The rod receives a horizontal blow at its free end, find the position of the particle so that it may start to move with its maximum velocity.

In this case show that the kinetic energies communicated to the rod and mass are equal.

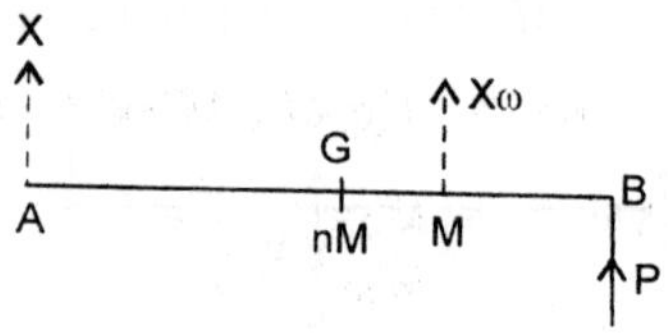

Fig. 8.17

Let AB be the rod of length $2a$, particle of mass M be placed at a distance x from the fixed end A.

Suppose an impules P is applied at the free end B, so as to give an angular velocity ω.

The velocity of the particle M is $x\omega$. Hence, momentum equation gives

$$(nM + M)k^2(\omega - 0) = P.\,2a$$

But

$$(nM + M)k^2 = nM\frac{4a^2}{3} + Mx^2$$

$$\therefore \quad M\left[n\frac{4a^2}{3}+x^2\right]\omega = P.\,2a$$

$$\therefore \quad \omega = \frac{6ap}{M(4na^2+3x^2)}$$

$$\therefore V = \text{velocity of the particle} = x\omega = \frac{6aPx}{M(4na^2+3x^2)}.$$

For maxima or mimina

$$\frac{dV}{dx} = \frac{6ap}{M}\left[\frac{(4na^2+3x^2)-x.6x}{(4na^2+3x^2)^2}\right]=0$$

$$\therefore \quad 4na^2+3x^2-6x^2 = 0$$

$$\therefore \quad x = 2a\sqrt{\frac{n}{3}}.$$

Kinetic energy of the rod is $\frac{1}{2}nMk^2\omega^2 = \frac{1}{2}nM\frac{4a^2}{3}\omega^2 = \frac{2nMa^2\omega^2}{3}$.

Kinetic energy of the mass $M = \frac{1}{2}Mx^2\omega^2 = \frac{1}{2}M.4a^2\frac{n}{3}\omega^2 = \frac{2nMa^2\omega^2}{3}$.

2. Two uniform rods AB, BC are freely jointed at B and laid on a horizontal table. AB is struck by a horizontal blow P in a direction perpendicular to AB at a distance c from the centre, the lengths of AB, BC being $2a$ and $2b$ and their massses M and M'; find motion immediately after the impulse.

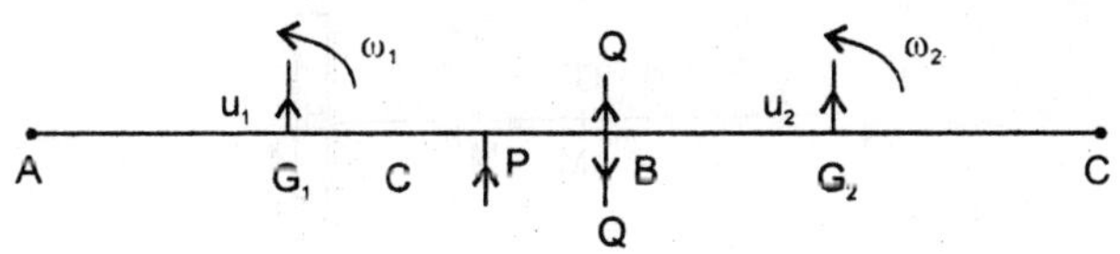

Fig. 8.18

Let G_1 and G_2 be the Centres of inertia of AB and BC. Let u_1 and ω_1 be the linear and angular velocities of G_1 ad u_2 and ω_2 be the linear and angular velocities of G_2 just after the blow. There will be an action between the two rods at B when the blow is struck, let its impulse be Q in opposite directions on the two rods. Initially rods were at rest, hence, we have for rod AB

$$Mu_1 = P - Q \tag{1}$$

$$M\frac{a^2}{3}\omega_1 = P.c - Q.a \tag{2}$$

So for *BC*

$$M' u_2 = Q \tag{3}$$

$$M' \frac{b^2}{3} \omega_2 = Qb \tag{4}$$

Now rods are connected at *B*, the motion of *B*, as deducted from each rod, must be same.

$$\therefore \quad u_1 + a\omega_1 = u_2 - b\omega_2 \tag{5}$$

On solving these equations, we get

$$Q = \frac{PM'}{4(M+M')}\left(1+\frac{3c}{a}\right), u_1 = \frac{P}{M}\left[1-\frac{M'}{4(M+M')}\left(1+\frac{3c}{a}\right)\right],$$

$$\omega_1 = \frac{3P}{Ma}\left[\frac{c}{a}-\frac{M'}{4(M+M')}\left(1+\frac{3c}{a}\right)\right], u_2 = \frac{P}{4(M+M')}\left(1+\frac{3c}{a}\right)$$

$$\omega_2 = -\frac{3P}{4(M+M')b}\left(1+\frac{3c}{a}\right).$$

3. A uniform rod *AB* of mass $2m$ is freely jointed at *B* to a second rod *BC* of mass *m*. The rods lie on a smooth horizontal plane at right angles to each other and an impulse *I* is applied to *AB* at *A* in direction parallel to *BC*. Find the initial velocity of *BC* and prove that K.E. created is $\frac{5I^2}{6m}$.

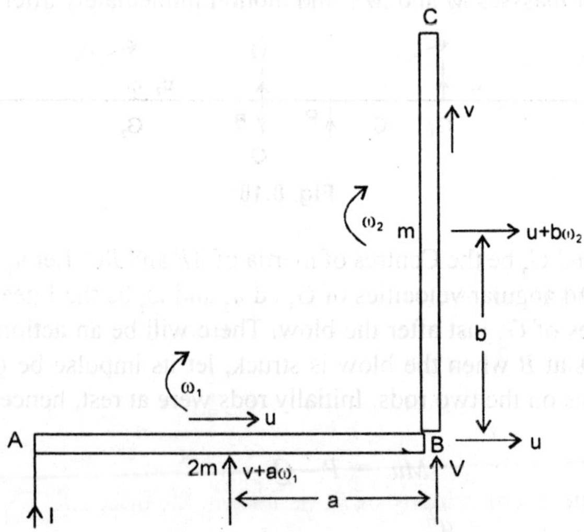

Fig. 8.19

Let B has velocity components u, v along, AB, BC and let ω_1, ω_2 be the angular velocities of the rods. The velocity of the centroid G_1 of the rod AB has the components $[u, v + a\omega_1]$ and G_2 of BC has velocity components $[u + b\omega_2, v]$, where $2a$, $2b$ are the lengths of the rods AB and BC respectively.

Since, there is no impulse along AB, hence, total linear momentum is conserved along AB.

$$\therefore \quad 2mu + m(u + b\omega_2) = 0$$

$$\therefore \quad 3u + b\omega_2 = 0 \tag{1}$$

The equation of motion in the direction BC is

$$2m(v + a\omega_1) + mv = I$$

$$\therefore \quad m(3v + 2a\omega_1) = I \tag{2}$$

An impulsive action is set up at B. By taking moments about B for the two rods, we need not introduce this impulse. Thuse, for the rod AB, the angular momentum generated about B is

$$2m(v + a\omega_1)a + \frac{1}{3}(2ma^2\omega_1) = m\left(2va + \frac{8}{3}a^2\omega_1\right)$$

and moment of I about B is $2aI$ in the same sense.

Equating these we have

$$m\left(v + \frac{4}{3}a\omega_1\right) = I \tag{3}$$

For the rod BC, the moment of momentum about B is $mb(u + b\omega_2)$ $+\frac{1}{3}mb^2\omega_2$.

Since, there is no impulse in this direction, we have

$$m\left(u + \frac{4}{3}b\omega_2\right) = 0$$

$$\therefore \quad u + \frac{4}{3}b\omega_2 = 0 \tag{4}$$

From (1) and (4), we get

$$u = 0, \omega_2 = 0.$$

From (2) and (3), we get

$$v = -\frac{1}{3m}, a\omega_1 = \frac{1}{m}.$$

Hence, the initial velocity of BC is $I/3m$ in the direction $\overrightarrow{CB}$.

The K.E. generated is

$$\frac{1}{2}(2m)(v+a\omega_1)^2+\frac{1}{2}(2m)\frac{a^2}{3}\omega_1^2+\frac{1}{2}mv^2$$

$$= m\left[\left(-\frac{I}{3m}+\frac{I}{m}\right)^2+\frac{I^2}{3m^2}+\frac{I^2}{18m^2}\right]=\frac{5I^2}{6m}.$$

Note: Since, the linear momentum gain is $\frac{I}{3}$ along CB, the impulsive action at B in BC is $\frac{I}{3}$ along CB and that on AB is equal, and opposite.

8.10 MOTION OF BODY INSIDE A BODY

Examples

1. A hollow cylinder of radius a, is fixed with its axis horizontal, inside it moves a cylinder of radius b, whose velocity in its lowest position is given; if the friction between the cylinders is sufficient to prevent any sliding. Discuss the motion.

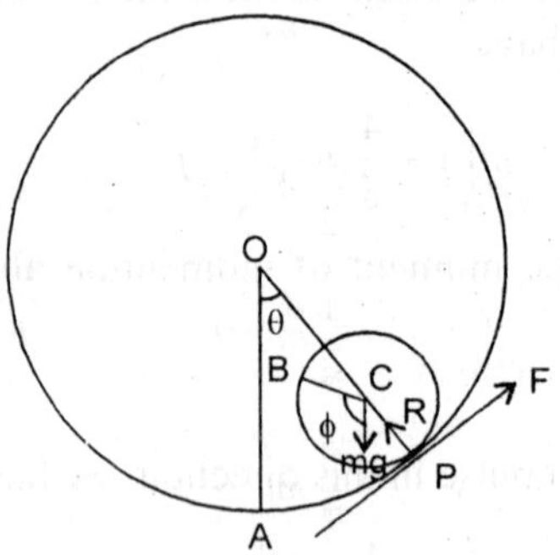

Fig. 8.20

The figure represents the vertical section through the centre of gravity C of the rolling cylinder.

Let ω be the angular velocity of the cylinder when it was projected from the lowest point. Suppose ϕ is the angle which a line CB (fixed in the moving body) makes with the vertical (a line fixed in space). Initially B coincides with A. Let F be the frictional force and R normal reaction at the point of contact P of two cylinders at time t. Let O be the centre of the outer cylinder and C that of the inner cylinder. Let $c = a - b$. Since, there is no slipping between the two cylinderes

$$\text{Arc } AP = \text{Arc } BP$$

$$\therefore \quad a\theta = b(\theta + \phi)$$

$$\therefore \quad (a-b)\theta = b\phi$$

$$c\theta = b\phi$$

$$\therefore \quad c\dot{\theta} = b\dot{\phi} \qquad (1)$$

C describes a circle of radius c with centre O.

Equations of motion of the inner cylinder are

$$mc\dot{\theta}^2 = R - mg\cos\theta \qquad (2)$$

$$mc\ddot{\theta} = F - mg\sin\theta \qquad (3)$$

where m is the mass of the inner cylinder.

Let us take horizontal and vertical lines through O as the co-ordinate axes.

The co-ordinates of C are $(c\sin\theta, c\cos\theta)$

Hence,

$$(\text{Velocity of } C)^2 = (c\cos\theta\dot{\theta})^2 + (c\sin\theta\dot{\theta})^2 = c^2\dot{\theta}^2.$$

Kinetic energy of the moving cylinder at time t is

$$= \frac{1}{2}mk^2\dot{\phi}^2 + \frac{1}{2}mc^2\dot{\theta}^2$$

$$= \frac{1}{2}m\frac{b^2}{2}\dot{\phi}^2 + \frac{1}{2}mc^2\dot{\theta}^2$$

$$= \frac{1}{4}mc^2\dot{\theta}^2 + \frac{1}{2}mc^2\dot{\theta}^2, \quad \text{using (1).}$$

$$= \frac{3}{4}mc^2\dot{\theta}^2$$

Kinetic energy at the time of projection is $\frac{3}{4}mc^2\Omega^2$ ($\because \dot{\theta} = \Omega$, initially)

From energy equation, we get

$$\frac{3}{4}mc^2(\dot{\theta}^2 - \Omega^2) = -mgc(1-\cos\theta)$$

$$\therefore \quad c\dot{\theta} = c\Omega^2 - \frac{4g}{3}(1-\cos\theta) \qquad (4)$$

Differentiating and dividing by $2\dot{\theta}$, we get

$$c\ddot{\theta} = -\frac{2g}{3}\sin\theta \qquad (5)$$

From (2) and (3), we get

$$R = mg\cos\theta + Mc\Omega^2 - \frac{4mg}{3}(1-\cos\theta)$$

$$= mc\Omega^2 + \frac{mg}{3}(7\cos\theta - 4) \qquad (6)$$

From (3) and (5), we get

$$F = mg\sin\theta - \frac{2mg}{3}\sin\theta$$

$$= \frac{mg}{3}\sin\theta. \qquad (7)$$

The equations (4), (6) and (7), determine the motion.

(i) When the cylinder makes complete revolution *i.e.*, when cylinder rolls round completely

$$R = 0,\ \theta = \pi.$$

From (6), we get

$$0 = mc\Omega^2 + \frac{mg}{3}(7\cos\pi - 4)$$

$$\therefore \qquad \Omega^2 = \frac{11g}{3c} = \frac{11g}{3(a-b)}.$$

Hence, the velocity of centre $C = (a-b)\Omega$

$$= (a-b)\sqrt{\frac{11g}{3(a-b)}} = \sqrt{\frac{11g(a-b)}{3}}.$$

This gives the position when two bodies separate.

(ii) Two bodies separate When cylinder does not roll completely, it will leave the outer cylinder when $R = 0$, hence, in that case we get from (4),

$$0 = mc\Omega^2 + \frac{mg}{3}(7\cos\theta - 4)$$

$$\therefore \qquad \cos\theta = \frac{4g - 3c\Omega^2}{7g} = \frac{4g - 3\Omega^2(a-b)}{7g}.$$

This is the least velocity of projection in order that cylinder may roll completely.

(iii) The cylinder will make small oscillations. If the rolling cylinder makes small oscillations about the lowest point of the fixed cylinder, then θ is always small. Then taking θ small, we get from (5),

$$c\ddot{\theta} = -\frac{2g}{3}\theta$$

$$\therefore \qquad \ddot{\theta} = -\frac{2g}{3c}\theta = -\frac{2g\theta}{3(a-b)}.$$

Hence, the time of oscillation is

$$\frac{2\pi}{\sqrt{\dfrac{2g}{3(a-b)}}} = 2\pi\sqrt{\frac{3(a-b)}{2g}}.$$

2. A circular cylinder of radius a and radius of gyration k rolls without slipping inside a fixed hollow cylinder of radius b. Show that the plane through their axes moves like a circular pendulum of length $(b-a)\left(1+\dfrac{k^2}{a^2}\right)$.

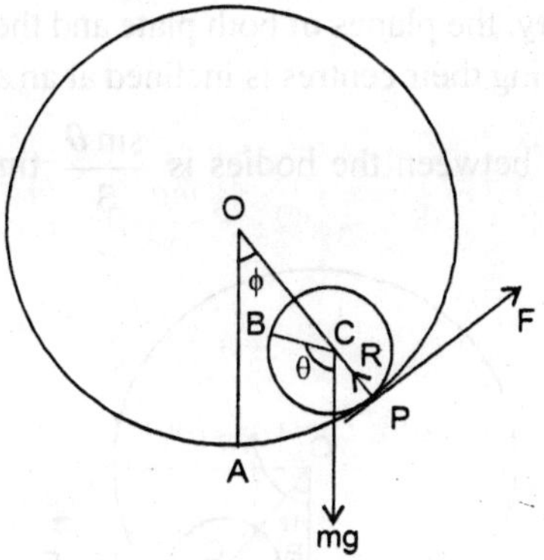

Fig. 8.21

The centre O of the outer cylinder is fixed. Let θ be the angle through which the plane of the axes turns, and ϕ the angle line fixed in moving cylinder makes with the vertical.

The outer cylinder is fixed. Equation of motion of the inner cylinder is

$$mk^2\ddot{\phi} = -Fa. \qquad (1)$$

Taking moments about C.G.

$$m(b-a)\ddot{\theta} = F - mg\sin\theta \qquad (2)$$

Since, there is no slipping, arc AP = Arc BP

$$\therefore \qquad b\theta = a(\theta+\phi)$$

$$\therefore \qquad a\ddot{\phi} = (b-a)\ddot{\theta} \qquad (3)$$

Eliminating F from (1) and (2), we get

$$m(b-a)\ddot{\theta} = -m\frac{k^2}{a}\ddot{\phi} - mg\sin\theta$$

$$\therefore \quad m\left(1+\frac{k^2}{a^2}\right)(b-a)\ddot{\theta} = -mg\sin\theta .$$

When θ is small, then we get

$$\therefore \quad \ddot{\theta} = \frac{-g}{\left(1+\frac{k^2}{a^2}\right)(b-a)}\theta.$$

Hence, the length of S.E.P. is

$$(b-a)\left(1+\frac{k^2}{a^2}\right).$$

3. A circular plate rolls down the inner circumference of a rough circle under the action of gravity, the planes of both plate and the circle being vertical. When the line joining their centres is inclined at an angle θ to the vertical, show that friction between the bodies is $\frac{\sin\theta}{3}$ times the weight of the plate.

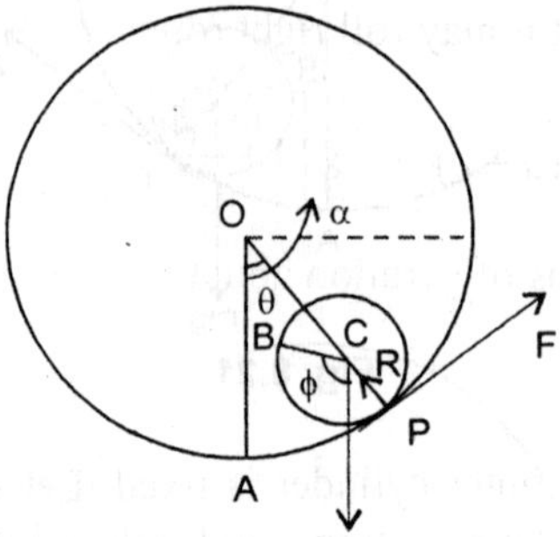

Fig. 8.22

Let O be the centre of the fixed circle and a the radius, C the centre and b the radius of the circular plate that rolls down.

At time t let the radius BC (a line fixed in the body) make an angle ϕ with the vertical OA (line fixed in the space), suppose initially B coincides with a fixed point Q on the fixed circle, where OQ is inclined at an angle α to the vertical OA.

Since, there is no slipping, arc QP = arc BP

$$\therefore \quad a(\alpha-\theta) = b[2\pi-(\theta+\phi)]$$

$$\therefore \quad -a\dot{\theta} = -b(\dot{\theta}+\dot{\phi})$$

$$\therefore \quad b\ddot{\phi} = (a-b)\ddot{\theta} \qquad (1)$$

Equation of motion of the plate are

$$m(a-b)\ddot{\theta} = F - mg\sin\theta \tag{2}$$

$$m\frac{b^2}{2}\ddot{\phi} = -Fb$$

$$\therefore \quad m\ddot{\phi} = -\frac{2F}{b} \tag{3}$$

From (1) and (3), we get

$$-2F = m(a-b)\ddot{\theta}.$$

With the help of (2), we get

$$-2F = F - mg\sin\theta$$

$$\therefore \quad 3F = mg\sin\theta$$

$$\therefore \quad F = \frac{W}{3}\sin\theta = \frac{\sin\theta}{3} \text{ times weight.}$$

4. A cylinder of radius a lies within a rough fixed cylindrical cavity of radius $2a$. The centre of gravity of the cylinder is at a distance c from the axis, and the initial state is that of stable equilibrium at the lowest point of the cavity. Show that the smallest angular velocity with which the cylinder must be started that it may roll right round the cavity is given by

$$\Omega^2(a+c) = g\left[1+\frac{4(a+c)^2}{(a-c)^2+k^2}\right],$$

where k is the radius of gyration about the centre of gravity.

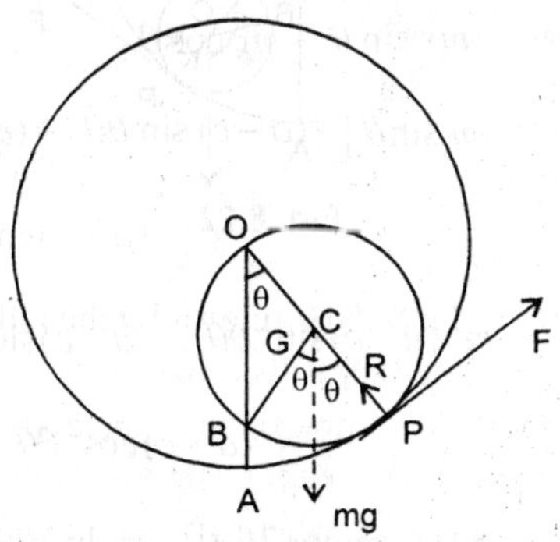

Fig. 8.23

Let O be the centre of fixed cavity whose radius is given $2a$, C the centre of moving cylinder whose radius is given a. Let at time t radius CB (a line fixed in moving body) make an angle θ with the vertical (a line fixed in space), then each of the two angles marked in the figure are also equal to θ. Initially B coincides with A; it is clear that B will also lie on the vertical OA.

Let the horizontal and vertical lines through the fixed point be taken as co-ordinate axes.

If (x, y) be the co-ordinates of centre of gravity G, then

$$x = (a-c)\sin\theta,\ y = (a+c)\cos\theta$$

$$\therefore \quad \dot{x} = (a-c)\cos\theta\dot{\theta},\ \dot{y} = -(a+c)\sin\theta\dot{\theta}$$

$$\ddot{x} = -(a-c)\sin\theta\dot{\theta}^2 + (a-c)\cos\theta\ddot{\theta},$$

$$\ddot{y} = -(a+c)\cos\theta\dot{\theta}^2 - (a+c)\sin\theta\ddot{\theta}.$$

$$\therefore \quad \dot{x}^2 + \dot{y}^2 = (a-c)^2\cos^2\theta\dot{\theta}^2 + (a+c)^2\sin^2\theta\dot{\theta}^2$$

$$= (a^2 + c^2 - 2ac\cos 2\theta)\dot{\theta}^2.$$

Hence, at time t

$$(\text{velocity})^2 \text{ of } C = (a^2 + c^2 - 2ac\cos 2\theta)\dot{\theta}^2.$$

At the lowest point $\theta = 0, \dot{\theta} = \Omega$.

Hence, at the lowest point $(\text{velocity})^2$ of $C = (a-c)^2\Omega^2$

From energy equation, we get

$$\frac{1}{2}m\left[k^2 + a^2 + c^2 - 2ac\cos 2\theta\right]\dot{\theta}^2 - \frac{1}{2}m[k^2 + (a-c)^2]\Omega^2$$

$$= -mg[(a+c) - (a+c)\cos 2\theta]$$

$$= -mg(a+c)(1 - \cos 2\theta) \qquad (1)$$

Equation of motion of the moving cylinder is

$$R - mg\cos\theta = -m\ddot{x}\sin\theta - m\ddot{y}\cos\theta$$

$$= -m\sin\theta\left[-(a-c)\sin\theta\dot{\theta}^2 + (a-c)\cos\theta\ddot{\theta}\right] - m\cos\theta \times \left[-(a+c)\cos\theta\dot{\theta}^2 - (a+c)\sin\theta\ddot{\theta}\right]$$

$$= m\left[(a-c)\sin^2\theta\dot{\theta}^2 - (a-c)\sin\theta\cos\theta\ddot{\theta} + (a+c)\cos^2\theta\dot{\theta}^2 + (a+c)\sin\theta\cos\theta\ddot{\theta}\right]$$

$$= m[(a + c\cos 2\theta)\dot{\theta}^2 + c\sin 2\theta\ddot{\theta}]$$

$$\therefore \quad R = mg\cos\theta + m\left[(a + c\cos 2\theta)\dot{\theta}^2 + c\sin 2\theta\ddot{\theta}\right] \qquad (2)$$

The cylinder will roll round if $R = 0$, when $\theta = \pi$.

From (2), we get

$$0 = -mg + m\left[(a+c)\dot{\theta}^2\right]$$

$$\therefore \qquad g = (a+c)\dot{\theta}^2.$$

and from (1)

$$\frac{m}{2}[k^2 + a^2 + c^2 - 2ac]\,\dot{\theta}^2 - \frac{m}{2}[k^2 + (a-c)^2]\Omega^2 = -2mg(a+c)$$

$$\therefore \left[k^2 + (a-c)^2\right]\dot{\theta}^2 - \left[k^2 + (a-c)^2\right]\Omega^2 = -4g(a+c).$$

Eliminating $\dot{\theta}^2$ from these equations, we get

$$[k^2 + (a-c)^2]\frac{g}{(a+c)} - [k^2 + (a-c)^2]\Omega^2 = -4g(a+c).$$

$$\therefore \quad \left[k^2 + (a-c)^2\right](a+c)\Omega^2 = \left[k^2 + (a-c)^2 + 4(a+c)^2\right]$$

$$\therefore \qquad (a+c)^2\,\Omega^2 = 1 + \frac{4(a+c)^2}{k^2 + (a-c)^2}.$$

8.11 COMPOUND PENDULUM

Definition : *A rigid body oscillating freely under gravity about a fixed horizontal axis is known as a compound pendulum.*

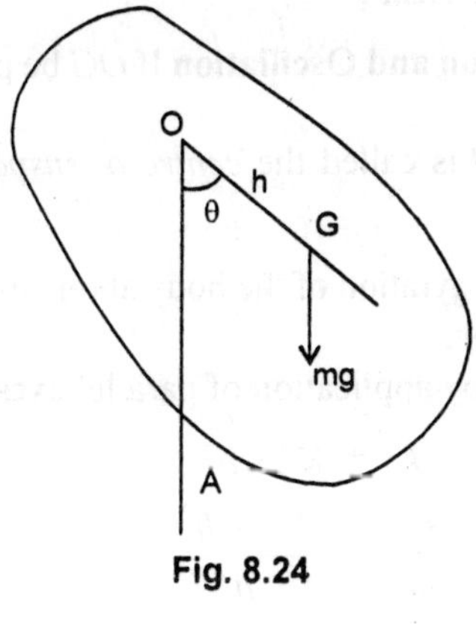

Fig. 8.24

(1) Let AOG be the vertical plane through G, the centre of gravity of the body which is at right angles to the axis of rotation through O. Let $OG(=h)$ make an angle θ with the vertical OA. If M be the mass and k the radius of gyration of the body about the axis of rotation, then the equation of motion is $Mk^2\dfrac{d^2\theta}{dt^2}$ = Sum of moments of the external forces about the axis of rotation + moment of the reaction at O about axis of rotation

$$= -Mgh\sin\theta + 0$$

If θ is small then, we get

$$\frac{d^2\theta}{dt^2} = -\frac{gh}{k^2}\theta.$$

This shows that the motion is Simple Harmonic and hence, period of oscillation is

$$2\pi\Big/\sqrt{\frac{gh}{k^2}} = 2\pi\sqrt{\frac{k^2}{gh}}.$$

(2) **Simple Equivalent Pendulum.** We know that the time of oscillation of simple pendulum of length l is $2\pi\sqrt{\frac{l}{g}}$. If the time of oscillation of the compound pendulum be equal to the time of oscillation of simple pendulum, then $\frac{k^2}{h} = l$.

A simple pendulum which oscillate in the same time as the compound pendulum, is called *Simple Equivalent Pendulum* (S.E.P.)

Hence, the length of S.E.P. $= \frac{k^2}{h}$.

(3) **Centre of Suspension and Oscillation** If OG be produced to O' such that $OO' = \frac{k^2}{h}$, then O is called the *centre of suspension* and O' is called *centre of oscillation.*

If K be the radius of gyration of the body about an axis through G parallel to the axis of rotation,

then since $OG = h$, by application of parallel axes theorem

$$k^2 = K^2 + h^2$$

$$\therefore \quad \frac{k^2}{h} = \frac{K^2 + h^2}{h}$$

$$\therefore \quad OO' = \frac{K^2 + h^2}{h}$$

$$\therefore \quad OO'.OG = K^2 + h^2 = K^2 + (OG)^2$$

$$\therefore \quad OG\,(OO' - OG) = K^2$$

$$\therefore \quad OG.OG' = K^2.$$

The symmetry of this relation shows that if the body were suspended from a parallel axis through O', then O would be the centre of oscillation. Hence, the centre of suspension and oscillation are interchangeable.

8.12 PRESSURE ON THE AXIS OF ROTATION

Let us assume that the body and the forces are symmetrical with respect to the plane through the centre of gravity perpendicular to the axis of rotation *i.e.*, with respect to the plane of the paper and let weight be the only force.

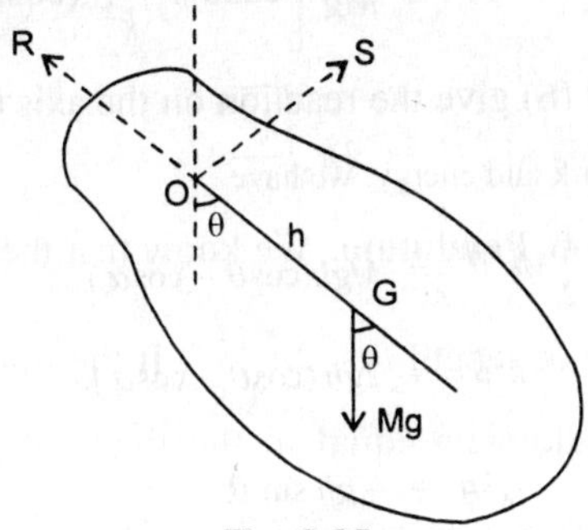

Fig. 8.25

Let R, S be the components of the reaction at O along and perpendicular to GO.

Since, G describes a circle of radius $OG = h$ and centre O, the components of acceleration are $h\dot{\theta}^2$ along GO and $h\ddot{\theta}$ perpendicular to GO in the direction of θ increasing.

Hence, equations of motion of G along and perpendicular to GO are

$$Mh\dot{\theta}^2 = R - Mg\cos\theta \qquad (1)$$

$$Mh\ddot{\theta} = S - Mg\sin\theta \qquad (2)$$

Equation of motion of the body about the axis of ratation is

$$Mk^2\ddot{\theta} = -Mgh\sin\theta \qquad (3)$$

Substituting the value of $\ddot{\theta}$ from (3) in (2), we get

$$S = Mg\sin\theta\left(1 - \frac{h^2}{k^2}\right) \qquad (4)$$

Multiplying (3) by $2\dot{\theta}$ and integrating, we get

$$k^2\dot{\theta}^2 = 2\,gh\cos\theta + C.$$

Initially $\dot{\theta} = 0, \theta = \alpha$, where α is the amplitude of the rotation.

$$\therefore \qquad C = -2gh\cos\alpha.$$

$$\therefore \qquad k^2\dot{\theta}^2 = 2gh(\cos\theta - \cos\alpha) \qquad (5)$$

Substituting the value of $\dot{\theta}^2$ from (5) and in (1), we get

$$R = Mg\cos\theta + 2Mg\frac{h^2}{k^2}(\cos\theta - \cos\alpha)$$

$$= Mg\left[\cos\theta + \frac{2h^2}{k^2}(\cos\theta - \cos\alpha)\right] \quad (6)$$

Equations (4) and (6) give the reaction on the axis of rotation.

Note : By principle of work and energy, we have

$$\frac{1}{2}Mk^2\dot{\theta}^2 = Mgh(\cos\theta - \cos\alpha)$$

$$\therefore \quad k^2\dot{\theta}^2 = 2gh(\cos\theta - \cos\alpha).$$

Differentiating, we have

$$k^2\ddot{\theta} = -gh\sin\theta.$$

These equations are (5) and (3) respectively.

8.13 CENTRE OF PERCUSSION

If a body rotating about a given axis, is so struck that there is no impulsive reaction on the axis, then any point on the line of action of the force is called a *centre of percussion*. The line of action of the force is called the *line of percussion*.

Again when the line of action of the blow is given, the axis about which the body begins to turn is called *axis of spontaneous rotation*.

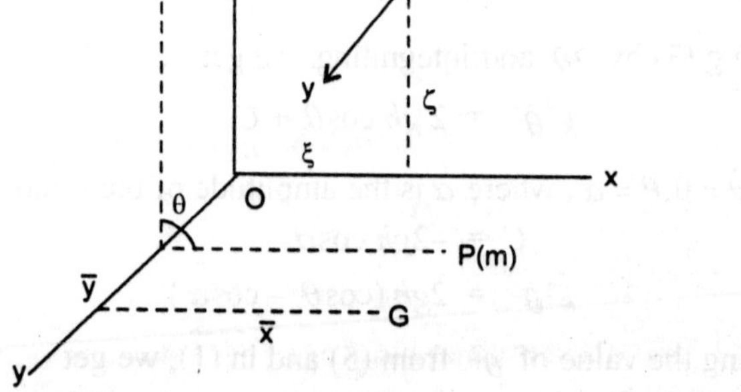

Fig. 8.26

Let fixed axis be taken as y axis, let the plane of xy pass through the instantaneous position of the centre of inertia G, and let the plane through the point of application Q of the blow perpendicular to the fixed axis be the plane of xz, so that G is a point $(\bar{x}, \bar{y}, 0)$ and Q be the point $((\xi, 0, \varsigma)$.

Let components of the blow parallel to the axes be X, Y and Z, and let us assume that there is no action on the axis of rotation.

Let u, v, w be the components of velocity before the blow and u', v', w', after the blow, ω is the angular velocity before the blow and ω' after the blow. Then equations of motion are

$$X = \sum m(u'-u) = \sum mz(\omega'-\omega), \text{ (since } u = z\omega, u' = z\omega')$$

$$= (\omega' - \omega) M\bar{z} = 0, \text{ as } \bar{z} = 0.$$

$$\therefore \quad X = 0 \tag{1}$$

$$Y = \sum m(v'-v) = 0 \tag{2}$$

$$\text{as } v = v' = 0.$$

$$Z = \sum m(w'-w) = -(\omega'-\omega) \sum m\bar{x} \text{ since } w = -x\omega, w' = -x\omega'.$$

$$= -(\omega'-\omega) M\bar{x} \tag{3}$$

$$-\varsigma Y = \sum m [y (w' - w) - z (v' - v)]$$

$$= -(\omega'-\omega) \sum mxy - F (\omega'-\omega)$$

$$\therefore \quad F = 0, \tag{4}$$

since, $Y = 0$.

$$\varsigma X - \xi Z = \sum m [z (u' - u) - x (w' - w)]$$

$$= (\omega'-\omega) \sum m (z^2 + x^2)$$

$$= (\omega'-\omega) Mk^2 \tag{5}$$

where Mk^2 is the moment of inertia of the body about y axis and

$$\xi Y = \sum m [x (v' - v) - y(u' - u)]$$

$$= -(\omega'-\omega) \sum myz$$

$$= -D(\omega'-\omega)$$

$$\therefore \quad D = 0, \tag{6}$$

since, $Y = 0$.

Equations (1) and (2) show that blow must have no components parallel to the axes of x and y, *i.e.*, it must be perpendicular to the plane through the fixed axis and instantaneous position of the centre of inertia (4) and (6) give F = 0 = D, so that the fixed axis must be a principal axis of the body at the origin, *i.e.*, at the point where the plane through the line of action of the blow perpendicular to the axis cuts it.

This is the essential condition for the existence of centre of percussion. Hence, if the fixed axis is not a principal axis at some point of its length, there is no centre of percussion. If it is a principal axis at only point of its length, then blow must act in the plane through this point perpendicular to the axis of rotation.

Finally (3) and (5) give $\xi = \dfrac{k^2}{\bar{x}}$.

It follows that when a centre of percussion does not exist, its distance from the fixed axis is the same as that of the centre of oscillation for the case when the body oscillates freely about the fixed axis as a horizontal axis of suspension.

Note : We note that the direction of the blow passes through centre of oscillation. Thus for a lamina suspended freely and capable of ratation about a principal axis the centre of oscillation is the centre of percussion.

Examples

1. Find the time of oscillation of a compound pendulum of a rod of mass m and length a, carrying at the end a sphere of mass M and radius b, the other end of the rod being fixed.

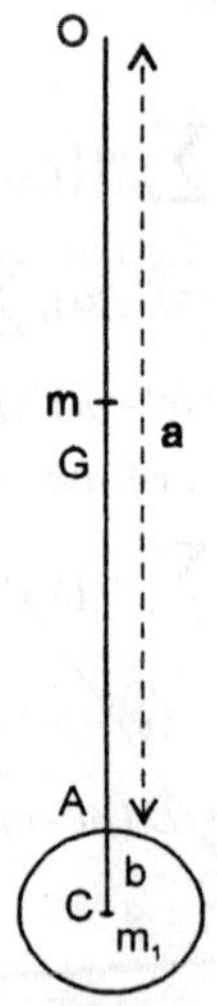

Fig. 8.27

Let OA be the rod of length a, carrying at A a sphere of radius b.

M. I. of the rod OA about the axis $OX = \frac{ma^2}{3}$.

M. I. of the sphere about the axis OX

$$= \text{M.I. about a parallel diameter} + M.(OB)^2$$

$$= M\frac{2}{5}b^2 + M(a+b)^2.$$

Thus if k be the radius of gyration of the compound pendulum, then

$$(M+m)k^2 = m\frac{a^2}{3} + M\left[\frac{2}{5}b^2 + (a+b)^2\right].$$

If G be the mid point of OA and G', the C.G. of the compound pendulum, then

$$h = OG' = \frac{mOG + MOB}{M+m} = \frac{m\frac{a}{2} + M(a+b)}{M+m}$$

Hence, the length of S.E.P. $l = \frac{k^2}{h} = \frac{m\frac{a^2}{3} + M\left[\frac{2}{5}b^2 + (a+b)^2\right]}{m\frac{a}{2} + M(a+b)}$

Time of oscillation $= 2\pi\sqrt{\frac{l}{g}}$.

2. A heavy homogeneous cube of weight w can swing about an edge which is horizontal, it starts from rest being displaced from its position of unstable equilibrium; when the perpendicular from the centre of gravity upon the edge has turned through an angle θ, find the components of reaction at the hinge along and right angle to this perpendicular.

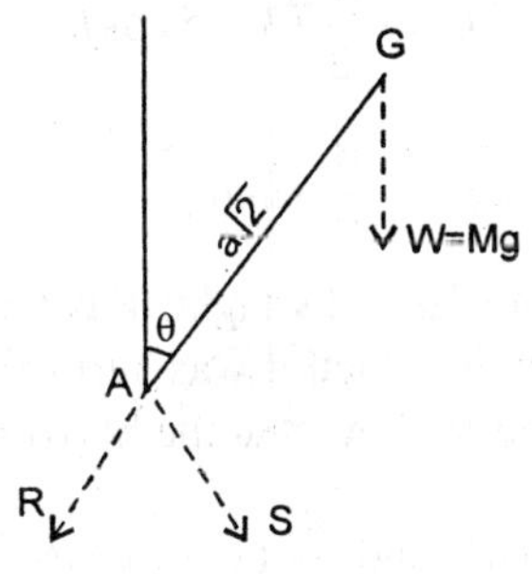

Fig. 8.28

If $2a$ be the edge of the cube, distance of centre of gravity G from the edge at A about which the cube rotates $= a\sqrt{2}$. Since, G describes a circle with centre A and radius AG, its components of acceleration are $a\sqrt{2}\dot{\theta}^2$ along GA and $a\sqrt{2}\ddot{\theta}$ perpendicular to GA. If R, S be the components of reaction at A, along and perpendicular to GA, then the equations of motion of G are

$$ma\sqrt{2}\dot{\theta}^2 = mg\cos\theta + R \tag{1}$$

$$ma\sqrt{2}\ddot{\theta} = mg\sin\theta + S \tag{2}$$

Equation of motion for the rotation about A is

$$m\left(2a^2 + \frac{2a^2}{3}\right)\ddot{\theta} = mga\sqrt{2}\sin\theta$$

$$\therefore \quad \ddot{\theta} = \frac{3\sqrt{2}g}{8a}\sin\theta \tag{3}$$

Multiplying (3) by $2\dot{\theta}$ and integrating, we get

$$\dot{\theta}^2 = -\frac{3\sqrt{2}g}{4a}\cos\theta + C$$

Initially, $\theta = 0$, $\dot{\theta} = 0$.

$$\therefore \quad C = \frac{3\sqrt{2}}{4a}g$$

$$\therefore \quad \dot{\theta}^2 = \frac{3\sqrt{2}g}{4a}(1 - \cos\theta) \tag{4}$$

Substituting the values of $\dot{\theta}^2$ and $\ddot{\theta}$ from (3) and (4) in (1) and (2), get

$$R = \frac{W}{2}(3 - 5\cos\theta)$$

$$S = \frac{W}{4}\sin\theta.$$

3. A uniform rod of length $2a$ and weight W is turning about its end O and starts from the position in which it was vertically above O. When it has turned through an angle θ, show that the horizontal and vertical reactions are $\frac{3}{4}W\sin\theta(2 - 3\cos\theta)$ and $\frac{W}{4}(1 - 3\cos\theta)^2$.

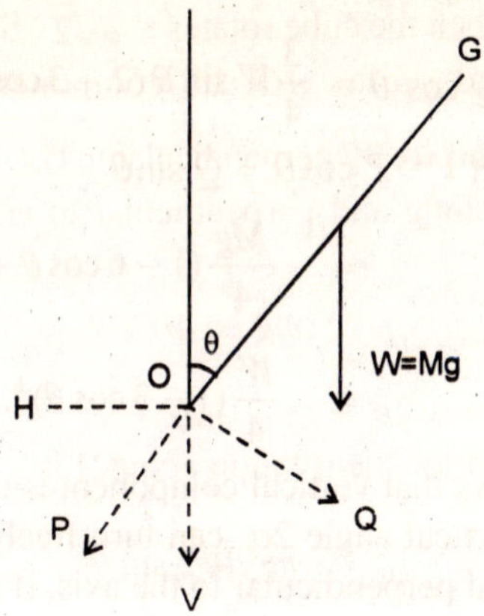

Fig. 8.29

Here Mk^2 = M.I. about axis of rotation = $\frac{4}{3}Ma^2$.

The C.G. is describing a circle of radius a about O. If P and Q be the normal reactions along GO and perpendicular to GO, then equations of motion of C.G. are

$$Ma\ddot{\theta} = Q + Mg\sin\theta \quad (1)$$

$$Ma\dot{\theta}^2 = P + Mg\cos\theta \quad (2)$$

Taking moments about the axis of rotation, we have

$$M\frac{4}{3}a^2\ddot{\theta} = Mga\sin\theta$$

$$\therefore \quad \ddot{\theta} = -\frac{3g}{4a}\sin\theta \quad (3)$$

Integrating and using initial conditons, $\theta = 0, \dot{\theta} = 0$, we get

$$\dot{\theta}^2 = \frac{3g}{2a}(1 - \cos\theta) \quad (4)$$

Substituting the value of $\ddot{\theta}$ from (3) in (1), we get

$$Q = \frac{Mg}{4}\sin\theta.$$

Similarly from (2) and (4), we get

$$P = \frac{Mg}{2}(3 - 5\cos\theta).$$

Now horizontal component

$$H = P\sin\theta - Q\cos\theta = \frac{3}{4}W\sin\theta(2 - 3\cos\theta).$$

Vertical component $V = P\cos\theta + Q\sin\theta$

$$= -\frac{Mg}{4}(1 - 6\cos\theta + 9\cos^2\theta)$$

$$= -\frac{W}{4}(1 - 3\cos\theta)^2.$$

Negative sign shows that vertical component is upwards.

4. A right cone of vertical angle 2α, can turn freely about an axis, passing through its base and perpendicular to the axis. If the cone starts from rest with its axis horizontal, show that when its axis is vertical, thrust on the fixed axis is to the weight of the cone as $1 + \frac{1}{2}\cos^2\alpha$ to $1 - \frac{1}{3}\cos^2\alpha$.

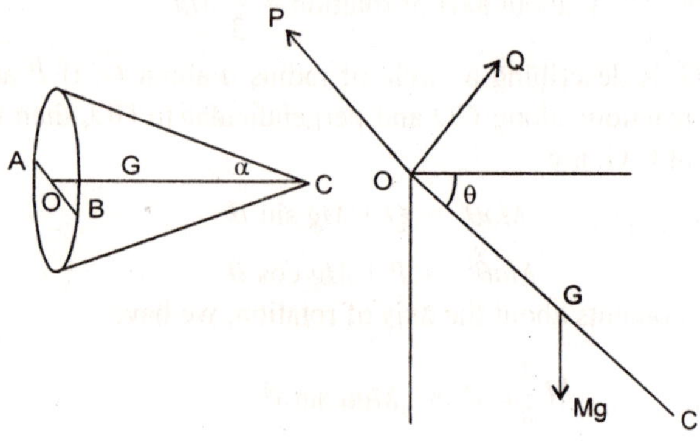

Fig. 8.30

The cone will turn around a horizontal diameter AB of the base

$$Mk^2 = \text{M. I. about } AB = \frac{3}{80}M\left(h^2 + 4a^2\right) + M\frac{h^2}{16},$$

where a is the radius of the base and h the height of the cone and $OG = \frac{h}{4}$.

$$\therefore \qquad k^2 = \frac{h^2}{20}(2 + 3\tan^2\alpha).$$

Let after time t the axis of the cone make an angle θ with the horizontal, then since, the centre of gravity G is describing a circle of radius $\frac{h}{4}$, the equations of motion of G are

$$M\frac{h}{4}\dot{\theta}^2 = P - Mg\sin\theta \qquad (1)$$

$$M\frac{h}{4}\ddot{\theta} = Mg\cos\theta - Q \qquad (2)$$

where P and Q are the components of reaction at O along and perpendicular to GO.

Taking moment about O, we get

$$Mk^2\ddot{\theta} = Mg\frac{h}{4}\cos\theta$$

$$\therefore \qquad k^2\ddot{\theta} = \frac{gh}{4}\cos\theta. \qquad (3)$$

Integrating and using initial conditions, $\theta = 0$, $\dot{\theta} = 0$, we get

$$k^2\dot{\theta}^2 = \frac{1}{2}gh\sin\theta \qquad (4)$$

From (1) and (4), we get

$$P = Mg\sin\theta\left(1 + \frac{h^2}{8k^2}\right)$$

From (2) and (3),

$$Q = Mg\cos\theta\left(1 - \frac{h^2}{16k^2}\right)$$

When the axis of the cone is vertical *i.e.*, when

$$\theta = \frac{\pi}{2},\ Q = 0,\ P = Mg\left(1 + \frac{h^2}{8k^2}\right) - Mg\left(\frac{8k^2 + h^2}{8k^2}\right)$$

Putting the value of k^2, we have

$$\frac{P}{Mg} = \left[\frac{8.\dfrac{h^2}{20}(2 + 3\tan^2\alpha) + h^2}{8.\dfrac{h^2}{20}(2 + 3\tan^2\alpha)}\right]$$

$$= \frac{9 + 6\tan^2\alpha}{4 + 6\tan^2\alpha}$$

$$= \frac{3}{2}\left[\frac{3\cos^2\alpha + 2\sin^2\alpha}{2\cos^2\alpha + 3\sin^2\alpha}\right]$$

$$= \frac{3(\cos^2\alpha + 2)}{2(2 + \sin^2\alpha)}$$

$$= \frac{1 + \frac{1}{2}\cos^2\alpha}{1 - \frac{1}{3}\cos^2\alpha}.$$

Note : If the cone turns about a fixed horizontal axis through the vertex, then distance from C. G. is $\frac{3h}{4}$. So writing $\frac{3h}{4}$ instead of $\frac{h}{4}$ and changed value of k^2 then we get same value of $\frac{P}{Mg}$.

Particular case if $2\alpha = \frac{\pi}{2}$,

$$\frac{P}{Mg} = \frac{1 + \frac{1}{4}}{1 - \frac{1}{6}} = \frac{3}{2}$$

Hence, the thrust on the fixed axis $= \frac{3}{2}Mg$.

5. Find the centre of percussion of a rod.
   Let a rod OA of length $2a$, suspended freely from one end O, be struck by a horizontal blow P at a point C, where $OC = x$.

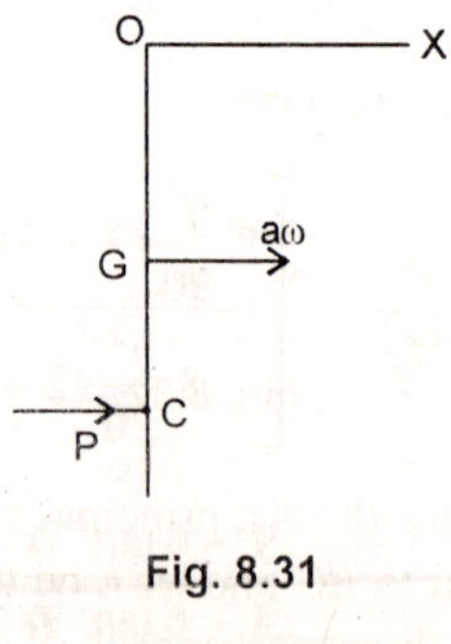

Fig. 8.31

Let ω be the instantaneous angular velocity communicated to the rod and

X the impulsive action at O. The velocity of the centre of gravity G after the blow is $a\omega$.

Hence, equating the change in momentum in the horizontal direction to the total external impulse in that direction, we get

$$M(a\omega - 0) = P + X$$

$$\therefore \quad Ma\omega = P + X \tag{1}$$

Moment equation gives

$$Mk^2\omega = P.x \tag{2}$$

where k is the radius of gyration of the rod about O, $\therefore\ k^2 = \dfrac{4a^2}{3}$

If the blow be given through the centre of percussion, there will be no impulse on the axis, *i.e.*, X = 0. Thus (1) gives

$$Ma\omega = P \tag{3}$$

Eliminating P from (2) and (3), we obtain

$$x = \frac{k^2}{a} = \frac{4a}{3} \text{ the length of S.E.P.}$$

If x be not equal to $\dfrac{k^2}{a}$, then

$$X = Ma\omega - M\frac{k^2}{x}\omega = Ma\omega\frac{x - \dfrac{k^2}{a}}{x} \tag{4}$$

It is positive or Negative according as x is greater then or less then $\dfrac{k^2}{a}$ *i.e.*, impulsive action at O on the body is in the same or opposite direction as the blow, according as the blow is applied at a point below or above the centre of percussion.

6. Find the centre of percussion of a triangle ABC which is free to turn about its side BC.

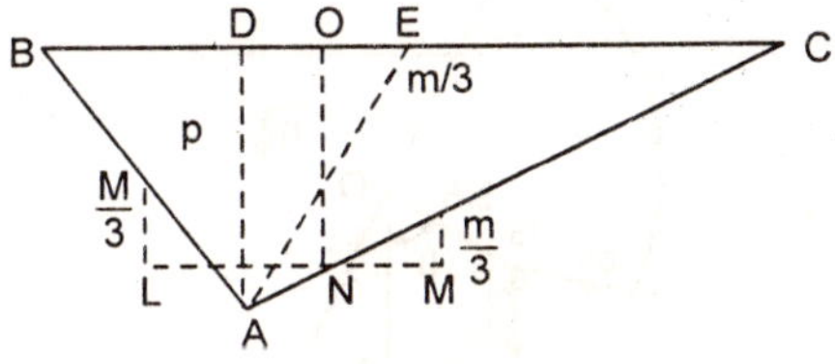

Fig. 8.32

Let us find a point where BC is a principal axis of the triangle.
Draw AD perpendicular to BC and let E be the middle point of BC.
Replace the triangle by three masses each of mass $m/3$ at the middle points of the sides of triangle, where m is the mass of the triangle.

Draw OYN perpendicular to BC, where O is the middle point of ED.
Now P.I. of the triangle about OC and OY

$$= \frac{m}{3}(-OE.O) + \frac{m}{3}(ON.NM) + \frac{m}{3}ON(-LN) = 0,$$

as $NM = LN$.

Thus we see that P. I. of the triangle vanishes about BC and a $\perp$ line through Q. Hence, BC is a principal axis of the triangle at O.

Let $AD = p$.

Now

$$Mk^2 = \text{M. I. of the triangle about } BC$$

$$= \frac{m}{3}\left[\left(\frac{1}{2}p\right)^2 + \left(\frac{1}{2}p\right)^2 + 0\right] = \frac{1}{6}mp^2.$$

$$\therefore \qquad k^2 = \frac{1}{6}p^2, h = \frac{p}{3}.$$

Hence, depth of centre of percussion below C along vertical through O is

$\frac{k^2}{h} = \frac{p}{2}$, which is the middle point of AE.

Miscellaneous Examples

1 A uniform heavy solid hemisphere of radius a is held at rest with its axis vertical and its curved surface in contact with horizontal table. If the hemisphere is released when the plane in rough to prevent sliding, show that the angle θ that the base makes with the horizontal at time t is such that

$$\dot{\theta}^2 = \frac{15g\cos\theta}{a(28 - 15\cos\theta)}.$$

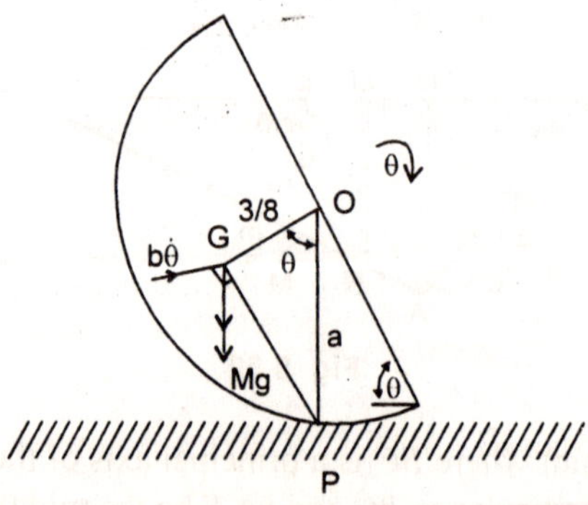

Fig. 8.33

Let M be the mass of hemisphere, the centroid G, O the centre of the base, so that $OG = \frac{3}{8}a$. Let $PG = b$. Since, $\dot{\theta}$ is the angular velocity of the hemisphere in the sense of θ increasing, and P is the instantaneous centre, $b\dot{\theta}$ is the velocity of G (perpendicular to PG), so the kinetic energy of the body is

$$T = \frac{1}{2}M(b\dot{\theta})^2 + \frac{1}{2}I_G\,\dot{\theta}^2.$$

Now

$$I_G = I_0 - M\left(\frac{3a}{8}\right)^2 = \frac{2}{5}Ma^2 - M\left(\frac{3a}{8}\right)^2$$

$$= \frac{83Ma^2}{320}.$$

Since,

$$b^2 = a^2 + \left(\frac{3a}{8}\right)^2 - 2a\left(\frac{3a}{8}\right)\cos\theta$$

$$= \frac{a^2}{64}(73 - 48\cos\theta).$$

$$\therefore \qquad T = \frac{Ma^2\dot{\theta}^2}{40}(28 - 15\cos\theta)$$

P.E. referred to the horizontal through P as zero level, is

$$V = Mga\left(1 - \frac{3}{8}\cos\theta\right).$$

From principle of conservation of energy

$$T + V = const.$$

$$\therefore a\dot{\theta}^2(28 - 15\cos\theta) + 40g\left(1 - \frac{3}{8}\cos\theta\right) = const.$$

Initially $\dot{\theta} = 0$, $\theta = \frac{\pi}{2}$.

$$\therefore const. = 40g.$$

$$\therefore a\dot{\theta}^2(28 - 15\cos\theta) + 40g\left(1 - \frac{3}{8}\cos\theta\right) = 40g.$$

$$\therefore \qquad \dot{\theta}^2 = \frac{15g\cos\theta}{a(28 - 15\cos\theta)}.$$

2. A frame made of thin wire of the line density ρ has the shape of a circle of radius a together with a diameter AB of the circle. The frame lies on a smooth horizontal table and is free to rotate about a fixed vertical axis through a point O on the circumference of the circle and on the diameter normal to AB. An insect of mass $\frac{1}{3}a\rho$, initially at rest at A begins to crawl along the diameter AB. If ϕ is the angle turned backwards by the frame in space after the insect has a crawled a distance x along AB, show that

$$\{n^2 a^2 + (a - x)^2\}\dot{\phi} = a\dot{x},$$

where
$n^2 = 12\pi + 9$. Prove also that when the insect reaches B, the frame will have rotated through an angle $\frac{1}{n}\tan^{-1}\left(\frac{2n}{n^2-1}\right)$.

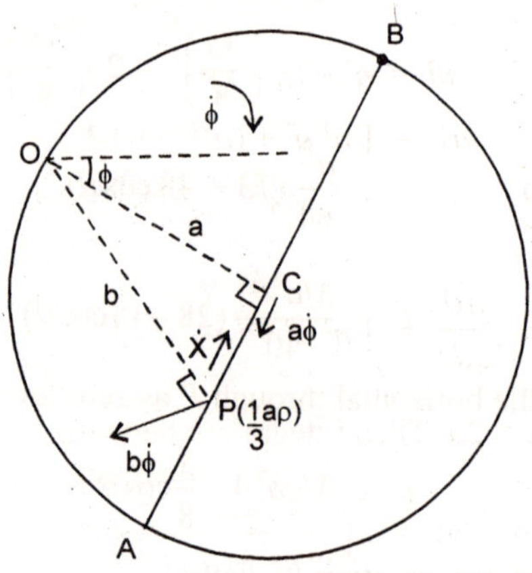

Fig. 8.34

Let the insect starts from A and after time t is at the point P, where $AP = x$. Since, spin of the frame is $\dot{\phi}$, so that the velocity of C, the centre of the circle is $a\dot{\phi}$ along $\overrightarrow{CA}$ while the insect has velocity $\dot{x}$ along $\overrightarrow{PC}$ together with $b\dot{\phi}$ along the direction perpendicular to OP in the sense of ϕ increasing ($OP = b$). Now the mass of the frame is $2a\rho(\pi + 1)$ and the M.I. about CO is

$$(2a\pi\rho)a^2 + \frac{1}{3}(2a\rho)a^2 = 2a^3\rho\left(\pi + \frac{1}{3}\right).$$

Hence, the angular momentum of the frame about O in the direction of ϕ increasing is

$$a\left[2a\rho(\pi+1)a\dot{\phi}\right]+2a^3\rho\left(\pi+\frac{1}{3}\right)\dot{\phi}=2a^3\rho\dot{\phi}\left(2\pi+\frac{4}{3}\right).$$

The angular momentum of the particle about O in the direction of ϕ increasing is

$$\frac{1}{3}a\rho(b^2\dot{\phi}-a\dot{x}) = \frac{a\rho}{3}\{(a-x)^2\dot{\phi}+a^2\dot{\phi}-a\dot{x}\}$$

Since, the moment about O of the external forces on the entire system comprising frame and the insect is zero, the total angular momentum about this point is constant and equal to zero, as the system is initially at rest.

$$\therefore\ 2a^3\rho\dot{\phi}\left(2\pi+\frac{4}{3}\right)+\frac{1}{3}a\rho\left[(a-x)^2+a^2\right]\dot{\phi}=\frac{1}{3}a^2\rho\dot{x}$$

since,

$$b^2 = a^2+(a-x)^2$$

$$\therefore\qquad a\dot{x} = [\,n^2a^2+(a-x)^2\,]\,\dot{\phi}$$

where $n^2 = 12\pi + 9$.

Now

$$\frac{d\theta}{adx} = [\,n^2a^2+(a-x)^2\,]^{-1}.$$

Let $\dot{\phi}=\phi_0$, when $x = 2a$. Then integrating between $x = 0$ and $x = 2a$

$$\phi_0 - 0 = \int_0^{2a}\frac{adx}{a^2n^2+(a-x)^2}.$$

$$= -\frac{a}{an}\left[\tan^{-1}\frac{a-x}{an}\right]_0^{2a}$$

$$= \frac{2}{n}\tan^{-1}\frac{1}{n}$$

$$= \frac{1}{n}\tan^{-1}\frac{\frac{2}{n}}{1-\frac{1}{n^2}}$$

$$= \frac{1}{n}\tan^{-1}\frac{2n}{n^2-1}$$

3. A sphere of radius a whose centre of gravity is at a distance c from its centre C is placed on a rough plane so that CG is horizontal, show that it will begin to roll or slide according as $\mu >$ or $< \dfrac{ac}{k^2 + a^2}$ where k is the radius of gyration about a horizontal axis through G. If μ is equal to this value, what happens ?

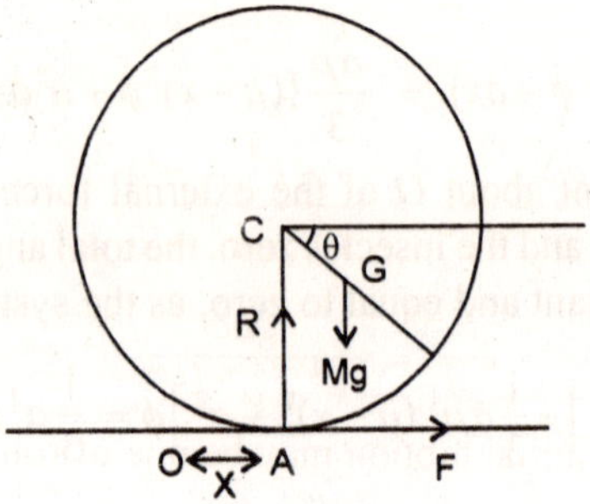

Fig. 8.35

When CG is inclined at an angle θ to the horizontal, let A, the point of contact, have moved through a horizontal distance x from its initial position O, and let $OA = x$.

Assume that sphere rolls so that friction F acts, since, the point of contact A is at rest,

$$\therefore \qquad \dot{x} = a\dot{\theta} \tag{1}$$

The equations of motion are

$$F = M\frac{d^2}{dt^2}[a\theta + c\cos\theta] = M[(a - c\sin\theta)\ddot{\theta} - c\cos\theta\dot{\theta}^2] \tag{2}$$

$$R - Mg = m\frac{d^2}{dt^2}[a - c\sin\theta] = M[-c\cos\theta\ddot{\theta} + c\sin\theta\dot{\theta}^2] \tag{3}$$

$$R.c\cos\theta - F(a - c\sin\theta) = Mk^2\ddot{\theta} \tag{4}$$

We only want the initial motion when $\theta = 0$, and then $\dot{\theta} = 0$, let $\ddot{\theta}$ is not equal to zero.

From (1) — (4), we get

$$F = Ma\ddot{\theta}$$

$$R = Mg - Mc\ddot{\theta}$$

$$Rc - Fa = Mk^2\ddot{\theta}$$

for initial values.

Substituting the values of F and R from first two equations in third, we have

$$(Mg - Mc\ddot{\theta})c - Ma^2\ddot{\theta} = Mk^2\ddot{\theta}$$

$$\therefore \quad (k^2 + a^2 + c^2)\ddot{\theta} = gc$$

$$\therefore \quad \ddot{\theta} = \frac{gc}{k^2 + a^2 + c^2}.$$

Hence, we have

$$\frac{R}{M} = g - \frac{gc^2}{k^2 + a^2 + c^2} = \frac{g(k^2 - a^2)}{k^2 + a^2 + c^2}$$

$$\frac{F}{M} = \frac{gac}{k^2 + a^2 + c^2}.$$

In order that the initial motion may be one of rolling, we must have

$$F < \mu R$$

$$\therefore \quad \mu > \frac{ac}{k^2 + a^2}$$

If $\mu <$ this value, the sphere will not roll, since, the friction is not sufficient.

Critical case

If $\mu = \dfrac{ac}{k^2 + a^2}$, it will be necessary to consider whether, when θ is small

but not absolutely zero, the value of $\dfrac{F}{R}$ is little greater or less than μ.

Eliminating F and R from (2)–(4), we get

$$(k^2 + a^2 + c^2 - 2ac\sin\theta)\ddot{\theta} - ac\cos\theta\dot{\theta}^2 = gc\cos\theta.$$

Integrating, and using initial conditions we get

$$(k^2 + a^2 + c^2 - 2ac\sin\theta)\dot{\theta}^2 = 2gc\sin\theta.$$

Since, θ is small, in the last two equations taking θ for $\sin\theta$ and unity for $\cos\theta$, retaining only first order terms, we get

$$(k^2 + a^2 + c^2)\dot{\theta}^2 = 2gc\theta.$$

$$(k^2 + a^2 + c^2 - 2ac\theta)\ddot{\theta} - \frac{ac.2gc\theta}{k^2 + a^2 + c^2} = gc$$

$$\therefore \quad (k^2 + a^2 + c^2)\ddot{\theta} = \left[gc + \frac{2gac^2\theta}{k^2 + a^2 + c^2}\right]\left[1 + \frac{2ac\theta}{k^2 + a^2 + c^2}\right]$$

$$= gc + \frac{4gac^2\theta}{k^2 + a^2 + c^2}$$

$$= gc\left[1 + \frac{4ac\theta}{k^2 + a^2 + c^2}\right].$$

Hence, to the first powers of θ, we have from (2) and (3)

$$\frac{F}{R} = \frac{(a - c\theta)\ddot{\theta} - c\dot{\theta}^2}{g - c\ddot{\theta}} = \frac{ac}{k^2 + a^2}\left[1 - \frac{(3k^2 - a^2)c\theta}{a(k^2 + a^2)}\right].$$

on substitution and simplification.

If $k^2 > \frac{a^2}{3}$, then $\frac{F}{R}$ is less than $\frac{ac}{k^2 + a^2}$ *i.e.*, $\frac{F}{R}$ is less than the coefficient of friction and the sphere rolls.

If $k^2 > \frac{a^2}{3}$, then $\frac{F}{R} >$ the coefficient of friction and the sphere slides.

4 A body of mass $m_1 + m_2$ is split into two parts of masses m_1 and m_2 by an impulsive explosion which generates K.E.T. Show that if after explosion the parts move in the same line as before, their relative speed is.

$$\sqrt{\frac{2T(m_1 + m_2)}{m_1 m_2}}.$$

Let u be the speed of the body before the expolsion and v_1, v_2 the speeds of m_1, m_2 after the explosion.
There is no change in the linear momentum,

$$\therefore \quad m_1 v_1 + m_2 v_2 = (m_1 + m_2)\, u.$$

K. E. exerted is T,

$$\therefore \quad \frac{1}{2}m_1 v_1^2 + \frac{1}{2}m_2 v_2^2 = \frac{1}{2}(m_1 + m_2)u^2 + T.$$

Eliminating u, we get

$$(m_1 v_1^2 + m_2 v_2^2)(m_1 + m_2) = (m_1 v_1 + m_2 v_2)^2 + 2(m_1 + m_2)T.$$

$$\therefore \quad m_1 m_2 (v_1 - v_2)^2 = 2(m_1 + m_2)T$$

$$\therefore \quad v_1 - v_2 = \sqrt{\frac{2T(m_1 + m_2)}{m_1 m_2}}.$$

5. A sphere of radius a, radius of gyration k about any axis through its centre, rolls with linear velocity v on a horizontal plane, the direction of motion being perpendicular to a vertical face of a fixed rectangular block of height h, where $h < a$. The sphere strikes the block, the sphere and block are perfectly rough and perfectly elastic. Show that the sphere will surmount

the block if

$$(a^2 + ah + k^2)^2 v^2 > 2gha^2 (a^2 + k^2).$$

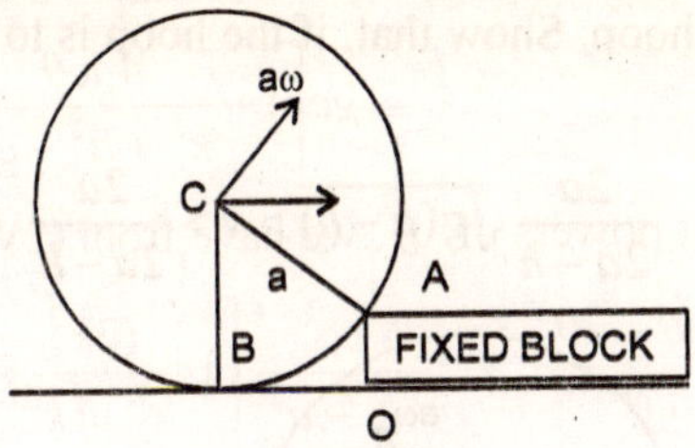

Fig. 8.36

Let C be the centre of the sphere and AO vertical face of fixed block; $AO = h$. The sphere strikes the block at A. Let ω' and ω be the angular velocities of the sphere just before and just after the impulse. After striking at A, the sphere begins to turn about A; so velocity of the centre C after the impulse is at right angles to CA and is equal to $a\omega$.

The impulse being at A, the moment of the impulse about A is zero, hence, moment of momentum about A must remain constant.

$$\therefore \quad mk^2\omega' + mv(a - h) = mk^2\omega + ma\omega.a$$

$$k^2 \frac{v}{a} + v(a - h) = (k^2 + a^2)\omega$$

$$\therefore \quad (k^2 + a^2)\omega = (a^2 - ah + k^2)\frac{v}{a}. \qquad (1)$$

This determines ω.

The condition that the sphere may surmount the block is that it must go on turning about A till the lowest point B comes in level with the upper face of the bolck *i.e.*, till the sphere rises a vertical heights h. This requires that when the sphere begins to turn up about A, it must have K.E. more than that would be required to do the work on the sphere to take it to a height h.

When sphere begins to rise, its K.E. $= \frac{1}{2}m(k^2\omega^2 + a^2\omega^2)$.

In order that the sphere may surmout the block, its kinetic energy must be more than that would be required for doing the work of taking it up to hight h,

$$\therefore \quad \frac{1}{2}m(k^2 + a^2)\omega^2 > mgh$$

$$\therefore \quad (k^2 + a^2)\omega^2 > 2hg$$

$$\therefore \quad (a^2 - ah + k^2)^2 \frac{v^2}{a^2} > 2gh(k^2 + a^2) \text{ from (1)}$$

$$\therefore \quad (a^2 - ah + k^2)^2 v^2 > 2gha^2(k^2 + a^2).$$

6. A hoop of radius a rolling on a horizontal road with velocity v comes into collision with a rough inelastic kerb of height h which is perpendicular to the plane of the hoop. Show that, if the hoop is to clear the kerb without jumping,

$$\frac{2a}{2a-h}\sqrt{g(a-h)} > v > \frac{2a}{2a-h}\sqrt{gh}\,.$$

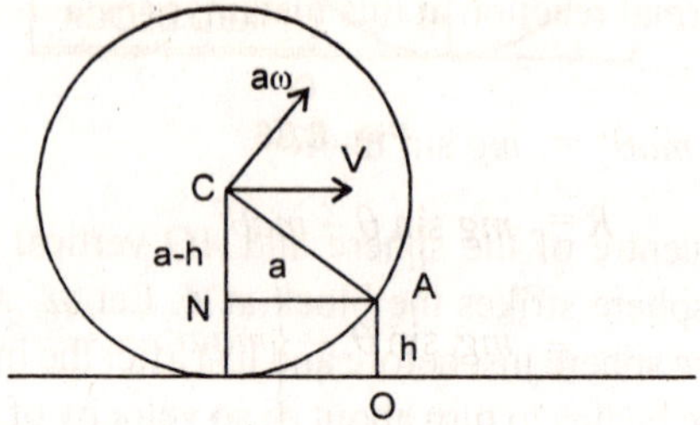

Fig. 8.37 (a)

Let AO be the kerb of height h. When the hoop strikes the kerb, let Ω be its angular velocity and velocity of the centre is v.

$$\therefore \qquad v = a\,\Omega. \tag{1}$$

After striking the kerb let ω be its angular velocity, so that the velocity of the centre C will be $a\omega$ at right anlges to the radius AC.

Since, the impulse acts at A, so its moment about A is zero, hence angular moment about A is constant.

Hence, we have

$$ma^2\Omega + mv\,(a-h) = ma^2\,\omega + ma.\,\omega\, a$$

$$\therefore \qquad mav + mv\,(a-h) = 2ma^2\,\omega, \text{ using (1).}$$

$$\therefore \qquad 2a^2\omega = v\,(2a-h) \tag{2}$$

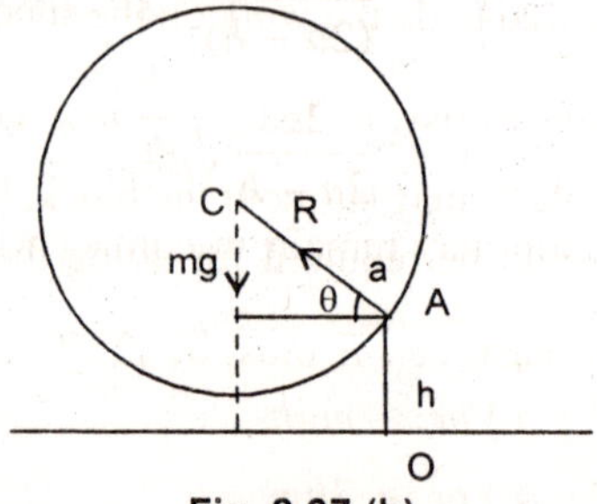

Fig. 8.37 (b)

After striking the kerb, when the hoop is turning about A and rising, let θ be the angle which the radius CA makes with the horizontal; then energy equation gives

$$\frac{1}{2}m(a^2\dot{\theta}^2 + a^2\dot{\theta}^2) - \frac{1}{2}m(a^2\omega^2 + a^2\omega^2) = -mg(h + a\sin\theta - a).$$

$$\therefore \quad a^2(\dot{\theta}^2 - \omega^2) = -g(h + a\sin\theta - a)$$

$$\therefore \quad \dot{\theta}^2 = \omega^2 - \frac{g}{a^2}(h + a\sin\theta - a) \tag{3}$$

Let R be the normal reaction at this instant; considering the motion of the hoop, we have

$$ma\dot{\theta}^2 = mg\sin\theta - R \tag{4}$$

$$\therefore \quad R = mg\sin\theta - ma\dot{\theta}^2$$

$$= mg\sin\theta - \left\{ma\omega^2 - \frac{mg}{a}(h + a\sin\theta - a)\right\}$$

$$= \frac{mg}{a}(h - a + 2a\sin\theta) - ma\omega^2 \tag{5}$$

In order that the hoop may surmount the kerb, $\dot{\theta}$ must vanish before the hoop goes to the highest point, *i.e.*, $\dot{\theta}^2$ must be positive, when $\theta = \frac{\pi}{2}$.

Hence, from (3), we get

$$\omega^2 > \frac{g}{a^2}(h + a - a)$$

$$\therefore \quad \omega^2 > \frac{gh}{a^2}$$

$$\therefore \quad v^2 > \frac{4a^2}{(2a - h)^2}gh, \text{ from (2)}$$

$$\therefore \quad v > \frac{2a}{2a - h}\sqrt{gh}.$$

In order that the hoop may remain in contact with the kerb, R must be positive when it is least, *i.e.*, R must be positive, when $\sin\theta = \frac{a - h}{a}$, from Fig. 8.37 (b).

Hence from (5),

$$\frac{mg}{a}\left(h - a + 2a\frac{a - h}{a}\right) = ma\omega^2 > 0.$$

$$\therefore \quad \frac{g}{a}(a - h) - a\omega^2 > 0.$$

$$\therefore \quad a\omega^2 < \frac{g}{a}(a-h)$$

$$\therefore \quad \omega < \frac{1}{a}\sqrt{g(a-h)}$$

$$\therefore \quad \frac{v(2a-h)}{2a^2} < \frac{1}{a}\sqrt{g(a-h)} \text{ from (2)}$$

$$\therefore \quad v < \frac{2a}{2a-h}\sqrt{g(a-h)}.$$

Hence, in order that the hoop may clear the kerb without jumping

$$\frac{2a}{2a-h}\sqrt{g(a-h)} > v > \frac{2a}{2a-h}\sqrt{gh}.$$

7. For equal rods, each of mass m and length $2a$, are freely jointed at their ends to form a rhombus. The rhombus falls with a diagonal vertical and is moving with velocity V when it hits a fixed horizontal inelastic plane. Find the motion of the rods immediately after the impact and show that their angular velocities are equal to $\frac{3V\sin\alpha}{2a(1+3\sin^2\alpha)}$, where α is the angle that each rod makes with the vertical.

Show also that the impact destroys the friction $\frac{3\sin^2\alpha}{1+3\sin^2\alpha}$ of the kinetic energy just before the impact.

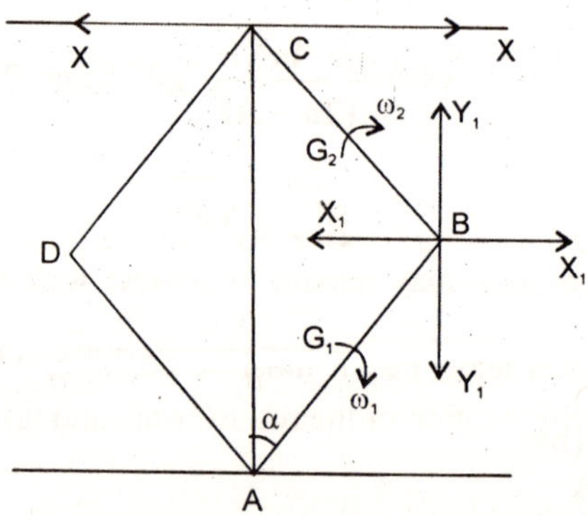

Fig. 3.38

Due to symmetry the motion of the rod DC is the same as the of BC and motion of the rod AD is the same as that of AB. Hence, we will consider the motion of AB and BC only.

After the impact at A with the horizontal plane, it is evident that the rod AB turns about A with angular velocity ω_1 (say) and BC turns about B with angular velocity ω_2 (say).

Due to symmetry the impulsive action at A will be horizontal, it be equal to X and let X_1 and Y_1 be the horizontal and vertical impulses at B in opposite direction on the two rods.

Also after impact, C moves (by symmetry) in vetical direction.

Therefore horizontal velocity of the point C= 0.

Hence, velocity of B + horizontal velocity of C relative to B = 0.

$$\therefore \quad 2a\omega_1 \cos\alpha + 2a\omega_2 \cos\alpha = 0$$

$$\therefore \quad \omega_1 = -\omega_2 \qquad (1)$$

This shows that the angular velocities of AB and BC are same, but in opposite sense.

Horizontal and vertical velocitis of G are $a\omega_1 \cos\alpha$ and $a\omega_1 \sin\alpha$ respectively.

Horizontal velocity of G_2 = horizontal velocity of B + horizontal velocity of G_2 relative to B.

$$= 2a\omega_1 \cos\alpha + a\omega_2 \cos\alpha$$
$$= 2a\omega_1 \cos\alpha - a\omega_1 \cos\alpha$$
$$= a\omega_1 \cos\alpha .$$

Similarly vertical velocity of G_2

$$= 2a\omega_1 \sin\alpha - a\omega_2 \sin\alpha$$
$$= 2a\omega_1 \sin\alpha + a\omega_1 \sin\alpha$$
$$= 3\,a\omega_1 \sin\alpha .$$

For combined motion of AB and BC, taking moment baout A, we have

$$m\frac{4a^2}{3}\omega_1 + m\left(\frac{a^2}{3}\omega_2 + a\omega_1 \cos\alpha.3a\cos\alpha + 3a\omega_1 \sin\alpha.\,a\sin\alpha\right) - 2maV\sin\alpha = X.4a\cos\alpha.$$

$$\therefore m\frac{4a^2}{3}\omega_1 - m\omega_1\left(\frac{a^2}{3} - 3a^2\cos^2\alpha - 3a^2\sin^2\alpha\right) - 2maV\sin\alpha = X.4a\cos\alpha.$$

$$\therefore \quad 2\omega_1 = \frac{V\sin\alpha}{a} + \frac{2X\cos\alpha}{ma} \qquad (2)$$

Also considering motion of the rod BC only and taking moments about B, we get

$$m\left[\frac{a^2}{3}\omega_2 + a\omega_1 \cos\alpha.a\cos\alpha - 3a\omega_1 \sin\alpha.a\sin\alpha\right] + m\,(V)\,a\sin\alpha = X.\,2a\cos\alpha.$$

$$\therefore \quad \omega_1 \left(\frac{2}{3} - 4 \sin^2 \alpha \right) = \frac{-V \sin \alpha}{a} + \frac{2X \cos \alpha}{ma} \qquad (3)$$

Subtracting (3) from (2), we get

$$\omega_1 \left(\frac{4}{3} + 4 \sin^2 \alpha \right) = \frac{2V \sin \alpha}{a}$$

$$\therefore \quad \omega_1 = \frac{3V \sin \alpha}{2a (1 + 3 \sin^2 \alpha)}.$$

Original K.E. (*i.e.*, K.E. just before the impulse) $= \frac{1}{2} 4mV^2 = 2mV^2$.

Final K.E. (*i.e.*, K.E. just after the impulse) = twice of (K.E. of AB + K.E. of BC).

$$= 2 \left[\frac{1}{2} m. \frac{4a^2 \omega_1^2}{3} + \frac{1}{2} m \left(\frac{a^2 \omega_2^2}{3} + a^2 \omega_1^2 \cos^2 \alpha + 9a^2 \omega_1^2 \sin^2 \alpha \right) \right]$$

$$= \frac{8ma^2 \omega_1^2}{3} (1 + 3 \sin^2 \alpha).$$

$$= \frac{8ma^2}{3} (1 + 3 \sin^2 \alpha) . \frac{9V^2 \sin^2 \alpha}{4a^2 (1 + 3 \sin^2 \alpha)^2},$$

Putting the value of ω_1

$$= \frac{3 \sin^2 \alpha}{1 + 3 \sin^2 \alpha} \cdot 2mV^2$$

$$= \frac{3 \sin^2 \alpha}{1 + 3 \sin^2 \alpha} \times \text{original K.E.}$$

EXERCISE

1. A uniform beam lies on a rough horizontal table at right angles to the edge, and is held so that one third of its length is in contact with the table. Prove that after it is released it will begin to slide over the edge of the table when it has turned through an angle $\tan^{-1} \left(\frac{\mu}{2} \right)$, μ being the co-efficient of friction between the table and the beam.
2. A hoop is projected with velocity V down an inclined plane of elevation α, the co-efficient of friction being μ (> $\tan \alpha$). It has initially such a backward spin Ω that after a time t_1, it starts moving uphill and continuous

to do so for a time t_2 after which it once more descends. The motion being in a vertical plane, show that

$$(t_1 + t_1)\, g \sin \alpha = a\,\Omega - V.$$

3. A thin napking ring, of radius a, is projected up a plane inclined at an angle α to the horizontal with velocity v, and an initial angular velocity ω in the sense which would cause the ring to move down the plane. If $v > 5\,a\omega$ and $\mu = \frac{1}{4}\tan\alpha$, show that the ring will never roll and will cease to ascend at the end of a time $\frac{4\,(2v - a\omega)}{9g\sin\alpha}$, and will slide back to the point of projection.
4. A uniform beam of mass M and length l stands upright on perfectly rough ground on the top of it, which is flat, rests a weight of mass m, the co-efficient of friction between the beam and the weight being μ. If the beam is allowed to fall to the ground its inclination θ to the vertical when the weight slips is given by

$$\left(\frac{4}{3}M + 3m\right)\cos\theta - \frac{M}{6\mu}\sin\theta = M + 2m.$$

5. If a uniform semi-circular wire be placed in a vertical plane with one extremity on a rough horizontal plane, and the diameter through that extremity vertical, show that the semi-circle will begin to roll or slide according as $\mu > or < \frac{\pi}{\pi^2 - 2}$. If μ has this value, prove that it will roll.
6. A heavy uniform sphere, of mass M, is resting on a perfectly rough horizontal plane, and a particle of mass m, is gently placed on it at an angular distance α from a highest point. Show that the particle will at once slip on the sphere if $\mu < \frac{\sin\alpha\,[7M + 5m\,(1 + \cos\alpha)]}{7M\cos\alpha + 5m\,(1 + \cos\alpha)^2}$, where μ is the co-efficient of friction between the sphere and the particle.
7. A homogeneous solid hemisphere, of mass M and radius a rests with its curved surface on a rough horizontal plane, and a particle of mass m is placed on its base, which is smooth, at a distance c from its cenre. Show that the hemi-sphere will commence to roll or slide as the co-efficient of friction.

$$\mu > \text{or} < \frac{25mac}{26\,(M + m)a^2 + 4amc^2}.$$

8. A sphere, of radius a, whose centre of gravity G is not at its cenre C is placed on a rough horizontal table, so that CG is inclined at an angle α to

the upwards drawn vertical, show that it will commence to slide along the table if the coefficient of friction $\mu < \dfrac{c \sin \alpha \ (a + c \cos \alpha)}{k^2 + (a + c \cos \alpha)^2}$, where $OG = c$ and k is the radius of gyration about a horizonal axis through G.

9. A rough solid cicular cylinder rolls down a second rough circular cylinder which is fixed with its axis horizontal. If the plane through their axes make an angle α with the vertical when the first cylinder is at rest, show that the bodies will separate when the angle of friction is $\cos^{-1}\left(\dfrac{4 \cos \alpha}{7}\right)$.

10. A disc rolls on the inside of a fixed hollow circular cylinder whose axis is horizontal, the plane of disc being vertical and perpendicular to the axis of the cylinder; if when in the lowest position, its centre is moving with a velocity $\sqrt{\dfrac{8}{3} g \ (a - b)}$, show that the centre of the disc will describe angle ϕ about the centre of the cylinder in time $\sqrt{\dfrac{3 \ (a - b)}{2g}} \ \log\left(\dfrac{\pi}{4} + \dfrac{\phi}{4}\right)$.

11. A solid homogeneous sphere is rolling on the inside of a fixed hollow sphere, the two centres being always in the same vertical plane. Show that the smaller sphere will make complete revolution, if it is in the lowest position, the pressure on it is greater than $\dfrac{34}{7}$ times its weight.

12. A uniform rough ball is at rest within a hollow cylindrical garden roller, and the roller is then drawn along a level path with velocity V. If $V^2 > \dfrac{27}{7} g \ (b - a)$, show that the ball will roll completely round the inside of the roller, a and b being the radii of the ball and the roller.

13. A hemisphere of mass M is free to slide with its base on a smooth horizontal table. A particle of mass m is placed on the hemisphere at an angular distance α from the vertex; show that the radius to the point of contact at which the particle leaves the surface, makes with the vertical an angle θ given by the equation

$$m \cos^3 \theta - (M + m) (3 \cos \theta - 2 \cos \alpha) = 0.$$

14. A rough cylinder, of mass $2nm$, is capable of motion about its axis, has a particle of mass m and co-efficient of friction μ placed on it vertically above the axis. The system is then slightly disturbed. Show that the particle will slip on the cylinder after it has moved through an angle θ given by

$$(n + 3) \cos \theta - 2 = \frac{n \sin \theta}{\mu}.$$

Hence, find the value of θ when the particle will leave the cylinder.

15. Two equal perfectly rough cylinders are placed in an unstable equilibrium, one on the top of the other, the latter resting on a perfectly smooth table. A slight disturbance being given, show that the spheres will continue to touch each other at the same point and that, if θ be the inclination to the vertical of the straight line joining the centres,

$$(k^2 + a^2 + a^2 \sin^2 \theta)\, \dot{\theta}^2 = 2ga\,(1 - \cos \theta),$$

where k is the radius of gyration of each sphere about an axis through its centre.

16. A uniform circular cylinder of radius a moves on the inside of a fixed straight tube of inner radius $c + a$ whose axis is horizontal. The cylinder starts by rolling from the lowest position in the tube with an angular velocity $(ngc)^{1/2}\, a$. Prove that, if the cylinder is to roll completely round the table, $n > \dfrac{11}{3}$ and the coefficient of friction between the cylinder and the table > $[(3n - 1)(3n + 3)]^{-1/2}$.

17. A uniform solid sphere of radius a, rotating about a horizontal diameter, is gently placed on the interior surface of a fixed rough circular cylinder of radius $a(1 + \sqrt{2})$ whose axis is horizontal and parallel to the axis of rotation of the sphere. The co-efficient of friction between the sphere and cylinder is $1/\sqrt{2}$ and the angle made by he plane of the axis of the cylinder and the centre of the sphere with the downward vertial is θ. If the initial value of θ is $\pi/6$ and the sense of rotation is such as to tend to make the sphere roll upwards, show that the centre of the sphere does begin to rise and that while slipping continues

$$\frac{2a\dot{\theta}^2}{g} = 2 \sin \theta - \exp\left\{\left(\theta - \frac{\pi}{6}\right)\sqrt{2}\right\}.$$

18. A thin, uniform solid sphere of mass M and radius b rolls on a rough horizontal table, and inside it is a rough uniform solid cylinder of mass m and radius a. The axes of the cylinders are parallel and their centres of gravity lie in the same cross section. Show that, when inclination to the downward vertical of the plane through the axis is θ, the kinetic energy of the system is

$$Mv^2 + \frac{1}{4} m\,(v + u)^2 + \frac{1}{2} m\,(u^2 + 2uv \cos \theta + v^2),$$

where v is the speed of the axis of M and $u = (b - a)\,\dot{\theta}$.

Prove also, that, for small oscillations of the system, the length of S.E.P. is $\dfrac{6M\,(b - a)}{4M + 3m}$.

19. A uniform solid sphere balanced on the top of a fixed rough sphere, not necessarily of the same radius, is slightly disturbed. Show that the initial motion is one of rolling and that the slipping commences before the spheres separate. If θ is the inclination of the line of centres of the spheres to the upwards vertical, and the co-efficient of friction between the spheres is 1/5, show that the slipping commences when $\theta = \cos^{-1}\left(\frac{340}{389}\right)$.

20. The rod AB, BC each of length a are freely jointed at B and placed on a horizontal table. The end A is freely pivoted to a fixed point on the table and the angle ABC is $\pi/3$. The system is then set in motion in such a way that the end C begins to move with velocity V and acceleration V^2/a, both in direction $\overline{CB}$. Find the initial angular velocities of the rods and show that their initial angular velocities are $\frac{8V^2}{3\sqrt{3}a^2}$ and $\frac{10V^2}{3\sqrt{3}\,a^2}$.
If the rods are uniform and each of mas m, find the initial kinetic energy generated.

21. Two uniform rods AB, AC, each of mass m and length $2a$ are freely hinged at A and held in vertical plane with A above BC and each rod inclined at angle α to the downward vertical. The system is then released from rest and, after falling through h, B and C strike a smooth inelastic horizontal table. Show from energy consideration, or otherwise, that neither rod rotates before the impact. Show also that immediately after the impact each rod has an angular velocity $3(2hg \sin^2 \alpha)^{1/2}/4a$ and find the impulsive reaction at the joint A.

22. A uniform circular hoop in a vertical plane is projected with velocity v down a rough plane inclind at an angle α to the horizontal. The hoop has an initial angular velocity ω tending to make it to roll up the plane. Show that the hoop will turn back if

$$a\omega(\mu - \tan\alpha) > v\mu,$$

where μ (> $\tan \alpha$) is the co-efficient of friction and a the radius of the hoop.
Show that the least value of ω for which the hoop will return to the point of projection satisfies the equation

$$\omega^2 - 2\omega \, . \frac{v\mu}{a(\mu - \tan\alpha)} + \frac{v^2}{a^2} = 0.$$

23. A uniform rod AB of length $2a$ can turn freely about one end A, and at time $t = 0$ is hanging from A in equilibrium under gravity. The end A is then set in motion in a horizontal straight line so that, when $t > 0$,

$OA = vt + \frac{1}{2}ft^2$, where O is a fixed point in the line and v, f are constants. Show that the initial angular velocity of the rod is $3v/4a$ and it will make complete revolutions about A if

$$3v^2 > 8a\left[g + \sqrt{g^2 + f^2}\right].$$

24. A uniform rod of length $2a$ and mass m is freely hinged to a fixed point at A and to an equal rod BC at B. The system is moving in a vertical plane about A and when ABC is vertically downwards, the angular velocity of BC is 2ω that of BC is ω in the same sense. At this instant a horizontal impulse I, opposing the motion, is applied to BC at a point $2ax$ below B, where $x < 1$. If the ratio of the angular velocities is unchanged by the blow, show that $x = \frac{1}{2}$.

Show that the initial kinetic energy of the system is $28ma^2\omega^2/3$ and find I if this is unchanged by the impulse.

25. A uniform rod PQ of length $2a$ and mass m freely hinged at P to a point on the circumference of a uniform circular disc of radius a, centre O and mass m. The system rests on a smooth horizontal table with O, P, Q collinear in order. An impulse I is applied at Q in a direction perpendicular to PQ. Show that the initial angular velocity of the disc is of magnitude $4I/7ma$. Show further that if the rod is rigidly fixed to the disc at P instead of being hinged, the angular velocity is $12I/17ma$.

26. A light rigid rod ABC has three particles each of mass m attached to it at A, B, C. The rod is struck by a blow P at right angles to it at a point distance from A equal to BC. Prove that kinetic energy set up is

$$\frac{P^2a^2 - ab + b^2}{2ma^2 + ab + b^2},$$

where $AB = a$, $BC = b$.

27. A uniform rod AB is freely movable on a rough inclined plane whose inclination to the horizon is α and whose coefficient of friction is μ, about a smooth pin fixed through end A; the rod is held in a horizontal position. If θ be the angle through which it falls from rest, show that

$$\frac{\sin\theta}{\theta} = \mu \cot\alpha.$$

28. A bent lever, whose arms are of lengths a and b, the angle between the being α, makes small oscillations in its own plane about the fulcrum, show that the length of corresponding simple pendulum is

$$\frac{2}{3} \cdot \frac{a^3 + b^3}{\sqrt{a^4 + 2a^2b^2 \cos\alpha + b^4}}.$$

29. A simple circular pendulum is formed of a mass M suspended from a fixed point by a weightless wire of length l, if a mass m, very small compared to M, be knotted to the wire at a point distant a from the point of suspension, show that the time of small vibrations of the pendulum is approximately diminished by

$$\frac{ma}{2Ml}\left(1 - \frac{a}{l}\right) \text{ of itself.}$$

30. Two bodies can move freely and independently under the action of gravity about the same horizontal axis, their masses are m, m' and distances of their centres of gravity from the axes are h and h'. If the lengths of their equivalent simple pendulum are l, l', prove that when fastened together the length of the equivalent simple pendulum will be

$$\frac{mhl + m'h'l'}{mh + m'h'}.$$

31. A circular area can turn freely about a horizontal axis which passes through a point O of its circumference and is perpendicular to its plane. If motion commences when the diameter through O is vertically above O, show that, when the diameter has turned through an angle θ, the components of the strain at O along and perpendicular to this diameter are respectively

$\frac{W}{3}(7\cos\theta - 4)$ and $\frac{W}{3}\sin\theta$.

32. A straight rod, of mass m and length $2l$; swinging about one end as a compound pendulum, starts from rest in a horizontal position and when vertical is struck a blow at its middle point which reverses and halves its angular velocity. Prove that the impulse of the blow is $m\sqrt{6gl}$.

33. Find the position of centre of percussion of sector of a circle, axis in the plane of the sector, perpendicular to its symmetrical radius, and passing through the centre of the circle.

34. A rod, of length $2a$, revolvs with uniform angular velocity ω about a vertical axis through a smooth joint at one extremity of the rod so that it describes a cone of semi-vertical angle α, show that

$$\omega^2 = \frac{3g}{4a\cos\alpha}.$$

Prove also that the direction of the reaction at the hinge makes with the vertical the angle $\tan^{-1}\left(\frac{3}{4}\right)$.

35. A uniform beam of mass m and length $2l$, is horizontal and can turn freely about its centre which is fixed. A ball of mass m' and moving with velocity u, hits the beam at one end. If the coefficient of restitution for the impulse be e, show that the angular velocity of the beam after the impact is

$$3m'(1 + e)\,u/(3m' + m)l,$$

and that the vertical velocity of the ball is then

$$u\,(em - 3m')/(m + 3m').$$

36. A pendulum is constructed of a solid sphere of mass M and radius a, which is attached to one end of the rod, of mass m and length b. Show that there will be no strain on the axis if the pendulum be struck at a distance

$$\frac{M\left[\frac{2}{5}a^2 + (a+b)^2\right] + \frac{1}{3}mb^2}{\left[M(a+b) + \frac{1}{2}mb\right]}.$$

37. A uniform rod AB can turn about its end A and is in equilibrium; find the point of its length where a blow must be applied to it so that the impulse at A may be in each case $\frac{1}{n}$ of that of the blow.

9

THREE DIMENSIONAL MOTION OF A RIGID BODY

9.1 INTRODUCTION

In previous chapter we applied the principles of rigid body dynamics to two dimensional problems. In this chapter, we extend the principles to the motion of a rigid body or system of particles in three dimensions. This application resembles to the corresponding applications in two dimensional problems, but it is often more convenient to formulate three dimensional problem in vector form before solving them. In discussion of the motion of a rigid body in three dimensions, we shell use two co-ordinate systems—one of which is fixed in space, *i.e.*, an inertial frame, and the other fixed in the body. The latter system is referred to as the body co-ordinate system and is non-inertial frame of reference. This is because this system undergoes a translational and rotational motion along with body.

9.2 EULER'S DYNAMICAL EQUATIONS OF MOTION OF A RIGID BODY ABOUT A FIXED POINT

Let O be a point fixed both in body and space, and body be turning about O. Let OX, OY, OZ be the three mutually perpendicular axes fixed in the body and $\vec{i}, \vec{j}, \vec{k}$ the unit vectors along these axes. If at any time t the body has angular velocity $\vec{\omega} = \omega_1\vec{i} + \omega_2\vec{j} + \omega_3\vec{k}$, then angular momentum $\vec{H}$ about O is given by

$$\vec{H} = A\omega_1\vec{i} + B\omega_2\vec{j} + C\omega_3\vec{k} \tag{1}$$

where A, B, C are the principal moment of inertia about these axes.

Since $\vec{\omega}$ is the spin of the frame,

$$\therefore \qquad \frac{d\vec{H}}{dt} = \frac{\partial\vec{H}}{\partial t} + \vec{\omega} \times \vec{H}.$$

$$\therefore \qquad \frac{d\vec{H}}{dt} = A\dot{\omega}_1\vec{i} + B\dot{\omega}_2\vec{j} + C\dot{\omega}_3\vec{k} + \vec{\omega}\times\vec{H} \qquad (2)$$

Let $\vec{L} = L_1\vec{i} + L_2\vec{j} + L_3\vec{k}$ be the moment of momentum of the external forces about O. Since O is fixed point,

$$\therefore \qquad \frac{d\vec{H}}{dt} = \vec{L}. \qquad (3)$$

Using (2) and (3) and equating the co-efficients of $\vec{i}, \vec{j}, \vec{k}$ we have

$$A\dot{\omega}_1 - (B-C)\omega_2\omega_3 = L_1,$$
$$B\dot{\omega}_2 - (C-A)\omega_3\,\omega_1 = L_2,$$
$$C\dot{\omega}_3 - (A-B)\omega_1\,\omega_2 = L_3.$$

These are called Euler's dynamical equations of motion of a rigid body about a fixed axies.

Note : The axies of H is fixed in direction and is called *invariable line*. A perpendicular plane is called an *invariable plane*. If the body were set in motion by an impulsive couple H, the plane of this couple would be an invariable plane and the axis through O at right angles to it the invariable line.

9.3 INTEGRALS OF ENERGY AND ANGULAR MOMENTUM

Prove that in absence of external forces K.E. of the body is constant during the motion.

In absence of external force Euler's dynamical equations reduce to

$$A\dot{\omega}_1 - (B-C)\omega_2\omega_3 = 0, \qquad (1)$$
$$B\dot{\omega}_2 - (C-A)\omega_3\,\omega_1 = 0, \qquad (2)$$
$$C\dot{\omega}_3 - (A-B)\omega_1\,\omega_2 = 0. \qquad (3)$$

Multiplying these equations by ω_1, ω_2, ω_3 respectively and adding, we get

$$A\dot{\omega}\,\omega_1 + B\dot{\omega}_2\,\omega_2 + C\dot{\omega}_3\,\omega_3 = 0$$

Integrating, we get

$$A\omega_1^2 + B\omega_2^2 + C\omega_3^2 = const.$$

But we know that K.E.

$$T = \frac{1}{2}(A\omega_1^2 + B\omega_2^2 + C\omega_3^2)$$

$$\therefore \qquad 2T = const.$$

$$\therefore \qquad T = const.$$

This is obvious since, in the absence of external forces no work is being done during the motion.

(*ii*) Prove that in absence of external forces the magnitude of angular momentum of the body remains constant during the motion.

Multiplying equations (1), (2), (3) by $A\omega_1$, $B\omega_2$, $C\omega_3$ respectively and adding, we get

$$A^2\omega_1\dot{\omega}_1 + B^2\omega_2\dot{\omega}_2 + C^2\omega_3\dot{\omega}_3 = 0$$

Integrating, we get

$$A^2\omega_1^2 + B^2\omega_2^2 + C^2\omega_3^2 = const.$$

But we know that magnitude of the angular momentum H is given by

$$H^2 = A^2\omega_1^2 + B^2\omega_2^2 + C^2\omega_3^2 .$$

$$\therefore \quad H = const.$$

This is obvious, because in the absence of external forces the moment of forces about O is zero.

Note : We shell see that these relations enable us to trace the path in space of any point of the body, but not its position at any time, because these are only first integrals of the equations of motion. We should remember that we are using moving axes fixed in the body and we have established that the axis of resultant moment of momentum is fixed in space (the invariable line), and that its direction cosines referred to the moving axes are proportional to $A\omega_1, B\omega_2, C\omega_3$.

9.4 EULER'S EQUATIONS OF MOTION FOR IMPULSIVE FORCES

Let a rigid body be rotating about a fixed point O and let OA, OB and OC be the principal axes fixed in the body moving along with the body.

Let the body be give an impulse such that the moments of the impulse about OA, OB and OC are respectively L_1, L_2, L_3. Let the components of the angular velocity be $\omega_1, \omega_2, \omega_3$ before impulse and $\omega'_1, \omega'_2, \omega'_3$ after the impulse.

Components of angular momentum about OA, OB, OC before impulse are $A\omega_1, B\omega_2, C\omega_3$ and after the impulse are $A\omega'_1, B\omega'_2, C\omega'_3$.

We know that change in angular momentum about any line is equal to the moment of the impulse about the same line, therefore taking moment about OA, OB. OC, we have

$$A(\omega'_1 - \omega_1) = L_1,$$
$$B(\omega'_2 - \omega_2) = L_2,$$
$$C(\omega'_3 - \omega_3) = L_3.$$

These are *Euler's equations of motion for impulsive forces.*

9.5 EULERIAN ANGLES (θ, φ, Ψ)

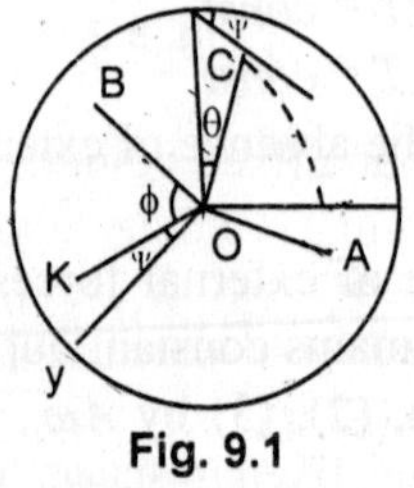

Fig. 9.1

Definition : Let a rigid body be free to move about a point O fixed in the body. Take a set of rectangular axes Ox, Oy, Oz fixed in space and OA, OB. OC the principal axes of the body about O. Since axes OA, OB, OC are fixed in the body, so they move when the body moves. Initially OA, OB, OC coincide with Ox, Oy, Oz.

Now suppose the body turns about O. In order to know the position of the body at any time, with reference to the initial position, three angles θ, ϕ, Ψ are chosen to define the position of principal axes and therefore of the body itself. the angles θ, ϕ, Ψ are called *Eulerian angles*.

Geometrical Concept of θ, ϕ, Ψ

Construct a spherical surface of radius unity with centre at O. Let the axes Ox, Oy, Oz which are fixed in space meet the spherical surface at point x, y, z. At any time t, when the body is rotating about O, let the moving axes meet the surface in points A, B, C. Then at time t,

θ is the angle between Oz and OC,

ϕ is the angle between the plane zOC and the plane COA.

Ψ is the angle between the plane zOC and the plane zOx.

To express the values of ϕ and Ψ more precisely, draw a line OK perpendicular to the plane zOC. Normals to the planes COA and zOx are respectively OB and Oy.

Since, angle between any two planes is the same as the angle between the normals to the planes, therefore it is obvious that φ is the angle between OK and OB. Ψ is the angle between OK and Oy.

9.6 EULER'S GEOMETRICAL EQUATIONS

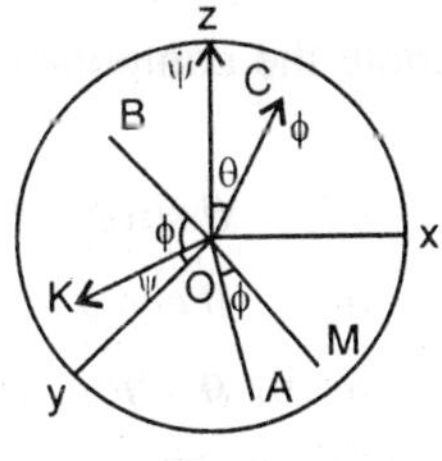

Fig. 9.2

Let a rigid body turn about a fixed point O, fixed in the body. If ω_1, ω_2, ω_3 are the components of angular velocity about OA, OB, OC respectively. Determination of ω_1, ω_2, ω_3 in terms of θ, φ, Ψ and $\dot{\theta}$, $\dot{\phi}$, $\dot{\Psi}$ are called *Euler's Geometrical equations*.

Let Ox, Oy, Oz be the axes fixed in space. Suppose θ, ϕ, ψ are the Eulerian angles defining the position of the body at time t.

Angle θ is in zOC plane, therefore angular velocity $\dot{\theta}$ is about the line OK, which is normal to the plane. Now line OK is perpendicular to both Oz and OC, lies in the xOy plane and as well as in AOB plane.

Angle φ is in AOB plane, therefore angular velocity $\dot{\phi}$ is about OC which is normal to the plane AOB.

Angle Ψ is in xOy plane, therefore angular veloicity $\dot{\Psi}$ is about Oz, which is normal to the plane xOy.

Now we will find the direction cosines of the lines OK, OC, and Oz with respect to the principal axes OA, OB, OC.

Since, OK is perpendicular to OC and is in the plane AOB making an angle φ with OB, therefore direction cosines of OK are $\sin\phi, \cos\phi, 0$ and direction cosines of OC are 0, 0, 1. To find the direction cosines of Oz, draw a line OM perpendicular to OC in the plane zOC. The line OM being in the plane, zOC is perpendicular to OK, and in virtue of its being perpendicular to OC, lies in the plane AOB. Thus both OM, OK lie in the plane AOB and are mutually perpendicular, and as OK makes the angle φ with OB, therefore OM must make angle φ with OA.

Now the direction cosines of Oz about the co-planar lines OC and OM are $\cos\theta, \sin\theta$.

But direction cosines of OM (since it is perpendicular to OC and lies in the plane AOB, making an angle φ with OA) about OA, OB are $-\cos\phi, \sin\phi$.

$\therefore$ Direction co-sines of Oz about OA, OB, OC are

$$-\sin\theta\cos\phi,\ \sin\theta\sin\phi,\ \cos\theta.$$

Hence, components of $\dot{\theta}, \dot{\phi}, \dot{\psi}$ about OA, OB, OC are $\dot{\theta}\sin\phi$, $\dot{\theta}\cos\phi$; 0, 0, $\dot{\phi}$; $\dot{\psi}\sin\theta\cos\phi$, $\dot{\psi}\sin\theta\sin\phi$, $\dot{\psi}\cos\theta$.

Now, if ω_1, ω_2, ω_3 denote the components of angular velocity about OA, OB, OC, then we have

$$\omega_1 = \dot{\theta}\sin\phi - \dot{\Psi}\sin\theta\cos\phi,$$

$$\omega_2 = \dot{\theta}\cos\phi + \dot{\Psi}\sin\theta\sin\phi,$$

$$\omega_3 = \dot{\theta} + \dot{\Psi}\cos\theta.$$

These are called *Euler's Geometrical equations.*

9.7 POINSOT'S REPRESENTATION OF MOTION BY ROLLING OF MOMENTAL ELLIPSOID ON AN INVARIABLE PLANE

Let the momental ellipsoid referred to the principal axex at O, be

$$Ax^2 + By^2 + Cz^2 = M\varepsilon^4 \quad (1)$$

Let the instantaneous axis of rotation meet this ellipsoid in P (x, y, z); if $OP = r$ and ω is the angular velocity, then

$$\frac{x}{\omega_1} = \frac{y}{\omega_2} = \frac{z}{\omega_3} = \frac{r}{\omega}. \tag{2}$$

Substituting this in (1), we have

$$A\omega_1^2 + B\omega_2^2 + C\omega_3^2 = \frac{M\varepsilon^4\omega^2}{r^2};$$

But we have

$$A\omega_1^2 + B\omega_2^2 + C\omega_3^2 = 2T,$$

$$\therefore \quad \frac{M\varepsilon^4\omega^2}{r^2} = 2T,$$

$$\therefore \quad \omega = \sqrt{\frac{2Tr^2}{M\varepsilon^4}} \tag{3}$$

Using ξ, η, ζ as current co-ordinates, the tangent plane to the ellipsoid at the point P is

$$Ax\xi + By\eta + Cz\zeta = M\varepsilon^4. \tag{4}$$

It follows from (2) that the direction co-sines of the perpendicular on the tangent plane are proportional to $A\omega_1$, $B\omega_2$, $C\omega_3$ *i.e.*, this tangent plane is an invariable plane.

Again if p denotes the distance of the tangent plane from O,

$$\frac{1}{p^2} = \frac{A^2x^2 + B^2y^2 + C^2z^2}{M^2\varepsilon^8};$$

$$= \frac{r^2}{\omega^2}[A^2\omega_1^2 + B^2\omega_2^2 + C^2\omega_3^2] / M^2\varepsilon^8.$$

$$= \frac{H^2r^2}{\omega^2 M^2\varepsilon^8}$$

$$\therefore \quad p^2 = \frac{\omega^2 M^2\varepsilon^8}{H^2r^2}$$

$$\therefore \quad p = \frac{\omega M\varepsilon^4}{Hr}.$$

Using (3)

$$p = \varepsilon^2\sqrt{2MT} / H. \tag{5}$$

Thus p is constant, so that the tangent plane to the ellipsoid at P is fixed in direction and at a fixed distance from O, therefore fixed in space.

But OP is the instantaneous axis of rotation, so at the instant under consideration the point P has no velocity, and the tangent plane at P being fixed in space, it follows that the motion is such that the momental ellipsoid rolls on this plane with its centre fixed at O, with an angular velocity which, by (3) is proportional to the radius vector of the ellipsoid in the direction of the instantaneous axis.

9.8 POLHODE, HERPOLHODE AND INVARIABLE CONE

As the ellipsoid rolls on the plane, the point of contact traces curves on the ellipsoid and on the plane. The former is called *Polhode* and the latter the *Herpolhode*, one is fixed in the body and the other fixed in space.

The equation of any polhode may be found from the fact that it is the locus of a point on the momental ellipsoid such that the tangent plane at that point is at a constant distance from the centre.

Now the distance p from the centre of the ellipsoid

$$Ax^2 + By^2 + Cz^2 = M\varepsilon^4 \tag{1}$$

to the tangent plane at (x, y, z) is given by

$$A^2x^2 + B^2y^2 + C^2z^2 = \frac{M^2\varepsilon^8}{p^2}$$

putting

$$p = \varepsilon^2 \sqrt{2MT} / H$$

we get

$$A^2x^2 + B^2y^2 + C^2z^2 = \frac{M\varepsilon^4 H^2}{2T} \tag{2}$$

Equations (1) and (2) represent for different values of H and T the polhodes on the ellipsoid (1).

Since, the instantaneous axis passes through the point of contact of the plane and ellipsoid, it always intersects the polhode, and the cone which it describes in the body may be called *polhode cone*.

The equation of the polhode cone may be found by eliminating $M\varepsilon^4$ from (1) and (2); it is

$$A\left(A - \frac{H^2}{2T}\right)x^2 + B\left(B - \frac{H^2}{2T}\right)y^2 + C\left(C - \frac{H^2}{2T}\right)z^2 = 0 \tag{3}$$

This will be real if, and only if , p lies between the greatest and least semi-axes of the ellipsoid (1) or if $\frac{H^2}{2T}$ lies between the greatest and least of the principal moments A, B, C.

The projection of the polhode on co-ordinate planes from (1) and (2), are

$$B(A-B)y^2 + C(A-C)z^2 = M\varepsilon^4\left(A - \frac{H^2}{2T}\right) \quad (4)$$

$$A(A-B)x^2 + C(C-B)z^2 = M\varepsilon^4\left(\frac{H^2}{2T} - B\right) \quad (5)$$

$$A(A-C)x^2 + B(B-C)y^2 = M\varepsilon^4\left(\frac{H^2}{2T} - C\right). \quad (6)$$

Suppose, that $A > B > C$, then (4) and (6) are ellipses and (5) is hyperbola; so that the polhodes are closed curves surrounding the axes of greatest and least moment.

Now, consider the polhode cone for different values of $H^2/2T$. If $H^2/2T = A$, the cone is the line $y = 0$, $z = 0$, *i.e.*, the axis of the greatest moment is the instantaneous axis and the distance of the plane from the centre does not permit of contact at any point but the end of this axis.

If $A > H^2/2T > B > C$, then the cone cuts the plane $x = 0$, in imaginary lines and the planes $y = 0$, $z = 0$ in real lines, so that in this case the cone surrounds the axis of the greatest moment, therefore so does the polhode.

If $H^2/2T = B$, the mean moment, the cone is in two planes passing through the axis of the mean moment cutting the ellipsoid in two ellipses which separate the polhodes round the axis of moment A from the polhode round the axis of the moment C.

If $B > H^2/2T > C$, the cone cuts the planes $x = 0$, $y = 0$ in real lines and $z = 0$ in imaginary lines, so that the cone surrounds the axis of the least moment and so does the polhode.

Finally, if $H^2/2T = C$, the cone is the line $x = 0$, $y = 0$, *i.e.*, the axis of the least moment is the instantaneous axis and the only possible position for it.

Herpolhode Cone : Since, Herpolhode is traced out by the points of contact of an ellipsoid turning about its centre and rolling on a fixed plane, it is obvious that it must be between two concentric circles which it touches alternatively; the centre of the circles being the foot of perpendicular from the centre of the ellipsoid.

If L be the centre and P a point on the herpolhode, we have

$$LP^2 = r^2 - p^2,$$

$$\therefore \quad LP^2 = \frac{M\varepsilon^4}{2T}\left(\omega^2 - \frac{4T^2}{H^2}\right) \quad (1)$$

Thus the radii of the circles touched by the herpolhode are got from (1) by giving to ω its least and greatest values.

The cone traced in space by the instantaneous axis passes through herpolhode, and may be called *herpolhode cone.*

Invariable Cone : Though the invariable lines if fixed in space, as the body moves, the line traces a cone in the body and this is called *invariable cone.*

To find the equation of invariable cone, we have

$$A^2\omega_1^2 + B^2\omega_2^2 + C^2\omega_3^2 = (A\omega_1^2 + B\omega_2^2 + C\omega_3^2)H^2/2T, \qquad (1)$$

and the direction cosines of the invariable line are proportional to $A\omega_1, B\omega_2, C\omega_3$, so that the equation of the cone is

$$x^2\left(1-\frac{H^2}{2AT}\right)+y^2\left(1-\frac{H^2}{2BT}\right)+z^2\left(1-\frac{H^2}{2CT}\right) = 0. \qquad (2)$$

Thus the general form resembles that of the polhode cone.

9.9 DEDUCE EULER'S EQUATIONS FROM LAGRANGE'S EQUATIONS

Let a rigid body turns about a fixed point O under the action of external forces.

Let OA, OB, OC be the principal axes at O and at time t let θ, ϕ, Ψ be the Eulerian angles which are the generalized co-ordinates of the body. Let ω_1, ω_2, ω_3 be the angular velocities and A, B, C be the moments of inertia of the body about OA, OB, OC respectively. We know that

K.E.

$$T = \frac{1}{2}(A\omega_1^2 + B\omega_2^2 + C\omega_3^2)$$

Where

$$\omega_1 = \dot{\theta}\sin\phi - \dot{\Psi}\sin\theta\cos\phi,$$

$$\omega_2 = \dot{\theta}\cos\phi + \dot{\Psi}\sin\theta\sin\phi,$$

$$\omega_3 = \dot{\phi} + \dot{\Psi}\cos\theta.$$

Hence, T can be regarded as function of generalized co-ordinates θ, ϕ, Ψ and of generalized velocities, $\dot{\theta}, \dot{\phi}, \dot{\Psi}$.

$$\therefore \quad \frac{\partial T}{\partial \dot{\phi}} = \frac{\partial T}{\partial \omega_1}\frac{\partial \omega_1}{\partial \dot{\phi}} + \frac{\partial T}{\partial \omega_2}\frac{\partial \omega_2}{\partial \dot{\phi}} + \frac{\partial T}{\partial \omega_3}\frac{\partial \omega_3}{\partial \dot{\phi}}$$

$$= \frac{\partial T}{\partial \omega_3}\frac{\partial \omega_3}{\partial \dot{\phi}} \text{ (since } \omega_1, \omega_2 \text{ are independent of } \dot{\phi}\text{)}$$

$$= C\omega_3.$$

and

$$\frac{\partial T}{\partial \phi} = \frac{\partial T}{\partial \omega_1}\frac{\partial \omega_1}{\partial \phi} + \frac{\partial T}{\partial \omega_2}\frac{\partial \omega_2}{\partial \phi} + \frac{\partial T}{\partial \omega_3}\frac{\partial \omega_3}{\partial \phi}$$

$$= A\omega_1 \frac{\partial \omega_1}{\partial \phi} + B\omega_2 \frac{\partial \omega_2}{\partial \phi} + C\omega_3.0$$

$$= A\omega_1 (\dot{\theta}\cos\phi + \dot{\Psi}\sin\theta\sin\phi) + B\omega_2 (-\dot{\theta}\sin\phi + \dot{\Psi}\sin\theta\cos\phi)$$

$$= A\omega_1\omega_2 - B\omega_2\omega_1 = (A - B)\,\omega_1\omega_2.$$

If W is the work function of the body and N be the moment of the external forces about OC, then

$$\frac{\partial W}{\partial \varphi} = N$$

Lagrange's φ equation is

$$\frac{d}{dt}\left(\frac{\partial T}{\partial \dot{\phi}}\right) - \frac{\partial T}{\partial \phi} = \frac{\partial W}{\partial \phi}.$$

$$\therefore \quad \frac{d}{dt}(C\omega_3) - (A - B)\,\omega_1\omega_2 = N,$$

$$\therefore \quad C\dot{\omega}_3 - (A - B) = N.$$

This is Euler's equation. Similary, we can deduce two other equations.

N.B. This article can be omitted at this stage and can be done after chapter *X*.

Examples

1. A rectangular parallelopiped whose edges are a, $2a$, $3a$, $3a$ can turn about its centre and is set rotating about a line perpendicular to the mean axis and making an angle $\cos^{-1}\frac{5}{8}$ with the least axis. Prove that ultimately body will rotate about the mean axis.

   Let ω be the angular velocity

   Initially,

$$\omega_1 = \frac{5}{8}\omega,\ \omega_2 = 0,\ \omega_3 = \sqrt{\omega^2 - \omega_2^2 - \omega_1^2} = \frac{\omega\sqrt{39}}{8}.$$

Now

$$A = \frac{b^2+c^2}{3} = \frac{(2a)^2+(3a)^2}{3} = \frac{13a^2}{3},$$

Similarly,

$$B = \frac{10a^2}{3} \text{ and } C = \frac{5a^2}{3}.$$

Since, body is moving under no forces, therefore Euler's equations of motion are

$$A\dot{\omega}_1 - (B-C)\,\omega_2\omega_3 = 0,$$

$$B\dot{\omega}_2 - (C-A)\,\omega_3\omega_1 = 0,$$

$$C\dot{\omega}_3 - (A-B)\,\omega_1\omega_2 = 0.$$

Substituting the values of A, B, C, we have

$$13\dot{\omega}_1 = 5\omega_2\omega_3, \quad (1)$$

$$10\dot{\omega}_2 = -8\omega_3\omega_1, \quad (2)$$

$$5\dot{\omega}_3 = 3\omega_1\omega_2. \quad (3)$$

Dividing (1) by (2), we have

$$\frac{13\dot{\omega}}{10\dot{\omega}_2} = \frac{-5\omega_2}{8\omega_1}$$

$$\therefore \quad 52\omega_1\dot{\omega}_1 + 25\omega_2\dot{\omega}_2 = 0.$$

Integrating we get

$$52\omega_1^2 + 25\omega_2^2 = const. = C_1.$$

Initially, $\omega_1 = \frac{5}{8}\omega,\ \omega_2 = 0,$

$$\therefore \quad C_1 = \frac{325\omega^2}{16}.$$

$$\therefore \quad 52\omega_1^2 + 25\omega_2^2 = \frac{325\omega^2}{16}. \quad (4)$$

Again dividing (1) by (3), we have

$$\frac{13\dot{\omega}_1}{5\dot{\omega}_3} = \frac{5\omega_3}{3\omega_1},$$

$$\therefore \quad 39\omega_1\dot{\omega}_1 - 25\omega_3\dot{\omega}_3 = 0.$$

Integrating, we have

$$39\omega_1^2 - 25\omega_3^2 = const.$$

Initially,

$$\omega_1 = \frac{5\omega}{8}, \omega_3 = \frac{\sqrt{39}\omega}{8},$$

$$\therefore \quad C_2 = 0.$$

$$\therefore \quad 39\omega_1^2 - 25\omega_3^2 = 0. \tag{5}$$

From (2) and (5),

$$10\dot{\omega}_2 = \frac{-8\sqrt{39}\omega_1^2}{5}$$

$$= \frac{-8\sqrt{39}}{5\sqrt{52}}\left(\frac{325\omega^2}{16} - 25\omega_2^2\right) \text{ by (4)}$$

$$= -\frac{4\sqrt{3}}{5}\left(\frac{325\omega^2}{16} - 25\omega_2^2\right).$$

Integrating, we have

$$\frac{4\sqrt{3}t}{5} = -10\int \frac{d\omega_2}{\left(\frac{325\omega^2}{16} - 25\omega_2^2\right)}$$

$$= -10.\frac{4}{5\sqrt{325}\omega} \tan h^{-1} \frac{5\omega_2}{\frac{\sqrt{325}}{4}\omega}.$$

$$= \frac{-8}{5\sqrt{13}\omega} \tan h^{-1} \frac{20\omega_2}{\sqrt{325}\,\omega}.$$

$$\therefore \quad -\frac{\sqrt{39}\omega t}{2} = \tan h^{-1} \frac{20\omega_2}{\sqrt{325}\omega}.$$

$$\therefore \quad \omega_2 = -\frac{\omega\sqrt{325}}{20} \tan \text{h} \frac{\sqrt{39}\omega t}{2},$$

$$= -\frac{\omega\sqrt{13}}{4} \tan \text{h} \frac{\sqrt{39}\omega t}{2}.$$

But when $t \to \infty$, $\tan h \dfrac{\sqrt{39}\omega t}{2} \to l$,

$$\therefore \qquad \omega_2 \to -\frac{\omega\sqrt{13}}{4}.$$

From (4) and (5), we get

$$\omega_1 = 0, \omega_3 = 0.$$

Hence, the body ultimately rotates about the mean axis.

2. A lamina rotating with uniform angular velocity n about an axis passing through its centre of gravity perpendicular to its plane has an additional angular velocity λn impressed upon it about the axis of least moment; $(A < B < C)$, where

$$\lambda^2 = \frac{(B+A)}{(B-A)}.$$

Prove that at time t, its angular velocities are

$$\lambda n \sec hnt, \lambda n \tan hnt, n \sec hnt.$$

It is given that $A < B < C$, therefore OA is the axis of least moment. Initially

$$\omega_1 = \lambda n, \omega_2 = 0, \omega_3 = n.$$

In the absence of external forces Euler's equations of motion are

$$A\dot{\omega}_1 - (B-C)\,\omega_2\omega_3 = 0$$

$$B\dot{\omega}_2 - (C-A)\,\omega_3\omega_1 = 0$$

$$C\dot{\omega}_3 - (A-B)\,\omega_1\omega_2 = 0.$$

For lamina $C = A + B$.

Substituting this in the above equations, we have

$$B\dot{\omega}_2 - (C-A)\,\omega_3\omega_1 = 0, \; A\dot{\omega}_1 = -A\omega_2\omega_3$$

$$\therefore \qquad \dot{\omega}_1 = -\omega_2\omega_3 \qquad (1)$$

and

$$\dot{\omega}_2 = \omega_3\omega_1, \qquad (2)$$

$$(A+B)\,\dot{\omega}_3 = (A-B)\,\omega_1\omega_2$$

$$\lambda^2\dot{\omega}_3 = -\,\omega_1\omega_2 \qquad (3)$$

Dividing (1) by (2), we get

$$\frac{\dot{\omega}_1}{\dot{\omega}_2} = -\frac{\omega_2}{\omega_1}$$

$$\therefore \qquad \omega_1\dot{\omega}_1 + \omega_2\dot{\omega}_2 = 0.$$

Integrating, we get

$$\omega_1^2 + \omega_2^2 = const. = C_1.$$

Initially,

$$\omega_1 = \lambda n, \omega_2 = 0,$$

$$\therefore \quad C_1 = \lambda^2 n^2.$$

$$\therefore \quad \omega_1^2 + \omega_2^2 = \lambda^2 n^2. \tag{4}$$

Dividing (1) by (3), we get

$$\frac{\dot{\omega}_1}{\lambda^2 \dot{\omega}_3} = \frac{\omega_3}{\omega_1},$$

$$\therefore \quad \omega_1 \dot{\omega}_1 - \lambda^2 \dot{\omega}_3 \omega_3 = 0.$$

Integrating, we get

$$\omega_1^2 - \lambda^2 \omega_3^2 = const. = C_2.$$

Initially,

$$\omega_1 = \lambda n, \omega_3 = n,$$

$$\therefore \quad C_2 = 0,$$

Hence,

$$\omega_1^2 = \lambda^2 \omega_3^2$$

$$\therefore \quad \omega_1 = \lambda \omega_3 \tag{5}$$

From (2), we have

$$\dot{\omega}_2 = \omega_3 \omega_1 = \frac{\omega_1^2}{\lambda}, \text{ using (4).}$$

$$= \frac{1}{\lambda}(\lambda^2 n^2 - \omega_2^2), \text{ using (5).}$$

Integrating, we get

$$t = \lambda \int \frac{d\omega_2}{\lambda^2 n^2 - \omega_2^2}$$

$$= \frac{1}{n} \tan h^{-1} \frac{\omega_2}{\lambda n} + C_3$$

Initially,

$$t = 0, \omega_2 = 0,$$

$$\therefore \quad C_3 = 0,$$

Hence,

$$t = \frac{1}{n} \tan h^{-1} \frac{\omega_2}{\lambda n}.$$

$$\therefore \quad \omega_2 = \lambda n \tan hnt.$$

Putting value of ω_2 in (4) we get

$$\omega_1^2 = \lambda^2 n^2 - \lambda^2 n^2 \tan h^2 nt = \lambda^2 n^2 \sec h^2 nt.$$

$$\therefore \quad \omega_1 = \lambda n \sec hnt.$$

Thus at time t ω_1, ω_2, ω_3 are given by

$$\lambda n \sec hnt, \lambda n \tan hnt, n \sec hnt.$$

3. A uniform elliptic disc is free to move about a focus and is set rotating with angular velocity Ω about an axis perpendicular to the corresponding latus rectum and making an angle θ with the plane of the disc. If $\cos 2\theta = \frac{A}{B}$ where A, B, are M.I. of the disc about major axis and latus rectum respectively, prove that after time t the angular velocity of the disc about the major axis will be

$$\Omega \sqrt{\frac{(B-A)}{2B}} \sec h \left[\Omega t \sqrt{\frac{(B-A)}{2B}} \right].$$

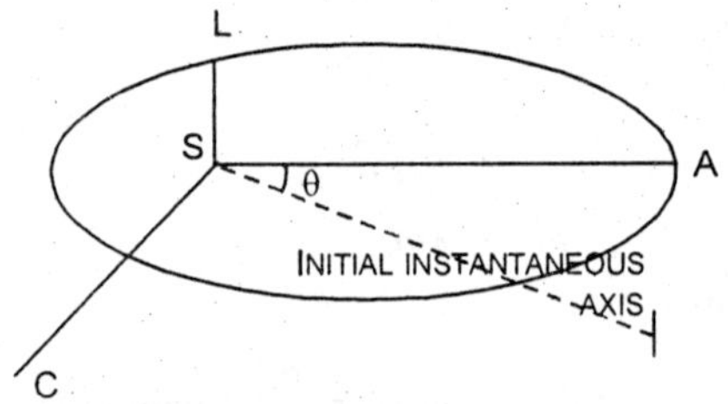

Fig. 9.3

Let S be the focus and SL the semi-latus rectum. Thus SA, SL, SC are the principal axes at S. Let SI be the initial axis of the rotation, which is given perpendicular to SL, therefore it lies in the plane CSA, making an angle θ with SA, so that direction cosines of SI are

$$\cos\theta, \cos(90° - \theta), \cos 90° \text{ i.e., } \cos\theta, \sin\theta, 0.$$

In the absence of external forces Euler's equations of motion are

$$A\dot{\omega}_1 - (B - C)\,\omega_2\omega_3 = 0,$$

$$B\dot{\omega}_2 - (C - A)\,\omega_3\omega_1 = 0,$$

$$C\dot{\omega}_3 - (A - B)\,\omega_1\omega_2 = 0.$$

For lamina $C = A + B$.

Substituting this in the above equations, we have

$$A\dot{\omega}_1 = -A\,\omega_2\omega_3 \;\therefore\; \dot{\omega}_1 = -\omega_2\omega_3, \tag{1}$$

Similarly, we get

$$\dot{\omega}_2 = \omega_3\omega_1, \tag{2}$$

$$\lambda^2 \dot{\omega}_3 = -\omega_1\omega_2 \tag{3}$$

where

$$\lambda^2 = \frac{A+B}{B-A}.$$

Dividing (1) By (2), we get

$$\frac{\dot{\omega}_1}{\dot{\omega}_2} = -\frac{\omega_2}{\omega_1}$$

$$\therefore \qquad \omega_1\dot{\omega}_1 + \omega_2\dot{\omega}_2 = 0.$$

Integrating, we get

$$\omega_1 + \omega_2^2 = \text{const.} = C_1$$

Initially,

$$\omega_1 = \Omega\cos\theta, \omega_2 = 0,$$

$$\therefore \qquad C_1 = \Omega^2\cos^2\theta,$$

$$\therefore \qquad \omega_1^2 + \omega_2^2 = \Omega^2\cos^2\theta. \tag{4}$$

Dividing (1) by (3), we get

$$\frac{\dot{\omega}_1}{\lambda^2\dot{\omega}_3} = \frac{\omega_3}{\omega_1},$$

$$\therefore \qquad \omega_1\dot{\omega}_1 - \lambda^2\dot{\omega}_3\omega_3 = 0.$$

Integrating, we get

$$\omega_1^2 - \lambda^2\omega_3^2 = \text{const.} = C_2.$$

Initially,

$$\omega_1 = \Omega\cos\theta, \omega_3 = \Omega\sin\theta,$$

$$\therefore \qquad C_2 = \Omega^2(\cos^2\theta - \lambda^2\sin^2\theta) = \Omega^2\left(\cos^2\theta - \frac{B+A}{B-A}\sin^2\theta\right)$$

$$= \Omega^2\left[\frac{B(\cos^2\theta - \sin^2\theta) - A(\cos^2\theta + \sin^2\theta)}{B-A}\right]$$

$$= \Omega^2\left(\frac{B\cos 2\theta - A}{B-A}\right) = 0,$$

Since $\qquad \cos 2\theta = \dfrac{A}{B}.$

Hence,

$$\omega_1^2 = \lambda^2\omega_3^2 \tag{5}$$

Now equation (2) gives

$$\dot{\omega}_2 = \omega_3\omega_1.$$

Using (5), we get

$$\dot{\omega}_2 = \frac{\omega_1^2}{\lambda} = \frac{1}{\lambda}(\Omega^2\cos^2\theta - \omega_2^2), \text{ using (4).}$$

Integrating, we get

$$t = \lambda\int\frac{d\omega_2}{\Omega^2\cos^2\theta - \omega_2^2}$$

$$= \frac{\lambda}{\Omega\cos\theta}\tan h^{-1}\frac{\omega_2}{\Omega\cos\theta} + C_3.$$

Initially,

$$t = 0, \omega_2 = 0,$$

$$\therefore \quad C_3 = 0.$$

Hence,

$$t = \frac{\lambda}{\Omega\cos\theta}\tan h^{-1}\frac{\omega_2}{\Omega\cos\theta},$$

$$\therefore \quad \omega_2 = \Omega\cos\theta\tan h\frac{\Omega t\cos\theta}{\lambda}$$

$$= \Omega\sqrt{\frac{(A+B)}{2B}}\tan h\,\Omega t\sqrt{\frac{B-A}{2B}}$$

Since,

$$\frac{\cos\theta}{\lambda} = \sqrt{\frac{A+B}{2B}}\sqrt{\frac{B-A}{B+A}} = \sqrt{\frac{B-A}{2B}}.$$

Now from (4), we have

$$\omega_1^2 = \Omega^2\cos^2\theta - \omega_2^2$$

$$= \Omega^2\cos^2\theta - \Omega^2\cos^2\theta\tan h^2\frac{\Omega t\cos\theta}{\lambda}$$

$$= \Omega^2\cos^2\theta\left(1 - \tan h^2\frac{\Omega t\cos\theta}{\lambda}\right)$$

$$= \Omega^2\cos^2\theta\sec h^2\frac{\Omega t\cos\theta}{\lambda}$$

$$\therefore \quad \omega_1 = \Omega \cos\theta \sec h \frac{\Omega t \cos\theta}{\lambda}$$

$$= \Omega \sqrt{\frac{A+B}{2B}} \sec h\Omega t \sqrt{\frac{B-A}{2B}}$$

$$\therefore \quad \frac{\cos\theta}{\lambda} = \sqrt{\frac{B-A}{2B}}.$$

Now from (5), we have

$$\omega_3 = \frac{\omega_1}{\lambda} = \sqrt{\frac{B-A}{A+B}} \Omega \sqrt{\frac{A+B}{2B}} \sec h\Omega t \sqrt{\frac{B-A}{2B}}$$

$$= \Omega \sqrt{\frac{B-A}{2B}} \sec h\Omega t \sqrt{\frac{B-A}{2B}}.$$

4. The principal moments of inertia of a body at the centre of mass are A, $3A$, $6A$. The body is so rotated that its angular velocities about the axes are $3n$, $2n$, n respectively. If in the subsequent motion under no forces, ω_1, ω_2, ω_3 denote the angular velocities about the principal axes at time t, prove that

$$\omega_1 = 3\omega_3 = \frac{9n}{\sqrt{5}} \sec hu \text{ and } \omega_3 = 3n \tan hu,$$

where

$$u = 3nt + \frac{1}{2} \log 5.$$

In the absence of external forces Euler's equations of motion are

$$A\dot{\omega}_1 - (B-C)\,\omega_2\omega_3 = 0$$

$$B\dot{\omega}_2 - (C-A)\,\omega_3\omega_1 = 0$$

$$C\dot{\omega}_3 - (A-B)\,\omega_1\omega_2 = 0.$$

It is given that

$$A = A,\ B = 3A, C = 6A.$$

Hence, Euler's equation reduce to

$$A\dot{\omega}_1 = (3A - 6A)\,\omega_2\omega_3.$$

$$\therefore \quad \dot{\omega}_1 = -3\,\omega_2\omega_3. \tag{1}$$

Similarly, we get

$$3\dot{\omega}_2 = 5\,\omega_3\omega_1 \tag{2}$$

$$3\dot{\omega}_3 = -\omega_2\omega_1 \quad (3)$$

Dividing (1) by (3), we get

$$\frac{\dot{\omega}_1}{3\dot{\omega}_3} = \frac{3\omega_3}{\omega_1}$$

$$\therefore \quad \omega_1\dot{\omega}_1 = 9\omega_3\dot{\omega}_3$$

Integrating, we get

$$\omega_1^2 = 9\omega_3^2 + C_1 .$$

Initially,

$$\omega_1 = 3n,\ \omega_3 = n,$$

$$\therefore \quad C_1 = 0.$$

$$\therefore \quad \omega_1^2 = 9\omega_3^2$$

$$\therefore \quad \omega_1 = 3\omega_2 \quad (4)$$

Again dividing (2) by (3), we get

$$\frac{\dot{\omega}_2}{\dot{\omega}_3} = -\frac{5\omega_3}{\omega_2}$$

$$\therefore \quad \omega_2\dot{\omega}_2 = -5\omega_3\,\dot{\omega}_3$$

Integrating, we get

$$\omega_2^2 + 5\omega_3^2 = C_2.$$

Initially,

$$\omega_2 = 2n,\ \omega_3 = n,$$

$$\therefore \quad C_2 = 9n^2.$$

$$\therefore \quad 5\omega_3^2 = 9n^2 - \omega_2^2 \quad (5)$$

Now equation (2) gives

$$3\dot{\omega}_2 = 5\omega_3\omega_1$$

$$= 15\omega_3^2 \text{, by (4).}$$

$$\therefore \quad \dot{\omega}_2 = 5\omega_3^2 = 9n^2 - \omega_2^2 \ \text{ by (5).}$$

Integrating, we get

$$t = \int \frac{d\omega_2}{9n^2 - \omega_2^2}$$

$$= \frac{1}{3n} \tan h^{-1} \frac{\omega_2}{3n} + C_3 .$$

Initially,

$$t = 0, \omega_2 = 2n,$$

$\therefore$

$$C_2 = -\frac{1}{3} n \tan h^{-1} \frac{2}{3}$$

$$= -\frac{1}{3n} . \frac{1}{2} \log \frac{3+2}{3-2}$$

$$= -\frac{1}{3n} \log \sqrt{5}$$

Hence,

$$t = \frac{1}{3n} \tan h^{-1} \frac{\omega_2}{3n} - \frac{1}{3n} \log \sqrt{5} .$$

$$\therefore \quad 3nt + \log \sqrt{5} = \tan h^{-1} \frac{\omega_2}{3n},$$

$$\therefore \quad \omega_2 = 3n \tan h \,(3nt + \log \sqrt{5})$$

$$= 3n \tan h\, u,$$

where

$$u = 3nt + \log \sqrt{5} .$$

Also, we have

$$5\omega_3^2 = 9n^2 - \omega_2^2$$

$$= 9n^2 - 9n^2 \tan h^2 u$$

$$\therefore \quad \omega_3 = \frac{3n \sec hu}{\sqrt{5}} .$$

Now from (4), we get

$$\omega_1 = 3\omega_3 = \frac{9n}{\sqrt{5}} \sec\ hu.$$

5. A uniform right circular cone of vertical angle 2α moves under no forces except at its vertex which is fixed. It is set rotating about a generator. Show that its axis describes in space a right cone of angle 2β, where

$$\tan \beta = \frac{1}{2} \tan \alpha + 9 \cot \alpha$$

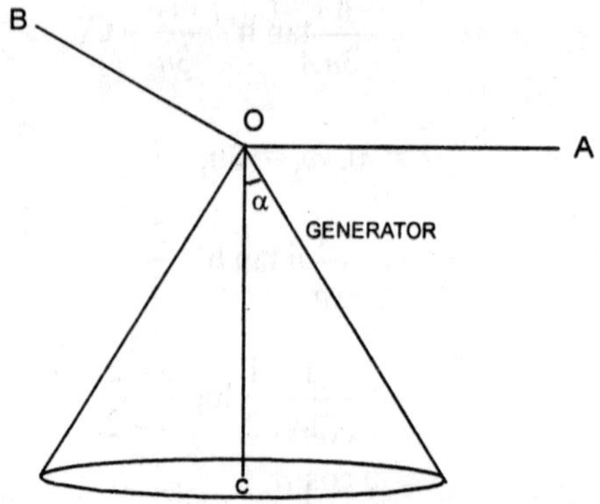

Fig. 9.4

Let O be the vertex of the cone, about which it rotates. The principal axes at the vertex O are the axis of the cone, and two lines OA, OB through O and at right angles to the axis OC of the cone.

Let ω be the initial angular velocity, with which the cone is set rotating about the generator. Thus initially,

$$\omega_1 = \omega \sin \alpha,\ \omega_2 = 0,\ \omega_3 = \omega \cos \alpha.$$

Here,

$$A = B = \frac{3M}{80}(h^2 + 4a^2) + M\left(\frac{3h}{4}\right)^2 = \frac{3M}{20}(a^2 + 4h^2)$$

where $h = \cot \alpha$.

$$C = \frac{3Ma^2}{10}$$

In the absence of external forces Euler's equations of motion are

$$A\dot{\omega}_1 - (B - C)\,\omega_2\omega_3 = 0,$$

$$B\dot{\omega}_2 - (C - A)\,\omega_3\omega_1 = 0,$$

$$C\dot{\omega}_3 - (A - B)\,\omega_1\omega_2 = 0.$$

Here $B = A$.

Hence, Euler's equations reduce to

$$A\dot{\omega}_1 = (A - C)\,\omega_2\omega_3$$

$$\therefore \quad \dot{\omega}_1 = \frac{(A - C)}{A}\,\omega_2\omega_3 \qquad (1)$$

$$B\dot{\omega}_1 = (C - A)\,\omega_3\omega_1$$

$$\therefore \qquad \dot{\omega}_2 = \frac{-(A-C)}{A}\omega_3\omega_1 \qquad (2)$$

$$C\dot{\omega}_3 = 0$$

$$\therefore \qquad \dot{\omega}_3 = 0 \qquad (3)$$

Integrating, we get

$$\omega_3 = const = C_1$$

Initially,

$$\omega_3 = \omega \cos \alpha$$

$$\therefore \qquad C_1 = \omega \cos \alpha$$

$$\therefore \qquad \omega_3 = \omega \cos \alpha. \qquad (4)$$

Dividing (1) by (2), we get

$$\frac{\dot{\omega}_1}{\dot{\omega}_2} = -\frac{\omega_2}{\omega_1}$$

$$\therefore \qquad \omega_1\dot{\omega}_2 + \omega_2\dot{\omega}_2 = 0.$$

Integrating, we get

$$\omega_1\omega_2^2 = const. = C_1$$

Initially,

$$\omega_1 = \omega \sin \alpha,\ \omega_2 = 0,$$

$$\therefore \qquad C_2 = \omega^2 \sin^2 \alpha,$$

$$\therefore \qquad \omega_1^2 + \omega_2^2 = \omega^2 \sin^2\alpha. \qquad (5)$$

The invariable line is fixed in space and its direction cosines are proportional to $A\omega_1$, $B\omega_2$, $C\omega_3$.

If θ be the angle the invariable line makes with axis OC, then

$$\cos \theta = \frac{C\omega_3}{\sqrt{A^2\omega_1^2 + B^2\omega_2^2 + C^2\omega_3^2}}$$

$$= \frac{C\omega_3}{\sqrt{A^2(\omega_1^2 + \omega_2^2) + C^2\omega_3^2}}$$

Using (4) and (5), we get

$$\cos \theta = \frac{C\omega \cos\alpha}{\sqrt{A^2\omega^2 \sin^2 \alpha + C^2\omega^2 \cos^2 \alpha}}$$

$$= \frac{C}{\sqrt{A^2 \tan^2 \alpha + C^2}}$$

$$\therefore \qquad \tan\theta = \frac{A\tan\alpha}{C}$$

$$= \frac{1}{2}\left(\frac{a^2+4h^2}{a^2}\right)\tan\alpha$$

$$= \frac{1}{2}(1+4\cot^2\alpha)\tan\alpha$$

$$= \frac{1}{2}\tan\alpha + 2\cot\alpha,$$

which is constant.

Now $\qquad \theta = \beta$

$$\therefore \qquad \tan\beta = \frac{1}{2}\tan\alpha + 2\cot\alpha$$

Hence, axis of the cone describes about the invariable line a right cone of vertical angle 2β, where

$$\tan\beta = \frac{1}{2}\tan\alpha + 2\cot\alpha\cdot$$

6. A thin elliptic plate of semi-axes a, b $(a > b)$ can turn freely about its centre which is fixed. It is set in motion with an angular velocity about an axis in its plane equally inclined to the axes of the ellipse. Show that the instantaneous axis will again be in the plane of the plate, after time

$$2\int_0^{\lambda} (\lambda^4 - x^4)^{-1/2}\,dx$$

where

$$\lambda^2 = \frac{n^2(a^2-b^2)}{2(a^2+b^2)}.$$

Initially,

$$\omega_1 = n\cos 45° = \frac{n}{\sqrt{2}},\ \omega_2 = n\sin 45° = \frac{n}{\sqrt{2}},\ \omega_3 = 0$$

Euler's equations in the absence of external forces reduce to

$$A\dot{\omega}_1 = (B-C)\,\omega_2\omega_1,$$

$$B\dot{\omega}_2 = (C-A)\,\omega_3\omega_1,$$

$$C\dot{\omega}_3 = (A-B)\,\omega_1\omega_2.$$

For the elliptic plate, we have

$$A = \frac{Mb^2}{4},\ B = \frac{Ma^2}{4},\ C = A + B \text{ for lamina.}$$

Substituting these values in the above equations, we have

$$A\dot{\omega}_1 = -A\omega_2\omega_3$$

$$\therefore \quad \dot{\omega}_1 = -\omega_2\omega_3. \qquad (1)$$

Similarly, we get

$$\dot{\omega}_2 = \omega_3\omega_1 \qquad (2)$$

$$\dot{\omega}_3 = \frac{-2\lambda^2}{n^2}\omega_1\omega_2 \qquad (3)$$

where

$$\lambda^2 = \frac{n^2}{2}\cdot\frac{a^2-b^2}{a^2+b^2} = \frac{n^2(B-A)}{2(B+A)}$$

Dividing (2) by (3), we get

$$\frac{\dot{\omega}_2}{\dot{\omega}_3} = -\frac{n^2}{2\lambda^2}\cdot\frac{\omega_3}{\omega_2},$$

$$\therefore \quad 2\lambda^2\omega_2\dot{\omega}_2 + n^2\omega_3\dot{\omega}_3 = 0.$$

Integrating, we get

$$2\lambda^2\omega_2^2 = -n^2\omega_3^2 + C_1.$$

Initially,

$$\omega_2 = \frac{n}{\sqrt{2}},\ \omega_3 = 0,$$

$$\therefore \quad C_1 = \lambda^2 n^2.$$

$$\therefore \quad 2\lambda^2\omega_2^2 = \lambda^2 n^2 - n^2\omega_3^2$$

$$\therefore \quad 2\lambda^2\omega_2^2 = n^2(\lambda^2 - \omega_3^2). \qquad (4)$$

Dividing (1) by (3) we get

$$\frac{\dot{\omega}_1}{\dot{\omega}_3} = \frac{n^2}{2\lambda^2}\cdot\frac{\omega_3}{\omega_1},$$

$$\therefore \quad 2\lambda^2\dot{\omega}_1\omega_1 = n^2\omega_3\dot{\omega}_3$$

Integrating, we get

$$2\lambda^2\omega_1^2 = n^2\omega_3^2 + C_2.$$

Initially,

$$\omega_1 = \frac{n}{\sqrt{2}},\ \omega_3 = 0$$

$$\therefore \quad C_2 = \lambda^2 n^2,$$

$$\therefore \quad 2\lambda^2\omega_1^2 = n^2(\lambda^2 + \omega_3^2). \tag{5}$$

From (4) and (5) we get

$$4\lambda^4\omega_1^2\omega_2^2 = n^4(\lambda^4 - \omega_3^4)$$

$$\therefore \quad \frac{2\lambda^2}{n^2}\omega_1\omega_2 = \pm(\lambda^4 - \omega_3^4),$$

$$\therefore \quad \dot{\omega}_3 = \pm\sqrt{\lambda^4 - \omega_3^4} \text{ by (3).} \tag{6}$$

ω_3 will be maximum, when $\dot{\omega}_3 = 0$,

or when

$$\lambda^4 - \omega_3^4 = 0,$$

$$\therefore \quad \omega_3 = \lambda$$

Integrating (6) we get

$$t_1 = (\text{time from } \omega_3 = 0 \text{ to } \omega_3 = \lambda) = \int_0^\lambda \frac{d\omega_3}{(\lambda^4 - \omega_3^4)}$$

(Since ω_3 increases from 0 to λ, therefore $\dot{\omega}_3$ is taken positive).

and

$$t_2 = (\text{time from } \omega_3 = \lambda \text{ to } \omega_3 = 0) = -\int_\lambda^0 \frac{d\omega_3}{\sqrt{\lambda^4 - \omega_3^4}} = \int_0^\lambda \frac{d\omega_3}{\sqrt{\lambda^4 - \omega_3^4}}$$

(Since ω_3 decreases from λ to 0, therefore $\dot{\omega}_3$ is taken negative).

Hence, the total time when the instantaneous axis will be again in the plane of plate is

$$t_1 + t_2 = 2\int_0^\lambda \frac{d\omega_3}{\sqrt{\lambda^4 - \omega_3^4}} = 2\int_0^\lambda \frac{dx}{\sqrt{\lambda^4 - x^4}}.$$

7. A body is rotating about its centre of mass under no forces and A, B, C ($A > B > C$) are the principal moments of inertia at the centre of mass. If initially ω_3 is positive and ω_1 is negative and if $\Omega^2 = 2BT$, where Ω is the

angular momentum and T is the kinetic energy, then show that

$$\omega_2 = \frac{\Omega}{B}\tan h\,\tau,$$

$$\omega_1 = -\frac{\Omega}{B}\sqrt{\frac{B(B-C)}{A(A-C)}}\sec h\,\tau,\ \omega_3 = \frac{\Omega}{B}\sqrt{\frac{B(A-B)}{C(A-C)}}\sec h\tau,$$

where

$$\tau = \frac{\Omega}{B}\sqrt{\frac{(A-B)(B-C)}{AC}}(t-t_0)$$

What happens when t increases indefinitely ?

Euler's equations in the absence of external forces reduce to

$$A\dot{\omega}_1 = (B-C)\omega_2\omega_3, \tag{1}$$

$$B\dot{\omega}_2 = (C-A)\omega_3\omega_1, \tag{2}$$

$$C\dot{\omega}_3 = (A-B)\omega_1\omega_2, \tag{3}$$

When $\Omega^2 = 2BT$

We know that

$$A\omega_1^2 + B\omega_2^2 + C\omega_3^2 = 2T. \tag{4}$$

$$A^2\omega_1^2 + B^2\omega_2^2 + C^2\omega_3^2 = \Omega^2 = 2BT. \tag{5}$$

Equations (4) and (5) can be written as

$$A\omega_1^2 + C\omega_3^2 - (2T - B\omega_2^2) = 0,$$

$$A^2\omega_1^2 + C^2\omega_3^2 - B(2T - B\omega_2^2) = 0.$$

Solving these, we get

$$\frac{\omega_1^2}{C(B-C)} = \frac{\omega_3^2}{A(A-B)} = \frac{2T - B\omega_2^2}{AC(A-C)} \tag{6}$$

Substituting values of ω_1^2 and ω_3^2 from (6) into (2), we get

$$\dot{\omega}_2^2 = \left(\frac{C-A}{B}\right)^2 AC(A-B)(B-C).\left\{\frac{2T - B\omega_2^2}{AC(A-C)}\right\}^2$$

$$= \frac{(A-B)(B-C)}{B^2 AC}.(2T - B\omega_2^2)^2$$

$$\therefore \quad \dot{\omega}_2 = \sqrt{\frac{(A-B)(B-C)}{AC}}.\frac{(2T - B\omega_2^2)}{B}$$

$$= \sqrt{\frac{(A-B)(B-C)}{AC}} \frac{2BT}{B^2}\left(1 - \frac{B^2\omega_2^2}{2BT}\right)$$

$$= \sqrt{\frac{(A-B)(B-C)}{AC}} \frac{\Omega^2}{B^2}\left(1 - \frac{B^2\omega_2^2}{\Omega^2}\right).$$

Integrating, we get

$$\frac{\Omega}{B}\sqrt{\frac{(A-B)(B-C)}{AC}} \int_{t_0}^{t} dt = \int \frac{\frac{B}{\Omega}}{1 - \frac{B^2\omega_2^2}{\Omega^2}} d\omega_2$$

$$\therefore \quad \frac{\Omega}{B}\sqrt{\frac{(A-B)(B-C)}{AC}} (t - t_0) = \frac{1}{2}\log \frac{1 + \frac{B\omega_2}{\Omega}}{1 - \frac{B\omega_2}{\Omega}}$$

$$\tau = \frac{1}{2}\log \frac{1 + \frac{B\omega_2}{\Omega}}{1 - \frac{B\omega_2}{\Omega}}$$

$$\therefore \quad 1 + \frac{B\omega_2}{\Omega} = \left(1 - \frac{B\omega_2}{\Omega}\right)e^{2\tau}$$

$$\therefore \quad \frac{B\omega_2}{\Omega}(1 + e^{2\tau}) = e^{2\tau} - 1$$

$$\therefore \quad \omega_2 = \frac{\Omega}{B}\cdot\frac{e^{2\tau} - 1}{1 + e^{2r}}$$

$$= \frac{\Omega}{B}\cdot\frac{e^{\tau} - e^{-\tau}}{e^{\tau} + e^{-\tau}}$$

$$= \frac{\Omega}{B}\tan h\tau$$

Then from (6), we have

$$\omega_1^2 = \frac{(B-C)}{A(A-C)}.(2T - B\omega_2^2)$$

$$= \frac{(B-C)}{A(A-C)}\left\{2T - \frac{B.\Omega^2}{B^2}\tan h^2\tau\right\}$$

$$= \frac{(B-C)\Omega^2}{AB(A-C)}(1-\tan h^2\tau)$$

$$= \frac{(B-C)\Omega^2}{AB(A-C)}\sec h^2\tau$$

$$\therefore \qquad \omega_1 = -\frac{\Omega}{B}\sqrt{\frac{B(B-C)}{A(A-C)}}\sec h\tau,$$

since ω_1 is given negative.

Similarly,

$$\omega_3 = \frac{\Omega}{B}\sqrt{\frac{B(A-B)}{A(A-C)}}\sec h\tau$$

when $t \to \infty$, $\omega_2 \to \frac{\Omega}{B}$ *i.e.* $\sqrt{\frac{2T}{B}}$ and ultimately

$$\omega_2 = \sqrt{\frac{2T}{B}} = \frac{\Omega}{B},$$

from (6)

$$\omega_1 = \omega_3 = 0.$$

8. If T be the K.E. and $\vec{G}$ be the moment of extrnal forces about the instantaneous axis and $\vec{\omega}$ the resulant angular velocity, then prove that

$$\frac{dT}{dt} = \vec{G}.\vec{\omega}.$$

We have

$$\vec{G} = G_1\vec{i} + G_2\vec{j} + G_3\vec{k},$$

$$\vec{\omega} = \omega_1\vec{i} + \omega_2\vec{j} + \omega_3\vec{k}$$

$$\therefore \qquad \vec{G}.\vec{\omega} = G_1\omega_1 + G_2\omega_2 + G_3\omega_3.$$

Now

$$T = \frac{1}{2}(A\omega_1^2 + B\omega_2^2 + C\omega_3^2)$$

$$\therefore \quad \frac{dT}{dt} = A\omega_1\dot{\omega}_1 + B\omega_2\dot{\omega}_2 + C\omega_3\dot{\omega}_3$$

$$= \omega_1[G_1 + (B-C)\omega_2\omega_3] + \omega_2[G_2 + (C-A)\omega_3\omega_1] + \omega_3[G_3 + (A-B)\omega_1\omega_2]$$

by Euler's equations.

$$= \omega_1 G_1 + \omega_2 G_2 + \omega_3 G_3 = \vec{G}.\vec{\omega}\,.$$

Examples of use of fixed axes

1. (a) *Motion of a sphere on rough plane, under the action of forces, the resultant of which passes through the centre of the sphere*

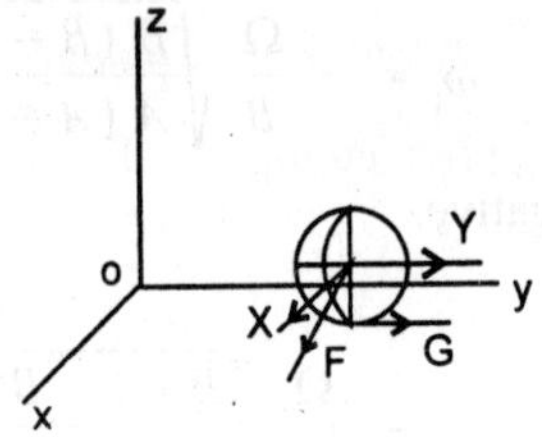

Fig. 9.5

Let us take the axes Ox, Oy in the plane and Oz at right angles to it. If x, y, z are the co-ordinates of the centre of the sphere and m its mass, the linear equations of motion are

$$m\ddot{x} = X + F,\ m\ddot{y} = Y + G,\ m\ddot{z} = Z + R. \qquad (1)$$

where X, Y, Z are the components of the forces; F, G the components of the friction and R is the normal reaction.

If $\omega_1, \omega_2, \omega_3$ be the components of the angular velocity, the equations of moments about axes through the centre parallel to the co-ordinate axes are

$$\frac{2}{5}ma^2\dot{\omega}_1 = Ga,\ \frac{2}{5}ma^2\dot{\omega}_2 = -Fa,\ \frac{2}{5}ma^2\dot{\omega}_3 = 0 \qquad (2)$$

The last equation gives

ω_3 = constant, so that spin of the sphere about an axis normal to the plane does not change.

Now, we will express the conditions for rolling, assuming that the friction prevents sliding. They are point of contact of the sphere when the plane is at rest, *i.e.*,

$$\dot{x} - a\omega_2 = 0, \quad \dot{y} + a\omega_1 = 0. \qquad (3)$$

From (1)–(3), we find that

$$m\ddot{x} = \frac{5}{7}X,\ m\ddot{y} = \frac{5}{7}Y, \qquad (4)$$

which determine the motion of the sphere.

(b) If the plane be made to revolve about Oz with angular velocity Ω, provided we retain the same fixed axes, the equations of motion of the sphere are unaltered but instead of (3), we have equations expressing the condition that the point of contact of the sphere and the plane have the same velocity, *i.e.*

$$\dot{x} - a\omega_2 = -\Omega y,\ \dot{y} + a\omega_1 = \Omega x. \tag{5}$$

In the case in which there are no forces except the reactions of the plane, taking moments about the lines parallel to the axes through the point of contact, we get

$$-ma\ddot{y} + \frac{2}{5}ma^2\dot{\omega}_1 = -Ya,\ ma\ddot{x} + \frac{2}{5}ma^2\dot{\omega}_2 = Xa. \tag{6}$$

From (5) and (6), taking $X = Y = 0$ we get

$$\ddot{x} + \frac{2}{7}\Omega\dot{y} = 0,\ \ddot{y} - \frac{2}{7}\Omega\dot{x} = 0.$$

Let $n = \frac{2}{7}\Omega$.

$$\therefore \qquad \ddot{x} + n\dot{y} = 0,\ \ddot{y} - n\dot{x} = 0$$

Multiplying second of these equations by i and adding to the first, we get

$$D(D - in)(x + iy) = 0,$$

where $D = \dfrac{d}{dt}$.

The general solution of this equation is

$$x + iy = A + Be^{int},$$

where A and B are complex constants, Taking

$$A = \alpha + i\beta,\ B = \lambda + i\mu,$$

where $\alpha, \beta, \lambda, \mu$ are real constants, Now we get

$$\begin{aligned} x + iy &= \alpha + i\beta + (\lambda + i\mu)(\cos nt + i \sin nt) \\ &= a + i\beta + \lambda \cos nt - \mu \sin nt + i(\mu \cos nt + \lambda \sin nt) \end{aligned}$$

Equating real and imaginary parts, we get

$$x = \alpha + \lambda \cos nt - \mu \sin nt,$$

$$y = \beta + \mu \cos nt + \lambda \sin nt.$$

$$\begin{aligned} \therefore (x - a)^2 + (y - \beta)^2 &= (\lambda \cos nt - \mu \sin nt)^2 + (\mu \cos nt + \lambda \sin nt)^2 \\ &= \lambda^2 + \mu^2. \end{aligned}$$

Thus, we have the circle in the plane $z = 0$, for the locus of O.

2. *Motion of a heavy sphere on the interior rough surface of a vertical circular cylinder*

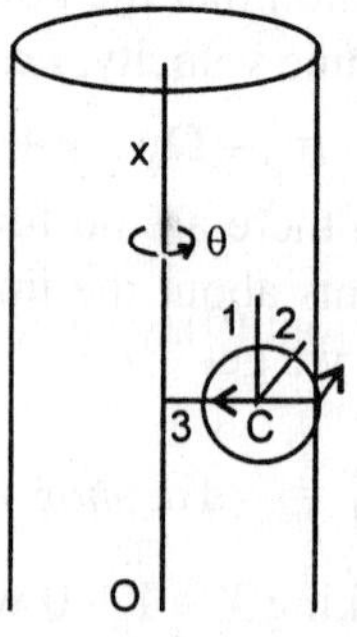

Fig. 9.6

Let us take axes through the centre C of the sphere, $C1$, verticaly upwards, $C3$ to intersect the axis of the cylinder and $C2$ horizontal forming a right handed screw system. Let a, b be the radii of the sphere and cylinder and θ the angle which the plane through the axis of the cylinder and the centre of the sphere makes with a fixed vertical plane. The angular velocities of the axes are

$$\theta_1, \theta_2, \theta_3 = \dot{\theta}, 0, 0. \tag{1}$$

If x be measured vertically upwards along the axis of the cylinder, the components of velocity of C are

$$u, v, w = \dot{x}, (b-a)\theta, 0, \tag{2}$$

and if ω_1, ω_2, ω_3 are the components of angular velocity of the sphere, the equations of motion for no sliding of the point of contact are

$$u - a\omega_2 = 0, v + a\omega_1 = 0$$

$$\therefore \quad a\omega_2 = \dot{x}, a\omega_1 = -(b-a)\,\dot{\theta} \tag{3}$$

In order to avoid introducing the forces of friction, we may take moments about the axes parallel to $C1$, $C2$, $C3$ through the point of contact. Thus we have

$$-ma(\dot{v} - w\theta_1 + u\theta_3) + A(\dot{\omega}_1 - \omega_2\theta_3 + \omega_3\theta_2) = 0, \tag{4}$$

$$ma(\dot{u} - v\theta_3 + w\theta_2) + A(\dot{\omega}_2 - \omega_3\theta_1 + \omega_1\theta_3) = mga, \tag{5}$$

$$A(\dot{\omega}_3 - \omega_1\theta_2 + \omega_2\theta_1) = 0 \tag{6}$$

where

$$A = \frac{2}{5}ma^2.$$

On substituting from (1) – (3) and (4) – (6), we have

$$\ddot{\theta} = 0 \tag{7}$$

$$\frac{7}{5}\ddot{x} - \frac{2}{5}a\omega_3\dot{\theta} = -g \tag{8}$$

$$a\dot{\omega}_3 + \dot{x}\dot{\theta} = 0. \tag{9}$$

From (7), we have

$\dot{\theta} = const. = n$, (say), then from (9), we get

$$a\dot{\omega}_3 + n\dot{x} = 0,$$

$$\therefore \quad a\omega_3 + nx = \mu. \tag{10}$$

where μ is a constant.
Then (8) reduces to

$$\frac{7}{5}\ddot{x} - \frac{2}{5}(\mu - nx)\,n = -g.$$

$$\therefore \quad \ddot{x} + \frac{2}{7}n^2 x = \frac{2}{7}\mu n - \frac{5}{7}g. \tag{11}$$

so that the sphere rises and falls in the cylinder with a vertical simple harmonic motion of the period $\sqrt{14\pi}\,/\,n$.
If initially,

$$x = 0,\ \dot{x} = 0,\ \dot{\theta} = n,\ \omega_3 = \mu$$

which means that the sphere has a spin μ about its radius drawn to the axis of the cylinder and the centre has a horizontal velocity $(b - a)n$, then the solution of (11) is

$$x = \left(\frac{\mu}{n} - \frac{5}{2}\cdot\frac{g}{n^2}\right)\left(1 - \cos\sqrt{\frac{2}{7}}nt\right)$$

so that the sphere rises above or descends below its initial level, according as

$$\mu n > \text{or} < \frac{5}{2}g.$$

9.10 MOTION OF SYMMETRICAL TOP

Definition : A top is defined a material body, which is symmetrical about an axis and terminate at one end of the axis in a sharp point, called the *apex* or *vertex*.

(i) Derive Equations of Motion of a Top from Euler's Equations

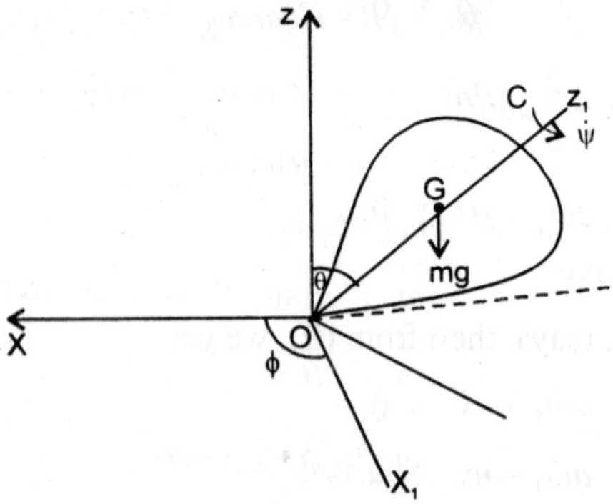

Fig. 9.7

Let top spin with its vertex O fixed in contact with a plane rough enough to prevent slipping. Let OC be the axis of the top and G its C.G., where $OG = h$.

Let Ox, Oy, Oz be the fixed axes and OA, OB, OC the principal axes at O, A the M. I. about OA and C the M.I. about OC. Since top is symmetrical about OC, therefore $A = B$.

Initially, OC is in the plane, zOx and let at any time t, OC be inclined at an angle θ to Oz and the plane zOC making the anle ψ with the fixed plane zOx. Let ω_1, ω_2 ω_3, be the angular velocities of the top about OA, OB, OC respectively, and L_1, L_2, L_3 the moments of the external forces about these axes. The external forces acting on the top are its weight mg acting vertically downward through G in the direction parallel to zO and the reaction acting at O.

The d.c's of Oz with respect to OA, OB, OC are

$$-\sin\theta\cos\phi, \sin\theta\sin\phi, \cos\theta.$$

In case of top $\varphi = 0$.

Therefore d.c.'s of Oz are $-\sin\theta$, 0, $\cos\theta$.

Hence, the resolved parts of $-mg$ acting at G (0, 0, h) are

$mg\sin\theta, 0, -mg\cos\theta$ parallel to OA, OB, OC respectively

the reactions are at the origin O (0, 0, 0)

Hence, we have

$$L_1 = \sum(y_1Z_1 - z_1Y_1) = 0,$$

$$L_2 = \sum(z_1X_1 - x_1Z_1) = mgh\sin\theta,$$

$$L_3 = \sum(x_1Y_1 - y_1X_1) = 0$$

Hence, Euler's equations of motion are

$$A\dot{\omega}_1 - (B-C)\omega_2\omega_3 = 0,$$

$$B\dot{\omega}_2 - (C-A)\omega_3\omega_1 = mgh \sin \theta,$$

$$C\dot{\omega}_3 - (A-B)\omega_1\omega_2 = 0.$$

Euler's *G*eometrical equations are

$$\omega_1 = \dot{\theta}\sin\phi - \dot{\psi}\sin\theta\cos\phi,$$

$$\omega_2 = \dot{\theta}\cos\phi + \dot{\psi}\sin\theta\sin\phi,$$

$$\omega_3 = \dot{\phi} + \dot{\psi}\cos\theta.$$

But for top $\varphi = 0$. Therefore we have

$$\omega_1 = -\dot{\psi}\sin\theta,\ \omega_2 = \dot{\theta},\ \omega_3 = \dot{\psi}\cos\theta.$$

Now $B = A$, hence from third Euler's equation, we get

$$\dot{\omega}_3 = 0$$

$$\therefore \qquad \omega_3 = n\ (const.)$$

$$\therefore \qquad \dot{\psi}\cos\theta = n.$$

Substituting these values of ω_1, ω_2, ω_3. Euler's first equation reduces to

$$A\frac{d}{dt}(-\dot{\psi}\sin\theta) - (A-C)\dot{\theta}\dot{\psi}\cos\theta = 0,$$

$$\therefore \qquad -A\ddot{\psi}\sin\theta - 2A\dot{\theta}\dot{\psi}\cos\theta + C\dot{\theta}\dot{\psi}\cos\theta = 0$$

$$-A\ddot{\psi}\sin\theta - 2A\dot{\theta}n + Cn\dot{\theta} = 0 \qquad (1)$$

Euler's second equation gives

$$A\frac{d\dot{\theta}}{dt} + (C-A)\,\dot{\psi}^2\sin\theta\cos\theta = mgh\sin\theta.$$

$$\therefore \qquad A\ddot{\theta} - A\dot{\psi}^2\sin\theta\cos\theta + C\dot{\psi}^2\sin\theta\cos\theta = mgh\sin\theta.$$

Putting

$$\dot{\psi}\cos\theta = n,$$

we get

$$A\ddot{\theta} - A\dot{x}^2\sin\theta\cos\theta + Cn\dot{\psi}\sin\theta = mgh\sin\theta. \qquad (2)$$

Equations (1) and (2) are equations of motion of a top and are of second order.

Deduction of these equations to first order

Multiplying (1) by sin θ and integrating, we get

$$A\dot{\psi}\sin^2\theta + Cn\cos\theta = D\,(const.) \quad (3)$$

Also $(2)\times 2\dot{\theta} - (1)\times 2\dot{\psi}\sin\theta$, gives

$$2A\dot{\theta}\ddot{\theta} + 2A\dot{\theta}\dot{\psi}^2\sin\theta\cos\theta + 2A\dot{\psi}\ddot{\psi}\sin^2\theta = 2mgh\sin\theta.$$

Integrating, we get

$$A(\dot{\theta}^2 + \dot{\psi}^2\sin^2\theta) + 2mgh\cos\theta = E\,(const). \quad (4)$$

Equations (3) and (4) are equations of motion of the top but are of first order.

The motion due to change in θ is called *nutation* and due to change in Ψ is called *precession*. The general motion of the top about fixed vertex is co-motion of these two.

***(ii)* Derive Equations of Motion of a Top from Lagrange's Equations**

In case of top $A = B$, $\phi = 0$ and

$$\omega_1 = -\dot{\psi}\sin\theta, \omega_2 = \dot{\theta}, \omega_3 = n = \dot{\psi}\cos\theta.$$

Hence, the K.E. T, is given by

$$T = \frac{1}{2}[A(\omega_1^2 + \omega_2^2) + C\omega_3^2]$$

$$= \frac{1}{2}[A(\dot{\psi}^2\sin^2\theta + \dot{\theta}^2) + C\dot{\psi}^2\cos^2\theta]$$

The work function W is given by $W = -mgh\cos\theta$.

Here, the generalized co-odinates are θ and ϕ.

Lagrange's θ equation is

$$\frac{d}{dt}\left(\frac{\partial T}{\partial\dot{\theta}}\right) - \frac{\partial T}{\partial\theta} = \frac{\partial W}{\partial\theta}.$$

$$\therefore \quad \frac{d}{dt}(A\dot{\theta}) - (A\dot{\psi}^2\sin\theta\cos\theta - C\dot{\psi}^2\sin\theta\cos\theta) = mgh\sin\theta$$

$$\therefore \quad A\ddot{\theta} - A\dot{\psi}^2\sin\theta\cos\theta + C\dot{\psi}^2\sin\theta\cos\theta = mgh\sin\theta$$

Similarly, Lagrange's ϕ equation is

$$\frac{d}{dt}\left(\frac{\partial T}{\partial\dot{\psi}}\right) - \frac{\partial T}{\partial\psi} = \frac{\partial W}{\partial\psi}.$$

$$\frac{d}{dt}(A\dot{\psi}\sin^2\theta + C\dot{\psi}\cos^2\theta) = 0$$

$$\therefore \quad A\dot{\psi}\sin^2\theta + Cn\cos\theta = D\,(const). \quad (2)$$

Equations (1)and (2) are equations of motion of a top.

Note : This article can be omitted at this stage and can be done after chapter 10.

(iii) Derive Equations of Motion of a Top from Principles of Energy and Momentum.

In case of top $A = B, \phi = 0$ and

$$\omega_1 = -\dot{\psi}\sin\theta, \omega_2 = \dot{\theta}, \omega_3 = n = \dot{\psi}\cos\theta.$$

Hence, the K.E. T is given by

$$T = \frac{1}{2}[A(\omega_1^2 + \omega_2^2) + C\omega_3^2]$$

$$= \frac{1}{2}[A\dot{\psi}^2\sin^2\theta + \dot{\theta}^2) + C\dot{\psi}^2\cos^2\theta]$$

$$\text{P.E.} = mgh\cos\theta.$$

By principle of conservation of energy

$$\text{K.E.} + \text{P.E.} = \text{constant}$$

$$\therefore \quad \frac{1}{2}[A(\dot{\psi}^2\sin^2\theta + \dot{\theta}^2) + Cn^2] + mgh\cos\theta = const.$$

$$\therefore \quad A(\dot{\psi}^2\sin^2\theta + \dot{\theta}^2) + 2mgh\cos\theta = const. = E. \tag{1}$$

Now external force mg is parallel to Oz and the reactions at O, cut Oz, so the sum of their moments baout Oz is zero. Hence, from principle of momentum, the moment of momentum about O is constant throughout the motion.

$A\omega_1$, $A\omega_2$, $C\omega_3$ are the angular momentum of the top about OA, OB, OC respectively, and the d.c.'s of Oz along these axes are

$$-\sin\theta, 0, \cos\theta. \text{ (since, for top } \phi = 0).$$

Resolving these angular momentum about Oz, we have moment of momentum about Oz as

$$-A\omega_1\sin\theta + C\omega_3\cos\theta,$$

Hence, by principle of moment of momentum, we have

$$-A\omega_1\sin\theta + C\omega_3\cos\theta = const.$$

$$\therefore \quad -A\sin\theta(-\dot{\psi}\sin\theta) + Cn\cos\theta = const.$$

$$\therefore \quad A\dot{\psi}\sin^2\theta + Cn\cos\theta = D\ (const.) \tag{2}$$

Equations (1) and (2) are equations of motion of a top.

(i) Steady Motion

Equations of motion of the top are

$$A\ddot{\theta} - A\dot{\psi}^2\sin\theta\cos\theta + Cn\sin\theta\dot{\psi} = mgh\sin\theta, \tag{1}$$

$$A\dot{\psi}\sin^2\theta + Cn\cos\theta = D. \quad (2)$$

When the top executes steady motion, then $\theta = \alpha$, $\dot{\psi} = \Omega$, hence, equations (1) becomes

$$A\Omega^2 \cos\alpha - Cn\Omega = -mgh \quad (3)$$

Equation (3) shows that there is a pair of real and distinct processional angular velocities Ω_1, Ω_2 provided that

$$C^2n^2 > 4\,Amgh\cos\alpha. \quad (4)$$

Thus, for a sufficiently great spin n about the axis of the symmetry, the top can execute a steady motion $\theta = \alpha$ about the vertical under two possible processional angular velocities Ω_1, Ω_2.

***(ii)* Stability Investigation.**

Equation (1) can be written as

$$A\ddot{\theta} + f(\theta) = 0 \quad (5)$$

where

$$f(\theta) = -A\dot{\psi}^2 \sin\theta\cos\theta + Cn\dot{\psi}\sin\theta - mgh\sin\theta.$$

Now, from (3)

$$f(\alpha) = 0.$$

Let us put $\theta = \alpha + \varepsilon$,
where ε is a small, then

$$f(\alpha+\varepsilon) = -A\,\dot{\psi}^2 \sin(\alpha+\varepsilon)\cos(\alpha+\varepsilon) + Cn\,\dot{\psi}\sin(\alpha+\varepsilon) - mgh\sin(\alpha+\varepsilon)$$

$$= \sin(\alpha+\varepsilon)\,[-A\dot{\psi}^2\cos(\alpha+\varepsilon) + Cn\,\dot{\psi} - mgh]$$

$$\approx (\sin\alpha + \varepsilon\cos\alpha)\,[-A\dot{\psi}^2(\cos\alpha - \varepsilon\sin\alpha) + Cn\dot{\psi} - mgh]$$

$$\approx \sin\alpha\,[-A\dot{\psi}^2\cos\alpha + Cn\dot{\psi} - mgh] + \varepsilon\cos\alpha \times [-A\dot{\psi}^2\cos\alpha + Cn\dot{\psi} - mgh] + \varepsilon A\dot{\psi}^2$$

$$\approx \varepsilon f'(\alpha).$$

Hence, equation (5) becomes

$$A\ddot{\varepsilon} + \varepsilon f'(\alpha) = 0, \quad (6)$$

If we can show $f'(\alpha) > 0$, this equation will reduce to the form

$$A\ddot{\varepsilon} + \omega^2\varepsilon = 0,$$

showing stability about $\theta = \alpha$ for small perturbations.

From (2)

$$\dot{\psi} = (D - Cn \cos \theta)/A \sin^2 \theta.$$

Hence,

$$Af(\theta) \sin^3 \theta = - \cos \theta (D - Cn \cos \theta)^2 + Cn \sin^2 \theta \times (D - Cn \cos \theta) - Amgh \sin^4 \theta.$$

Differentiating both sides w.r.t. θ, putting $\theta = \alpha$ and using $f(\alpha) = 0$, we get

$$Af'(\alpha) \sin^3 \alpha = \sin \alpha (D - Cn \cos \alpha)^2 - 2Cn \sin \alpha \cos \alpha (D - Cn \cos \alpha) + 2Cn \sin \alpha \cos \alpha (D - Cn \cos \alpha) + C^2n^2 \sin^3 \alpha - 4 Amgh \sin^3 \alpha \cos \alpha$$

$$= (D - Cn \cos \alpha)^2 \sin \alpha + C^2n^2 \sin^3 \alpha - 4Amgh \sin^3 \alpha \cos \alpha$$

From (2) and (3), we get

$$D - Cn \cos \alpha = A\Omega \sin^2 \alpha,$$

$$Cn = A\Omega \cos \alpha + \frac{mgh}{\Omega}.$$

Hence,

$$Af'(\alpha) = A^2 \Omega^2 \sin^2 \alpha + \left(A\Omega \cos \alpha + \frac{mgh}{\Omega} \right)^2 - 4 Amgh \cos \alpha$$

$$= A^2\Omega^2 - 2 Amgh \cos \alpha + \left(\frac{mgh}{\Omega} \right)^2$$

Thus (6) gives

$$\ddot{\varepsilon} + \left[\Omega^2 - \frac{1}{A} . 2mgh \cos \alpha + \left(\frac{mgh}{A\Omega} \right)^2 \right] \varepsilon = 0.$$

Now, the co-efficient of ε

$$\left[\Omega - \left(\frac{mgh}{A\Omega} \right) \right]^2 + \frac{2mgh}{A} (1 - \cos \alpha)$$

> 0, when $a \neq 0$.

Thus, for $a \neq 0$, the motions is S.H.M. type showing that the position $\theta = \alpha$ is one of the stability

Examples

1. If initially the axis of the top is horizontal and it is set spinning with angular velocity ω in a horizontal plane, prove that axis will start to rise if $Cn\omega > Mgh$ and that when $Cn\omega = 2Mgh$, the axis will rise to an angular distance

$$\cos^{-1}\left(\frac{A\omega}{Cn}\right),$$

provided that $A\omega < Cn$, and will be at an instantaneous rest.
Equations of motion of the top are

$$A\ddot{\theta} + Cn\dot{\psi}\sin\theta - A\dot{\psi}^2\sin\theta\cos\theta = Mgh\sin\theta \quad (1)$$

$$A\dot{\psi}\sin^2\theta + Cn\cos\theta = D.$$

Initially, $\theta = \frac{\pi}{2}, \dot{\psi} = \omega,$

$$\therefore \quad D = A\omega.$$

Hence,

$$A\dot{\psi}\sin^2\theta + Cn\cos\theta = A\omega. \quad (2)$$

Initially, equation (1) becomes

$$A\ddot{\theta} = Mgh - Cn\omega$$

R.H.S is negative, if $Cn\omega > Mgh$. In other words, the axis will rise, if $Cn\omega > Mgh$. From (2)

$$\dot{\psi} = \frac{D - Cn\cos\theta}{A\sin^2\theta}.$$

Substituting the value of $\dot{\psi}$ in (1), we have

$$A\ddot{\theta} + Cn.\frac{(D - Cn\cos\theta)}{A\sin\theta} - \frac{A(D - Cn\cos\theta)^2\cos\theta}{A^2\sin^3\theta} = Mgh\sin\theta.$$

Putting value of $Mgh = \frac{1}{2}Cn\omega$ (given), we get

$$A\ddot{\theta} + Cn\frac{(D - Cn\cos\theta)}{A\sin\theta} - \frac{A(D - Cn\cos\theta)^2\cos\theta}{A^2\sin^3\theta} = \frac{1}{2}Cn\omega\sin\theta.$$

$$\therefore \quad A\ddot{\theta} = \frac{(C^2n^2 + A^2\omega^2)\cos\theta}{A\sin^3\theta} + Cn\omega\left\{\frac{1}{2}\sin\theta - \left(\frac{1}{\sin\theta} + \frac{2\cos^2\theta}{\sin^3\theta}\right)\right\}.$$

Integrating, we get

$$A\dot{\theta}^2 = -\frac{A^2\omega^2 + C^2n^2}{A\sin^2\theta} + Cn\omega\left\{-\cos\theta + \frac{2\cos\theta}{\sin^2\theta}\right\} + K.$$

Initially, $\theta = \frac{\pi}{2}, \dot{\theta} = 0,$

$$\therefore \qquad K = \frac{A^2\omega^2 + C^2n^2}{A}.$$

$$A\dot{\theta}^2 = -\frac{A^2\omega^2 + C^2n^2}{A\sin^2\theta} + Cn\omega.\frac{(-\sin^2\theta\cos\theta + 2\cos\theta)}{\sin^2\theta} + \frac{A^2\omega^2 + C^2n^2}{A}$$

$$\begin{aligned} \therefore \qquad A^2\dot{\theta}^2\sin^2\theta &= -(A^2\omega^2 + C^2n^2)\cos^2\theta + Cn\omega A(-\sin^2\theta + 2)\cos\theta \\ &= -(A^2\omega^2 + C^2n^2)\cos^2\theta + Cn\omega A(1 + \cos^2\theta)\cos\theta \\ &= \cos\theta\,(A\omega - Cn\cos\theta)\,(Cn - A\omega\cos\theta) \end{aligned}$$

Hence, $\dot{\theta} = 0$, when

$$\cos\theta\,(A\omega - Cn\cos\theta)\,(Cn - A\omega\cos\theta) = 0,$$

i.e., when $\cos\theta = 0$, $\therefore\ \theta = \dfrac{\pi}{2}$, which is initial position,

and when $A\omega - Cn\cos\theta = 0$, $\therefore \cos\theta = \dfrac{A\omega}{Cn}$

$\therefore \theta = \cos^{-1}\dfrac{A\omega}{Cn}$, provided that $A\omega < Cn$.

and when $Cn - A\omega\cos\theta = 0$,

$\therefore \qquad \theta = \cos^{-1}\dfrac{Cn}{A\omega}$, which is impossible, when $A\omega < Cn$.

Now,

$$\dot{\psi} = \frac{A\omega - Cn\cos\theta}{A\sin^2\theta} = 0, \text{ when } \cos\theta = \frac{A\omega}{Cn}.$$

Thus we see that in the position $\ddot{\theta} = \cos^{-1}\dfrac{A\omega}{Cn}$, we have $\dot{\theta} = 0$ and $\dot{\psi} = 0$,

i.e., the top comes to rest when it is in the position $\theta = \cos^{-1}\dfrac{A\omega}{Cn}$. But we see from (1) that at time $\theta = \dfrac{Mgh\sin\theta}{A}$, which is clearly positive, hence, top begins to fall (moving towards θ increasing). Hence, this position of rest is instantaneous.

2. Prove that a top can be set spinning steadily with its axis of the figure horizontal, by an impulsive couple about an axis making with the vertical an angle

$$\tan^{-1}\left(\frac{C^2n^2}{AMgh}\right).$$

Let β be the inclination to the vertical, of the axis about which an impulsive couple of moment J (say) is applied.

The motion is steady with axis OC of the figure horizontal, hence, J is equivalent to two couples namely $J\cos\beta$ about the vertical Oz, and $J\sin\beta$ about the horizontal (i.e., about OC).

We know that the angular momentum about Oz is

$$A\dot{\psi}\sin^2\theta + Cn\cos\theta,$$

and angular momentum about OC is Cn.

Now, we know that

Change in angular momentum = Sum of moments of the impulsives.

Hence, taking moments about Oz and OC, we get

$$A\dot{\psi}\sin^2\theta + Cn\cos\theta = J\cos\beta \qquad (1)$$

$$Cn = J\sin\beta, \qquad (2)$$

Dividing (2) by (1), we get

$$\tan\beta = \frac{Cn}{A\dot{\psi}\sin^2\theta + Cn\cos\theta}$$

$$= \frac{Cn}{A\dot{\psi}}, \text{ because } \theta = \frac{\pi}{2} \; (constant)$$

is steady with axis horizonal.

$$= \frac{Cn}{A\omega}, \text{ because precessional velocity } \dot{\psi} = \omega$$

(*constant*) for steady motion.

$$= \frac{Cn}{A(Mgh / Cn)}$$

$$= \frac{C^2n^2}{AMgh}.$$

$$\therefore \qquad \beta = \tan^{-1}\frac{C^2n^2}{AMgh}.$$

EXERCISE

1. A uniform solid sphere rolls on the inside of a rough circular cylinder whose axis is vertical. Show that if the cylinder be made to rotate with constant angular velocity ω about a generating line in the plane through the axis from which the azimuth θ is measured, the azimuth motion of the centre of the sphere relative to the cylinder is given by

$$7(b-a)\ddot{\theta} = 5b\,\omega^2 \sin\theta,$$

where a, b are the radii of the sphere and the cylinder.

2. A uniform solid sphere of radius a rolls without sliding inside a fixed cylinder of radius b whose axis is inclined at an angle α to the horizon. Obtain the equations of motion

$$(b-a)\dot{\theta}^2 = \frac{10}{7} g\cos\alpha\cos\theta + const.,$$

$$\dot{x}^2 + \frac{2}{7}a^2\omega_3^2 = \frac{10}{7} gx\sin\alpha + const.,$$

$$\alpha\dot{\theta}\left(\frac{d^2\omega_3}{d\theta^2} + \frac{2}{7}\omega_3\right) = \frac{5}{7} g\sin\alpha,$$

where x is the distance which the sphere has moved parallel to the axis of the cylinder, θ the angle between the axial plane through the centre and the vertical axial plane, and ω_3 the angular velocity of the sphere about the radius of the point of contact.

3. A uniform heavy square lamina, free to turn about its centre which is fixed is set rotating with angular velocity Ω about an axis inclined at an angle 45° to its plane. Show that in subsequent motion, the axis of the figure of lamina describes with uniform velocity, a right circular cone whose axis is inclined at $\tan^{-1}(1/3)$ with the invariable line. Find the semi-vertical angle of the cone and the uniform angular velocity with which it is described.

4. A uniform square lamina is rotating under no forces about one corner O, which is fixed. If ω_1, ω_2, ω_3 are the components of the angular velocity about the principal axes at O (in the order of increasing moments of inertia), show that $\omega_1^2 + \omega_2^2$ and $3\omega_2^2 + 4\omega_3^2 = const.$

5. A plane lamina is pivoted at a point O and moves under the action of no forces other than the reaction at O. The principal axis in the plane of the lamina are Ox and Oy and the corresponding principal moments of inertia are A and B $(B > A)$. Initially lamina is rotating with angular velocity Ω about a fixed line (not necessarily in the plane of the lamina) which makes a small angle with Ox. Show that, during the subsequent motion, the direction of the component of the angular velocity in the plane of the lamina oscillates about Ox with period approximately

$$(2\pi/\Omega)(B+A)^{1/2}/(B-A)^{1/2}.$$

6. An ellipsoid, free to move about its centre, is set in rotation at time $t = 0$, with angular velocity $\vec{\Omega}$. If n, 0, $3n$ are the components of $\vec{\Omega}$ along the

principal axes at the centre, and $6A$, $3A$, A are the moments of inertia about these axes, find the components of the angular velocity at time t, and show that for large value of t, the magnitude of the angular velocity is practically $n\sqrt{5}$.

7. A uniform circular disc is set rotating with an angular velocity ω about an axis passing through the centre making an angle α with the normal. Prove that the semi-vertical angle θ of the cone described by the axis of the disc is given by

$$\tan\theta = \frac{1}{2}\tan\alpha.$$

If ω be the angular velocity, prove that the above cone is described in the period

$$\frac{2\pi}{\omega\sqrt{1+3\cos^2\alpha}}.$$

8. A rigid body is free to rotate about its centroid G, the principal moments of inertia are 7, 25, 32 units respectively. The body is given an angular velocity Ω about a line through G whose direction ratios are 4, 0, 3. Show that after time t the components of angular velocity about the principal axes of inertia at G are

$$\frac{4}{5}\Omega\cos\phi, \frac{4}{5}\Omega\sin\varphi, \frac{3}{5}\Omega\cos\phi,$$

where $\tan(\phi/2) = \tan h\,(3\Omega t/10)$.

Deduce that ultimately the body rotates about the principal axis of intermediate moment of inertia.

9. A perfectly rough plane is made to rotate about a vertical axis in its plane with angular velocity Ω. A uniform solid sphere rolls on the plane under the action of gravity. Initially, it is at rest relative to the plane. Prove that the centre never descends more than $5g/\Omega^2$ below its original level.

10. A uniform sphere of radius a rolls on the outside of a fixed rough cylinder of the same radius whose axis is horizontal. Initially the sphere stands on the top of the cylinder and spins about its vertical axis with angular velocity Ω. When the normal at the point of the contact makes an angle θ with the vertical, show that $7a\dot{\theta}^2 = 5g(1-\cos\theta)$. Show also that the angular velocity of the sphere about the normal is then $\Omega\cos n\theta$, where $7n^2 = 2$.

11. A uniform solid right circular cone of height h, mass M and semi-vertical angle $\pi/4$ moves freely about its vertex O, under no forces than the reaction at O. If the cone is set spinning with angular velocity Ω about a generator $O1$, find the components of the angular velocity at any subsequent time t

in the right handed orthogonal frame OA, OB, OC fixed in the cone, where OC is the axis of the cone and $O1$ bisects the angle between OA and OC.

Find the reaction at O and show that its magnitude is $3\sqrt{29}\ Mh\Omega^2 / 40$.

12. A rigid body rotates freely about its centre of mass under no forces. The body is rotating with angular velocity Ω about the axis of the greatest moment of inertia (C), when a small angular momentum Ap is applied about the axis of the moment of inertia A. Show that instantaneous axis moves in the body with period $2\pi/\Omega$, where $\Omega^2 = \dfrac{(C-B)(C-A)\omega^2}{AB}$.

13. Show that in the free motion of the body, with an axis of the symmetry C, about the centre of gravity, if n denotes the spin about the axis C and ϕ denotes Euler's third angle

$$A\dot{\phi} = (A - C)n.$$

14. A rigid straight light rod, of length $5a$, rigidly and normally attached at one end to a point O on the surface of a uniform solid sphere of radius a. To the other end of the rod is fixed a particle of mass equal to that of the sphere such that $OA = 8a$. Prove that at time t later, the components of angular about OA is

$$(\Omega/25)\ \{16 + a\cos(2\Omega t/3\}.$$

Show that the instantaneous axis of the system describes in a space a cone of angle

$$2\tan^{-1}(6/7).$$

15. A uniform rectangular lamina of sides $2a$, $3a$ is freely hinged to a horizontal axis along one of its shortest edges. This axis is fixed to a vertical shaft which passes through mid point of the hinged edge, and the shaft is forced to rotate with angular velocity ω. If θ is the inclination of the plate to the downward vertical at time t, show that one of the Euler's equations can be written as

$$a\ddot{\theta} - a\omega^2 \sin\theta\cos\theta = -\frac{1}{2} g\sin\theta.$$

Show that if ω is sufficiently large there is a position of relative equilibrium in which the lamina makes an acute angle with the downward vertical. Prove that this position is stable and find the period of small oscillations when the lamina is slightly disturbed from this position.

16. A uniform circular disc of radius a and mass m is free to turn about a fixed point A of the circumference in any way. It is rotating with angular velocity Ω about the vertical diameter through A. A moving particle of mass m hits the disc at one end of the horizontal diameter and adheres to the disc. When the particle hits the disc it has the velocity perpendicular to the disc

and is moving in the same direction as the point where it hits. Find the velocity of the particle after impact, and show that the impulsive reaction at A is

$$m\,(u - a\Omega)/29.$$

17. A uniform solid sphere rolls on the inside of a fixed rough circular cylinder whose axis is vertical. Prove that the curve traced out on the cylinder by the point of the contact is

$$x = a\cos\sqrt{\frac{2}{7}}\,\theta.$$

where a depends on the circumstances of projection.

18. A rough material plane turns with angular velocity ω about a horizontal aixs fixed in its self. At the instant when the plane is horizontal, a uniform solid sphere of radius a having an angular velocity Ω about a vertical diameter is gently placed on the plane, touching at a point on the axis of rotation. Show that so long as the sphere remains in contact with the plane the displacement of the point of contact after time t is given by

$$x = \sqrt{\frac{7}{5}}\left(\frac{5g}{12\omega^2} - a\right)\sin h\sqrt{\frac{5}{7}}\omega t - \frac{5g}{12\omega^2}\sin\omega t,$$

$$y = \frac{\Omega a}{\omega}\left(1 - \cos\sqrt{\frac{2}{7}}\,\omega t\right),$$

where x is measured perpendicular and y parallel to the axis of rotation.

19. A uniform circular cylinder of mass M is free to turn about its axis which is horizontal. The cylinder is at rest, when a uniform solid sphere of mass m and radius a, is placed in contact with the highest generator, sphere having initially a spin n about its vertical diameter. Show that if the system is slightly disturbed, the sphere will leave the cylinder when the plane through the axis of the cylinder and the centre of the sphere makes an angle α with the vertical, where $\cos\alpha = \dfrac{10M + 8m}{17M + 12m}$, and that at the instant of separation, the component of velocity of the centre of the sphere in the direction of the axis of the cylinder is $\sqrt{\frac{2}{7}}\,an\sin\sqrt{\frac{2}{7}}\alpha$.

20. A circular cylinder is fixed with its axis horizontal, and a rough uniform sphere is projected horizontally from a point of the lowest generator in a direction making an angle α with this generator. Prove that for the sphere to come in contact with the highest generator the initial velocity must be

greater than $\sqrt{\dfrac{27\,(b-a)g}{7\sin^2\alpha}}$, where a and b are the radii of the sphere and the cylinder respectively.

21. A hoop of radius a, rolls without slipping on a rough horizontal surface. Obtain the equations of motion in the form

$$3\ddot{\theta} - \dot{\phi}^2 \sin\theta\cos\theta + 4\omega\dot{\phi}\sin\theta = -2(g/a)\cos\theta,$$

$$\frac{d}{dt}(\dot{\varphi}\sin^2\theta) - 2\omega\dot{\theta}\sin\theta = 0,$$

$$2\dot{\omega} - \dot{\theta}\dot{\phi}\sin\theta = 0,$$

where θ is the angle the plane of the hoop makes with the horizontal, $\dot{\phi}$ is the angular velocity of the vertical plane containing the diameter through the point of contact, and ω is the angular velocity of the hoop abot its axis of symmetry.

If the hoop rolls along a straight line with its plane vertical, show that the motion is stable if

$$\omega > \frac{1}{2}\sqrt{\frac{g}{a}}.$$

22. If the top be started when its axis makes an angle of 60° with the upward drawn vertical, so that initial spin about its axis is $\dfrac{A}{C}\sqrt{\dfrac{3Mgh}{A}}$, and the velocity of its axis in azimuth is $2\sqrt{\dfrac{Mgh}{3A}}$, its angular velocity in the meridian plane being initially zero, show that inclination θ of its axis of the vertical at any time is given by the equation

$$\sec\theta = 1 + \sec h\left\{\frac{\sqrt{Mgh}}{A}t\right\},$$

so that the axis continuously approaches to the vertical without ever reaching it.

23. Show that vertical pressure of the top on the point of support is equal to its weight when the inclination of its axis to the vertical is given by the least root of the equation

$$3AMgh\cos^2\theta - \cos\theta\,[C^2n^2 + 2Amgbh] + C^2 + n^2 - AMgh = 0,$$

where a and b are constants depending on initial circumstances of the motion.

24. A symmetrical top is set in motion on a rough horizontal plane with an angular motion about its axis of the figure, the axis being inclined at an angle i to the vertical. Show that between the greatest approach to the recess from the vertical, the centre of gravity describes an arc

$$h \tan^{-1}\left(\frac{\sin i}{p - \cos i}\right),$$

where p and h are given by $OG = h$ and $c^2n^2 = 4\,Amghp$.

25. A top, of vertex O, is set in motion so that angular momentum Cn is equal to that about the vertical drawn upwards through O. The inclination of the axis of the vertical was initially i and every point on the axis was moving horizontally. Determine if the axis will become vertical in a finite tme, and, if it does, show that the velocity of the centre of gyration p, at that instant is equal to

$$2\sqrt{gl}\,\tan\frac{i}{2}\sqrt{p - \cos^2\frac{i}{2}},$$

where $4A^2\,gp = C^2n^2l$, and $l = OP$.

what is the nature of motion if $p < \cos^2\dfrac{i}{2}$?

26. When the axis of the symmetrical top is stationary and then spin is large and eqal to n, and a blow J is applied perpendicular to the axis at a distance c from the fixed point. Prove that maximum angular deviation of the axis is approximately, $2\tan^{-1}\left(\dfrac{Jc}{Cn}\right)$, C being the M.I. of the top about its axis of symmetry.

10

GENERALIZED CO-ORDINATES AND LAGRANGIAN FORMULATION

10.1 DYNAMIC MEHODS

Dynamical system consists of particle or system of particles; it may include rigid bodies, since they are made up of particles. There are two dynamic methods:

1. **Newtonian formulation.** In Newtonian formulation two different approaches are used for the solution of mechanical problems. In first approach, the laws of motion (linear and/or rotational) are involved, such an approach often leads to many equations containing unknown as unwanted internal forces. In second approach, however, where the forces are conservative, conservation principles are used, thereby avoiding the introduction of unwanted forces. The first method is essentially vectorial and derives directly from the work of Newton. In second method, scalar quantities, *i.e.* potential and kinetic energies are calculated and energy equation is set up. Of two methods; the first is more cumbersome, but it is more general, because many problems involve non-conservative fields of force or else impulsive motion to which principle of conservation of energy is not applicable, where both approaches are applicable, energy method is preferred because in certain cases it leads to the desired results much more quickly than the vectorial method.
2. **Lagrangian formulation.** While solving specific problems, generally we write equations of motion in terms of cartesian, cylindrical, or spherical polar co-ordinates. For example, the motion of particle in a central field of force, the polar co-ordinates are used. For the motion of projectile, cartesian co-ordinates are used. It is seen that a particular co-ordinate system chosen from the symmetry of the physical system helps to simplify the problem. Such a dependence on the co-ordinate system, however, is undesirable and we should be able to write equations of motion without reference to any specific co-ordinate system used. This is the approach in the Lagrangian formulation. This approach has a number of advantages

over the Newtonian formulation. because here we need only Kinetic and Potential energies, which simplifies the work.
This approach may be compared to the use of vectors in describing physical quantities. When expressed through vectors, these quantities do not depend on any specific co-ordinate system and the physical significance of the equation is also clearly seen. This is, however, a very limited comparison. The Lagrangian formulation is of a very general nature and as we will see that it makes use of generalized co-ordinates and velocities which are independent of the co-ordinate system.

10.2 CONSTRAINTS

In practice the motion of particle or system of particles is generally restricted in some way. The limitations or geometrical restrictions on the motion of a particle or system of particles are known as *constraints*.

Classification of constraints. The constraints can be classified in following ways:

(*a*) **Holonomic and Non-Holonomic Constraints:** A holonomic constraint is one that may be expressed in the form of a equation relating the co-ordinate of the system and time. The general form of such equations for a system of N particles is

$$F_i(x_1, y_1, z_1, x_2, y_2, z_2 \ldots\ldots x_N, y_N, z_N, t) = 0 \quad (1)$$

where $i = 1, 2, 3, \ldots.\ k$ and F_i is some funtion of the co-ordinates and time. In this equation i denotes ith constraint. It should be remembered that all the variables may not enter in the equation of the constraints.

A Non-Holonomic constraint can not be expressed in the form of equation (1). It may be in the form of inequality. If there are k equations of constraint, the number of degree of freedom is reduced to $(3N - k)$. Hence instead of $3N$ co-ordinates (x_i, y_i, z_i), where $i = 1, 2, 3 \ldots.., N$, we can assign $(3N - k)$ independent variable $q_1, q_2, q_3, \ldots\ldots\ldots, q_{3N-k}$ to describe the system. These variables need not necessarily have dimensions of length or angles, they may be even charges.

It should be remebered that the constraints are consequences of the forces exerted on the system by constraining mechanism. This mechanism may not be generally known. Hence, it is necessary to eliminate the forces of constraints. We mostly come across systems in which holonomic constraints have been introduced and such systems are called *holonomic systems*. In this type of system any generalized co-ordinate can be changed without changing other co-ordinate.

The *Non-Holonomic constraints* can be eliminated by any general method. Each problem involving such constraints needs to be solved individually. Moreover, the constraints need not always be written as in the form of equation connecting co-ordinates, but they can be expressed as well in terms of the velocities. For example, a disc of radius a rolling down an inclind plane along the line of greatest slope will have its constraint like

$\frac{dl}{dt} = a\dot{\theta}$ or $dl = ad\theta$, after integration gives $l - a\theta =$ constant.

This is holonomic constraint obtained from the relationship between velocities.

Examples of Holonomic constraints.

1. The constraints involved in the motion of rigid bodies, in which the distance between any two particular points is always fixed, because the conditions of constraints are expressed as $(\vec{r}_i - \vec{r}_j)^2 - C_{ij}^2 = 0$.
2. The constraints involved in the motion of the point mass of simple pendulum, because the point mass remains at constant distance l from the point of suspension, whose position vector is $\vec{a}$. Hence, constraints are expressed as $(\vec{r} - \vec{a})^2 = l^2$, where $\vec{r}$ is the position vector of the point mass.
3. When a particle is restricted to move along a curve or surface.

Examples of Non-Holonomic constraints.

1. The constraints involved in the motion of the particle placed on the surface of solid sphere, the conditions of constraints are expressed as $r^2 - a^2 \leq 0$, where a is the radius of the sphere. This is an ineqaulity and not in the form of equation (1).
2. The constraints involved in the motion of molecules in a gas container.
3. An object rolling on the rough surface without slipping.

(*b*) **Scleronomic and Rhenomic Constraints:** If the constraints are independent of time, they are called *Scleronomic* and if they contain time explicitly, they are called *Rhenomic*.

A bead sliding on a moving wire is an example of rhenomic constraint, while constraint in case of rigid body is scleronomous. If, however, we construct a simple pendulum whose length changes with time, then constraint expressed in the equation form will be time dependent. Similarly, if the radius of sphere is changing with time, is also rheonomic.

10.3 GENERALIZED CO-ORDINATES

Suppose the system is made up of N particles of masses m_i ($i = 1, 2, 3, ..., N$) and that at any time the position of each particle may be specific by means of n independent variables q_j ($j = 1, 2, ..., n$). These variables are called *generalized co-ordinates* of the system and in number they are equal to the number of degree of freedom of the system. The position of a free rigid body is determined by three co-ordinates x, y, z of an assigned point of the body together with three angles, *e.g.* the Eulerian angle θ, ϕ, ψ, which determine the orientation. These six variables are called *generalized co-ordinates* of a body. The generalized co-ordinates can be chosen in number of ways.

In order to chose a suitable set of generalized co-ordinates in a given problem, the following three principles must be taken into account.

(*a*) They specify the configuration of the system.

(*b*) They may be varied arbitrarily and independent of each other, without violating the constraints of the system.

(*c*) They must simplify mathematical calculations without effecting the information and enlightening solution.

The number of co-ordinates satisfying conditions (1) and (2) is the number of degree of freedom. Condition (3) then implies that after ascertaining about the degree of freedom, we chose a suitable co-ordinate system to achieve mathematical simplicity in setting up equations of motion as well as in their solutions. Unlike, Cartesian co-ordinates, generalized co-ordinates are not divided into three convenient groups of co-ordinates, which may form vectors.

Examples

1. Let threre be a particle of mass m constrained to move on a circular wire in X-Y plane having co-ordinates (x, y) at any instant. Let r be the radius of the circle; then we have

$$x = r \cos \theta, y = r \sin \theta.$$

Thus, the motion of the particle can be simplified by use of generalized co-ordinates alone, since r is constant.

2. Let us consider the motion of two particles system. Three co-ordinates specify the position of the one particle and another three those of the other particle, making a total of six. Hence, there are six generalized co-ordinates. But if the points are connected by rigid weightless rod of length a or they lie on a rigid body, then distance between two points remain constant, hence, number of generalized co-ordinates is reduced from six to five.

3. If system consists of three particles, the number of generalized co-ordinates is 9 but if there is rigid connection between three particles number of generalized co-ordinates reduces to 9 – 3 = 6.

10.4 TRANSFORMATION EQUATIONS

If x, y, z are rectangular co-ordinates of a particle of mass m, then we shell denote generalized co-ordinates by q_i (j = 1, 2, 3, ..., n), if t is time, then we have

$$x = x\,(q_1, q_2, \ldots\ldots\ldots, q_n, t)$$

$$y = y(q_1, q_2, \ldots\ldots\ldots q_n, t) \qquad (1)$$
$$z = z(q_1, q_2, \ldots\ldots\ldots q_n, t)$$

If $\vec{r}$ is the position vector of mass m, then

$$\vec{r} = \vec{r}(q_1, q_2, \ldots\ldots, q_n, t) \qquad (2)$$

The equations of type (1) and (2) are called transformation equations. The equations also contain the constraints explicitly. The inverse transformation equations, wherever they exist, can be written as

$$q_1 = q_i(\vec{r}_1, \vec{r}_2, \ldots\ldots \vec{r}_n, t)$$

10.5 CONFIGURATION SPACE

The motion of single particle can be described by a set of three co-ordinates *i.e.*, the space for the description of motion of a single particle must be three dimensional. In case of two particles the motion of each particle will be described by a set of three co-ordinates, *i.e.* for the description of the motion of two particles we need six dimensional space. Alternatively, we can say that the problem of two particles can be considered as that of a single particle in six dimensional space. Similarly, if system consists of N particles, it can be considered as a problem of single particle in $3N$ dimensional space. In case of l constraints, the problem of N particles in $3N$-l dimensional space. This $3n$ dimensional space in the absence of constraints and $3N$-l dimensional space in the presence of constraints is called *configuration space.*

10.6 GENERALIZED VELOCITIES AND GENERALIZED MOMENTUM

The quantities $\dot{q}_i$ are called *generalized velocities.*

It T is kinetic energy of the system, the quantities $p_i = \dfrac{\partial T}{\partial \dot{q}_i}$ are called *generalized momentum.*

Let $\vec{r}$ be the position of the particle at time t, so that

$$\vec{r} = \vec{r}(q_1, q_2, \ldots\ldots, q_n, t),$$
$$\dot{\vec{r}} = \frac{\partial \vec{r}}{\partial t} + \sum \frac{\partial \vec{r}}{\partial q_i} \cdot \dot{q}_j .$$

we shall through out take the variation of j from l to n and of i from l to N.

Now, we regard the new variables $\dot{q}_1, \dot{q}_2, \ldots\ldots, \dot{q}_n$, then

$$\frac{\partial \vec{r}}{\partial q_j} = \frac{\partial \dot{\vec{r}}}{\partial \dot{q}_j}$$

10.7 VIRTUAL WORK AND GENERALIZED FORCES

Suppose particles of a dynamical system undergo a small instantaneous displacement independent of time, consistent with constraints of the system and such that all external and internal forces remain unchanged in magnitude and direction during the displacement, such a displacement is called *virtual displacement* because displacement is of hypothetical nature. Although purely fictitious, the notion of virtual displacement is essential to the elucidation of principles of analytical dynamics.

Let its particle m_i at position $\vec{r}_i$ at time t undergoes a virtual displacement to position.

$\vec{r}_i + \delta \vec{r}_i$. Let $\vec{F}_i$ and $\vec{F}_i'$ be the external and internal forces acting on m_i.

Then the virtual work done on m_i is $(\vec{F}_i + \vec{F}_i').\delta \vec{r}_i$ and so that total work done on all the particles when similar displacement are given, is

$$\delta W = \sum (\vec{F}_i + \vec{F}_i').\delta \vec{r}_i = \sum \vec{F}_i.\delta \vec{r}_i + \sum \vec{F}'_i.\delta \vec{r}_i.$$

But in many cases $\sum \vec{F}_i'.\delta \vec{r}_i$, *i.e.*, work done by internal forces is zero, *e.g.* when the particles are connected by rigid constraints, but in case of elastic connections, the work done is not zero. When internal forces do not work in virtual displacement, then

$$\delta W = \sum \vec{F}_i.\delta \vec{r}_i,$$

and W is called virtual work function.

Now, suppose the system is holonomic and specified by n generalized co-ordinats q_j we can change q_j to $q_j + \delta q_j$ witout changing remaining co-ordinates. Let this displacement take place instantaneously and suppose the corresponding work done on the system to be $Q_j \delta q_j$. Then

$$Q_j \delta q_j = \vec{F}_i.\delta \vec{r}_i$$

If now we make similar variations in each q_j, then

$$\delta W = \sum_j Q_j \delta q_j = \sum_i \vec{F}_i.\delta \vec{r}_i$$

Thus, we see that Q_j is the co-efficient of δq_j in the virtual function δW which has been constructed solely from external forces acting on the system. Q_j is called the *generalized force.*

10.8 KINETIC ENERGY

Kinetic energy of the system is given by

$$T = \sum m_i \dot{r}_i^2 / 2 = \sum m_i v_i^2 / 2. \quad (1)$$

Since, $\vec{r}_i = \vec{r}_i(q_j, t)$, we can write

$$\dot{\vec{r}}_i = \frac{\partial \vec{r}_i}{\partial t} + \sum \frac{\partial \vec{r}_i}{\partial q_j} . \dot{q}_j \tag{2}$$

Now, in order to get kinetic energy we have to square the expression (2), so we will get three types of terms: the term involving square of velocity, the term involving velocity and the term independent of velocity. Squaring expression (2), we get

$$\dot{\vec{r}}_i^2 = \sum_j \sum_k \frac{\partial \vec{r}_i}{\partial q_j} \cdot \frac{\partial \vec{r}_i}{\partial q_k} \dot{q}_j . \dot{q}_k + 2 \sum \frac{\partial \vec{r}_i}{\partial q_j} \cdot \frac{\partial \vec{r}_i}{\partial t} \dot{q}_j + \left(\frac{\partial \vec{r}_i}{\partial t} \right)^2 \tag{3}$$

Substituting this value of $\dot{\vec{r}}^2$ in (1), we get

$$T = \sum_j \sum_k a_{jk} \dot{q}_j \dot{q}_k + \sum_j b_j \dot{q}_j + C \tag{4}$$

where

$$a_{jk} = \frac{1}{2} \sum_i m_i \frac{\partial \vec{r}_i}{\partial q_j} \cdot \frac{\partial \vec{r}_i}{\partial q_k}$$

$$b_j = \sum_i m_i \frac{\partial \vec{r}_i}{\partial t} \cdot \frac{\partial \vec{r}_i}{\partial q_j} \tag{5}$$

$$C = \frac{1}{2} \sum_i m_i \left(\frac{\partial \vec{r}_i}{\partial t} \right)^2 .$$

Thus, kinetic energy is a quadratic function of the generalized velocity components, in which the co-efficients are functions of the generalized co-ordinates and time.

The case when transformation equations are independent of time is of particular interest. The constraints are scleronomous and the partial derivatives with respect to time vanish. Hence, $b_j = C = 0$. In this case, the kinetic energy is a homogeneous quadratic function of the generalized velocities. Thus,

$$T = \sum_j \sum_i a_{ji} \dot{q}_j \dot{q}_i \tag{6}$$

Let us now form the sum $\dot{q}_i \dfrac{\partial T}{\partial \dot{q}_i}$

From (6)

$$\frac{\partial T}{\partial \dot{q}_i} = \sum_j \sum_i a_{ij}\dot{q}_j + \sum_i \sum_j a_{ij}\dot{q}_j$$

$$\sum_i \dot{q}_i \frac{\partial T}{\partial \dot{q}_i} = \sum_j \sum_i a_{ij}\dot{q}_i\dot{q}_j + \sum_i \sum_j a_{ij}\dot{q}_i\dot{q}_j \cdot$$

Since, all suffixes are dummy and are symmetrical in i and j, the two sums on R H.S. are identical. Hence,

$$\sum_i \dot{q}_i \frac{\partial T}{\partial \dot{q}_i} = 2\sum_j \sum_i a_{ij}\dot{q}_i\dot{q}_j = 2T. \qquad (7)$$

The result of this equation follows immediately from Euler's theorem, which states that if $f(x)$ is a homogeneous function of the nth degree of a set of variables, then

$$\sum x_i \frac{\partial f}{\partial x_i} = nf. \qquad (8)$$

The co-efficients a's — are called *co-efficients of inertia* of the system. They must be such that the kinetic energy is positive for all values of velocities.

10.9 LAGRANGE'S EQUATIONS FOR HOLONOMIC DYNAMICAL SYSTEM

The quation of motion of the ith particle is

$$\vec{F}_i + \vec{F}_i' = m_i \ddot{\vec{r}}_i \qquad (1)$$

$$\therefore \sum_i (\vec{F}_i + \vec{F}_i').\delta\vec{r}_i = \sum_i m_i \ddot{\vec{r}}_i.\delta\vec{r}_i$$

We know that

$$\delta W = \sum_i \vec{F}_i.\delta\vec{r}_i = \sum_i (\vec{F}_i + \vec{F}_i').\delta\vec{r}_i,$$

since, in case of rigid constraints the work done by internal forces is zero.

Now

$$\ddot{\vec{r}}_i.\left(\frac{\partial \vec{r}_i}{\partial q_j}\right) = \frac{d}{dt}\left\{\dot{\vec{r}}_i.\left(\frac{\partial \vec{r}_i}{\partial q_j}\right)\right\} - \dot{\vec{r}}_i.\frac{\partial \dot{\vec{r}}_i}{\partial q_j}. \qquad (2)$$

Since,

$$\frac{d}{dt}\left(\frac{\partial \vec{r}_i}{\partial q_j}\right) = \frac{\partial \dot{\vec{r}}_i}{\partial q_j}.$$

We now that

$$\frac{\partial \vec{r}_i}{\partial q_j} = \frac{\partial \dot{\vec{r}}_i}{\partial \dot{q}_j}$$

Taking dot product of both sides of (1) with $\frac{\partial \vec{r}_i}{\partial q_j}$ and using (2), we get

$$\frac{d}{dt}\left\{\frac{\partial}{\partial \dot{q}_j}\left(\frac{1}{2}m_i\dot{\vec{r}}_1^2\right)\right\} - \frac{\partial}{\partial q_j}\left(\frac{1}{2}m_i\dot{\vec{r}}_i^2\right) = (\vec{F}_i + \vec{F}_i').\frac{\partial \vec{r}_i}{\partial q_j} \tag{3}$$

$$\therefore \frac{d}{dt}\sum_i \frac{\partial}{\partial \dot{q}_j}\left(\frac{1}{2}m_i\dot{\vec{r}}_i^2\right) - \sum_i \frac{\partial}{\partial q_j}\left(\frac{1}{2}m_i\dot{\vec{r}}_i^2\right) = \sum_i (\vec{F}_i + \vec{F}_i').\frac{\partial \vec{r}_i}{\partial q_j}$$

Now total energy of the system is

$$T = \frac{1}{2}\sum_i m_i\dot{\vec{r}}_i^2$$

$$\therefore \frac{d}{dt}\left(\frac{\partial T}{\partial \dot{q}_j}\right) - \frac{\partial T}{\partial q_j} = \sum_i (\vec{F}_i + \vec{F}_i').\frac{\partial \vec{r}_i}{\partial q_j} \tag{4}$$

When all generalized co-ordinates are constants except q_i, the virtual work function is

$$\delta W = \sum_j Q_j \delta q_j = \sum_i \vec{F}_i.\frac{\partial \vec{r}_i}{\partial q_j}\delta q_j = \sum_i (\vec{F}_i + \vec{F}_i').\frac{\partial \vec{r}_i}{\partial q_j}\delta q_j$$

$$\therefore \qquad Q_j = \sum_i (\vec{F}_i + \vec{F}').\frac{\partial \vec{r}_i}{\partial q_j}$$

Using this result (4) becomes

$$\frac{d}{dt}\left(\frac{\partial T}{\partial \dot{q}_j}\right) - \frac{\partial T}{\partial q_j} = Q_j. \tag{5}$$

These are called *Lagrange's equations*. Since, j varies from l to n, hence, we get n equations.

Case of conservative field of forces. When field of force is conservative, then there exists a potential function

$$V = V(q_1, q_2, \ldots, q_n),$$

such that

$$\delta W = -\,\delta V, \text{ hence } \sum Q_j \cdot Sq_i = -\sum \frac{\partial V}{\partial q_{ij}} \sum q_j$$

hence $Q_j = -\dfrac{\partial V}{\partial q_j}$ so that Lagrange's equation for conservative holonomic system becomes

$$\frac{d}{dt}\left(\frac{\partial T}{\partial \dot{q}_j}\right) - \frac{\partial T}{\partial q_j} = -\frac{\partial V}{\partial q_j}$$

$$\therefore \quad \frac{d}{dt}\left(\frac{\partial T}{\partial \dot{q}_j}\right) - \frac{\partial}{\partial q_j}(T - V) = 0. \tag{6}$$

Let $L = T - V$

$$\frac{\partial L}{\partial \dot{q}_j} = \frac{\partial T}{\partial \dot{q}_j}, \frac{\partial L}{\partial q_j} = -\frac{\partial L}{\partial q_j},$$

since, T is a function of q's $\dot{q}$'s and V is a function of q's only.

Therefore equation (6) becomes

$$\frac{d}{dt}\left(\frac{\partial L}{\partial \dot{q}_j}\right) - \frac{\partial L}{\partial q_j} = 0 \tag{7}$$

These are called *Lagrange's equations for conservative holonomic system*, L is called *Lagrangian* or *kinetic potential.*

Note :

1. By using generalized co-ordinates and Lagrange's equations, solution of the problem can be obtained without reference to inertial reactions.
2 Q's are not necessarily of the dimensions of force. When q_j is a length, then Q_j is of the dimension of force but if q_j is an angle then Q_j is of the dimension of couple.

Examples

1. A particle is sliding on a uniformly rotating wire in a force free space. Discuss the motion.

   Let m be the mass of the particle and T kinetic energy, then

$$T = m(\dot{r}^2 + r^2\dot{\theta}^2)/2, \tag{1}$$

Since,

$$\theta = \omega t, \therefore \dot{\theta} = \omega$$

$$\therefore \qquad T = \frac{1}{2}m(\dot{r}^2 + r^2\omega^2)$$

$$\therefore \qquad \frac{\partial T}{\partial r} = mr\omega^2, \frac{\partial T}{\partial \dot{r}} = m\dot{r}.$$

The Lagrange's equation in the absence of force is

$$\frac{d}{dt}\left(\frac{\partial T}{\partial \dot{r}}\right) - \frac{\partial T}{\partial r} = 0.$$

$$\therefore \qquad \frac{d}{dt}(m\dot{r}) - mr\omega^2 = 0.$$

$$\therefore \qquad \ddot{r} = r\omega^2,$$

This represents the centripetal acceleration due to which the particle moves outwards.

2. Derive equations of motion of a particle in spherical polar co-ordinates.

Let the co-ordinates of particle of mass m be, r, θ, φ. Now the velocity components are $\dot{r}, r\dot{\theta}, r\sin\theta\dot{\phi}$.

$$\therefore \qquad 2T = m(\dot{r}^2 + r^2\dot{\theta}^2 + r^2\sin^2\theta\dot{\phi}^2) \qquad (1)$$

Let mF, mG, mH denote the components of force in the direction of elements, δr, $r\delta\theta$, $r\sin\theta\delta\phi$, the potential is given by

$$-\delta V = m(F\delta r + Gr\delta\theta + Hr\sin\theta\delta\varphi) \qquad (2)$$

Now Lagrange's equations are

$$\frac{d}{dt}\left(\frac{\partial T}{\partial \dot{r}}\right) - \frac{\partial T}{\partial r} = mF,$$

$$\frac{d}{dt}\left(\frac{\partial T}{\partial \dot{\theta}}\right) - \frac{\partial T}{\partial \theta} = mG,$$

$$\frac{d}{dt}\left(\frac{\partial T}{\partial \dot{\phi}}\right) - \frac{\partial T}{\partial \phi} = mH.$$

$$\therefore \qquad \frac{d}{dt}(m\dot{r}) - mr\dot{\theta}^2 - mr\dot{\phi}^2\sin^2\theta = mF,$$

$$\therefore \qquad \frac{d}{dt}(mr^2\dot{\theta}) - m\dot{\phi}^2r^2\sin\theta\cos\theta = mGr,$$

$$\frac{d}{dt}(mr^2\dot{\phi}\sin^2\theta) = mHr\sin\theta.$$

$$\therefore \qquad \ddot{r} - r\dot{\theta}^2 - r\dot{\phi}^2\sin^2\theta = F,$$

$$\therefore \qquad r^2\ddot{\theta} + 2r\dot{r}\dot{\theta} - r^2\dot{\phi}^2\sin\theta\cos\theta = Gr,$$

$$\therefore r^2\ddot{\phi}\sin^2\theta + 2r\dot{r}\dot{\phi}\sin^2\theta + 2r^2\dot{\phi}\dot{\theta}\sin\theta\cos\theta = Hr\sin\theta.$$

$$\therefore \qquad \ddot{r} - r\dot{\theta}^2 - r\dot{\phi}^2\sin^2\theta = F,$$

$$\therefore \qquad r\ddot{\theta} + 2\dot{r}\dot{\theta} - r\dot{\phi}^2\sin\theta\cos\theta = G,$$

$$\therefore \qquad r\ddot{\phi}\sin\theta + 2\dot{r}\dot{\phi}\sin\theta + 2r\dot{\theta}\dot{\phi}\cos\theta = H.$$

3. A uniform circular wire has radius R, enter C and is free to rotate about a vertical axis through a point O in its plane distant d from C. The wire carries a smooth particle P and $\angle OCP = \theta$ at time t. If ω is the angular velocity of the wire, show that

$$R\ddot{\theta} + \dot{\omega}(R - d\cos\theta) = d\omega^2\sin\theta.$$

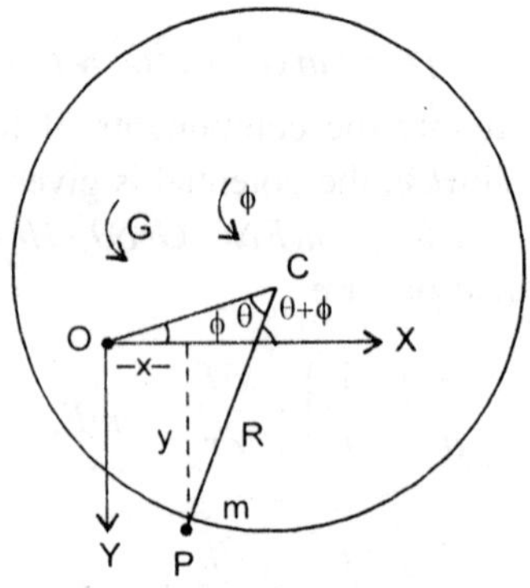

Fig. 10.1

At time t, let OC make an angle ϕ with the fixed line OX, so that PC makes an angle $(\theta + \varphi)$ with OX. Let horizontal line Ox be x-axis and vertically downward OY be y-axis then co-ordinates of P (x, y) are given as

$$x = d\cos\varphi - R\cos(\theta + \varphi),$$
$$y = R\sin(\theta + \varphi) - d\sin\varphi$$
$$\dot{x} = -d\dot{\phi}\sin\phi + R(\dot{\theta} + \dot{\phi})\sin(\theta + \phi),$$
$$\dot{y} = R(\dot{\theta} + \dot{\phi})\cos(\theta + \phi) - d\dot{\phi}\cos\phi.$$

Hence, if m be the mass of the particle at P, then K.E. is given by

$$2T = m(\dot{x}^2 + \dot{y}^2)$$

$$= m\{-d\dot{\phi}\sin\phi + R(\dot{\theta}+\dot{\phi})\sin(\theta+\phi)\}^2 + m\{R(\dot{\theta}+\dot{\phi})\cos(\theta+\phi) - d\dot{\phi}\cos\phi\}^2$$

$$= m[R^2(\dot{\theta}+\dot{\phi})^2 + d^2\dot{\phi}^2 - 2Rd\dot{\phi}(\dot{\theta}+\dot{\phi})\cos\theta]$$

Let G be the couple applied in the direction of ϕ increasing, then the virtual work function is

$$\delta W = G\delta\varphi,$$

so generalized forces are

$$Q_\theta = 0, Q_\varphi = G.$$

Lagrange's equation for co-ordinate θ is

$$\frac{d}{dt}\left(\frac{\partial T}{\partial \dot{\theta}}\right) - \frac{\partial T}{\partial \theta} = Q_\theta$$

$$\therefore \quad \frac{d}{dt}\{R^2(\dot{\theta}+\dot{\phi}) - Rd\dot{\phi}\cos\theta\} - Rd\dot{\phi}(\dot{\theta}+\dot{\phi})\sin\theta = 0,$$

$$\therefore \quad R(\ddot{\theta}+\ddot{\phi}) - d\ddot{\phi}\cos\theta + d\dot{\phi}\dot{\theta}\sin\theta - d\dot{\phi}(\dot{\theta}+\dot{\phi})\sin\theta = 0.$$

Putting $\dot{\phi} = \omega, \ddot{\phi} = \dot{\omega}$, we get

$$R\ddot{\theta} + \dot{\omega}(R - d\cos\theta) = d\omega^2\sin\theta.$$

4. Motion of an insect which crawls at a uniform rate on a uniform heavy rod which is turning about a fixed end.

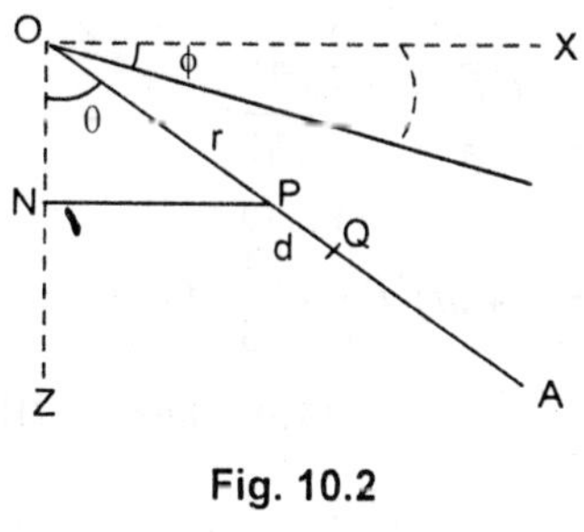

Fig. 10.2

Let $2a$ be the length of the rod, M its mass, O its fixed end, OZ vertical, θ the inclination of the rod to the vertical and φ the angle, which the vertical plane through the rod makes with a fixed vertical plane. Consider an element $Mdr/2a$ of the rod at P wher $OP = r$. If PN be drawn perpendicular to the vertical through the end O of the rod, then P is moving perpendicular to NP with velocity $\dot{\phi} NP$, *i.e.*, $\dot{\phi}\, r \sin\theta$ and it is moving perpendicular to OP in the plane ZOA with velocity $r\dot{\theta}$. Hence, K.E. is given by

$$2T = \int_0^{2a} \frac{Mdr}{2a}(r^2\dot{\phi}^2 \sin^2\theta + r^2\dot{\theta}^2)$$

$$= \frac{M}{2a}(\dot{\theta}^2 + \dot{\phi}^2 \sin^2\theta)\int_0^{2a} r^2 dr$$

$$= \frac{4Ma^2}{3}(\dot{\theta}^2 + \dot{\phi}^2 \sin^2\theta)$$

Let m be the mass of the insect and u the speed at which it crawls and suppose it is at the middle point of the rod when $t = 0$ and crawls downward. Then its co-ordinates x, y, z at time t are given by
$x = (a + ut) \sin\theta \cos\varphi$, $y = (a + ut) \sin\theta \sin\varphi$, $z = (a + ut) \cos\theta$
and square of its velocity is

$$\dot{x}^2 + \dot{y}^2 + \dot{z}^2 = u^2 + (a + ut)^2 (\dot{\theta}^2 + \dot{\phi}^2 \sin^2\theta).$$

$$\therefore\ 2T = \frac{4}{3}Ma^2 (\dot{\theta}^2 + \dot{\phi}^2 \sin^2\theta) + m\{u^2 + (a+ut)^2(\dot{\theta}^2 + \dot{\phi}^2 \sin^2\theta\}$$

$$-V = Mga \cos\theta + mg(a + ut)\cos\theta + const.$$

Lagrange's equations are

$$\frac{d}{dt}\left(\frac{\partial T}{\partial \dot{\theta}}\right) - \frac{\partial T}{\partial \theta} = -\frac{\partial V}{\partial \theta},$$

$$\frac{d}{dt}\left(\frac{\partial T}{\partial \dot{\phi}}\right) - \frac{\partial T}{\partial \phi} = -\frac{\partial V}{\partial \phi}.$$

$$\therefore\ \frac{d}{dt}\left\{\frac{4}{3}Ma^2\dot{\theta} + m(a+ut)^2\dot{\theta}\right\} - \left\{\frac{4}{3}Ma^2 + m(a+ut)^2\right\}\dot{\phi}^2 \sin\theta\cos\theta$$

$$= -\{Ma + m(a + ut)\}\, g \sin\theta$$

$$\therefore\ \frac{d}{dt}\left[\left\{\frac{4}{3}Ma^2 + m(a+ut)^2\right\}\dot{\phi}\sin^2\theta\right] = 0. \quad (2)$$

We observe that (2) integrates at once, which gives

$$\left\{\frac{4}{3}Ma^2 + m\,(a+ut)^2\right\}\dot{\phi}\sin^2\theta = \text{const.} \quad (3)$$

which shows that the moment of momentum about the vertical is constant Eliminating $\dot{\phi}$ from (1) and (3) we obtain a differential equation for θ.

5. A uniform rod, of length $2a$, can turn freely about one end, which is fixed. Initially it is inclined at an acute angle α to the downward drawn vertical and it is set rotating about a vertical axis through its fixed end with angular velocity ω. Show that, during the motion, the rod is always inclined at an angle which is > or < α, according as ω^2 > or < $3g/4\,a\cos\alpha$, and that in each case its motion is included between the inclination α and $\cos^{-1}[-n + \{1 - 2n\cos\alpha + n^2\}^{1/2}]$, where

$$n = a\omega^2 \sin^2\alpha/3g.$$

If it be slightly disturbed when revolving steadily at constant inclination α, show that the time of small oscillation is $2\pi\sqrt{\dfrac{4a\cos\alpha}{3g(1+3\cos^2\alpha)}}$.

Let m be the mass of the rod proceeding as in Example (4), we get kinetic energy of the rod given by

$$2T = \frac{4ma^2}{3}(\dot{\theta}^2 + \dot{\phi}^2\sin^2\theta).$$

$$-V = mga\cos\theta + \text{constant}.$$

Lagrange's equations are

$$\frac{d}{dt}\left(\frac{\partial T}{\partial\dot{\theta}}\right) - \frac{\partial T}{\partial\theta} = -\frac{\partial V}{\partial\theta},$$

$$\frac{d}{dt}\left(\frac{\partial T}{\partial\dot{\phi}}\right) - \frac{\partial T}{\partial\phi} = -\frac{\partial V}{\partial\phi}.$$

$$\therefore \quad \frac{d}{dt}\left(\frac{4}{3}ma^2\dot{\theta}\right) - \frac{4ma^2}{3}\dot{\phi}^2\sin\theta\cos\theta = -mga\sin\theta,$$

$$\frac{d}{dt}\left(\frac{4}{3}ma^2\dot{\phi}\sin^2\theta\right) = 0.$$

$$\therefore \quad \ddot{\theta} - \dot{\phi}^2\sin\theta\cos\theta = -\frac{3g}{4a}\sin\theta, \tag{1}$$

$$\dot{\phi}\sin^2\theta = \text{const.} = \omega\sin^2\alpha. \tag{2}$$

Eliminating $\dot{\phi}$, we get

$$\ddot{\theta} - \frac{\omega^2\sin^4\alpha\cos\theta}{\sin^3\theta} = -\frac{3g}{4a}\sin\theta. \tag{3}$$

Steady motion : The rod goes round at a constant inclination α to the vertical if $\ddot{\theta}$ be zero when $\theta = \alpha$, *i.e.*, if

$$\omega^2 = 3g/4\,a\cos\alpha \tag{4}$$

When ω has not this particular value, integrating equation (3), we get

$$\dot{\theta}^2 + \frac{\omega^2 \sin^4\alpha}{\sin^2\theta} = \frac{3g}{2a}\cos\theta + C = \omega^2\sin^2\alpha + \frac{3g}{2a}(\cos\theta - \cos\alpha)$$

from the initial conditions.

$$\dot{\theta}^2 = \frac{3gn}{a}\left[1 - \frac{\sin^2\alpha}{\sin^2\theta}\right] + \frac{3g}{2a}(\cos\theta - \cos\alpha),$$

$$= \frac{3gn}{a}\frac{(\cos^2\alpha - \cos^2\theta)}{\sin^2\theta} + \frac{3g}{2a}(\cos\theta - \cos\alpha)$$

$$= \frac{3g}{2a}(\cos\alpha - \cos\theta)\left[-1 + \frac{2n(\cos\theta + \cos\alpha)}{\sin^2\theta}\right]$$

$$= \frac{3g\,(\cos\alpha - \cos\theta)}{2a\sin^2\theta}[-\sin^2\theta + 2n\cos\alpha + 2n\cos\theta]$$

$$= \left\{\frac{3g}{2a\sin^2\theta}(\cos\alpha - \cos\theta)\right\}\{\cos^2\theta + 2n\cos\theta - 1 + 2n\cos\alpha\}$$

Hence, $\dot{\theta}$ is zero when $\theta = \alpha$, *i.e.* initially, or when

$$\cos^2\theta + 2n\cos\theta - 1 + 2n\cos\alpha = 0$$

i.e. when $\cos\theta = -n + \sqrt{\{1 - 2n\cos\alpha + n^2\}}$ (5)

[The +sign must be taken; for the –sign would give a value of $\cos\theta$ numerically greater than unity.]

The motion is therefore included between the values $\theta = \alpha$ and $\theta = \theta_1$ where $\cos\theta_1$ is equal to the right of (5).

Now, when $\theta_1 >$ or $< \alpha$ *i.e.* the rod will rise higher than or fall below its initial position, according as $\cos\theta_1 < \cos\alpha$ or $\cos\theta_1 > \cos\alpha$,

i.e., according as $\sqrt{[1 - 2n\cos\alpha + n^2]} <$ or $> \cos\alpha + n$

i.e., according as $1 - 2n\cos\alpha + n^2 <$ or $\cos^2\alpha + n^2 + 2n\cos\alpha$.

i.e., according as $\sin^2\alpha <$ or $> 4n\cos\alpha$

i.e., according as $\sin^2\alpha/4\cos\alpha <$ or $> 4n$

i.e., according as $\sin^2\alpha/4\cos\alpha <$ or $> 4a\omega^2\sin^2\alpha/3g$

i.e., according as $\omega^2 >$ or $< 3g/4\,a\cos\alpha$.

i.e., according as the initial angular velocity is greater or less than that for steady motion at the inclination α.

Small oscillations about the steady motion.

From (4) the value of ω^2 is $3g/4a \cos \alpha$ for steady motion. If ω^2 has this value, then (3) gives

$$\ddot{\theta} = \frac{3g\sin^4 \alpha \cos\theta}{4a\sin^3 \theta \cos\alpha} - \frac{3g\sin\alpha}{4a} \qquad (6)$$

Here put $\theta = \alpha + x$, where x is small, and therfore

$$\sin\theta = \sin\alpha + x \cos \alpha, \cos \theta = \cos \alpha - x \sin \alpha$$

Hence, (6) gives on neglecting squares and higher powers of x

$$\ddot{x} = 3g \sin \alpha \left[(1 - x \tan \alpha)(1 + x \cot \alpha)^{-3} - 1 - x \cot \alpha\right] / 4a.$$

$$= -\frac{3g\sin\alpha}{4a}[4\cot\alpha + \tan\alpha]x.$$

$$= -\frac{3g(1+3\cos^2\alpha)}{4a\cos\alpha}x.$$

This is S.H.M. of period

$$2\pi\sqrt{\frac{4a\cos\alpha}{3g(1+3\cos^2\alpha)}}.$$

6. A homogeneous rod OA of length $2a$ and mass m_1, is freely hinged to a point, at its other end is freely attached another homogeneous rod AB of mass m_2 and length $2b$; the system moves under gravity, find equations of motion.

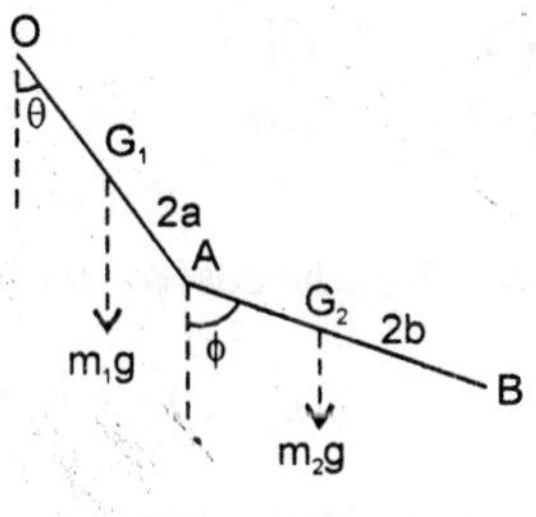

Fig. 10.3

Let G_1 and G_2 be the centres of masses of the rods, and θ, ϕ their inclination to the vertical at time t. Referred to horizontal and vertical lines through O as axes of reference, the co-ordinates of G_1 are

$$x_1 = 2a\dot{\theta}\cos\theta + b\dot{\phi}\cos\phi,$$

$$\therefore \quad \dot{x}_1 = a\,\dot{\theta}\,\cos\theta$$

$$\dot{y}_1 = -a\,\dot{\theta}\,\sin\theta$$

$$\dot{x}_1^2 + \dot{y}_1^2 = a^2\dot{\theta}^2.$$

The K.E. of the rod OA about $G_1 = \frac{1}{2}m_1\frac{a^2\dot{\theta}^2}{3}$

The K.E. of the rod OA about O is $= \frac{1}{2}\left[m_1\frac{a^2\dot{\theta}}{3} + m_1a^2\dot{\theta}^2\right] = \frac{2}{3}m_1a^2\dot{\theta}^2$

Similarly co-ordinates of G_2 are

$$x_2 = 2a\sin\theta + b\sin\phi,$$
$$y_2 = 2a\cos\theta + b\cos\phi$$

Hence,

$$\dot{x}_2^2 + \dot{y}_2^2 = (2a\dot{\theta}\cos\theta + b\dot{\phi}\cos\phi)^2 + (2a\dot{\theta}\sin\theta + b\dot{\phi}\sin\phi)^2$$
$$= 4a^2\dot{\theta}^2 + b^2\dot{\phi}^2 + 4ab\dot{\theta}\dot{\phi}\cos(\theta-\phi)$$

Also the K.E. of the rod AB about G_2 is $= \frac{1}{2}\left(\frac{m_2b^2\dot{\phi}^2}{3}\right)$

Hence, K.E. of the system is given by

$$T = \frac{2}{3}m_1a^2\dot{\theta}^2 + \frac{1}{2}m_2\left[4a^2\dot{\theta}^2 + b^2\dot{\phi}^2 + 4ab\dot{\theta}\dot{\phi}\cos(\theta-\phi) + \frac{b^2\dot{\phi}^2}{3}\right]$$

$$= \frac{1}{2}\left(\frac{m_1}{3} + m_2\right)4a^2\dot{\theta}^2 + \frac{2}{3}m_2b^2\dot{\phi}^2 + 2m_2ab\dot{\theta}\dot{\phi}\cos(\theta-\phi) \quad (1)$$

$$V = -m_1\,ga\cos\theta - m_2g\,(2a\cos\theta + b\cos\phi) \quad (2)$$

Lagrange's θ equation is

$$\frac{d}{dt}\left(\frac{\partial T}{\partial\dot{\theta}}\right) - \frac{\partial T}{\partial\theta} = -\frac{\partial V}{\partial\theta}$$

$$\therefore \frac{d}{dt}\left[\left(\frac{m_1}{3} + m_2\right)4a^2\dot{\theta} + 2m_2ab\dot{\phi}\cos(\phi-\theta)\right] - 2m_2ab\dot{\theta}\dot{\phi}\sin(\phi-\theta)$$
$$= -(m_1 + 2m_2)\,ga\sin\theta$$

$$\therefore \left(\frac{m_1}{3} + m_2\right)4a\ddot{\theta} + 2m_2b\,[\ddot{\phi}\cos(\phi-\theta) - \dot{\phi}^2\sin(\phi-\theta)]$$
$$= -g\,(m_1 + 2m_2)\sin\theta \quad (3)$$

Lagrange's ϕ equations is

$$\frac{d}{dt}\left(\frac{\partial T}{\partial\dot{\phi}}\right) - \frac{\partial T}{\partial\phi} = -\frac{\partial V}{\partial\phi}$$

$$\therefore \frac{d}{dt}\left[m_2\left\{\frac{4b^2\dot{\phi}}{3} + 2ab\dot{\theta}\cos(\phi-\theta)\right\}\right] + m_2 2ab\dot{\theta}\,\dot{\phi}\sin(\phi-\theta)$$
$$= -m_2\,bg\sin\phi$$

$$\therefore \quad \frac{4}{3}b\ddot{\phi} + 2a\ddot{\theta}\cos(\phi - \theta) + 2a\dot{\theta}^2 \sin(\phi - \theta) = -g\sin\phi \qquad (4)$$

Multiplying (3) by a $\dot{\theta}$ (4) by $m_2 b\dot{\phi}$, adding and integrating, we get

$$\frac{1}{2}\left(\frac{m_1}{3} + m_2\right)4a^2\dot{\theta}^2 + \frac{1}{2}m_2.\frac{4b^2}{3}\dot{\phi}^2 + \frac{1}{2}m_2 4ab\dot{\theta}\dot{\phi}\cos(\phi - \theta)$$
$$= (m_1 + 2m_2)\, ga \cos\theta + m_2 gb \cos\phi + C \qquad (5)$$

This is energy equation

Again multiplying (3) by a (4) by $m_2 b$ and adding, we have

$$\frac{d}{dt}\left[\left(\frac{m_1}{3} + m_2\right)4a^2\dot{\theta} + m_2\left\{4b^2\frac{\dot{\phi}}{3} + 2ab(\dot{\theta} + \dot{\phi})\cos(\varphi - \theta)\right\}\right]$$
$$= -ag\,(m_1 + 2m_2)\sin\theta - m_2 gb\sin\varphi.$$

This is the equation of moment of momentum about O.

7. A bead, of mass M slides on a smooth fixed wire, whose inclination to the vertical is α, and has hinged to it a rod of mass M and length $2b$, which can move freely in the vertical plane through wire. If the system starts from rest with the rod hanging vertically, show that

$$[4M + m\,(1 + 3\cos^2\theta)]\,l\dot{\theta}^2 = 6(M + m)g\sin\alpha\,(\sin\alpha - \sin\theta),$$

where θ is the angle between the rod and lower part of the wire.

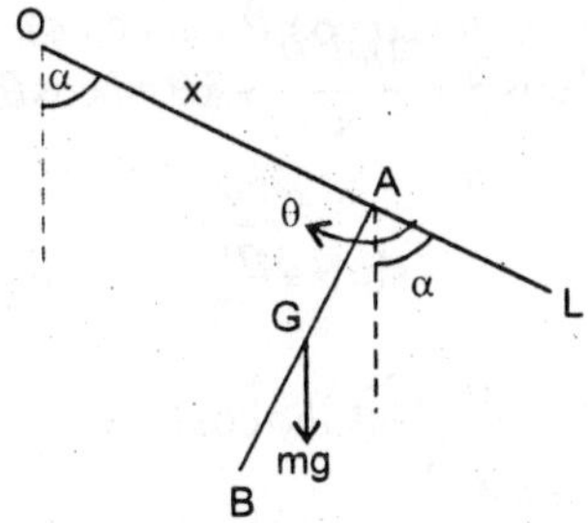

Fig. 10.4

OL is fixed wire. At any time t let the bead be at A, where $OA = x$, at that time rod AB is making an angle θ with the lower part of the fixed wire. G is the centre of gravity of the rod.

Taking O as origin and fixed wire OL as x-axis co-ordinates of G are $(x + l\cos\theta, l\sin\theta)$.

$$\therefore \qquad (\text{velocity})^2 \text{ of } G = (\dot{x} - l\dot{\theta}\sin\theta)^2 + (l\dot{\theta}\cos\theta)^2.$$

Let T be the K.E. and V the potential function of the system, then

$$T = M\frac{\dot{x}^2}{2} + \frac{1}{2}m\left[\frac{l^2\dot{\theta}^2}{3} + (\dot{x} - l\dot{\theta}\sin\theta)^2 + (l\dot{\theta}\cos\theta)^2\right]$$

$$= \left[(M+m)\dot{x}^2 - 2ml\,\dot{x}\dot{\theta}\sin\theta + \frac{4ml^2\dot{\theta}^2}{3} \right] \Big/ 2$$

$$V = -Mg\,x\cos\alpha - mg\,\{x\cos\alpha + l\cos(\theta-\alpha)\}$$

Lagrange's x equation is

$$\frac{d}{dt}\left(\frac{\partial T}{\partial \dot{x}}\right) - \frac{\partial T}{\partial x} = -\frac{\partial V}{\partial x},$$

$$\therefore \quad \frac{d}{dt}[(M+m)\dot{x} - ml\dot{\theta}\sin\theta] = (M+m)g\cos\alpha.$$

$$\therefore \quad (M+m)\ddot{x} - ml\ddot{\theta}\sin\theta - ml\dot{\theta}^2\cos\theta = (M+m)\,g\cos\alpha. \quad (1)$$

Lagrange's θ equation is

$$\frac{d}{dt}\left(\frac{\partial T}{\partial \dot{\theta}}\right) - \frac{\partial T}{\partial \theta} = -\frac{\partial V}{\partial \theta}$$

$$\therefore \frac{d}{dt}\left[-ml\dot{x}\sin\theta + \frac{4ml^2\dot{\theta}}{3}\right] + ml\dot{x}\dot{\theta}\cos\theta = -mgl\sin(\theta-\alpha),$$

$$\therefore \; -m\ddot{x}l\sin\theta - \dot{x}\dot{\theta}\,ml\cos\theta + \frac{4ml^2\ddot{\theta}}{3} + \dot{x}\dot{\theta}\,ml\cos\theta = -mgl\sin(\theta-\alpha),$$

$$\therefore \quad -\ddot{x}\sin\theta + \frac{4l\ddot{\theta}}{3} = -g\sin(\theta-\alpha) \quad (2)$$

Eliminating $\ddot{x}$ from (1) and (2), we have

$$\ddot{\theta}\left[-ml\sin^2\theta + \frac{4(M+m)l}{3}\right] - ml\dot{\theta}^2\sin\theta\cos\theta$$

$$= (M+m)g\,[\cos\alpha\sin\theta - \sin(\theta-\alpha)]$$

$$\therefore \; l\ddot{\theta}\left[4M + m(1+3\cos^2\theta)\right] - 3ml\dot{\theta}^2\sin\theta\cos\theta$$

$$= 3\,(M+m)\,g\cos\theta\sin\alpha.$$

Integrating, we get

$$l\dot{\theta}^2\left[4M + m(1+3\cos^2\theta)\right] = 6\,(M+m)g\sin\theta\sin\alpha + D.$$

When $\theta = \alpha,\ \dot{\theta} = 0$

$$\therefore \quad D = -6\,(M+m)\,g\sin^2\alpha.$$

Hence,

$$l\dot{\theta}^2\left[4M + m(1+3\cos^2\theta)\right] = 6\,(M+m)g\sin\alpha\,(\sin\theta - \sin\alpha).$$

8. A uniform rod OA of length $2a$ and mass M is smoothly pivoted at one end to a fixed point O of the rod makes an angle θ with the downward vertical OZ and the plane AOZ makes an angle ϕ with a fixed vertical plane. A bead of mass m slides on the smooth rod and is connected to O by a light elastic string of modulus nmg and a natural length a. Show that the K.E. of the system is given by

$$2T = \frac{4}{3}ma^2(\dot{\theta}^2 + \dot{\phi}^2\sin\theta) + \lambda m(\dot{x}^2 + x^2\dot{\theta}^2 + x^2\dot{\phi}^2\sin^2\theta),$$

where x is the stretched length of the string, and derive the equations of motion.

Show that, if a steady motion with $\theta = \pi/3$ and $x = 4a/3$ is possible, then

$$\dot{\phi}^2 = \frac{3g}{2a} \text{ and } n = 6\lambda.$$

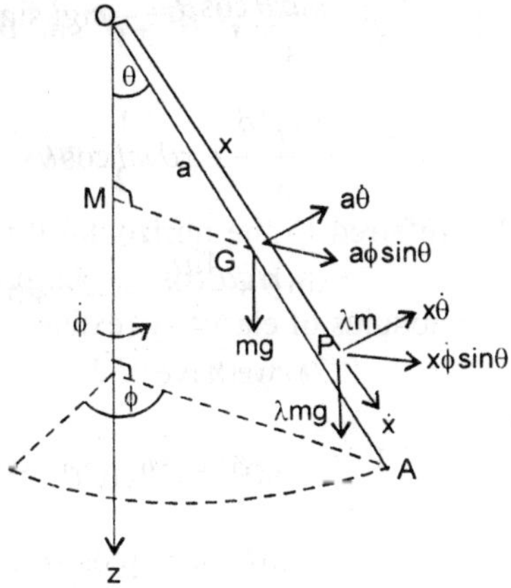

Fig. 10.5

Let G be the centroid of the rod and P is the position of the particle at time t, such that $OP = x$. Draw perpendicular GM on the vertical OZ through O. Now G is moving perpendicular to GM with velocity $\dot{\phi}$. GM *i.e.*, $\dot{\phi}$ a sin θ and it is moving perpendicular to OG in the plane AOZ with velocity $a\dot{\theta}$

Now the rod has spin components $\dot{\theta}$, $\dot{\phi}\sin\theta$, $\dot{\phi}\cos\theta$ about the Principal axes and corresponding Principal M.I. are $ma^2/3$, $ma^2/3$, 0

Hence, K.E. of the rod due to rotation is $\frac{1}{2}\left(\frac{1}{3}ma^2\right)(\dot{\theta}^2+\dot{\phi}^2\sin^2\theta)$.

$$\therefore \quad \text{K.E. of the rod is} = \frac{1}{2}m(a^2\dot{\theta}^2+a^2\dot{\phi}^2\sin^2\theta)$$

$$+\frac{1}{2}\left(\frac{1}{3}ma^2\right)(\dot{\theta}^2+\dot{\phi}^2\sin^2\theta)$$

$$=\frac{2ma^2}{3}(\dot{\theta}^2+\dot{\phi}^2\sin^2\theta).$$

Similarly, K.E. of the particle is

$$=\frac{1}{2}\lambda m[\dot{x}^2+(x\dot{\phi}\sin\theta)^2+(x\dot{\theta})^2]$$

$\therefore$ K.E. T of the system is given by

$$T=\frac{2}{3}ma^2(\dot{\theta}^2+\dot{\phi}^2\sin^2\theta)+$$

$$\frac{1}{2}\lambda m(\dot{x}^2+x^2\dot{\theta}^2+x^2\dot{\phi}^2\sin^2\theta).$$

Gravitational P.E. (referred to the horizontal through O as zero level)

$$=-mga\cos\theta-\lambda mgx\cos\theta.$$

Now elastic P.E. = modulus of elasticity (extension)^2/2 natural length

$$=nmg(x-a)^2/2a.$$

Thus total P.E. is

$$V=-mga\cos\theta-\lambda mgx\cos\theta+mng(x-a)^2/2a.$$

$$\therefore \quad L=T-V$$

$$=\frac{2}{3}ma^2(\dot{\theta}^2+\dot{\phi}^2\sin^2\phi)+\frac{1}{2}\lambda m(\dot{x}^2+x^2\dot{\theta}^2+x^2\dot{\phi}^2\sin^2\theta)$$

$$+mga\cos\theta+\lambda mgx\cos\theta-\frac{nmg(x-a)^2}{2a}.$$

Lagrange's equations for x, θ are

$$\frac{d}{dt}\left(\frac{\partial L}{\partial \dot{x}}\right)-\frac{\partial L}{\partial x}=0,$$

$$\frac{d}{dt}\left(\frac{\partial L}{\partial \dot{\theta}}\right)-\frac{\partial L}{\partial \theta}=0.$$

First equation becomes

$$\frac{d}{dt}(\lambda m\dot{x}) - \lambda m(x\dot{\theta}^2 + x\dot{\phi}^2 \sin^2\theta) - \lambda mg\cos\theta + nmg(x-a)/a = 0,$$

$$\therefore\ \lambda\ddot{x} - \lambda x(\dot{\theta}^2 + \dot{\phi}^2\sin^2\theta) - \lambda g\cos\theta + \frac{ng(x-a)}{a} = 0 \qquad (1)$$

Similarly θ– equation becomes

$$\frac{4}{3}a^2\ddot{\theta} + \lambda x^2\ddot{\theta} + 2\lambda\dot{x}\dot{\theta}x - \frac{4}{3}a^2\dot{\phi}^2\sin\theta\cos\theta - \lambda x^2\dot{\phi}^2\sin\theta\cos\theta$$

$$+ ag\sin\theta + \lambda gx\sin\theta = 0. \qquad (2)$$

For steady motion with $\theta = \pi/3$, and $x = \dfrac{4a}{3}$ equations (1) and (2) become

$$\therefore \qquad \lambda a\dot{\phi}^2 - \frac{1}{2}\lambda g + \frac{1}{3}ng = 0. \qquad (3)$$

$$-\frac{2}{3}a\dot{\phi}^2 - \frac{8}{9}\lambda a\dot{\phi}^2 + g + \frac{4}{3}\lambda g = 0. \qquad (4)$$

(4) gives

$$\dot{\phi}^2 = \frac{3g}{2a}$$

and (3) gives

$$n = 6\lambda.$$

10.10 CONSERVATION THEOREMS

(*a*) ***Conservation Theorem for Generalized Momentum.*** Prove that the generalized momentum corresponding to a cyclic co-ordinate of ignorable co-ordinate is constant.

If a co-ordinate q_j is absent in the expression for L, then q_j is called *ignorable* or *cyclic co-ordinate.*

Lagrange's equations are

$$\frac{d}{dt}\left(\frac{\partial L}{\partial \dot{q}_j}\right) - \frac{\partial L}{\partial q_j} = 0$$

If q_j is a cyclic co-ordinate, then $\dfrac{\partial L}{\partial q_j} = 0$, hence, we get

$$\frac{d}{dt}\left(\frac{\partial L}{\partial \dot{q}_j}\right) = 0$$

$$\therefore \qquad \frac{\partial L}{\partial \dot{q}_j} = const.$$

$$\therefore \qquad p = const.$$

(*b*) ***Conservation Theorem for Energy.*** Prove that, if the Lagrangian L does not contain time explicitly, the total energy of the conservative system is constant.

Let us consider a conservative system in which L is independent of time. Now Lagrange's equations are

$$\frac{d}{dt}\left(\frac{\partial L}{\partial \dot{q}_j}\right) - \frac{\partial L}{\partial q_j} = 0,$$

$$\therefore \qquad \frac{d}{dt}\left(\frac{\partial L}{\partial \dot{q}_j}\right) = \frac{\partial L}{\partial q_j}. \qquad (1)$$

We have

$$L = L(q_1, q_2, \ldots\ldots, q_n, \dot{q}_1, \dot{q}_2, \ldots\ldots, \dot{q}_n)$$

$$\therefore \qquad \frac{dL}{dt} = \sum_j \frac{\partial L}{\partial q_j} \dot{q}_j + \sum_j \frac{\partial L}{\partial \dot{q}_j} \ddot{q}_j$$

Using (1), we get

$$\frac{dL}{dt} = \sum_j \frac{d}{dt}\left(\frac{\partial L}{\partial \dot{q}_j}\right)\dot{q}_j + \sum_j \frac{\partial L}{\partial \dot{q}_j} \ddot{q}_j$$

$$= \sum_j \frac{d}{dt}\left(\dot{q}_j \frac{\partial L}{\partial \dot{q}_j}\right)$$

$$\therefore \quad \frac{d}{dt}\left[L - \sum_j \dot{q}_j \frac{\partial L}{\partial \dot{q}_j}\right] = 0$$

$$\therefore \qquad L - \sum_j \dot{q}_j \frac{\partial L}{\partial \dot{q}_j} = \text{const.} \qquad (2)$$

But

$$\frac{\partial L}{\partial \dot{q}_j} = \frac{\partial}{\partial \dot{q}_j}(T - V) = \frac{\partial T}{\partial \dot{q}_j}$$

$$= \frac{\partial}{\partial \dot{q}_j} \sum \frac{1}{2} m_j \dot{q}_j^2$$

$$= m_j \dot{q}_j$$

because $$T = \sum_j \frac{1}{2} m_j q_j^2$$

Substituting the value of $\frac{\partial L}{\partial \dot{q}_j}$ in (2) we get

$$L - \sum m_j \dot{q}_j^2 = \text{const.}$$

$$\therefore \quad L - 2T = \text{const.}$$

$$\therefore \quad T - V - 2T = \text{const.}$$

$$\therefore \quad -(T + V) = \text{const.}$$

$$\therefore \quad T + V = \text{const.}$$

10.11 LAGRANGE'S EQUATIONS FOR IMPULSIVE FORCES

Lagrange's equations for a holonomic system are

$$\frac{d}{dt}\left(\frac{\partial T}{\partial \dot{q}_j}\right) - \frac{\partial T}{\partial q_j} = Q_j$$

Now

$$\frac{\partial T}{\partial \dot{q}_j} = p_j$$

Hence, Lagrange's equations become

$$\frac{dp_j}{dt} - \frac{\partial T}{\partial q_j} = Q_j$$

Integrating from $t = 0$ to $t = \tau$, we obtain

$$(p_j)_{t=\tau} - (p_j)_{t=0} = \int_0^\tau \frac{\partial T}{\partial q_j} dt + \int_0^\tau Q_j dt$$

Let $Q_j \to \infty$, $\tau \to 0$ in such a way that $\int_0^\tau Q_j dt \to \text{finite} = I_j$.

as the co-ordinates do not change abruptly,

Let $\Delta p_j = \lim \left[(p_j)_{t=\tau} - (p_j)_{t=0}\right]$, we obtain equations in impulsive form

$$\Delta p_j = I_j$$

These are *Lagrange's equations for Impulsive forces.*

Hence, these equations can be stated as that generalized momentum increment is equal to the generalized impulsive force associated with each co-ordinate q_j.

Examples

1. Three equal uniform rods *AB*, *BC*, *CD* are freely jointed at *B* and *C* and the ends *A* and *D* are fastened to smooth fixed pivots whose distance apart is equal to the length of either rod. The frame being at rest in the form of a square, a blow *I* is given perpendicular to *AB* at its middle point and in the plane of the square. Show that energy set up is $3I^2/40m$ where *m* is the mass of each rod. Find also the blows at the joints *B* and *C*.

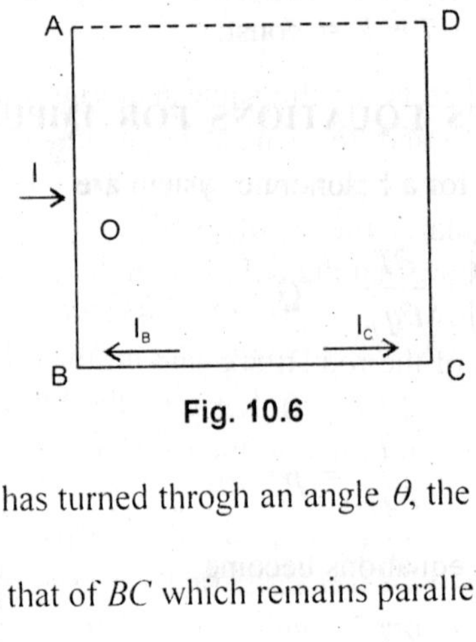

Fig. 10.6

When *AB* or *CD* has turned throgh an angle θ, the energy of either rod is $\frac{1}{2}m.\frac{4a^2\dot{\theta}^2}{3}$, and that of *BC* which remains parallel to *AD*, is $\frac{1}{2}m(2a\dot{\theta})^2$.

$\therefore$ K.E. of the system is given by

$$T = 2.\frac{1}{2}m.\frac{4a^2\dot{\theta}^2}{3} + \frac{1}{2}m4a^2\dot{\theta}^2 = \frac{10}{3}ma^2\dot{\theta}^2$$

Generalized momentum is

$$p_\theta = \frac{\partial T}{\partial \dot{\theta}} = \frac{20ma^2\dot{\theta}}{3}.$$

Impulsive virtual work function is

$$\delta U = Ia\delta\theta.$$

Lagrange's equation for impulsive force is

$$\Delta p = I.a; \text{ i.e. } 20\, ma^2 \dot{\theta} / 3 = 1.a.$$

$$\therefore \qquad \dot{\theta} = 3I/20ma.$$

$$\therefore \qquad T = [(10ma^2)/3].[9I^2/400\, m^2a^2] = 3I^2/40\, m.$$

If Y_1, Y_2 be the blows at the points B and C, then by taking moments about A and D for the rods AB and DC, we have

$$m.\frac{4a^2}{3}\dot{\theta} = I.a - Y_1.2a$$

$$m.\frac{4a^2}{3}\dot{\theta} = Y_2.2a$$

$$\therefore \qquad Y_1 = \frac{1}{2a}\left(I.a - \frac{m4a^2}{3}\dot{\theta}\right) = \frac{I}{2} - \frac{I}{10} = \frac{2I}{5},$$

$$Y_2 = m.\frac{2a}{3}\dot{\theta} = \frac{I}{10}.$$

2. Four freely jointed rods, of the same material and thickness, form a rectangle of sides $2a$ and $2b$ of mass m. When lying in this form on a horizontal plane an elastic particle of mass M moving with velocity V in a direction perpendicular to the rod of length $2a$ impinges on it at a distance c from its center. Show that the K.E. lost in the impact is

$$[(V^2)/2] \div [1/M + 1/m\,\{1 + (3a + 3b)c^2/(a + 3b)a^2\}].$$

Let m_1 be the mass of the rod struck, and m_2 that of the adjacent rod, so that $m_1/a = m_2/b = m/2\,(a + b)$. Let u be the linear velocity and ω the angular velocity communicated to the rod that is struck.

Then

$$T = 2m_1\,(u^2 + a^2\,\omega^2/3)/2 + m_2[(u + a\omega)^2 + (u - a\omega)^2]/2$$

$$= (m_1 + m_2)\,u^2 + (m_1/3 + m_2)a^2\omega^2$$

$$= mu^2/2 + ma^2\,\omega^2\,(a + 3b)/6\,(a + b) \qquad (1)$$

Also the below

$$X = M\,(V - u - c\omega) \qquad (2)$$

and

$$\delta U = M\,(V - u - c\omega)\,(\delta x + c\,\delta\theta) \qquad (3)$$

$$= I_x \delta x + I_\theta$$

where $\qquad u = x,\ \omega = \theta$

$$\therefore \qquad I_x = M(V - u - c\omega),\ I_\theta = Mc\,(V - u - c\omega)$$

Generalized momenta are

$$p_u = \frac{\partial T}{\partial u} = mu,\ p_\theta = \frac{\partial T}{\partial \omega} = ma^2\,\omega(a + 3b)/3(a + b).$$

Lagrange's equations for impulsive forces are

$$\Delta p_x = I_x \text{ and } \Delta p_\theta = I_\theta$$

$$\therefore \quad mu = M(V - u - c\omega) \tag{4}$$

$$\frac{ma^2\omega(a+3b)}{3(a+b)} = Mc\,(V - u - c\omega) \tag{5}$$

Let $\lambda = 3c^2(a+b)/a^2(a+3b)$

then we have

$$u = MV/[M(1+\lambda) + m] \text{ and } c\omega = MV\lambda/[M(1+\lambda)+m] \tag{6}$$

Total loss of K.E.

$$= X[V + (u + c\omega)]/2 - X(u + c\omega)/2 = XV/2 = MV(V - u - c\omega)/2$$

$$= MV[V - MV/\{M(1+\lambda)+m\} - MV\lambda/\{M(1+\lambda)+m\}]/2$$

$$= MV^2[1 - M(1+\lambda)/\{M(1+\lambda)+m\}]/2 = MV^2\,[m/\{M(1+\lambda)+m\}]/2$$

$$= V^2\,[1/\{(1+\lambda)/m + 1/M\}]/2$$

$$= \frac{1}{2}V^2 \div [1/M + \{1 + 3(a+b)\,c^2/a^2\,(a+3b)\}/m]$$

10.12 EQUILIBRIUM CONFIGURATIONS FOR CONSERVATIVE HOLONOMIC DYNAMICAL SYSTEMS

For a conservative holonomic dynamical system specified by n generalized co-ordinates q_j with time explicitly absent, we know that K.E. is a homogeneous quadratic function of velocities of the form

$$T = \frac{1}{2}\sum_{r=1}^{n}\sum_{s=1}^{n} a_{rs}\dot{q}_r\dot{q}_s,$$

where $a_{rs} = a_{sr}$ and these co-efficients are each functions of q_i for all r, s. Further the P.E. is of the form

$$V = V(q_1, q_2, \ldots\ldots\ldots, q_n)$$

Thus Lagrange's equations for such a system are

$$\frac{d}{dt}\left(\frac{\partial T}{\partial \dot{q}_j}\right) - \frac{\partial T}{\partial q_j} = -\frac{\partial V}{\partial q_j} \tag{1}$$

Now

$$\frac{\partial T}{\partial q_j} = \frac{1}{2}\sum_{r=1}^{n}\sum_{s=1}^{n} \frac{\partial a_{rs}}{\partial q_j}\dot{q}_r\dot{q}_s,$$

$$\frac{\partial T}{\partial \dot{q}_j} = \frac{1}{2}\sum_{r=1}^{n}\sum_{s=1}^{n} a_{rs}\left(\dot{q}_r\frac{\partial \dot{q}_s}{\partial \dot{q}_j} + \dot{q}_s\frac{\partial \dot{q}_r}{\partial \dot{q}_j}\right)$$

$$= \frac{1}{2}\sum_{r=1}^{n} a_{rj}\dot{q}_r + \frac{1}{2}\sum_{s=1}^{n} a_{js}\dot{q}_s = \sum_{r=1}^{n} a_{rj}\dot{q}_r;$$

$$\therefore \frac{d}{dt}\left(\frac{\partial T}{\partial \dot{q}_j}\right) = \sum_{r=1}^{n}\left\{a_{rj}\ddot{q}_r + \dot{q}_r\left(\dot{q}_1\frac{\partial a_{rj}}{\partial \dot{q}_1} + \dot{q}_2\frac{\partial a_{rj}}{\partial \dot{q}_2} + \ldots + \dot{q}_n\frac{\partial a_{rj}}{\partial \dot{q}_n}\right)\right\}$$

Now suppose that the system is in equilibrium for a configuration specified by the co-ordinates $q_j = \alpha_j$. For these values $\dot{q}_j = 0 = \ddot{q}_j$.

Hence, it follows that $\frac{\partial T}{\partial q_j} = 0;\ \frac{d}{dt}\left(\frac{\partial T}{\partial \dot{q}_j}\right) = 0.$

Thus Lagrange's equations (1) reduce to

$$\frac{\partial V}{\partial q_j} = 0, \tag{2}$$

for $q_j = \alpha_j$. It follows that the equilibrium configuration for a system specified by n generalized co-ordinates q_j, we express the potential energy V as a function of these co-ordinates and then solve n equations (2) for the co-ordinates specifying the equilibrium configuration or configurations.

Having found the configuration according to this rule, it is important to investigate whether these positions are stable or unstable. We know that for conservative system $T + V =$ constant and if small perturbations are made from an equilibrium position, T must decrease or increase according to whether the position is stable or unstable, *i.e.* V must increase or decrease, respectively. Thus at a stable position of equilibrium V is minimum and at an unstable position V must be a maximum.

10.13 NORMAL OR PRINCIPAL CO-ORDINATES

The quantities X, Y which are such that the corresponding equations each contain only X or Y, are called *Principal* or *Normal co-ordinates.*

Let the K.E. and the work function W be expressible in terms of θ, ϕ, $\Psi, \dot{\theta}, \dot{\phi}, \dot{\psi}$.

In the case of small oscillations about a position of equilibrium, so that

$$T = A_{11}\dot{\theta}^2 + A_{22}\dot{\phi}^2 + A_{33}\dot{\psi}^2 + 2A_{12}\dot{\theta}\dot{\phi} + 2A_{23}\dot{\phi}\dot{\psi} + 2A_{13}\dot{\theta}\dot{\psi} \tag{1}$$

$$W = C + B_1\theta + B_2\phi + B_3\psi + B_{11}\theta^2 + B_{22}\phi^2 + B_{33}\psi^2 \tag{2}$$

If θ, ϕ, Ψ be expressed in tems of X, Y, Z by linear equations of the form

$$\theta = \lambda_1 X + \lambda_2 Y + \lambda_3 Z,$$
$$\varphi = \mu_1 X + \mu_2 Y + \mu_3 Z,$$

$$\psi = \nu_1 X + \nu_2 Y + \nu_3 Z,$$

and $\lambda_1, \lambda_2, \lambda_3, \mu_1, \mu_2, \mu_3, \nu_1, \nu_2, \nu_3$ be so chosen that on substitution in (1) and (2) there are no terms in T containing $\dot{Y}\dot{Z}$, $\dot{Z}\dot{X}$, $\dot{X}\dot{Y}$ and YZ, ZX, XY, in W; then X, Y, Z are *Principal* or *Normal co-ordinates.* Then

$$T = A'_{11}\dot{X}^2 + A'_{22}\dot{Y}^2 + A'_{33}\dot{Z}^2$$

$$W = C'_1 + B'_1 X + B'_2 Y + B'_3 Z + B'_{11}X^2 + B'_{22}Y^2 + B'_{33}Z^2.$$

The typical Lagrange's equation is then

$$2A'_{11}\ddot{X} = B'_1 + 2B'_{11}X$$

i.e. an equation containing only X.

On solving this and the two similar equations for Y and Z, we have θ given by sum of three simple harmonic motions. We can generalize for the case, if the original equations contain more than three co-ordinates.

Examples

1. A uniform rod AB, of mass $2m$ and length $2a$, swings freely about a horizontal axis through the end A. One end of the elastic string is attached to the end B of the rod and carries a particle of mass m at its other end. When the system is in stable equilibrium the string is of length $4a/3$, its extension beings ε. If the system performs small oscillations in the vertical plane through AB, the rod and string making angles θ, φ respectively with the downward vertical, and the length of string being $x + 4a/3$, show that x, $\varphi + 2\theta$, $2\varphi - 3\theta$ are normal co-ordinates for the system and the lengths of the corresponding being ε, $8a/3$, $a/3$.

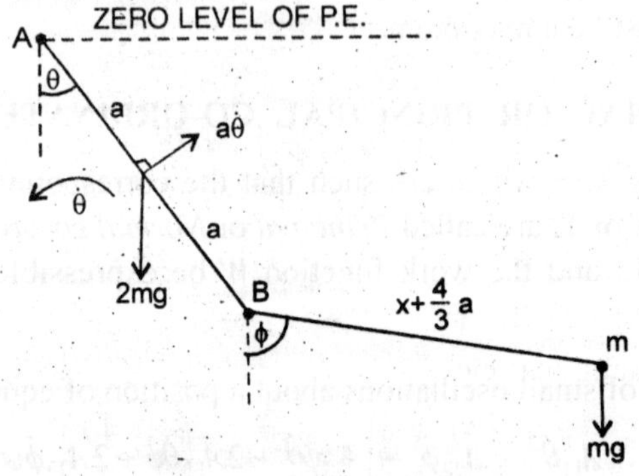

Fig. 10.7

Let λ be the modulus of elasticity of the string. In the equilibrium position mg is the tension in the string,

$$\therefore \qquad mg = \lambda\varepsilon/(4a/3 - \varepsilon)$$

$$\therefore \qquad \lambda = mg\,(4a - 3\varepsilon)/3\varepsilon \qquad (A)$$

Let G be the C.G. of the string. Referred to A as origin, let horizontal and vertical lines be taken as axes of co-ordinates. Co-ordinates of G are $(a \sin \theta, - a \cos \theta)$.

$\therefore$ velocity components of G are $a\dot{\theta} \cos \theta$, $a\dot{\theta} \sin \theta$.

$\therefore$ (velocity)^2 of G = $a^2\dot{\theta}^2$

$$\text{K.E. of } G \text{ due to rotation} = \frac{1}{2}\frac{(2m)a^2\dot{\theta}^2}{3} = \frac{1}{3}ma^2\dot{\theta}^2$$

$$\therefore \text{ K.E. of the rod} = \frac{1}{2}(2m)a^2\dot{\theta}^2 + \frac{1}{3}ma^2\dot{\theta}^2 = \frac{4}{3}ma^2\dot{\theta}^2.$$

Let BC be the string, co-ordinates of C are $[2a \sin \theta + (x + 4a/3) \sin \varphi]$, $[2a \cos\theta + (x + 4a/3) \cos \varphi]$

Velocity components of C are

$$2a\dot{\theta}\cos\theta + \dot{\phi}\left(x + \frac{4a}{3}\right)\cos\phi + \dot{x}\sin\phi,$$

$$- 2a\dot{\theta}\sin\theta + \dot{x}\cos\phi - \dot{\phi}\left(x + \frac{4a}{3}\right)\sin\phi.$$

$\therefore$ (Velocity)^2 of C

$$= \left[2a\dot{\theta}\cos\theta + \dot{\phi}\left(x + \frac{4a}{3}\right)\cos\phi + \dot{x}\sin\phi\right]^2$$

$$+ \left[-2a\dot{\theta}\sin\theta + \dot{x}\cos\theta - \dot{\phi}\left(x + \frac{4a}{3}\right)\sin\phi\right]^2$$

$$= 4a^2\dot{\theta}^2 + \dot{\phi}^2\left(x + \frac{4a}{3}\right)^2 + \dot{x}^2$$

$$+ 4a\dot{\theta}\dot{\phi}\left(x + \frac{4a}{3}\right)\cos(\phi - \theta) + 4a\dot{\theta}\,\dot{x}\sin(\phi - \theta).$$

K.E. due to particle

$$= \frac{1}{2}m\left[4a^2\dot{\theta}^2 + \dot{\phi}^2\left(x + \frac{4a}{3}\right)^2 + \dot{x}^2\right.$$

$$\left. + 4a\dot{\theta}\dot{\phi}\left(x + \frac{4a}{3}\right)\cos(\phi - \theta) + 4a\dot{\theta}\,\dot{x}\sin(\phi - \theta)\right]$$

Hence, K.E. of the system is given by

$$T = \frac{4}{3}ma^2\dot{\theta}^2 + \frac{1}{2}m\left[4a^2\dot{\theta}^2 + \dot{\phi}^2\left(x+\frac{4a}{3}\right)^2 + \dot{x}^2\right.$$

$$\left. + 4a\dot{\theta}\dot{\phi}\left(x+\frac{4a}{3}\right)\cos(\phi-\theta) + 4a\dot{\theta}\,\dot{x}\sin(\phi-\theta)\right]$$

$$= \frac{10}{3}ma^2\dot{\theta}^2 + \frac{1}{2}m\dot{\phi}^2\left(x+\frac{4a}{3}\right)^2 + \frac{1}{2}m\dot{x}^2$$

$$+ 2am\dot{\theta}\dot{\phi}\left(x+\frac{4a}{3}\right)\cos(\phi-\theta) + 2am\dot{\theta}\,\dot{x}\sin(\phi-\theta)$$

$$= \frac{10}{3}ma^2\dot{\theta}^2 + \frac{1}{2}m\dot{x}^2 + \frac{8}{9}ma^2\dot{\phi}^2 + \frac{8}{3}ma^2\dot{\theta}\dot{\phi}$$

(to second order)

Gravitational P.E. referred to the horizontal through A as zero level is

$$= -2mga\cos\theta - mg\,[2a\cos\theta + (x + 4a/3)\cos\varphi]$$

$$= -mgx + 2mga\theta^2 + \frac{2}{3}mga\phi^2 + \text{const.}$$

(to second order)

The elastic P.E. of the string is $= \lambda\,(x+\varepsilon)^2/2(4a/3-\varepsilon)$

Substituting value of λ from (A) this energy becomes

$$= mg\,(x^2/\varepsilon + 2x)/2 + \text{const.}$$

Thus, to the second order the total P.E. is

$$V = mgx^2/2\varepsilon + 2mga\theta^2 + 2mga\varphi^2/3 + const.$$

The Lagrangian

$$L = T - V$$

$$= \frac{10}{3}ma^2\dot{\theta}^2 + \frac{1}{2}m\dot{x}^2 + \frac{8}{9}ma^2\dot{\phi}^2 + \frac{8}{3}ma^2\dot{\theta}\dot{\phi}$$

$$-\frac{mgx^2}{2\varepsilon} - 2mga\theta^2 - \frac{2}{3}mga\phi^2 + \text{const.}$$

Lagrange's equations are

$$\frac{d}{dt}\left(\frac{\partial L}{\partial \dot{x}}\right) - \frac{\partial L}{\partial x} = 0,$$

$$\frac{d}{dt}\left(\frac{\partial L}{\partial \dot{\theta}}\right) - \frac{\partial L}{\partial \theta} = 0,$$

$$\frac{d}{dt}\left(\frac{\partial L}{\partial \dot{\phi}}\right) - \frac{\partial L}{\partial \phi} = 0.$$

$$\therefore \quad \ddot{x} + \frac{gx}{\varepsilon} = 0, \tag{1}$$

$$\therefore \quad \frac{20}{3}a\ddot{\theta} + \frac{8}{3}a\ddot{\phi} + 4g\theta = 0, \tag{2}$$

$$\therefore \quad \frac{8}{3}a\ddot{\theta} + \frac{16}{9}a\ddot{\phi} + \frac{4}{3}g\phi = 0. \tag{3}$$

From (1) it is clear that x is a normal co-ordinate, then length of the corresponding simple equivalent pendulum is ε.

Putting $\theta = \theta_0 \sin(\omega t + \varepsilon)$ and $\phi = \phi_0 \sin(\omega t + \varepsilon)$ in (2) and (3) we get

$$(5a\omega^2 - 3g)\theta_0 + 2a\omega^2 \varphi_0 = 0 \tag{4}$$

$$6a\omega^2\theta^2 + (4a\omega^2 - 3g)\varphi_0 = 0 \tag{5}$$

Eliminating θ_0 and φ_0, we get

$$(5a\omega^2 - 3g)(4a\omega^2 - 3g) - 12a^2\omega^4 = 0$$

$$\therefore \quad 8a^2\omega^4 - 27a\omega^2 g + 9g^2 = 0,$$

$$\therefore \quad (8a\omega^2 - 3g)(a\omega^2 - 3g) = 0,$$

Its roots are given by

$$\omega_1^2 = 3g/a, \omega_2^2 = 3g/8a$$

From (4)

$$\theta_0/(2a\omega^2/g) = -\phi_0/(5a\omega^2/g - 3)$$

Let $\omega = \omega_1$, the solutions are

$$\theta_0 = C_1, \phi_0 = -2C_1,$$

when $\omega = \omega_2$, the solutions are

$$\theta_0 = 2C_2, \varphi_0 = 3C_2$$

Hence, general solution is

$$\theta = C_1 \sin(\omega_1 t + \varepsilon_1) + 2C_2 \sin(\omega_2 t + \varepsilon_2) \tag{6}$$

$$\varphi = -2C_1 \sin(\omega_1 t + \varepsilon_1) + 3C_2 \sin(\omega_2 t + \varepsilon_2) \tag{7}$$

From (6) and (7) we get

$$2\theta + \varphi = -7C_2 \sin(w_2 t + \varepsilon_2)$$

$$2\phi - 3\theta = -7C_1 \sin(w_2 t + \varepsilon_1)$$

Thus $(2\theta + \varphi)$, $(2\varphi - 3\theta)$ can be expressed as single sinusoidal terms, therefore they are normal co-ordinates. Their periods are

$$2\pi/\omega_2 = 2\pi(8a/3g)^{1/2} \text{ and } 2\pi/\omega_1 = 2\pi(a/3g)^{1/2}.$$

Hence length of S.E.P. are $8a/3$ and $a/3$.

2. A hollow cylindrical garden roller is pulled with a counterpoise which can turn about the axis of the cylinder, the system is placed on rough horizontal plane and oscillates under gravity, if $2\pi/p$ be the time of small oscillation, show that p is given by the equation

$$p^2 [(2M+m)k^2 - mh^2] = 2(M+m)\, gh,$$

where M and m are the masses of the roller and counterpoise, k is the radius of gyration of m about the axis of the cylinder and h is the distance of the center of mass m from the axis.

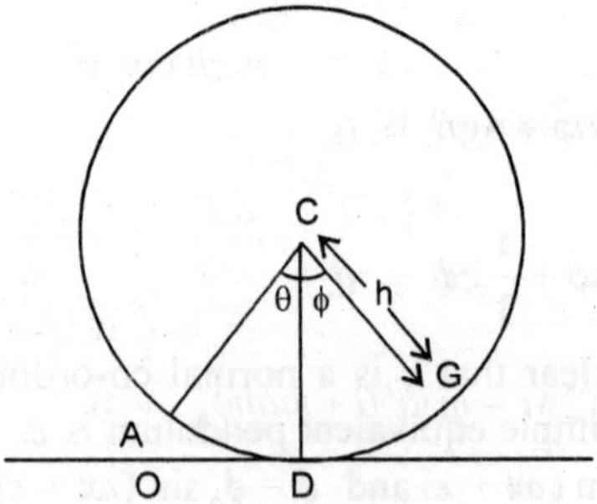

Fig. 10.8

Let C be the center of cylindrical roller and a its radius. Let G be the center of gravity of counterpoise, whose mass is m.

The radius CA is a line fixed in the roller and CG fixed in the counterpoise; these lines were initially vertical and let they make angles θ and ϕ with the vertical at time t. O is the initial point of contact and D is the point of contact at time t.

Since, there is no slipping between the cylinder and the horizontal plane, therefore arc $AD = a\theta$. Taking O as origin and the horizontal and vertical through O as co-ordinate axes.

Co-ordinates of G are

$$x = a\theta + h \sin \varphi,\ y = a - h \cos \varphi,$$

where $CG = h$.

(velocity)^2 of G

$$\dot{x}^2 + \dot{y}^2 = (a\dot{\theta} + h\dot{\phi}\cos\phi)^2 + (h\dot{\phi}\sin\phi)^2$$

$$= a^2\dot{\theta}^2 + h^2\dot{\phi}^2 + 2ah\dot{\theta}\dot{\phi}\cos\phi$$

$$= a^2\dot{\theta}^2 + h^2\dot{\phi}^2 + 2ah\dot{\theta}\dot{\phi},$$

neglecting higher powers of θ and φ.

Now k is the radius of gyration of the counterpoise about C, therefore its square of radius of gyration about C is $(k^2 - h^2)$

Co-ordinates of C are $(a\theta, a)$

$\therefore$ (velocity)^2 of $C = a^2\dot{\theta}^2$

Let T be the K.E. and V the potential of the system, then

$$T = \frac{1}{2}M(a^2\dot{\theta}^2 + a^2\dot{\theta}^2) + \frac{1}{2}m\left\{(k^2 - h^2)\dot{\phi}^2 + a^2\dot{\theta}^2 + h^2\dot{\phi}^2 + 2ah\dot{\theta}\dot{\phi}\right\}$$

$$= \frac{1}{2}(2M + m)a^2\dot{\theta}^2 + \frac{1}{2}mk^2\dot{\phi}^2 + mah\dot{\theta}\dot{\phi}$$

$$V = -mgh\cos\phi$$

Lagrange's θ equation is

$$\frac{d}{dt}\left(\frac{\partial T}{\partial \dot{\theta}}\right) - \frac{\partial T}{\partial \theta} = -\frac{\partial V}{\partial \theta},$$

$$\therefore \quad \frac{d}{dt}\left\{(2M + m)a^2\dot{\theta} + mah\dot{\phi}\right\} = 0.$$

Integrating it and using $\dot{\theta} = 0 = \dot{\phi}$ at $t = 0$, we get

$$(2M + m)\ a^2\dot{\theta} + mah\dot{\phi} = 0. \tag{1}$$

Lagrange's ϕ equation is

$$\frac{d}{dt}\left(\frac{\partial T}{\partial \dot{\phi}}\right) - \frac{\partial T}{\partial \phi} = -\frac{\partial V}{\partial \phi},$$

$$\therefore \quad \frac{d}{dt}(mk^2\dot{\phi} + mah\dot{\theta}) = -mgh\phi,$$

neglecting higher powers of φ.

$$\therefore \quad k^2\ddot{\phi} + ah\ddot{\theta} = -gh\phi. \tag{2}$$

eliminating $\ddot{\theta}$ from (1) and (2), we get

$$\left[mh^2 - (2M + m)k^2\right]\ddot{\phi} = -(2M + m)\ gh\phi.$$

$$\therefore \left[(2M + m)k^2 - mh^2\right]\ddot{\phi} = (2M + m)\ gh\phi. \tag{3}$$

If $2\pi/p$ be the time of small oscillations, then $\phi = A\cos(pt + B)$

$$\therefore \quad \dot{\phi} = -Ap\sin(pt + B),$$

$$\ddot{\phi} = -Ap^2\cos(pt + B) = -p^2\phi.$$

Substituting value of $\ddot{\phi}$ in (3), we get

$$-p^2\ [(2M + m)k^2 - mh^2] = -(2M + m)\ gh,$$

as $\varphi \neq 0$,

$$\therefore \quad p^2\left[(2M + m)k^2 - mh^2\right] = (2M + m)gh.$$

3. A ring slides on a smooth circular hoop of equal mass and radius a which can turn in vertical plane about a fixed point O on its circumference. If θ and φ be the inclinations to the vertical of radius through O and of radius through the ring, prove that the principal co-ordinates are $(2\theta + \varphi)$ and $(\varphi - \theta)$, and the periods of small oscillations are $2\pi\sqrt{(a/2g)}$ and $2\pi\sqrt{(2a/g)}$.

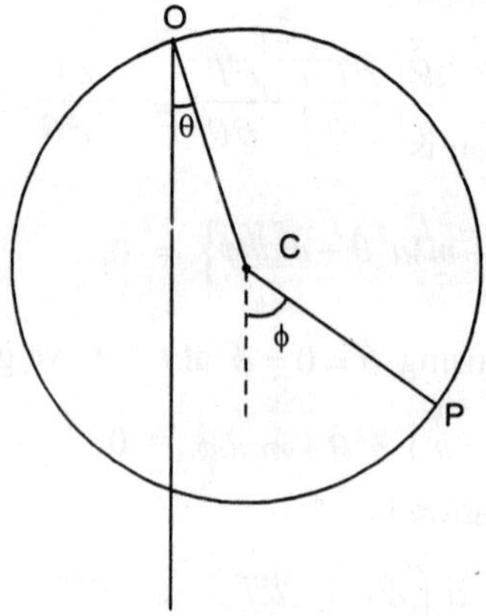

Fig. 10.9

Let m be the mass of the hoop and the ring. In a displaced position let C be the center of the hoop and P the position of the ring.
Referred to O co-ordinates of P are given by

$$x = a(\sin\theta + \sin\varphi),\ y = a(\cos\theta + \cos\varphi)$$

$$\therefore \qquad \dot{x}^2 + \dot{y}^2 = a^2(\dot{\theta}^2 + \dot{\phi}^2 + 2\dot{\theta}\dot{\phi}),$$

when θ and φ are small.
Co-ordinates of C are $(a\sin\theta, a\cos\theta)$

$$\therefore \qquad (\text{Velocity})^2 \text{ of } C = a^2\dot{\theta}^2$$

$\therefore$ K.E. of the system is given by

$$T = \frac{1}{2}m(k^2\dot{\theta}^2 + a^2\dot{\theta}^2) + \frac{1}{2}m(\dot{x}^2 + \dot{y}^2)$$

$$= \frac{1}{2}m(k^2\dot{\theta}^2 + a^2\dot{\theta}^2) + \frac{1}{2}ma^2(\dot{\theta}^2 + \dot{\phi}^2 + 2\dot{\theta}\dot{\phi})$$

$$= \frac{1}{2}ma^2(3\dot{\theta}^2 + \dot{\phi}^2 + 2\dot{\theta}\dot{\phi}).$$

Potential V is given by

$$V = -mga\cos\theta - mga(\cos\theta + \cos\varphi)$$

$$= -mga(2\cos\theta + \cos\varphi)$$

Lagrange's θ equation is

$$\frac{d}{dt}\left(\frac{\partial T}{\partial \dot{\theta}}\right) - \frac{\partial T}{\partial \theta} = -\frac{\partial V}{\partial \theta}$$

$$\therefore \quad \frac{d}{dt}(3\dot{\theta} + \dot{\phi}) = -\frac{2g}{a}\theta$$

$$\therefore \quad 3\ddot{\theta} + \ddot{\phi} = -\frac{2g}{a}\theta. \tag{1}$$

Lagrange's φ equation is

$$\frac{d}{dt}\left(\frac{\partial T}{\partial \dot{\phi}}\right) - \frac{\partial T}{\partial \phi} = -\frac{\partial V}{\partial \phi},$$

$$\therefore \quad \frac{d}{dt}(\dot{\theta} + \dot{\phi}) = -\frac{g}{a}\phi$$

$$\therefore \quad \ddot{\theta} + \ddot{\phi} = -\frac{g}{a}\phi. \tag{2}$$

Adding (1) and (2), we get

$$D^2(2\theta + \varphi) = -g(2\theta + \varphi)/2a \tag{3}$$

where $\qquad D = \dfrac{d}{dt}.$

Multiplying (2) by 2 and subtracting (1), we get

$$D^2(\varphi - \theta) = -2g(\varphi - \theta)/a$$

Putting $2\theta + \varphi = X$ and $\varphi - \theta = Y$, we get

$$D^2X = -gX/2a \text{ and } D^2Y = -2gY/a. \tag{4}$$

which represents two independent simple harmonic motions.

Thus X and Y are principal co-ordinates *i.e.* $2\theta + \varphi$ and $\varphi - \theta$ are principal co-ordinates. From (4) it is clear that periods of small oscillations are $2\pi\sqrt{(2a/g)}$ and $2\pi\sqrt{(a/2g)}$.

4. Four equal rods each of length $2a$, are hinged at their ends to form a rhombus $ABCD$. The angles B and D are connected by an elastic string and the lowest end A rests on a horizontal plane while the end C slides on a smooth vertical wire passing through A; in the position of equilibrium the string is stretched to twice its natural length and the angle BAD is 2α. Show that the time of small oscillations about this position is

$$2\pi\{2a(1 + 3\sin^2\alpha)\cos\alpha/3g\cos 2\alpha\}^{1/2}.$$

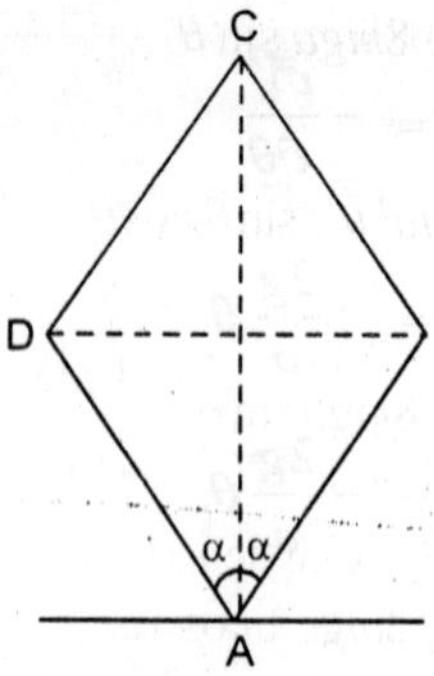

Fig. 10.10

When the rods are inclined at an angle θ to the vertical, the component velocities of the center of the upper rods are $\frac{d}{dt}(3a\cos\theta)$, $\frac{d}{dt}(a\sin\theta)$.

i.e. $-3a\dot{\theta}\sin\theta, a\dot{\theta}\cos\theta$.

Hence, total K.E. is

$$T = 2\times\frac{1}{2}m\left[\frac{4a^2}{3}\dot{\theta}^2 + (-3a\dot{\theta}\sin\theta)^2 + (a\dot{\theta}\cos\theta)^2 + \frac{a^2\dot{\theta}^2}{3}\right]$$

$$= 8ma^2\dot{\theta}^2\left(\frac{1}{3}+\sin^2\theta\right).$$

The potential V is given by

$$V = mg.2(a\cos\theta + 3a\cos\theta) + 2\lambda\int_c^{2a\sin\theta}\frac{x-c}{c}dx$$

$$= 8mga\cos\theta + \frac{\lambda}{c}(2a\sin\theta - c)^2,$$

where $2c$ is the stretched length of the string and λ the modulus of elasticity. Lagrange's equation is

$$\frac{d}{dt}\left(\frac{\partial T}{\partial\dot{\theta}}\right) - \frac{\partial T}{\partial\theta} = -\frac{\partial V}{\partial\theta}$$

$$\therefore \frac{d}{dt}\left[16ma^2\dot{\theta}\left(\frac{1}{3}+\sin^2\theta\right)\right] - 16ma^2\dot{\theta}^2\sin\theta\cos\theta$$

$$= 8mga\sin\theta - \frac{4a\lambda\cos\theta}{c}(2a\sin\theta - c)$$

$$\therefore\ 16ma^2\ddot{\theta}\left(\frac{1}{3}+\sin^2\theta\right)+16ma^2\dot{\theta}^2\sin\theta\cos\theta$$

$$= 8mga\sin\theta - \frac{4a\lambda}{c}\cos\theta(2a\sin\theta - c).$$

Now we are given that $\dot{\theta} = \ddot{\theta} = 0$ when $\theta = \alpha$, $c = a\sin\alpha$.

$\therefore$ $\lambda = 2mgc/a\cos\alpha$.

Putting $\theta = \alpha + \varepsilon$ in (1), where ε is small, and retaining first order terms of ε, we have

$$16ma^2\ddot{\varepsilon}\left(\frac{1}{3}+\sin^2\alpha\right) = 8mga(\sin\alpha + \varepsilon\cos\alpha) - \frac{8mg}{\cos\alpha}$$

$$\times(\cos\alpha - \varepsilon\sin\alpha)(a\sin\alpha + 2a\varepsilon\cos\alpha)$$

$$= -8mga\varepsilon(\cos\alpha - \sin\alpha\tan\alpha)$$

$$= -8mga\varepsilon\frac{\cos 2\alpha}{\cos\alpha}$$

$$\therefore\quad \ddot{\varepsilon} = \frac{3g\cos 2\alpha}{2a\cos\alpha(1+3\sin^2\alpha)}\varepsilon.$$

This is S.H.M. hence required time is

$$= 2\pi[2a\cos\alpha(1+3\sin^2\alpha)/3g\cos 2\alpha]^{1/2}$$

EXERCISE

1. Derive equations of planetary motion by using Lagrange's equations.
2. A uniform rod AB of mass m and length $2a$ attached to a fixed point by a light elastic string OA of natural length l and modulus of elasticity λ, moves in a vertical plane through O with string taut. At time t, OA, OB make angles θ, φ with the downward vertical and length of OA is $1 + x$. Prove that K.E. of the rod is

$$\frac{1}{2}m\left[\dot{x}^2 + (l+x)^2\dot{\theta}^2 + \frac{4}{3}a^2\dot{\phi}^2\right]$$

$$+ma\dot{\phi}\left[\dot{\theta}(l+x)\cos(\phi-\theta) - \dot{x}\sin(\phi-\theta)\right].$$

Derive Lagrange's equations of motion of the system, if in addition to gravity, an external force with horizontal and vertical components X, Y acts on the rod at B.

3. A uniform rod AB of mass m and length $2a$ is at rest at time $t = 0$ and is suspended from a fixed smooth horizontal rail by means of small ring at A of mass m. A force F is applied at A along the rail. If x is the distance moved by A from its initial position O and $\angle OBA = \pi/2 - \theta$, obtain the equations of motion of the system.
If A is constrained to move with constant acceleration g, show that the rod oscillates between vertical ($\theta = 0$) and the horizontal ($\theta = \pi/2$), and that
$$F = mg\,[19 - 12 \sin\theta + 9 \sin 2\theta - \cos 2\theta)]/8.$$
4. A spherical shell of mass M, whose outer surface is rough and of radius a, has its inner surface smooth and of radius b. A particle of mass m moves inside while the shell rolls on a rough table. Show that if the excursions of the particle be α on either side of the vertical, then
$$[M(a^2 + k^2) + ma^2 \sin^2\theta]b\,\dot{\theta}^2 = 2g[M(a^2 + k^2) + ma^2][\cos\theta - \cos\alpha]$$
5. Two uniform rods AB, AC each of mass m and length $2a$, are smoothly hinged together at A and move on horizontal plane. At time t the mass centre of the rods is at the point (ξ, η) referred to fixed perpendicular axes OX, OY in the plane, and the rods make angles $\alpha \pm \varphi$ with OX. Prove that the K.E. of the system is
$$m\left[\dot{\xi}^2 + \dot{\eta}^2 + \left(\frac{1}{3} + \sin^2\phi\right)a^2\dot{\theta}^2 + \left(\frac{1}{3} + \cos^2\phi\right)a^2\dot{\phi}^2\right]$$
and derive Lagrange's equations of motion for the system if the external force with components $[X, Y]$ along axes acts at A.
6. Two uniform rods AB, BC of masses m, m' are smoothly hinged at B and rest at smooth horizontal plane with the end A fixed. Prove that if an impulse I is applied at right angles to AB at the middle point, the K.E. created is
$$\frac{1}{2}I^2 \Big/ \left(\frac{4}{3}m + 4m' - 3m'\cos^2\alpha\right),$$
where α is the angle ABC
Prove also that, if the motion of BC is constrained by the presence of smooth peg at the middle point against which it is pressed by the impulsive action, the K.E. is
$$\frac{1}{2}I^2 \Big/ \left(\frac{4}{3}m + 4m' - \frac{8}{3}m'\cos^2\alpha\right).$$
7. Three uniform rods AB, BC, CD of masses m, M, m and lengths a, b, a are smoothly joined at B and C and placed in a straight line. They are set in motion by an impulse I applied at A at right angles to AB. Prove that the K.E. is

$$2l^2(2m^2 + 7Mm + 3M^2)/m(3m^2 + 8Mm + 4M^2),$$

if the rods are free; and

$$3l^2(4m^2 + 8Mm + 3M^2)/2m(3m^2 + 7Mm + 3M^2),$$

if the end D is fixed.

8. A square plate has mass M and side $2a$. It is freely suspended from a corner by a string of length $2\sqrt{2}a$ and receives a blow I in the plane of plate at its lowest point. Show that the plate will begin to move with angular velocity $3I/Ma\sqrt{2}$.
9. Two uniform rods AB, BC of masses m and m' and lengths $2a$, $2b$ are smoothly hinged at B and initially they lie at rest on a smooth table and in a straight line. AB receives a blow of impulse I at A perpendicular to AB. Construct the equations of motion of the system just after impact.
10. The ends of a uniform rigid rod of mass m are moving with velocities $\vec{u}, \vec{v}$. Prove that the K.E. of the rod is $\frac{1}{2}m(\vec{u}^2 + \vec{v}^2 + \vec{u}.\vec{v})$.

    There rods OA, AB, BC each of mass m, are smoothly jointed together at A, B hang in equilibrium under gravity from a fixed smooth joint at O. An impulsive couple G is applied to the rod AB in a vertical plane through O. Prove that the initial K.E. of the system is $\frac{57G^2}{26ml^2}$, where l is the length of AB.
11. A thin hollow cylinder has a particle suspended from its axis by a string and can roll on a rough table. Show that system oscillates about its equilibrium position in the same period as simple pendulum of lengh $2Mc/(2M+m)$, where M, m are the masses of the cylinder and the particle and c is the length of the string. If the cylinder is set in motion with velocity V, show that the angular velocity of the string is given by

$$c^2\dot{\phi}^2 = \frac{1}{2}\cdot\frac{\left[MV^2 - 2(2M+m)gc\sin^2\frac{\phi}{2}\right]}{\left[M + \frac{1}{2}m\sin^2\frac{\phi}{2}\right]}.$$

    where φ is the angle it makes with the vertical.
12. Four equal uniform rods, smoothly jointed at their extremities to form a rhombus, are suspended from one corner and rotate with uniform angular velocity round a vertical diagonal. Find the position of relative equilibrium, and, if the system be slightly displaced so that the diagonal remains vertical, show that the length of the simple equivalent pendulum for the consequent small oscillation is

$$2a\cos\alpha(1+3\sin^2\alpha)/3(1+3\cos^2\alpha),$$

$2a$ being the length of a rod, and 2α the angle between the two higher rods.

13. A uniform heavy rod AB of mass m and length $12a$ is suspended from a fixed point O by a light string OC of length $5a$ attached to a apoint C of the rod, so that $AC = 4a$. The system makes small oscillations about the position of equilibrium in a vertical plane through O, and OC, AB make angles q_1, q_2 respectively with the vertical at time t. Prove that the K.E. of the system is

$$\frac{1}{2}ma^2(25\dot{q}_1^2 + 20\dot{q}_1\dot{q}_2 + 16\dot{q}_2^2)$$

and that the normal periods of small oscillations are $2\pi/p_1$ and $2\pi/p_2$, where

$3ap_1^2 = g$ and $10ap_2^2 = g$.

14. A hollow circular roller of radius a and mass M has attached to the centre of its axis a particle of mass m by a light rod of length b. The particle is free to swing about the axis in the plane perpendicular to its axis. The roller is at rest on a horizontal plane rough enough to prevent slipping, with the rod held at angle α with the down vertical. If the rod is then released, prove that the centre of the roller will oscillate through a distance $2mba^2\sin\alpha\,[M(a^2+k^2)+ma^2]$,
Where k is the radius of gyration of the roller about the axis.

15. A uniform rod, of length $2a$, is freely movable about its centre and a particle of mass one third that of the rod is attached by a light inextensible string, of length a, to one end of the rod, show that one period of principal oscillation is

$$\pi\left(\sqrt{5}+1\right)\sqrt{(a/g)}.$$

16. A perfectly rough sphere lying inside of a hollow cylinder, which rests on a perfectly rough plane is slightly displaced from its position of equilibrium. Show that the time of small oscillations is

$$2\pi\sqrt{\frac{(a-b)}{g}\cdot\frac{14M}{10M+7m}},$$

where a is the radius of the cylinder, b that of the sphere, and M, m are the masses of the cylinder and sphere.

17. A plank of mass M, radius of gyration k and length $2b$, can swing like a see-saw across perfectly rough fixed cylinder of radius a. At its ends hang two particles each of mass m, by strings of length l. Show that as the system swings, the lengths of its simple equivalent pendulum are l and $(Mk^2+2mb^2)/(M+2m)a$

18. A uniform rod AB, of length $2a$, can turn freely about a point distance c from its centre, and is at rest at an angle α to the horizon when a particle is hung by a light string of length l from one end. If the particle be slightly displaced in the vertical plane of the rod, show that it will oscillate in the same time as a simple pendulum of length

$$l(a^2 + 3a\cos^2\alpha + 3c^2\sin^2\alpha)/(a^2 + 3ac).$$

19. At the lowest point of a smooth circular tube, of mass M and radius a is placed a particle of mass m, the tube hangs in a vertical plane from its highest point, which is fixed, and can turn freely in its own plane about this point. If the system be slightly displaced, show that the periods of two independent oscillations are $2\pi\sqrt{(2a/g)}$ and $2\pi\sqrt{[Ma/(M+m)g]}$.

11

HAMILTONIAN FORMULATION

11.1 PHASE SPACE

In Lagrangian formulation there are $3n$ equations of motion of the second order for a system containing n particles in the absence of harmonic constraints. For general consideration it is more convenient to solve 6n partial differential equations of the first order in place of $3n$ equations of second order. For this purpose there must be $6n$ degrees of freedom or 6n dimensional space known as phase space-sometime also called Γ-space. In phase space momenta are also regarded as independent variables. A single particle is specified by $6n$ co-ordinates in phase space, $3n$ position co-ordinates and $3n$ momenta co-ordinates. The 6 dimensional space is sometimes called μ-space. The Γ-space is superposition of μ-spaces

In short, phase space is a 6n dimensional space formed of co-ordinates q_1, $q_2, ...q_j, ..., q_{3n}, p_1, p_2, ..., p_j, ..., p_{3n}$.

We shall now obtain Hamilton's equations of motion for a system with n degrees of freedom. We shall assume that the constraints are harmonic and the forces are derivable from potential which depends either on position or velocity-dependent variables. Hamilton's equations make use of Phase space.

11.2 HAMILTON'S EQUATIONS OF MOTION

We know that

$$L = L(q_j, \dot{q}_j, t)$$

and

$$p_j = \frac{\partial L}{\partial \dot{q}_j} \tag{1}$$

These equations will serve to determine $\dot{q}$'s in terms of p's and q's, so that L may be regarded as a function of p's, q's and t. We may put

$$H = -L + \sum p_j \dot{q}_j \tag{2}$$

H is called *Hamiltonian Function.*
Then

$$\delta H = -\delta\left[L - \sum p_j \dot{q}_j\right]$$

$$= -\sum\left\{\frac{\partial L}{\partial q_j}\delta q_j + \frac{\partial L}{\partial \dot{q}_j}\delta \dot{q}_j + \frac{\partial L}{\partial t}\delta t - p_j \delta \dot{q}_j - \dot{q}_j \delta p_j\right\} \tag{3}$$

But from Lagrange's equations,

$$\frac{d}{dt}\left(\frac{\partial L}{\partial \dot{q}_j}\right) - \frac{\partial L}{\partial q_j} = 0,$$

$$\therefore \quad \frac{dp_j}{dt} = \frac{\partial L}{\partial q_j}.$$

$$\therefore \quad \dot{p}_j = \frac{\partial L}{\partial q_j}. \tag{4}$$

Using (1), (3) becomes

$$\delta H = \sum(\dot{q}_j \delta p_j - \dot{p}_j \delta q_j) - \frac{\partial H}{\partial t}\delta t \tag{5}$$

Hence, H can be regarded as a function of q's, p's and t,

$$\therefore \quad H = H(q_j, p_j, t)$$

We want to describe the motion of the system in terms of an equation of motion involving the Hamiltonian. This clearly becomes problem of transformation from set of variables $(q_j, \dot{q}_j)$ to (q_j, p_j). In order to achieve this transformation, we write the differential form

$$\delta H = \sum\left[\frac{\partial H}{\partial p_j}\delta p_j + \frac{\partial H}{\partial q_j}\delta q_j\right] + \frac{\partial L}{\partial t}\delta t. \tag{6}$$

Comparing the coefficients of δp_j and δq_j in (5) and (6), we get

$$\dot{q}_j = \frac{\partial H}{\partial p_j}, \quad \dot{p}_j = -\frac{\partial H}{\partial q_j} \tag{7}$$

On comparing the co-efficients of δt in (5) and (6), we get

$$-\frac{\partial L}{\partial t} = \frac{\partial H}{\partial t}. \tag{8}$$

Equations (7) are called *Hamilton's equations* or *Hamilton's canonical* equations of motion. They from a set of $2n$ first order differential equations and replace the n-Lagrangian differential equations of second order. Mostly the treatment of Hamilton's equations to dynamical problems is less convenient than Lagrange's, but there are some fields where former can be used with advantage.

11.3 PHYSICAL SIGNIFICANCE OF HAMILTONIAN FUNCTION H

Prove that if L does not involve t explicitly, Hamiltonian function H is constant and is equal to total energy of the system for conservative system.

Since, $H = H(q_j, p_j, t)$, the total time derivative of H is given by

$$\frac{dH}{dt} = \sum\left\{\frac{\partial H}{\partial q_j}\dot{q}_j + \frac{\partial H}{\partial p_j}\dot{p}_j\right\} + \frac{\partial H}{\partial t}. \qquad (1)$$

Substituting Hamilton's equations in this, we get

$$\frac{dH}{dt} = \frac{\partial H}{\partial t} = -\frac{\partial L}{\partial t}.$$

If L does not involve t explicitly, $\frac{\partial L}{\partial t} = 0.$

$$\therefore \quad \frac{dH}{dt} = 0.$$

$$\therefore \quad H = const.$$

Now for conservative system, the potential V is not a function of velocities

i.e., $\frac{\partial V}{\partial \dot{q}_j} = 0$ and K.E.T is given by

$$2T = \sum \dot{q}_j\left(\frac{\partial T}{\partial \dot{q}_j}\right).$$

Now

$$H = -L + \sum p_j \dot{q}_j$$

$$= -L + \sum \frac{\partial L}{\partial \dot{q}_j}\dot{q}_j$$

$$= -L + \sum\left(\frac{\partial T}{\partial \dot{q}_j}\right)\dot{q}_j, \text{ Since } \frac{\partial L}{\partial \dot{q}_j} = \frac{\partial T}{\partial \dot{q}_j}$$

$$= -L + 2T$$
$$= -(T - V) + 2T = T + V = h = const.$$

11.4 DERIVATION OF HAMILTON'S EQUATIONS FROM A VARIATIONAL PRINCIPLE

From Hamilton's Principle

$$\delta L = \delta \int_{t_1}^{t_2} L dt = 0 \qquad (1)$$

We have

$$L = -H(q_j, P_j, t) + \sum p_j \dot{q}_j.$$

Substituting value of L in (1), we get

$$\delta \int_{t_1}^{t_2} \left\{ \sum p_j \dot{q}_j - H(q_j, p_j, t) \right\} dt = 0 \qquad (2)$$

As a variational Principle in the phase space, this equation is known as modified *Hamilton's Principle*. Now this principle is exactly of the form of the variational problem in a space of 2n dimensions.

i.e.,

$$\delta L = \delta \int_{t_1}^{t_2} f(q_j, \dot{q}_j, p_j, \dot{p}_j, t) dt = 0, \qquad (3)$$

for which the $2n$ Euler equations are,

$$\left(\frac{d}{dt}\right)\left(\frac{\partial f}{\partial \dot{q}_j}\right) - \frac{\partial f}{\partial q_j} = 0, \qquad (4)$$

$$\frac{d}{dt}\left(\frac{\partial f}{\partial \dot{p}_j}\right) - \frac{\partial f}{\partial p_j} = 0. \qquad (5)$$

The integrand f as given in (3) contains $\dot{q}_j$ only through $p_j \dot{q}_j$ term and q_j only in H.

Hence, equation (4) leads to

$$\dot{p}_j + \frac{\partial H}{\partial \dot{q}_j} = 0. \qquad (6)$$

On the other hand there is not explicit dependence of the integrand in (5) on $\dot{p}_j$, hence equations (5) reduce to

$$\dot{q}_j - \frac{\partial H}{\partial p_j} = 0. \tag{7}$$

Equations (6) and (7) are Hamilton's equations of motion. Thus Euler-Lagrange's equations of the modified Hamilton's principle are Hamilton's equations of motion.

11.5 APPLICATION OF HAMILTON'S EQUATIONS OF MOTION

We have seen that dynamical problems can be solved easily by using generalized co-ordinates and Lagrange's equations. In fact there is hardily a mechanical problem where generalized co-ordinates are not applicable. For the solution of problem Hamilton's equations are generally not so convenient as the Lagrange's equations and for their relation to more advanced mechanics, particularly to three fields; celestial mechanics, statistical mechanics, and quantum mechanics. Here, we shall consider some examples with the dynamics of the particles to illustrate the application of Hamilton's equations.

Examples

1. Motion of a spherical pendulum.

   Let a particle of mass m be suspended by a fine string of length r, θ as the inclination of the string to the vertical and φ the azimuth of the vertical plane through the string. Then K.E.T is given by

$$2T = mr^2(\dot{\theta}^2 + \dot{\phi}^2 \sin^2\theta)$$

P.E.V is given by

$$V = -mgr\cos\theta.$$

$$\therefore \qquad \mathrm{L} = T - V = mr^2(\dot{\theta}^2 + \dot{\phi}^2 \sin^2\theta)/2 + mgr\cos\theta. \tag{1}$$

$$\therefore \qquad p_\theta = \frac{\partial L}{\partial \dot{\theta}} = mr^2\dot{\theta}, \qquad p_\phi = \frac{\partial L}{\partial \dot{\phi}} = mr^2\dot{\varphi}\sin^2\theta. \tag{2}$$

Hence,

$$\mathrm{H} = -L + p_\theta\dot{\theta} + p_\varphi\dot{\varphi}$$

$$= p_\theta^2/2mr^2 + p_\varphi^2/2mr^2\sin^2\theta - mgr\cos\theta. \tag{3}$$

Then Hamiltonian equations become

$$\dot{\theta} = \frac{\partial H}{\partial p_\theta} = p_\theta/mr^2, \dot{\phi} = \frac{\partial H}{\partial p_\varphi} = p_\varphi/mr^2\sin^2\theta, \tag{4}$$

and

$$\dot{p}_\theta = -\frac{\partial H}{\partial \theta} = p_\varphi^2\cos\theta/mr^2\sin^3\theta - mgr\sin\theta, \dot{p}_\varphi = -\frac{\partial H}{\partial \varphi} = 0. \tag{5}$$

From (5) $p_\varphi = const. = mr^2 c$ (say)

$\therefore \quad mr^2 \dot{\varphi} \sin^2\theta = mr^2 c$

$\therefore \quad \dot{\varphi} \sin^2\theta = c.$

The remaining equations give

$$\ddot{\theta} - c^2 \cos\theta / \sin^3\theta + g\sin\theta / r = 0.$$

On integrating we get

$$\dot{\theta}^2 + c^2 / \sin^2\theta - 2g\cos\theta / r = const.$$

Thus we have found $\dot{\theta}$ and φ, hence, motion is determined.

2. Motion of simple pendulum.

Let l be the length of the string, and m the mass of the bob, forming a simple pendulum. At time t, let the string be inclined at an angle θ to the vertical.

Velocity of the bob $= l\dot{\theta}$.

$\therefore \quad T = ml^2\dot{\theta}^2/2$ and $V = -mgl\cos\theta$

$\therefore \quad L = T - V = ml^2\dot{\theta}^2/2 + mgl\cos\theta$

Now

$$p_\theta = \frac{\partial L}{\partial \dot{\theta}} = ml^2\dot{\theta}.$$

Since, L does not involve t explicitly, therefore Hamiltonian H is

$$H = T + V = ml^2\dot{\theta}^2/2 - mgl\cos\theta,$$

$$= mp_\theta^2/2m^2l^2 - mgl\cos\theta,$$

$$= p_\theta^2/2ml^2 - mgl\cos\theta.$$

Hamilton's equations are

$$\dot{p}_\theta = -\frac{\partial H}{\partial \theta} = -mgl\sin\theta. \quad (1)$$

$$\dot{\theta} = \frac{\partial H}{\partial p_\theta} = p_\theta/ml^2 \quad (2)$$

From (1) and (2) we get

$$\ddot{\theta} = \dot{p}_\theta / ml^2 = -mgl\sin\theta / ml^2 = -g\sin\theta / l.$$

This is the usual equation of motion of simple pendulum.

3. Motion of Top.

For a top we have

$$2T = [A(\dot{\theta}^2 + \dot{\psi}^2\sin^2\theta) + Cn\dot{\psi}\cos\theta],$$

$$V = mgh\cos\theta.$$

$$\therefore \quad L = T - V = \{A(\dot{\theta}^2 + \dot{\psi}^2 \sin^2\theta) + cn\dot{\psi}\cos\theta\}/2 - mgh\cos\theta.$$

Now

$$p_\theta = \frac{\partial L}{\partial \dot{\theta}} = A\dot{\theta}$$

$$p_\psi = \frac{\partial L}{\partial \dot{\psi}} = A\dot{\psi}\sin^2\theta + Cn\cos\theta.$$

Since L does not involve t explicitly, Hamiltonian H is given by

$$\begin{aligned} H &= T + V \\ &= [A(\dot{\theta}^2 + \dot{\psi}^2 \sin^2\theta) + Cn\dot{\psi}\cos\theta]/2 + mgh\cos\theta \\ &= [p_\theta^2 / A + (p_\psi - Cn\cos\theta)^2 / A\sin^2\theta + Cn^2]/2 + mgh\cos\theta. \end{aligned}$$

Hamilton's equations are

$$\begin{aligned} \dot{p}_\theta &= -\frac{\partial H}{\partial \theta} \\ &= \frac{(p_\psi - Cn\cos\theta)^2 \cos\theta}{A\sin^3\theta} - \frac{(p_\psi - Cn\cos\theta)Cn\sin\theta}{A\sin^2\theta} + mgh\sin\theta \end{aligned} \tag{1}$$

$$\dot{\theta} = \frac{\partial H}{\partial p_\theta} = \frac{p_\theta}{A}, \tag{2}$$

$$\dot{p}_\psi = -\frac{\partial H}{\partial \psi} = 0, \tag{3}$$

$$\dot{\psi} = \frac{\partial H}{\partial p_\psi} = \frac{p_\psi - Cn\cos\theta}{A\sin^2\theta}. \tag{4}$$

From (3), we get

$$p_\psi = D \text{ (const.)},$$

and substituting value of p_ψ in (4), we get

$$\dot{\psi} = \frac{D - Cn\cos\theta}{A\sin^2\theta}$$

$$\therefore A\dot{\psi}\sin^2\theta + Cn\cos\theta = D. \tag{5}$$

From (1) and (2), we have

$$A\ddot{\theta} = \dot{p}_\theta$$

$$= \frac{(p_\psi - Cn\cos\theta)^2\cos\theta}{A\sin^3\theta} - \frac{(p_\psi - Cn\cos\theta)Cn\sin\theta}{A\sin^2\theta} + mgh\sin\theta$$

$$= \frac{A^2\dot{\psi}^2\sin^4\theta\cos\theta}{A\sin^3\theta} - \frac{A\dot{\psi}\sin^2\theta Cn\sin\theta}{A\sin^2\theta} + mgh\sin\theta \quad \text{[using (4)]}$$

$$= A\dot{\psi}^2\sin\theta\cos\theta - Cn\dot{\psi}\sin\theta + mgh\sin\theta,$$

$$\therefore A\ddot{\theta} - A\dot{\psi}^2\sin\theta\cos\theta + Cn\dot{\psi}\sin\theta = mgh\sin\theta. \quad (6)$$

Equation (5) and (6) are equations of motion of a top.

12

VARIATIONAL METHOD

In chapter X Lagrange's equations were obtained from D' Alembert's principle in which Newton's law of motion, viz. $F = mf$ was used. D' Alembert's principle is often called a differential principle, since we use it in the instantaneous state of the system with some infinitesimal virtual displacements from that position of the system. The variational principle, in that sense, must be looked upon as 'integral principle'. Hence, we consider the entire motion of the system between two instants t_1 and t_2 and consider small virtual variations of the entire motion of the system from the actual motion.

Here, we assume that the system point moves in configuration space tracing out a curve. This curve represents the path of motion of entire system. Here, by the 'motion of the system' we mean the motion of the system point along this path in the configuration space. We can consider time as parameter of the curve, since each point in the configuration space has one or more values of time associated with it. It should be clearly born in mind that the n-dimensional configuration space does not have any bearing with the three dimensional space which we can visualize physically and write the instead of the path of the system point in this n-dimensional configuration space, therefore in only a convenient way of visualizing the actual motion of the N-particles with n-degrees of freedom.

12.1 ACTUAL PATH AND VARIED PATH

Suppose a particle of mass m moves between two fixed pints A and B from time t_0 at A to time t_1 at B traveling along the curve APB. A curve along which the particle actually travels is called the *actual path of the particle.*

Any other curve AQB drawn through the points A and B and adjacent to the actual path APB, but not coinciding with it, is called varied path.

Since, a number of curves can be drawn through A and B which are adjacent to the actual path APB, hence, there will be a family of varied paths corresponding to the actual path of the particle. Similarly, a rigid body, moving from a configuration A at time t_0 to another configuration B at time t_1 has an actual path and varied path.

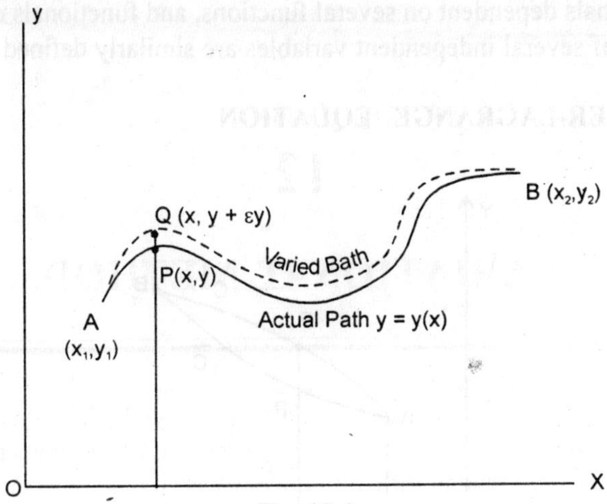

Fig. 12.1

12.2 VARIATIONAL METHOD (CALCULUS OF VARIATION)

Besides the problems in which it is necessary to determine the maximum and minimum values of certain function $z = f(x)$, there are often encountered problems where one has to find the maximal and minimal values of specific quantities called *functionals.*

Functionals are variable quantities whose values are determind by the choice of one or several functions.

Moments of inertia, Static moments, the co-ordinate of the center of gravity of a homogeneous curve or surface are some examples of functionals, since, their values are determined by the choice of the curve or surface, *i.e.* by the choice of the functions that enter into the equation of the curve or surface.

In all these examples, we have a relationship that is characteristics of functionals to a function (or vector function) there corresponds a number whereas when we specify a function $z = f(x)$, there corresponds a number.

The *calculus of variations* investigates methods that permit finding maximal and minimal values of functionals. Problems in which it is reqired to investigate a function for a maximal and minimal are called *variational problems*.

Methods of solving variational problems *I*, *i.e.*, problems involving the functionals for maximal and minimal, are extremly similar to the methods of investigating functions for maximal and minimal.

A variable quantity I is a functional dependent on a function $y(x)$. *i.e.* $I = I\{y(x)\}$ if to each function $y(x)$ of a certain class of functions there B corresponds a value I, *i.e.* we have a corespondence : to the function $y(x)$ there corresponds a number I.

Functionals dependent on several functions, and functionals dependent on functions of several independent variables are similarly defined.

12.3 EULER-LAGRANGE EQUATION

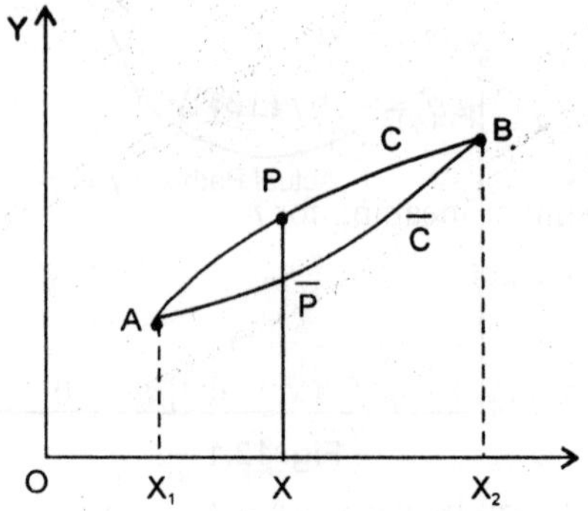

Fig. 12.2

Let A, B be fixed points (x_1, y_1), (x_2, y_2) in a Cartesian co-ordinate plane. Suppose $f(x, y, y')$ is a known functional form of a variable $y'\left(=\dfrac{dy}{dx}\right)$. Then if C is a curve joining A to B and having equation $y = y(x)$, the integral

$$I = \int_{x_1}^{x_2} f(x, y, y')dx \qquad (1)$$

has a definite value whenever the function $y(x)$ is prescribed. The value of I will change as we vary the form of the curve C through A and B. Hence, we may expect that in general there will be some curve $\overline{C}$ through these fixed points such that the value of I taken along it, is stationary (in general a maximum or minimum) compared with the value along neighbouring paths C.

Let the path $\overline{C}$ which renders I stationary have equation $y = \bar{y}(x)$ and let the equation of the neighbouring curve C be $y = \bar{y}(x) + \varepsilon\eta(x)$, where ε is small into, $\eta(x)$ is an arbitrary continuous differentiable function of x satisfying $\eta(x_1) = 0 = \eta(x_2)$, to ensure that the curve passes through A and B. In the figure $\overline{P}P = \varepsilon\eta(x)$ the value of I taken along C is thus a function of ε of the form

$$I(\varepsilon) = \int_{x_1}^{x_2} f(x, \bar{y} + \varepsilon\eta, \bar{y}' + \varepsilon\eta')\,dx \qquad (2)$$

Since, the curve $\bar{C}$ for which $\varepsilon = 0$, is to render I stationary, for which condition is $I'(0) = 0$. Differentiating (2) with respect to ε we have

$$I'(\varepsilon) = \int_{x_1}^{x_2} [\eta f_y(x, \bar{y} + \varepsilon\eta, \bar{y}' + \varepsilon\eta') + \eta' f_{y'}(x, \bar{y} + \varepsilon\eta, \bar{y}' + \varepsilon\eta')]\, dx,$$

where $\quad f_y(\bar{y} + \varepsilon\eta, \bar{y}' + \varepsilon\eta' = \dfrac{\partial}{\partial y} f(x, y. y') \quad$ when $\quad y = \bar{y} + \varepsilon\eta,$

$y' = \bar{y}' + \varepsilon\eta'$, with similar meaning for $f_{y'}$.

Putting $I'(0) = 0$, we get

$$\int_{x_1}^{x_2} [\eta f_y(x, \bar{y}, \bar{y}') + \eta' f_{y'}(x, \bar{y}, \bar{y}')]\, dx = 0 \tag{3}$$

Integrating second term by parts, we get

$$\int_{x_1}^{x_2} [\eta' f_{y'}(x, \bar{y}, \bar{y}')\, dx = [\eta f_{y'}(x, \bar{y}, \bar{y}')]_{x_1}^{x_2} - \int_{x_1}^{x_2} \eta \frac{d}{dx} f_y(x, \bar{y}, \bar{y}')\, dx$$

$$= -\int_{x_1}^{x_2} \frac{\eta d}{dx} f_y(x, \bar{y}, \bar{y}')\, dx$$

since, $\eta(x_1) = 0 = \eta(x_2)$.

Using this result in (3), we get

$$\int_{x_1}^{x_2} \eta(x) \left[f_y(x, \bar{y}, \bar{y}' - \frac{d}{dx} f_{y'}(x, \bar{y}, \bar{y}') \right] dx = 0 \tag{4}$$

Lemma : If $\int \eta(x)\, \phi(x)\, dx = 0$, where $\eta(x)$ is arbitrary, subject to its being continuously differentiable and satisfying conditions

$\eta(x_1) = \eta(x_2) = 0$, then $\phi(x) = 0$.

Applying this result in (4), we get

$$f_y(x, \bar{y}, \bar{y}') - \frac{d}{dx} f_{y'}(x, \bar{y}, \bar{y}') = 0 \tag{5}$$

since, $\eta \neq 0$ except at $x_1 x_2$.

This differential equation is called Euler Lagrange's equation. Its solution giBes the desired relation $y - y(x)$.

Particular cases :

(*i*) If y is absent, *B*hen $\frac{\partial f}{\partial y} = 0$ and the above equation reduces to

$$\frac{d}{dx}\left(\frac{\partial f}{\partial y'}\right) = 0.$$

$$\therefore \quad \frac{\partial f}{\partial y'} = \text{constant.}$$

(*ii*) Multiplying Euler Lagrange's equation by y' we get

$$y'\frac{\partial f}{\partial y} - y'\frac{d}{dx}\left(\frac{\partial f}{\partial y'}\right) = 0.$$

$$\therefore \quad y'\frac{\partial f}{\partial y} - \left[\frac{d}{dx}\left(y'\frac{\partial f}{\partial y'}\right) - \frac{\partial f}{\partial y'}\frac{dy'}{dx}\right] = 0.$$

Adding $\frac{\partial f}{\partial x}$ on both sides, we get

$$\left[\frac{\partial f}{\partial x} + y'\frac{\partial f}{\partial y} + \frac{\partial f}{\partial y'}\frac{dy'}{dx}\right] - \frac{d}{dx}\left(y'\frac{\partial f}{\partial y'}\right) = \frac{\partial f}{\partial x}.$$

$$\therefore \quad \frac{df}{dx} - \frac{d}{dx}\left(y'\frac{\partial f}{\partial y'}\right) = \frac{\partial f}{\partial x}.$$

$$\therefore \quad \frac{d}{dx}\left(f - y'\frac{\partial f}{\partial y'}\right) = \frac{\partial f}{\partial x}.$$

This is another form of Euler Lagrange's equation . If x is absent in f, then $\frac{\partial f}{\partial x} = 0$ and the above equation becomes

$$f - y'\frac{\partial f}{\partial y'} = \text{constant.}$$

12.4 GENERAL CASE

We can generalize the above result, if we consider the functional of more general form

$$I = \int_{x_1}^{x_2} f(x, y_1, y_2, \ldots. y_n, y_1', y_2', \ldots.. y_n')\, dx$$

and proceed as above, we get

$$\frac{\partial f}{\partial y_r} - \frac{d}{dx}\left(\frac{\partial f}{\partial y_r'}\right) = 0. \ (r = 1, 2, \ldots., n).$$

These are n differential equations, whose solutions are

$$y_r = y_r(x) \ (r = 1, 2, \ldots.., n).$$

when substituted in I will make it stationary.

Examples

1. Find the shortest curve joining two fixed points

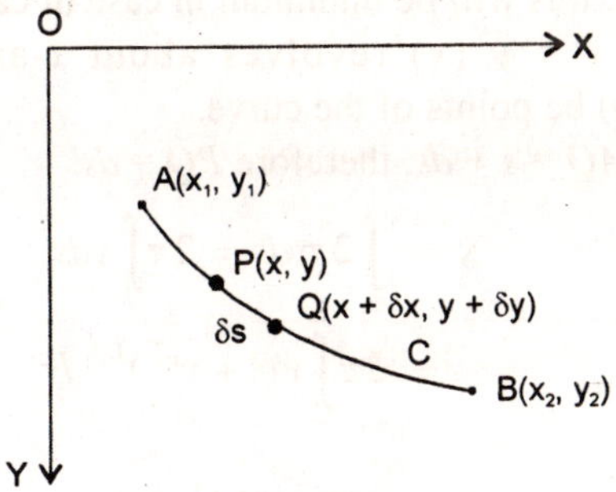

Fig. 12.3

Let the curve be given by $y = y(x)$ and two fixed points are $A(x_1, y_1)$ and $B(x_2, y_2)$.

Now for any curve

$$\frac{ds}{dx} = \sqrt{1 + \left(\frac{dy}{dx}\right)^2} = (1 + y'^2)^{1/2}$$

$$s = \int (1 + y'^2)^{1/2} dx,$$

Let $\qquad f(y') = (1 + y'^2)^{1/2},$

then $\qquad s = \int f(y')\, dx.$

Since y is absent in f, Euler Lagrange's equation is given by

$$\frac{\partial f}{\partial y'} = \text{constant}.$$

$$\therefore \quad \frac{\partial}{\partial y'}\left[(1+y'^2)^{1/2}\right] = \text{constant.}$$

$$\therefore \quad \frac{y'}{(1+y'^2)^{1/2}} = \text{constant.}$$

$$\therefore \quad y' = \text{constant} = b$$

$$\therefore \quad \frac{dy}{dx} = b,$$

$$\therefore \quad y = a + bx, \text{ where } a \text{ is a constant.}$$

Hence, the shortest curve joining A and B is a straight line

2. Show that the area of the surface of revolution of a curve $y = y(x)$ is $2\pi\int y(1 + y'^2)\,dx$.

Hence, show that this will be minimum in case of catenary.

Let the curve $y = y(x)$ revolves about x-axis, and $P(x, y)$, $Q(x + dx, y + dy)$ be points of the curve.

Let $AP = s$, and $AQ = s + ds$, therefore $PQ = ds$.

$$S = \int 2\pi y ds = 2\pi\int yds.$$

$$= 2\pi\int y(1+y'^2)^{1/2}dx$$

Let

$$f(y, y') = y(1+y'^2)^{1/2}.$$

$$\therefore \quad S = 2\pi\int f(y, y')\,dx.$$

Since x is absent in f, Euler Lagrange's equation is given by

$$f - y'\frac{\partial f}{\partial y'} = \text{constant.}$$

$$\therefore \quad y(1+y'^2)^{1/2} - yy'^2/(1+y'^2)^{1/2} = \text{constant}$$

$$\therefore \quad y/(1+y'^2)^{1/2} = \text{constant} = C$$

$$\therefore \quad y = C(1+y'^2)^{1/2}$$

But $y' = \dfrac{dy}{dx} = \tan\varphi$.

$$\therefore \quad y = C(1+\tan^2\varphi)^{1/2} = C\sec\varphi.$$

This is the equation of catenary. Hence, the minimum area of surface of revolution is that of catenary.

3. Show that the geodesics of spherical surface are great circles *i.e.* the circles whose centers lie at the center of the sphere.

A geodesic is a line which represents distance between two points, when the path is restricted to lie on the same surface. In the present case the surface is a spherical surface.
The element ds, on the surface of a sphere of radius r, in the spherical co-ordinates, is given by

$$ds = r\left[(d\theta)^2 + \sin^2\theta(d\phi)^2\right]^{1/2} \qquad (1)$$

The total distane between two points whose co-ordinates are (r_1, θ_1, ϕ_1) and (r_2, θ_2, ϕ_2) is given by

$$I = \int ds = \int r\left[1 + \sin^2\theta\left(\frac{d\phi}{d\theta}\right)^2\right]^{1/2} d\theta \qquad (2)$$

Here, the function f is given by

$$f = r\left[1 + \sin^2\theta\left(\frac{d\phi}{d\theta}\right)^2\right]^{1/2} \qquad (3)$$

If I is to be extremum, we must have

$$\frac{\partial f}{\partial \phi} - \frac{d}{d\theta}\left(\frac{\partial f}{\partial \phi'}\right) = 0, \text{ where } \phi' = \frac{d\phi}{d\theta}.$$

Since, $\frac{\partial f}{\partial \phi} = 0$, Euler Lagrange's equation becomes $\frac{d}{d\theta}\left(\frac{\partial f}{\partial \phi'}\right) = 0.$

$$\therefore \quad \frac{d}{d\theta}\left[\frac{\partial}{\partial \phi'}\left\{(1 + \phi'^2 \sin^2\theta)^{1/2}\right\}\right] = 0$$

$$\therefore \quad \frac{d}{d\theta}\left[\frac{\phi' \sin^2\theta}{(1 + \phi'^2 \sin^2\theta)^{1/2}}\right] = 0,$$

since, $r \neq 0$, and is constant.
Integrating this, we get

$$\frac{\phi' \sin^2\theta}{(1 + \phi^2 \sin^2\theta)^{1/2}} = C$$

$$\therefore \quad \phi'^2 \sin^2\theta = C^2(1 + \phi'^2 \sin^2\theta)$$

$$\phi' = \frac{C}{[\sin^2\theta(\sin^2\theta - C^2)]^{1/2}} = \frac{C \operatorname{cosec}^2\theta}{(1 - C^2 - C^2\cot^2\theta)^{1/2}}.$$

Substituting for cot θ leads to standard form of integral for ϕ and integration yields

$$\phi = \alpha - \sin^{-1}(k \cot \theta), \tag{4}$$

where α and $k = \dfrac{C}{(1-C^2)^{1/2}}$ are constants

This gives

$$k \cot \theta = \sin(\alpha - \phi),$$

$$\therefore \quad k \cos \theta = \sin(\alpha - \phi) \sin \theta. \tag{5}$$

Using the relations between Cartesian and spherical polar co-ordinates, we can write these equations as

$$zk = r \sin \theta \cos \phi \sin \alpha - r \sin \theta \sin \phi \cos \alpha.$$

$$\therefore \quad zk = x \sin \alpha - y \cos \alpha,$$

where

$$x^2 + y^2 + z^2 = r^2. \tag{6}$$

The equation (6) represents a plane passing through the origin and hence cutting the surface of the sphere in a great circle. The distance between two points on the surface of the sphere is an arc of a circle whose center lies at the center of the sphere.

12.5 HAMILTON'S PRINCIPLE

Statement : Of all the paths along which a dynamical system may move from one point to another point in the configuration space within a given interval of a time (consistant with constraint, if any) the actual path followed is that which minimizes the time integral of Lagrangian L.

Or

The motion of the system from instant t_1 to t_2 is such that the line integral

$$I = \int_{t_1}^{t_2} L dt \tag{1}$$

where $L = T - V$, extremum for the path of motion.

In terms of the calculus of variations this principle can be stated as

$$\delta L = \delta \int_{t_1}^{t_2} L dt. \tag{2}$$

with variations zero at $t = t_1$ and $t = t_2$.

This statement reveals that the line integral of L is extremum. But in most of the applications in dynamics, we obtain a minimum condition. This is incorporated in first statement.

Proof : We know that if $q_1, q_2, \ldots\ldots q_n$ are generalized co-ordinates of the conservative holonomic dynamical system, then $\dot{q}_s$ are generalized velocities,

K .E. is function of $q_s, \dot{q}_s$ and time t and P.E. is a function of position. But both position and velocity are functions of time implicitly as well as explicitly. Hence, L can be expressed as

$$L = L(q_1, q_2, \ldots\ldots q_n, \dot{q}_1, \dot{q}_2 \ldots\ldots \dot{q}_n, t).$$

Hence, equation (2) can be written as

$$\delta L = \delta \int_{t_1}^{t_2} L(q_r, \dot{q}_r, t)\, dt = 0. \ (r = 1, 2, \ldots\ldots, n) \tag{3}$$

Euler Lagrange's equations are

$$\frac{\partial L}{\partial q_r} - \frac{d}{dt}\left(\frac{\partial L}{\partial \dot{q}_r}\right) = 0,$$

$$\therefore \quad \frac{d}{dt}\left(\frac{\partial L}{\partial \dot{q}_r}\right) - \frac{\partial L}{\partial q_r} = 0, \ (r = 1, 2, \ldots\ldots, n) \tag{4}$$

These are familiar Lagrange's equations.

Hence, in terms of Lagrangian L, Hamilton's principle can be stated as:

Of all the possible paths, along which a dynamical system may move from one point to another point in the configuration space within a given interval of a time, the actual path followed is that for which the time integral of the Lagrangian L is an extremum.

Although suffix r of equation (4) is stated to take all values from 1 to n, but it will not be in case if some constraints are acting on the system. In that case the number of independent co-ordinates (*i.e.*, the degree of freedom) is decreased and the number of Lagrange's equations is also correspondingly reduced.

12.6 PRINCIPLE OF LEAST ACTION FOR CONSERVATIVE HOLONOMIC SYSTEM

The action A of a dynamical system over the interval $t_1 < t < t_2$ is defined as

$$A = \int_{t_1}^{t_2} 2T dt,$$

where T is the kinetic energy.

Suppose motion is not dependent on time t but on variable r which is to assume fixed values at the end points of the path. Then

$$A = \int_{r_1}^{r_2} 2T \frac{dt}{dr} dr \quad (r_1, r_2 \textit{ fixed})$$

We suppose that total energy is described

$$\therefore \qquad T + V = C$$

and T is independent of time. Then

$$L = T - V = 2T - C$$

And L is also independent of time.

Let when r varies from r to $r + \delta r$ to, then A varies from A to $A + \delta A$, so that

$$\delta A = \delta \int_{r_1}^{r_2} 2T \frac{dt}{dr} dr$$

$$= \int_{r_1}^{r_2} \left[\delta(2T) \frac{dt}{dr} + 2T\delta\left(\frac{dt}{dr}\right) \right] dr.$$

But

$$\delta\left(\frac{dt}{dr}\right) = \frac{d}{dr}(\delta t) = \frac{d}{dt}(\delta t)\frac{dt}{dr}.$$

and

$$\delta(2T) = \delta L$$

Therefore

$$\delta A = \int_{r_1}^{r_2} \left[\delta L + 2T \frac{d}{dt}(\delta t) \right] \frac{dt}{dr} dr.$$

Moreover

$$\delta L = \sum_1^n \frac{\partial L}{\partial q_k} \delta q_k + \sum_1^n \frac{\partial L}{\partial \dot{q}_k} \delta \dot{q}_k,$$

Now denoting total differential with respect to r by primes

$$\delta \dot{q}_k = \delta(q'_k / t') = \delta q'_k / t' - (q'_k / t'^2)\delta t',$$

$$= \frac{1}{t'}\frac{d}{dr}(\delta q_k) - \frac{q'_k}{t'^2}\frac{d}{dt}(\delta t)$$

$$= \frac{d}{dt}(\delta q_k) - (q'_k / t')\frac{d}{dt}(\delta t)$$

$$= \frac{d}{dt}(\delta q_k) - \dot{q}_k \frac{d}{dt}(\delta t).$$

$$\therefore \qquad \sum_1^n \frac{\partial L}{\partial \dot{q}_k} \delta \dot{q}_k = \sum_1^n \frac{\partial L}{\partial \dot{q}_k} \frac{d}{dt}(\delta q_k) - \sum_1^n \dot{q}_k \frac{\partial L}{\partial \dot{q}_k} \frac{d}{dt}(\delta t).$$

But

$$\frac{\partial L}{\partial \dot{q}_k} = \frac{\partial T}{\partial \dot{q}_k}$$

and T is a homogeneous quadratic function of the velocities.

$$\therefore \quad \sum_{k=1}^{n} \dot{q}_k \frac{\partial L}{\partial \dot{q}_k} \frac{d}{dt}(\delta t) = \sum_{k=1}^{n} \dot{q}_k \frac{\partial L}{\partial \dot{q}_k} \frac{d}{dt}(\delta t) = 2T \frac{d}{dt}(\delta t).$$

$$\therefore \quad \delta L = \sum_{k=1}^{n} \frac{\partial L}{\partial q_k} \delta q_k + \sum_{k=1}^{n} \frac{\partial L}{\partial \dot{q}_k} \delta \dot{q}_k$$

$$= \sum_{k=1}^{n} \left\{ \frac{\partial L}{\partial q_k} \delta q_k + \frac{\partial L}{\partial \dot{q}_k} \frac{d}{dt}(\delta q_k) - 2T \frac{d}{dt}(\delta t) \right\}.$$

$$\therefore \quad \delta A = \int_{r_1}^{r_2} \sum_{k=1}^{n} \left\{ \frac{\partial L}{\partial q_k} \delta q_k + \frac{\partial L}{\partial \dot{q}_k} \frac{d}{dt}(\delta q_k) \right\} \frac{dt}{dr} dr$$

$$= \int_{r_1}^{r_2} \sum_{k=1}^{n} \left[\delta q_k \frac{d}{dt}\left(\frac{\partial L}{\partial \dot{q}_k} \right) + \frac{\partial L}{\partial \dot{q}_k} \frac{d}{dt}(\delta q_k) \right] \frac{dt}{dr} dr$$

$$= \int_{r_1}^{r_2} \sum_{k=1}^{n} \frac{d}{dt} \left\{ \delta q_k \frac{\partial L}{\partial \dot{q}_k} \right\} \frac{dt}{dr} dr$$

$$= \left[\sum_{k=1}^{n} \delta q_k \frac{\partial L}{\partial \dot{q}_k} \right]_{r_1}^{r_2}.$$

If we allow $\delta q_k = 0$ ($k = 1, 2, \ldots\ldots, n$) at each end point $r = r_1 = r_2$, then $\delta A = 0$. this shows that A is stationary. This result is known as Principle of Least Action. Here, we have established only stationary character of A, but it can be shown to be a minimum. Here, total energy is constant, as constrasted with total time in Hamilton's Principle.

12.7 DISTINCTION BETWEEN HAMILTON'S PRINCIPLE AND PRINCIPLE OF LEAST ACTION

In Hamilton's principle the time interval is prescribed (fixed) as the body moves from one configuration to another configuration, but in principle of Least Action, however the time interval from one configuration to another configuration is not fixed. In principle of Least Action end points $r = r_1, r = r_2$ are fixed and the total energy $T + V$ is prescribed (fixed).

Examples

1. A particle moves on a smooth curve, joining two fixed points A and B, under the gravity, starting from rest from A. Find the form of the path in order that the time from A to B is minimum. (See fig. 12.3)
   Let us take the fixed point A as origin and horizontal and vertical through A as co-ordinates axes, and (x_1, y_1) the co-ordinates of B.
   At time t, let v be the velocity in a position $P(x, y)$ and actual distance is measured from the origin A.
   By energy equation, we have

$$\frac{1}{2}mv^2 = mgy.$$

$$\therefore \quad v = \sqrt{2gy},$$

$$\therefore \quad \frac{ds}{dt} = \sqrt{2gy}.$$

Integrating, we get

$$t = \int \frac{ds}{\sqrt{2gy}} = \int \frac{\sqrt{1+y'^2}}{\sqrt{2gy}}\,dx$$

since,

$$\frac{ds}{dx} = \sqrt{1+y'^2}$$

$$\therefore \quad t = \frac{1}{\sqrt{2g}}\int f(y, y')\,dx$$

where

$$f(y, y') = \sqrt{\frac{1+y'^2}{y}}.$$

Since, x is absent in f, therefore, for t to be stationary (here minimum), we should have

$$f - y'\frac{\partial f}{\partial y'} = \text{constant}.$$

$$\sqrt{\frac{1+y'^2}{y}} - \frac{y'}{\sqrt{y(1+y'^2)}} = \text{const.}$$

$$\therefore \quad \frac{1}{\sqrt{y(1+y'^2)}} = \text{constant}$$

$$\therefore \quad y(1+y'^2) = 2b \text{ (const.)}$$

$$\therefore \quad y(1+\tan^2\psi) = 2b$$

$$\therefore \quad y = 2b(\cos^2\Psi)$$

$$\therefore \quad y = b(1+\cos 2\Psi).$$

Now

$$\frac{dy}{dx} = \tan\Psi.$$

$$\therefore \quad dx = \cot\Psi\, dy$$

$$= \cot x\,(-4b\cos\Psi\sin\Psi\, d\Psi)$$

$$= -4b\cos^2\Psi$$

$$= -2b(1+\cos 2\Psi).$$

Integrating, we get

$$x = -2b\left(\psi + \frac{1}{2}\sin 2\psi\right) + a,$$

where a is a constant.

$$\therefore \quad x = a - b(2\psi + \sin 2\psi),\ y = b(1-\cos 2\psi).$$

2. A particle moves along a curve from one point to another point, if the rate of motion $v = x$. Find the curve, so that time spent is minimum.

It is given that

$$v = x,$$

$$\therefore \quad \frac{ds}{dt} = x,$$

$$\therefore \quad dt = \frac{ds}{x}.$$

$$\therefore \quad t = \int_{x_1}^{x_2} \frac{\sqrt{1+y'^2}}{x}\,dx$$

Let

$$f(x, y') = \frac{\sqrt{1+y'^2}}{x},$$

$$\therefore \qquad t = \int_{x_1}^{x_2} f(x, y')\, dx.$$

Since, y is absent in f, hence, Euler Lagrange's equation is given by

$$\frac{\partial f}{\partial y'} = C,$$

$$\therefore \qquad \frac{y'}{x\sqrt{1 + y'^2}} = C.$$

Putting

$$y' = \tan t$$

we get

$$x = \frac{1}{C} \sin t.$$

$$\therefore \qquad x = C_1 \sin t.$$

where

$$C_1 = \frac{1}{C}.$$

Now

$$\frac{dy}{dt} = \tan t,$$

$$\therefore \qquad dy = \tan t dx$$
$$= \tan t.\, C_1 \cos t dt$$
$$= C_1 \sin t dt,$$

Integrating, we get

$$y = -C_1 \cos t + C_2 .$$

Thus

$$x = C_1 \sin t, y - C_2 = -C_1 \cos t.$$

Eliminating t, we get

$$x^2 + (y - C_2)^2 = C_1^2$$

which represents a family of circles with centers on the axes of ordinates.

3. A particle of units mass moves along OX under a cosntant force f, starting from rest from the orogin O at time $t = 0$. If T and V are K.E. and P.E. of the particle, calculate

$$\int_0^{t_0} (T - V)\, dt.$$

Evaluate this for the varied path in which the position of the particle is given by

$$x = \frac{1}{2}ft^2 + \varepsilon ft(t - t_0),$$

where ε is a constant; and that the result is in agreement with Hamilton's principle. What are the essential features of the varied motion that ensures this agreement ?

The particle moves along OX with acceleration f starting from rest; thereore its actual path is

$$x = \frac{1}{2}ft^2,$$

$$\therefore \qquad \dot{x} = ft.$$

Hence, K.E.

$$T = \frac{1}{2}\dot{x}^2 = \frac{1}{2}f^2t^2.$$

Since, particle moves from $t = 0$ to t_0, hence, end points of the actual path are 0 and $\frac{1}{2}ft_0^2$.

The potential V at a distance x at time t is

$$V = -fx = -f\left(\frac{1}{2}ft^2\right) = -\frac{1}{2}f^2t^2.$$

$$\therefore \qquad T - V = \frac{1}{2}f^2t^2 - \left(-\frac{1}{2}f^2t^2\right) = f^2t^2.$$

Hence, along actual path

$$\int_0^{t_0}(T - V)\,dt = \int_0^{t_0} f^2t^2dt = \frac{1}{3}f^2t_0^3. \qquad (1)$$

The varied path is given by

$$x = \frac{1}{2}ft^2 + \varepsilon ft(t - t_0)$$

$$= \frac{1}{2}[(1 + 2\varepsilon)ft^2 - 2\varepsilon ftt_0].$$

Substituting $t = 0$ and $t = t_0$, we see that end points of varied path are also 0 and $\frac{1}{2}ft_0^2$.

Hence, the end pionts of actual and varied path are same.

Velocity along varied path is

$$\dot{x} = (1+2\varepsilon) ft - \varepsilon f t_0$$

Hence, along varied path

$$T = \frac{1}{2}[(1+2\varepsilon) ft - \varepsilon f t_0]^2.$$

$$= -\frac{1}{2}[(1+4\varepsilon+4\varepsilon^2) f^2 t^2 - (2\varepsilon+4\varepsilon^2) f^2 t_0 t + \varepsilon^2 f^2 t_0^2].$$

Potential, along varied path is

$$V = -f\left[\frac{1}{2}\left\{(1+2\varepsilon) ft^2 - 2\varepsilon f t_0 t\right\}\right]$$

$$= -\frac{1}{2}[(1+2\varepsilon) f^2 t^2 - 2\varepsilon f^2 t_0 t]$$

So that in the varied path,

$$T - V = \frac{1}{2}\ [\{(1+4\varepsilon+4\varepsilon^2) f^2 t^2 + (1+2\varepsilon)\} f^2 t^2 - (4\varepsilon + 4\varepsilon^2) f^2 t_0 t + \varepsilon^2 f^2 t_0^2].$$

$$= \frac{1}{2}[(2+6\varepsilon+4\varepsilon^2) f^2 t^2 - (4\varepsilon+4\varepsilon^2) f^2 t_0 t + \varepsilon^2 f^2 t_0^2]$$

Hence, in the varied path,

$$\int_0^{t_0} (T-V)\, dt$$

$$= \int_0^{t_0} \frac{1}{2}\ [(2+6\varepsilon^2+4\varepsilon^2) f^2 t^2 - (4\varepsilon+4\varepsilon^2) f^2 t_0 t + \varepsilon^2 f^2 t_0^2] dt$$

$$= \frac{1}{2}\ \{(2+6\varepsilon+4\varepsilon^2)\frac{1}{3} f^2 t_0^3 - (4\varepsilon+4\varepsilon^2)\ \frac{1}{2} f^2 t_0^3 + \varepsilon^2 f^2 t_0^3\}$$

$$= \frac{1}{3}\left(1+\frac{1}{2}\varepsilon^2\right) f^2 t_0^3.$$

which is minimum if $\varepsilon = 0$, and then the minimum value is $\frac{1}{3} f^2 t_0^3$, which is the same as in the actual path. Also we see that when $\varepsilon = 0$, the varied path coincides with the actual path.

Hence, the result is in agreement with Hamilton's principle.

4. A particle moves in a straight line with central acceleration $\omega^2 x$ between two points x_0 and x_1 in the prescribed time $t_1 - t_0$. Show that Hamilton's principal function S is

$$\frac{\omega}{2\sin\omega\,(t_1 - t_0)}\left[(x_1^2 + x_0^2)\cos\omega(t_1 - t_0) - 2x_1x_0.\right]$$

Here the motion is simple harmonic in a straight line

$$\therefore \qquad \ddot{x} = -\omega^2 x.$$

Hence, the path is

$$x = A\cos\omega t + B\sin\omega t,$$

$$\therefore \qquad \dot{x} = -A\,\omega\sin\omega t + B\,\omega\cos\omega t.$$

K.E.

$$T = \frac{1}{2}\dot{x}^2.$$

Potential V = work done against the force

$$= \int_0^x \omega^2 x\,dx = \frac{1}{2}\omega^2 x^2.$$

Let us consider the motion between two fixed points x_0, x_1 on the path from t_0 to t_1. Since, the points lie on the path, therefore

$$A\cos\omega t_0 + B\sin\omega t_0 - x_0 = 0,$$
$$A\cos\omega t_1 + B\sin\omega t_1 - x_1 = 0.$$

Solving these, we get

$$A = \frac{x_0\sin\omega t_1 - x_1\sin\omega t_0}{\sin\omega(t_1 - t_0)},$$

$$B = \frac{x_1\cos\omega t_0 - x_0\cos\omega t_1}{\sin\omega(t_1 - t_0)}.$$

$$\therefore\ B^2 - A^2 = [x_1^2\cos 2\omega t_0 + x_0^2\cos 2\omega t_1 - 2x_1x_0\cos\omega(t_1 + t_0)] \times \frac{1}{\sin^2\omega(t_1 - t_0)}.$$

$$2AB = -[x_1^2\sin 2\omega t_0 + x_0^2\sin 2\omega t_1 - 2x_1x_0\sin\omega(t_1 + t_0)]\,\frac{1}{\sin^2\omega(t_1 - t_0)}.$$

Hamilton's principal function S is

$$S = \int_{t_0}^{t_1} L\,dt$$

$$= \int_{t_0}^{t_1} (T - V)\, dt$$

$$= \int_{t_0}^{t_1} \left(\frac{1}{2} \dot{x}^2 - \frac{1}{2} \omega^2 x^2 \right) dt$$

$$= \frac{1}{2} \omega^2 \int_{t_0}^{t_1} [(-A \sin \omega t + B \cos \omega t)^2 - (A \cos \omega t + B \sin \omega t)^2]\, dt$$

$$= \frac{1}{2} \omega^2 \int_{t_0}^{t_1} [(B^2 - A^2) \cos 2\omega t - 2AB \sin 2\omega t]\, dt$$

$$= \frac{1}{4} \omega [(B^2 - A^2) \sin 2\omega t + 2AB \cos 2\omega t]_{t_0}^{t_1}$$

$$= \frac{1}{4} \omega [(B^2 - A^2)(\sin 2\omega t_1 - \sin 2\omega t_0) + 2AB(\cos 2\omega t_1 - \cos 2\omega t_0)]$$

$$= \frac{1}{2} \omega [(B^2 - A^2) \cos \omega (t_1 + t_0) \sin \omega (t_1 - t_0) - 2AB \sin \omega (t_1 + t_0) \times \sin \omega (t_1 - t_0)]$$

Substituting the values of $A^2 - B^2$ and $2AB$ and simplifying, we get

$$S = \frac{\omega}{2 \sin \omega (t_1 - t_0)} [(x_0^2 + x_1^2) \cos \omega (t_1 - t_0) - 2x_1 x_0)]$$

5. A heavy bead of mass m is freely movable on a smooth circular wire of radius a which is made to rotate about a vertical diameter with spin ω, prove that the action

$$A = ma^2 \int_{\theta_1}^{\theta_2} \left[\frac{2H}{ma^2} + \frac{2g}{a} \cos \theta + \omega^2 \sin^2 \theta \right]^{1/2} d\theta,$$

where H is Hamiltonian and θ the angle the radius through bead makes with the downward vertical.

At any time t let the particle m be in a position P whose distance from the fixed vertical diameter is y and angular distance θ.

Reduce the curve to rest by introducing an additional force $m\omega^2 y$ on the particle in the direction perpendicular to the vertical diameter AOB. Thus particle is treated moving on the fixed vertical circle under the forces namely *(i)* weight mg acting vertically downward

(ii) the force $m\omega^2 y$ $(= m\omega^2 a \sin\theta)$

Hence, potential due to these two forces is

$$V = -mga\cos\theta - \int_0^y m\omega^2 y\,dy$$

$$= -mga\cos\theta - \frac{1}{2}m\omega^2 y^2$$

$$= -mga\cos\theta - \frac{1}{2}m\omega^2 a^2\sin^2\theta$$

velocity of the particle = $a\dot{\theta}$

$$\therefore \quad T = \frac{1}{2}ma^2\dot{\theta}^2$$

The action

$$A = \int_{t_1}^{t_2} 2T\,dt$$

$$= \int_{t_1}^{t_2} ma^2\dot{\theta}^2\,dt$$

$$= ma^2\int_{\theta_1}^{\theta_2}\dot{\theta}^2\frac{dt}{d\theta}d\theta$$

$$= ma^2\int_{\theta_1}^{\theta_2}\dot{\theta}\,d\theta.$$

Since, the potential exists, therefore, the Hamiltonian H is constant and equal to sum of kinetic and potential energies, that is

$$H = T + V$$

$$= \frac{1}{2}ma^2\dot{\theta}^2 - \left(mga\cos\theta + \frac{1}{2}ma^2\omega^2\sin^2\theta\right)$$

$$\therefore \quad \dot{\theta}^2 = \frac{2H}{ma^2} + \frac{2g}{a}\cos\theta + \omega^2\sin^2\theta.$$

Hence,

$$A = ma^2 \int_{\theta_1}^{\theta_2} \left[\frac{2H}{ma^2} + \frac{2g}{a} \cos\theta + \omega^2 \sin^2\theta \right]^{1/2} d\theta.$$

Proved.

6. A particle of mass m falls a given distance z_0 in time $t_0 = \sqrt{\frac{2z_0}{g}}$ and the distance travelled in time is given by $z = at + bt^2$, where a and b are such that time t_0 is always same. Show that the integral $\int_0^{t_0} L dt$ is an extremum for real values of the co-efficients only when $a = 0$ and $b = g/2$.

Let z be the distanced travelled by the particle after time t; then K.E. of the particle

$$T = \frac{1}{2} m\dot{z}^2$$

P.E. of the particle

$$V = -mgz.$$

$$\therefore \quad L = T - V = \frac{1}{2} m\dot{z}^2 + mgz.$$

According to Hamilton's Principle, we have

$$\delta \int_0^{t_0} L dt = 0$$

$$\therefore \quad \delta \int_0^{t_0} \left(\frac{1}{2} m\dot{z}^2 + mgz \right) dt = 0$$

$$\therefore \quad \int_0^{t_0} \left(\frac{1}{2} m\dot{z}^2 + mgz \right) dt = \text{extremum.}$$

$$\therefore \quad \int_0^{t_0} f(z, \dot{z}, t)\, dt = \text{extremum}$$

where

$$f(z, \dot{z}, t) = \frac{1}{2}m\dot{z}^2 + mgz.$$

Euler Langrange's equation is given by

$$\frac{d}{dt}\left(\frac{\partial f}{\partial \dot{z}}\right) - \frac{\partial f}{\partial z} = 0.$$

$$\therefore \quad \frac{d}{dt}(m\dot{z}) - mg = 0.$$

$$\therefore \quad \ddot{z} - g = 0, (A)$$

Now

$$z = at + bt^2$$

$$\therefore \quad \ddot{z} = 2b.$$

Substituting the value of $\ddot{z}$ in (A), we get

$$2b - g = 0.$$

$$\therefore \quad b = \frac{1}{2}g.$$

When $z = z_0$, $t = t_0$ then

$$z_0 = at_0 + bt_0^2 = at_0 + \frac{1}{2}gt_0^2 \quad (B)$$

It is given that

$$t_0 = \sqrt{\frac{2z_0}{g}}$$

$$\therefore \quad z_0 = \frac{1}{2}gt_0^2.$$

Substituting the value of z_0 in (B), we get

$$\frac{1}{2}gt_0^2 = at_0 + \frac{1}{2}gt_0^2.$$

$$\therefore \quad a = 0.$$

Hence, we have $a = 0$, $b = \frac{1}{2}g$.

7. Derive Langrange's equations from Hamilton's principle.

Now

$$L = L(q_j, \dot{q}_j, t).$$

According to Hamilton's Principle

$$\int_{t_0}^{t} L dt = extremum$$

The resulting Euler-Langrange's equations are

$$\frac{d}{dt}\left(\frac{\partial L}{\partial \dot{q}_j}\right) - \frac{\partial L}{\partial q_j} = 0, (j = 1, 2, ..., n).$$

These are familar Langrange's equations.

8. Apply Principle of Least Action to prove that of all the possible motions between two points, the system for which K.E. is conserved moves along the curve for which time is extremum.

According to Principle of Least Action, we have

$$\delta \int_{t_0}^{t_1} 2T dt = 0.$$

If K.E. is conserved then we have

$$\delta \int_{t_0}^{t_1} dt = 0.$$

$$\therefore \quad \delta(t_1 - t_0) = 0.$$

which states that the system moves along the path for which transit time is extremum.

EXERCISES

1. A particle of unit mass is projected so that its energy is h in the field of force of which P.E. is $\phi(r)$ at a distance r from the origin. Deduce from the principle of energy and Least Action that the differential equation of the path is

$$c^2\left[r^2 + \left(\frac{dr}{d\theta}\right)^2\right] = r^4 [h - \phi(r)].$$

2. A mass m is attached to a coil spring having a constant k, oscillates along a smooth horizontal line with motion given by

$$x = A \sin \omega t,$$

where $\omega = \sqrt{\frac{k}{m}}$. Assuming that varied path is given by is

$$x = A \sin \omega t + \varepsilon \sin 2\omega t,$$

where ε is a small constant quantity, show that for actual path

$$\int_0^{\frac{\pi}{2\omega}} L dt = 0,$$

and for the varied path this integral is equal to $\frac{1}{8} 3m\pi\omega\varepsilon^2$.

3. A particle is projected in a vertical plane with a velocity whose components are u and v respectively. Calculate the value of the integral

$$\int_0^{\frac{n\pi}{\omega}} L dt$$

Assuming that a varied path is given by

$$x = ut, y = vt - \frac{1}{2} gt^2 + \varepsilon \sin \omega t,$$

where ε is small constant quantity, show that for varied path $\int_0^{\frac{n\pi}{\omega}} L dt$ is greater than for actual path, but the result is in agreement with Hamilton' Principle.

4. A particle moves on a plane curve, under the central acceleration $\omega^2 r$, between two fixed points (x_0, y_0), (x_1, y_1) in the prescribed time $t_1 - t_0$, prove that Hamilton's Principal function S is given by

$$\frac{\omega}{2\sin\omega(t_1 - t_0)}[(x_0^2 + y_0^2 + x_1^2 + y_1^2)\cos\omega(t_1 - t_0) - 2(x_1 x_0 + y_1 y_0)]$$

SHORT ANSWER QUESTIONS

1. Write down the resultant of the system of forces acting at given points of the body.
2. Write down the conditions of equilibrium of a rigid body.
3. Write short note on 'Constrained bodies'.
4. Find the conditions of equilibrium of a rigid body with one point fixed.
5. Write down conditions of equilibrium of a rigid body having two points fixed.
6. A circular uniform table of weight W, rests on four equal legs placed symmetrically round its edge, find the least weight that can be hung upon the edge of the table just to upset the table.
7. A square table stands on four legs placed respectively at the middle points of its sides; find the greatest weight that can be put at one of the corners without upsetting the table.
8. State the Principle of Virtual Work.
9. Write short note on 'Conservative system'.
10. Find the relation between kinetic energy, work done and potential energy.
11. Explain stable and unstable equilibrium, giving examples.
12. Write short note on 'Stability test of energy'.
13. Explain the term 'central axis' and write down its equation.
14. Prove that the moment of the resultant couple about a central axis is less than the moment of the resultant couple corresponding to any point O which is not on the central axis.
15. Define 'Wrench', 'Intensity of wrench', 'Pitch' and 'Screw'.
16. Derive the condition for a given system of forces to be compounded into a single force.
17. Explain Reciprocal Screw.
18. Prove that if a screw is reciprocal to each of two screws α and β, it is reciprocal to any screw δ on the cylindroid determined by α and β.
19. Write down equation of cylindroid.
20. Write short note on 'Frame of reference'.
21. Explain the terms 'Kinematics' and 'Kinetic'.
22. Define the follwing:
    Mass, Matter, Velocity, Acceleration.

23. Write down the radial and transverse components of velocity and acceleration when the particle is moving along a plane curve.
24. Write down the tangential and normal components of velocity and acceleration, when a particle is moving along a plane curve.
25. Write down components of velocity and acceleration for the motion in a circle.
26. Write short notes on *(i)* Relative velocity *(ii)* Angular velocity of the body
27. Derive the relation between linear and angular velocity of the body.
28. Prove that angular velocity of the body is independent of the position of the fixed point O.
29. Prove that general motion of a body is a screw motion.
30. Can vector angular velocities of a body about a fixed point be added? Give reason.
31. Show that two equal and opposite rotation of a rigid body about distinct parallel axes arc equivalent to a translation of the body.
32. If the axes of x and y rotate with angular velocities ω_1 and ω_2 respectively and Ψ is the angle between them, write down the components of acceleration of the point parallel to the axes.
33. Write down components of velocity and acceleration in cylindrical and spherical co-ordiantes.
34. State Newton's Laws of moton.
35. Prove that earth's gravitational field is conservative.
36. If, $\bar{F} = kd\bar{r}$ where k is constant, prove that field is non conservative.
37. Give the significance of conservative laws.
38. State principles of conservation of energy, conservation of linear momentum and conservation of angular momentum.
39. Explain S.H.M, its periodic time, frequency and amplitude of the oscillations.
40. Write short note on disturbed S.H.M.
41. Distinguish between 'Damped harmonic oscillations' and 'Forced oscillation' and 'Damped forced oscillations'.
42. State Hooke's Law.
43. Find work done in stretching the string.
44. Write down equations of motion of a particle in vector form, then derive for three dimensional motion.
45. Write down equation of projectile motion under gravity.
46. In case of projectile motion under gravity, write down co-ordinates of vertex and focus, length of latus rectum, equation of directrix, time of flight and range.

47. Show that the product of two times of flight from P to Q with given velocity of projection is $2PQ/g$.
48. Find the range on inclined plane through the point of projection.
49. Write down equation of the path of a particle projected in a resisting medium, whose resistance is $k \times$ velocity.
50. Find the equation of parabola of safety.
51. Write short note on constrained particle motion.
52. Prove that rate of change of linear angular momentum of the particle is equal to the resulting force which acts on it.
53. What do you mean by cycloid motion? Prove that it is periodic motion.
54. Write down equations of motion of a particle moving on a rough plane curve.
55. If a particle is describing a central orbit, prove that it must be a plane curve.
56. If a particle is describing a central orbit, prove that the rate of description of area by the radius vector drawn from the fixed point to the particle is constant.
57. Explain the terms 'Apse', 'Apse line', 'Apsidal angle', 'Perihelion', and 'Aphelion'.
58. Write down equations of the orbit of central force and expressions of velocity in
    (i) Reciprocal co-ordinates
    (ii) Pedal co-ordinates.
59. Write down the conditions for the path of particle moving under inverse square law to be an ellipse, parabola, or hyperbola.
60. If a particle is describing an ellipse under the inverse square law of attraction, find the periodic time.
61. Write short note on 'Velocity from infinity'.
62. Prove that principles of energy and momentum can be applied to central orbits.
63. Write shot note on 'Stability' of an orbit.
64. State Kepler's laws of planetary motion.
65. State Newton' Law of gravitation.
66. Derive Kepler's laws from Newton's law of gravitation.
67. Write short note on 'Disturbed elliptic motion'.
68. What is the necessary modification of Kepler's Law?
69. Explain Elliptic Harmonic Motion.
70. Find the central force when a particle describes the spiral $r\theta = a$.
71. Find the central force when a particle describes the lemniscate $r^2 = a^2 \cos 2\theta$.
72. Find the central force when a particle describes cardroid $r = a(1 + \cos \theta)$.

73. Find the law of force when a particle describes a circle $r = 2a \cos \theta$.

74. If the law of force is $f = -m\left(\frac{\mu}{r^2} + \frac{\eta}{r^3}\right)$, $\eta < h^2$, find the path.

75. A particle is attracted towards two fixed centres for each of which the force is proportional to the distance. Show that the resultant force is a central force, which is directly proportional to the distance.

76. If the maximum speed in a certain elliptic harmonic motion is 31/4 km. per sec., and the minimum speed is 3 km. per sec., what is the eccentricity of the ellipse ?

77. Prove that average value (with respect to time) of the K.E. in elliptic harmonic motion is eqaul to the average value of the P.E.

78. If the rate at which the radius vector sweeps area is the same for a circular obrit and parabolic orbit, show that the perihelion distance of the parabola is one-half of the radius of the circle.

79. Prove that the speed in a parabolic orbit at any point is $\sqrt{2}$ times the speed in the circular orbit which passes through that point.

80. A particle moves in an ellipse under the Newtonian law of gravitation, prove that the average value (with respect to time) of the K.E. is half the average value of the P.E.

81. Distinguish between external and internal forces for a system of particles.

82. Prove that for a system of particles the angular momentum of internal forces about any fixed point is zero.

83. Prove that total external impulse applied to the system of particles is equal to the total change of momentum produced.

84. Define physical moments and prove that the rate of change of angular momentum about any axis throgh a fixed point for such a system is eqaul to the sum of the physical moments of the external forces acting on the system about that line.

85. Prove that K.E. of a system of particles moving in space is equal to the sum of K.E. of a single particle whose mass is eqal to the total mass of all the particles concentrated at its centroid and moving with centroid's velocity, together with K.E. of the system in motion relative to the centroid.

86. Prove that the angular momentum of the system of particles about a fixed point O is equal to that about O of a single particle whose mass is equal to the total mass of all the particles concentrated at the centroid and moving with centroid's velocity, together with the angular momentum about the centroid of the system in its motion relative to the centroid.

87. Prove that the rate of change of angular momentum of the system about O is equal to the moment of the rate of change of the linear momentum about O of a single particle whose mass is equal to the total mass of all

the particles, concentrated at its centroid and having centroid's motion, together with the moment of the rate of change of momentum about the centroid of the system in its motion relative to the centroid. Further prove that this is equal to the total moment about O of the external forces acting on the system.

88. Write short note on 'Conservation laws'.
89. Explain the terms 'Moment of Inertia and 'Product of Inertia'.
90. Find the 'Moment of Inertia' of a rod about a line perpendicular to the rod and passing through one end.
91. Find the M.I. of a rod about a line perpendicular to the rod and passing through its centre.
92. Find the M.I. of a rectangle $ABCD$ about a line parallel to AB.
93. Find the M.I. of a rectangle $ABCD$ about a line parallel to BC.
94. Find the M.I. of a parallelopiped having 2a, 2b, 2c as length of its edges about a line passing through its centre and parallel to 2a.
95. Find the M.I. of a cube about a line passing through its centre and parallel to its edge.
96. Find the M.I. of a circumference of a circle about a diameter.
97. Find the M.I. of a circumference of a circle about an axis throgh its centre.
98. Find the M.I. of a circular disc about a diameter.
99. Find the M.I. of a circular disc about an axis through its centre.
100. Find the M.I. of an elliptic disc about major axis.
101. Find the M.I. of an elliptic disc about minor axis.
102. Find the M.I. of a hollow sphere about its diameter.
103. Find the M.I. of a solid sphere about its diameter.
104. State parallel and perpendicular axes theorems.
105. Write down the Inertia Matrix.
106. Write down the M.I. of a body about a given line.
107. Prove that sum of M.I. of a body about any three rectangular axes meeting at a point is constant and is eqaul to twice the M.I. about that point.
108. Define momental ellipsoid and write down its equation.
109. With usual notations prove that $A > 2D$.
110. Define ellipsoid of gyration and write down its equation.
111. Find the principal axes at the corner A of the rectangular lamina $ABCD$.
112. State necessary and sufficient conditions for equimomental systems.
113. Show that a uniform rod of mass m is equimomental to three masses $m/6$, $m/6$ placed at its ends and $2m/3$ at its middle point.
114. Write down Kinetic Energy of a rigid body in different forms.
115. Write down equation of angular momentum of a rigid body.
116. Write down equation determining directions of principal axes.

117. Explain two dimensional plane motion.
118. State D'Alembert's principle and advantage of its application to dynamical problems of rigid body.
119. Distinguish between 'Rolling' and 'Sliding'.
120. Distinguish between 'Perfectly rough surface' and 'Perfectly smooth surface'.
121. In the motion of two bodies in contact, what is the condition for the separation of the bodies?
122. Write short note on 'Motion about fixed axis'.
123. Write short note on 'Compound pendulum'.
124. Explain the terms 'Simple pendulum', 'Centre of suspension' and 'Centre of oscillation'.
125. Prove that the centres of suspension and oscillation are interchangeable.
126. A solid homogeneous cone of height h and vertical angle 2α oscillates about a horizontal axis through its vertex, find the length of simple equivalent pendulum.
127. A uniform triangular lamina oscillates in its own plane about the angle A. Find the length of simple equivalent pendulum.
128. An ellipse of axes a, b and a circle of radius b are cut from a sheet of thin uniform metal and fixed together their centres co-incident. The figure is free to move in its own vertical plane, find the length of simple equivalent pendulum.
129. A thin circular disc of radius a has a hole in it of radius b whose centre is at a depth c from the centre of the disc ($c < a - b$). The disc is free to oscillates in a vertical plane about a smooth horizontal circular rod of radius b passing through the hole. Find the length of simple equivalent pendulum.
130. Write down equation of motion about a fixed axis for impulsive forces.
131. A rod of mass m and of length $2a$, which is capable of free motion about one end A, falls from a vertical position, and when it is horizontal, strikes a fixed end A. Find the impu'se of the blow and impulse of the reaction at A.
132. The door of a railway carriage stands upon at right angles to the length of the door when latter starts to move with an acceleration f; the door being supposed to be smoothly hinged to the carriage and to be uniform and of length $2a$, find the angular velocity, when it has turned through an angle θ.
133. Find the time of oscillation of a compound pendulum, consisting of a rod of mass m and length a, carrying at one end a sphere, of mass m, and diameter $2b$, the other end being fixed.

134. An elliptic lamina is such that when it swings about one latus rectum as a horizontal axis, the other latus rectum passes through the centre of oscillation, find the eccentricity.
135. A uniform elliptic board swings about a horizontal axis at right angles to the board and passing through one focus. If the centre of oscillation be at the other focus, find the eccentricity of the ellipse.
136. Find the length of simple equivalent simple pendulum, in the following cases:
    (i) Circular disc; axis a tangent to it.
    (ii) A hemisphere; axis a diameter of the base.
137. Find the length of simple equivalent pendulum of a cube of side $2a$, axis being *(a)* diagonal of one face *(b)* an edge.
138. A uniform circular disc, of mass m and radius a is fixed at its ends to two points in the same vertical line, and is rotating with constant angular velocity. Find the horizontal thrust on the upper end.
139. Find the centre of percussion of a triangle ABC which is free to move about its side BC.
140. Find how an equilateral triangular lamina must be stuck that it may commence to rotate about a side.
141. Find the centre of percussion of a sector of a circle, axis in the plane of the sector, perpendicular to its symmetric radius and passing through the centre of the circle.
142. Find the centre of percussion of a uniform circular plate, axis a horizontal tangent.
143. Find the centre of percussion of a uniform rod with one end fixed.
144. Write down Euler's equations of motion of a rigid boyd about a fixed axis.
145. Prove that in absence of external forces, K.E. of the body is constant during the motion.
146. Prove that in absence of external forces the angular momentum of the body is constant during the motion.
147. Derive Euler's equations of motion for impulsive forces.
148. Write short note on 'Eulerian angles'.
149. Write down Eulrer's Geometrical equations.
150. Distinguish between 'Polhode' and 'Herpolhode'.
151. Distinguish between 'Polehode cone' and 'Herpolhode cone'.
152. Define symmetrical top.
153. Write down equations of motion of symmetrical top.
154. Derive the stability conditions for the motion of a symmetrical top.
155. Write short note on 'Invariable line' and 'Invariable plane'.
156. Explain the terms 'Generalized co-ordinates' and 'Degree of freedom'. Give examples.

157. Write short note on 'Constraints'.
158. State the conservation theorems for Generalized momentum and energy.
159. Distinguish between 'Holonomic' and 'Non-holonomic' dynamical systems. Give examples.
160. Write down Lagrange's equations for Non-conservative and conservative field of forces.
161. Write short note on 'Normal modes'.
162. Derive Lagrange's equations for impulsive forces.
163. Write short note on 'Dynamical methods'.
164. Using Lagrangian method, derive equations of planetary motion.
165. Write short note on 'Transformation equations'.
166. Write short note on 'Configuration space'.
167. Explain the terms 'Generalized velocity', 'Generalized momentum' and 'Generalized forces'.
168. Define cyclic or ignorable co-ordinates and state conservative theorems for generalised energy and momentum.
169. Define Hamiltonian function H and write down Hamilton's equations.
170. Define 'Cyclic' or 'Ignorable' co-ordinates and give physical significance of Hamiltonian function.
171. With references to Lagrangian function, state conservation theorem for energy.
172. Distinguish between Lagrange's equations and Hamilton's equations.
173. Explain the terms 'Actual path' and 'Varied path'.
174. Write different forms of Euler-Lagrange's equation.
175. State Hamilton's principle and principle of Least Action. Distinguish between two principles.
176. State the conditions, where principle of conservation of energy is not applicable.